W9-BVZ-785

Experimental Analysis of
Insect Behaviour

Edited by
L. Barton Browne

With 151 Figures

Springer-Verlag New York · Heidelberg · Berlin 1974

Dr. L. Barton Browne
CSIRO, Division of Entomology
Canberra City, A.C.T. 2601/Australia

ISBN 0-387-06557-1 Springer-Verlag New York Heidelberg Berlin
ISBN 3-540-06557-1 Springer-Verlag Berlin Heidelberg New York

Offsetprinting: Julius Beltz, Hemsbach/Bergstr. Bookbinding: Brühlsche Universitätsdruckerei, Gießen

PREFACE

This volume has come about as a direct result of a Symposium entitled "Experimental Analysis of Insect Behaviour" which was an important contribution to the 14 International Congress of Entomology held in Canberra, Australia, in August 1972 under the joint sponsorship of the Australian Academy of Science and the Australian Entomological Society. It is not, however, strictly Symposium proceedings. I have included, in this volume contributions from several workers who had to withdraw from the Symposium at a fairly late stage. Furthermore, quite intentionally, a number of the contributions bear only a general relationship to the papers given at the Congress. To permit this, the deadline for contributions was set at some six months after the Symposium.

I imposed no restrictions on the form of the contributions. I did, however, indicate that speculative reviews highlighting the author's own recent research or that of his immediate colleagues would be particularly acceptable, and a number of the contributors have taken the opportunity to write this kind of paper. Several contributors, notably those whose task it was to give more general papers in the Symposium itself, have written reviews of somewhat greater scope.

The resultant volume is a collection of papers representative of a great variety of approaches to the experimental analysis of insect behaviour. The approaches vary from the purely behavioural to the purely electrophysiological, from the identification of adequate stimuli to the study of behaviour patterns which are almost entirely of endogenous origin, and from pure physiology to the experimental analysis of behavioural strategies. The breadth of my understanding of insect behaviour, in particular, and even of behaviour in general has been greatly increased by my editorial activities and it is my hope that readers will benefit as much as I have.

Thanks are due to Mrs. P.K. Hicks who typed the entire camera-ready manuscript, to L.A. Marshall, C. Lourandos, J. Green and Mrs. G. Palmer who were involved in the preparation of illustrations, to A.C.M. van Gerwen who took responsibility for the checking of the references and helped in a number of ways, and to Dr. R.M.M. Traynier on whom I relied for advice during the construction of the subject index. I wish also to thank Dr. D.F. Waterhouse who supported the production of this volume by allowing me access to the typing, illustrating and photographic facilities of the CSIRO, Division of Entomology.

December 1973

L. Barton Browne
CSIRO, Division of Entomology
P.O. Box 1700,
Canberra City, A.C.T. 2601
Australia

CONTENTS

CHANGES OF RESPONSIVENESS IN THE PATTERNING OF BEHAVIOURAL SEQUENCES

J.S. Kennedy

Agricultural Research Council Insect Physiology Group: Department of Zoology, Imperial College Field Station, Silwood Park, Ascot, Berks., England

In 1970, Hoyle diagnosed the condition of behavioural physiology, or Neuroethology in his own term, as "ripe with promise" but "not quite off the mark". As far as published work goes that diagnosis still stands. Over the last ten years or so more and more evidence has been brought forward that what the ethologists called "fixed action patterns" are due, as Hoyle (1970) put it, to "neural activity whose programming is achieved by central nervous cells to a large extent independently of the detailed sensory input". This seems to be true not only of reiterative patterns - wing beating, ambulation, stridulation, ventilation - but also of some progressive ones such as courtship and eclosion. Going on from there to the next step of discovering the actual neural mechanism of a fixed action pattern has proved exceedingly difficult. We still do not know the nature of the neural oscillator for the simplest reciprocating motor pattern. Faced with this situation, as well as Hoyle's (1970) paradox that "the investigator suffers simultaneously from a dearth of information ... but a surfeit of recordings", one cannot but think twice about his tactic of the sustained frontal assault: "Record first, think later - and then record some more".

Fixed action patterns were by their nature excellent material for the historically necessary task of establishing that central patterning exists at all. The physiologist then knew just what to look for: he had only to find, in the output from a de-afferented ganglion preparation, a replica of the familiar, predictable behaviour pattern. It does not follow, however, that fixed action patterns are the best material for the next and quite different task of identifying an actual coordination mechanism. Being wholly central, that cannot be very easy of access experimentally. An alternative might be to look at a behavioural sequence that is not fixed but instead shows a large measure of sensory dependence, for that will facilitate experimental access to the coordination mechanism. The central contribution to such a coordination will show itself as changes of responsiveness to the sensory inputs, but there will of course be no ready-made model available of the pattern of central changes to be looked for. The model must first be derived from behavioural experiments designed to codify the changes of responsiveness. Unfortunately the study of this aspect of behaviour has much ground to make up after having been side-tracked by early ethological theory. Although there are many reported cases of a stimulus serving to switch on responsiveness to another stimulus, as when an odour switches on a visual orientation response, this kind of information has remained disconnected, non-quantitative and concerned with one-way effects only.

For some years we have therefore been looking for regularities in changes of responsiveness in an aphid, *Aphis fabae* Scopoli. For this we have used two reflex systems which are not fixed but vary in relative strength so that it is possible to study the shifting balance between them and how inputs affect it. In the first phase of these experiments the two systems were flight on the one hand, and settling down on a surface to feed, etc., on the other. More recently we have been looking at flight in relation to the landing-approach response.

The basic equipment for both phases of the work is a vertical wind tunnel. In its present form this has a working section 90 cm square and 60 cm high, the roof and floor being screens of brush nylon which serve to smooth the air flow. The speed of the flow, usually downwards but upwards if necessary, is controlled by varying the pressure difference between chambers above and below the working section. A strong light is projected downwards through the white centre part of the upper screen, the rest of this screen, the walls and floor being black. The aphid flies upward towards the brightly lit window but is prevented from reaching the roof screen and is held flying freely to and fro just below it by the downflow of air, which the operator adjusts continually so as to balance the aphid's upward flying toward the light. Continuous recording of that airflow provides a continuous record of the aphid's rate of climb as a sensitive measure of the flight response. For the earlier experiments flight was terminated after pre-set intervals by holding up a vertical leaf or other landing platform under the window, which evoked a visual landing response, and we recorded the responses of the insect after it had landed on the surface and also its rate of climb when it took off again. For the more recent experiments we have used a retractable landing target, a leaf-shaped yellow card which could be removed before an approaching flier landed on it.

The results of the earlier experiments on the interaction of flight and settling responses (Kennedy 1966 *et ante*) may be summarized as follows. Temporarily inhibiting either one of these antagonistic reflex systems, by the simple procedure of eliciting the other one with an appropriate stimulus, either increased or reduced its subsequent strength. For example, the flying aphid's rate of climb toward the overhead light could be strengthened, or in other circumstances weakened, by interrupting the flight with a brief landing. These two, opposite after-effects of temporary inhibition of a reaction by stimulating its antagonist were called antagonistic induction and antagonistic depression, respectively. Frequently repeated elicitation of the antagonistic activity had cumulative after-effects, boosting or depressing the protagonist progressively. Thus, interposing a brief landing after each successive minute of flight boosted the rate of climb of a fresh aphid well above the rate attained during continuous, uninterrupted flying. Now the after-effect of each spell of inhibition contained both excitatory and inhibitory components; it was the proportions in which these appeared that varied with the pretreatment and produced a net boost, or net depression. There was also evidence that the after-effect of inhibiting a given activity by stimulating the antagonistic one was due, not to any feedback from the performance of the antagonistic activity, but to the central inhibition itself. For landings on two very unequally stimulating surfaces had different after-effects on flight even when the settling performance on the two surfaces was apparently equal (Kennedy 1966).

The key point here, with respect to behavioural sequence-building, is that, thanks to the cumulative after-effects, the balance between the antagonists was unstable and shifted one way or the other according to the inputs received. If those inputs were such as to maintain uninterrupted flight, the balance shifted in favour of settling. If, on the other hand, the inputs were such as to induce frequent interruptions of flight, then the balance shifted the other way in favour of stronger flight and weaker settling responses. In this way one behaviour pattern could gain upon the other and even win, as it were, quite unlike the familiar rhythmic oscillations between lower-order reflex antagonists which either continue their balanced alternation together, or instead both stop together.

If the type of interaction between different behaviours seen there in the aphid is of general occurrence, it would mean, for instance, that the waxing and waning of responsiveness, the hallmark of normal behaviour, are not to be ascribed solely to peripheral events or to special physiological systems such as clocks and endocrines, but are generated in the normal working of the central nervous reflex machinery itself. That much is already evident from the large body of behavioural and neurophysiological work that has been done on the decremental or habituating effects of repeated stimulation of some one response, with which there are often associated incremental effects (Horn and Hinde 1970). The aphid results take us further, into the interaction between different responses requiring different stimuli as in normal behaviour, and show that incremental and decremental changes of responsiveness can be brought about by stimulating not the response in question but an antagonistic one that inhibits it. Centrally organized changes of response strength could then be a function of the constantly shifting reciprocal inhibition which underlies the 'singleness of action' characteristic of behaviour in general. There are some neurophysiological analogues of these aftereffects of inhibition, also. The most striking one to my mind is what Maynard (1961) called *paradoxical driving* when he was able to produce, by appropriately timed inhibitory stimuli only, a sustained series of rebound bursts from an otherwise silent pacemaker cell in the lobster cardiac ganglion. Indeed it seems to me strange that modellers of neural oscillator systems involving reciprocal inhibition as in wing-beating, etc. leave aside the possibility of such post-inhibitory rebound.

However, those earlier aphid experiments were concerned with two clear-cut, mutually exclusive antagonists, and one could reasonably question whether the results had any very general relevance because the behavioural repertoires of insects are not in fact made up of neat pairs of mutually exclusive, incompatible antagonists. There are also what Sherrington called "allied reflexes" and all kinds of intermediates. Do the principles of antagonistic induction and depression apply to these, too?

We have been looking at such a case in the landing-approach response of a flying aphid to the visual stimulus from a conspicuous object such as a leaf in its immediate vicinity. The approach response in certainly not incompatible with flight, for it occurs during and by means of flight. At the same time it appears to be allied to the settling responses of an aphid after landing. For example, flying aphids of many species, including the bean aphid, used in our experiments, are readily caught in yellow traps, because yellow is the most powerful attractant and arrestant for them when they have been flying for a while. And yellow is also the colour of substrate which most readily arrests a walking aphid and induces it to probe, the first of the settling responses (Moericke 1950, 1955; Kring 1972).

We have therefore examined the effects on flight excitability (measured by the rate of climb) of stimulating the landing-approach reaction in the flight chamber, without allowing any actual landing which would of course quickly arrest wing-beating and initiate the settling responses. During each bout of presentation the landing target, a small yellow card, is promptly swung up out of sight whenever the flying aphid approaches it closely and promptly dropped into view again as the aphid wheels back toward the central roof window. A single brief presentation bout (5-10 sec) induces 1-3 oriented approaches by the aphid which are associated with a single ripple or blip in the rate-of-climb trace: this most often takes the form of a drop first, then a rise to above the pre-presentation level and then a return to it. Ripples occur from time to time also in the absence of the target, which suggests that what the target stimulus does is not to elicit a once-for-all response but rather to give an impulse to a built-in pendulum or oscillator which, having been pushed one way, then swings back of itself and even overshoots. The opposite swings are seldom equal and their amplitude and frequency are so variable that the term "oscillator" seems hardly appropriate. But the idea that there is some kind of unstable, see-sawing mechanism affecting the flight excitability here gains support from many records where the target presentation induces a high-frequency fluctuation in the rate of climb (as well as slower changes in the net rate) and this continues after the target has been withdrawn.

When the interval between successive 10 sec presentations is reduced to 60 sec the effects on the rate of climb still remain discrete and reiterative. But when the target is presented for several minutes at a time, thus allowing a long series of approaches in rapid succession (i.e. at intervals of 3-4 sec, the time the aphid takes to complete a flight circuit under the illuminated roof window), then the rate of climb shows cumulative shifts one way or the other, recalling the shifts obtained when flight was broken up by actual landings at one-minute intervals, already mentioned. The pattern followed by these rate-of-climb shifts during and after a bout of target presentation varies greatly but the five types of sequence that are most frequently induced by a presentation bout lasting 200 sec and repeated at 10-min intervals, are as follows.

In sequence type A the rate of climb is depressed throughout the presentation period and remains below the pre-presentation level for at least a minute after the target has been withdrawn. This type of sequence may occur at any time but is the rule when an aphid is about to become negative to the light and "range out" from the illuminated area toward the black walls, as it always does sooner or later. In type B the rate of climb is depressed throughout the presentation period but begins rising again before it ends and there is no apparent after-effect, pre- and post-presentation levels being virtually the same. In type C, the rate of climb again remains somewhat depressed throughout the presentation period but when the presentations end it "rebounds" to above the pre-presentation level. In type D, there is some depression initially but then the rate of climb rises to above the pre-presentation level while the target is still present and stimulating repeated approaches. Moreover, the rate of climb remains boosted for more than a minute after the target has gone. Finally, in type E there is no depression of the rate of climb at all. It is boosted from the beginning of the target presentation and remains so, as in type D, until well after the presentations have entirely ceased.

Thus the effects of single bouts of target presentation on the rate of climb cover a continuous spectrum from depression alone to boosting alone, and there is every possible intermediate and combination of the two among the minority of sequences not described above. Hence, in those instances where the presentation produces no net rise or fall in the rate of climb but only fluctuation about an unchanged mean, this fluctuation is interpreted not as mere "noise" but as a balanced alternation between boosting and depression - and thus conceivably a model of the familiar rhythmic coordinations.

The longer the aphid has flown before being presented with the target for the first time, the more depressing is the target's effect on the aphid's rate of climb. This priming effect of uninterrupted flight on the approach response parallels the priming of the settling responses by uninterrupted flight that was demonstrated previously. However, the priming effect of flight can be counteracted for some time by repeated presentations of the target, just as it was counteracted by repeated actual landings in the earlier experiments. For example, a 200 sec presentation which followed five previous ones coming at 10 min intervals, depressed the rate of climb very much less than a 200 sec presentation after the same duration of flight but without any previous presentations.

In each of four separate series of flights using presentation bouts of 3-5 min duration at 7-10 min intervals, there was a clear if irregular shift, in the effect of successive presentation bouts on the rate of climb, in the direction from sequence type A toward type E, that is, toward less depression and more boosting. Eventually (again as in the earlier flight-settling experiments) the trend shifted toward depression again, and finally type A sequences predominated when the aphid was about to "range out".

Up to a point, then, the experiments have given a straightforward answer to the question posed. Although the approach response is obviously not antagonistic to upward flight in the incompatible way that the system of settling responses is antagonistic to it, nevertheless the approach system does interact with upward flight in much the same way as does the settling system. Stimulating it results in antagonistic induction or depression of the rate of climb which with repetition can be cumulative, so these principles of behavioural coordination appear to be of wider application than we might have supposed.

However, antagonistic induction and depression are now seen in a new light. Originally, these terms referred solely to after-effects. This was because, when reflex systems are incompatible like settling and flight in the aphid, any changes that may be occurring in the central excitatory state of either system while it is inhibited are invisible and immeasurable behaviourally. But when the two systems are not incompatible, like climbing flight toward a light and in-flight approach to a lateral target object, then the excitatory state of the former flight system can still be monitored behaviourally while it is actually being subjected to the inhibitory effect of stimulating the antagonistic approach system. It turns out that this stimulation has both inhibitory (depressing) and excitatory (boosting) effects on the upward flight system while the stimulation is coming in, and not only as after-effects.

When the immediate effect of the target stimulus on upward flight is excitatory, then plainly this stimulus cannot be called inhibitory or antagonistic. Nor can boosting of flight with no prior depression at all (sequence type E) be called post-inhibitory rebound. There was a complete spectrum of effects of this stimulus on the rate of climb in respect of the relative amounts and of the relative time-courses of depression and boosting. Indeed the simplest hypothesis consistent with the behavioural evidence would be that the inhibitory and excitatory effects of the same input are two separate and opposite processes with different time-courses, rather than one of them being a consequence of the other as assumed in the notion of post-inhibitory excitation or rebound. The results of the earlier experiments on flight and settling do not contradict this hypothesis; it is perhaps one for neurophysiology to consider.

ACKNOWLEDGEMENT

Mr. A.R. Ludlow's cooperation in the experiments and comments on the manuscript are gratefully acknowledged.

REFERENCES

HORN, G., HINDE, R.A.: "Short Term Changes in Neural Activity and Behaviour". London: Cambridge University Press. (1970).

HOYLE, G.: Cellular mechanisms underlying behavior - neuroethology. Adv. Insect Physiol. 7, 349-444 (1970).

KENNEDY, J.S.: The balance between antagonistic induction and depression of flight activity in *Aphis fabae* Scopoli. J. exp. Biol. 45, 215-228 (1966).
KRING, J.B.: Flight behavior of aphids. A. Rev. Ent. 17. 461-492 (1972).
MAYNARD, D.M.: Cardiac inhibition in decapod crustacea. *In* "Nervous Inhibition", (Ed., E. Florey) pp. 144-178. New York: Pergamon Press (1961).
MOERICKE, V.: Über das Farbsehen der Pfirsichblattlaus (*Myzodes persicae* Sulz.). Z. Tierpsychol. 7, 265 (1950).
MOERICKE, V.: Über die Lebensgewohnheiten der geflügelten Blattläuse (Aphidina) unter besonderer Berücksichtigung des Verhaltens beim Landen. Z. angew. Ent. 37, 29-91 (1955).

SEQUENTIAL ANALYSIS AND REGULATION OF INSECT REPRODUCTIVE BEHAVIOUR

G. Richard

Laboratoire d'Ethologie, Complexe Scientifique Universitaire, Rennes-Beaulieu, France

Even though the reproductive behaviour of insects has long been a popular subject for research by entomologists, detailed descriptions have been made only since the beginning of the 20th Century. However, descriptions in the period before 1940 were often marked by anthropomorphic interpretations or by Fabre's finalism despite some serious efforts to analyse the bases of behaviour. These efforts were in three directions:
(a) Investigations were carried out into external factors eliciting behaviour.
(b) Experiments were carried out on the nervous pathways and factors involved in the organization of various kinds of behaviour.
(c) The idea was developed that there was inter-dependence between internal and external factors which results in there being a complex relationship between the organism and its environment.

These efforts lead to modern developments in the field of reproductive behaviour, spearheaded by three fundamental publications: that of Baerends (1941) who made a detailed analysis of what can be called the parental sequences; of Grassé (1942) who set reproductive behaviour in an eco-ethological context characteristic of social insects; and of Tinbergen *et al.* (1943) whose paper remains a classic for its description of insect courtship.

In the general context of the evolution of ideas regarding behaviour, our group of researchers has, for the past few years, been interested in the analysis of the sequential succession of reproductive behaviour and in the problems of the physiological regulation of each sequence as well as with the succession of sequences.

It is well known that there is, in insects, following their moult to the stage in which they carry out reproductive functions (even neotenic individuals undergo a special moult: Grassé and Noiret (1946); Lüscher (1960); Lebrun (1966)), a period during which the genital products mature. This may be very short as in Ephemeroptera and Lepidoptera (Bombycidae, Saturniidae, etc.), or relatively long as in Odonata. When the period is long, genital maturation is preceded by a compulsory feeding phase and, in some insects, this pre-reproductive period is one of intense wide-ranging locomotor activity which results in migration (e.g., Odonata, Diptera, Homoptera, Heteroptera, etc.).

After the feeding phase, insects display the following sequences characteristic of reproductive behaviour:
(a) Courtship which consists mainly of locomotion and immobilization, and of antennal and wing movements, including vibration, which not only provide visual stimulation but also possess an important tactile component. These elements of behaviour, which are not always clearly separable from aggressive behaviours (Pajunen 1964), and are often combined with responses to sound at a distance (Orthoptera) and especially to sex pheromones, lead finally to the point at which contact between the partners' genital organs is achieved. The likelihood of this happening depends on the preparedness of the individuals and may be facilitated or hindered by specific abdominal postures.
(b) Copulation may occur once only, as in Bombycidae, or may be repeated,

as in Orthoptera, before each act of oviposition. From our point of view it is necessary to distinguish between the situation in which a female lays only one large batch of eggs and that in which she lays many small batches or even single eggs. In this latter case (e.g. Sphegidae) there is repetition of the same sequential series of behavioural elements.
(c) In some instances (Sphegidae, Dermaptera, social insects) females (or sterile females = workers in social insects) extend their behaviour by caring for their eggs or even the offspring.

It is of fundamental importance to attempt to understand the series of external and physiological events by which insects, having been prepared for their adult state by their larval existence, regulate the processes of the appearance, disappearance and integration of different phases of reproductive behaviour (Fig. 1).

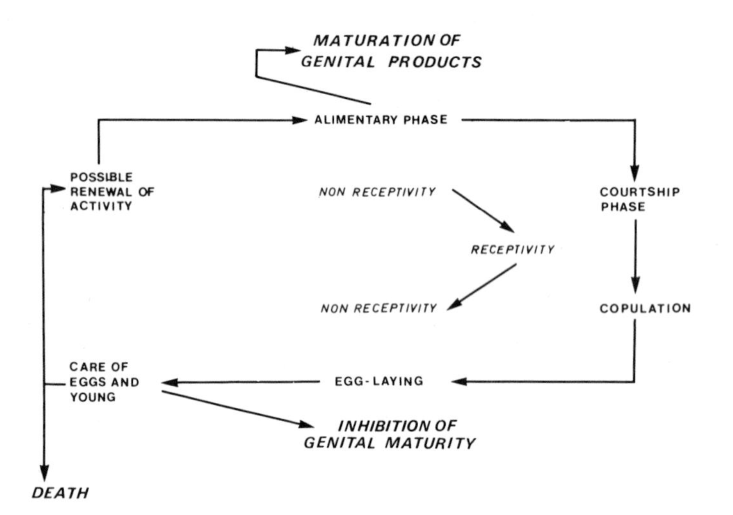

Fig. 1. Schematic diagram of the succession of phases of sexual behaviour superimposed on physiological events

SUCCESSION OF THE MAIN BEHAVIOURAL PHASES

Williams (1935), Larsen (1943), Shorey, Gaston and Fukuto (1964), Shorey and Gaston (1965) on noctuid moths; Haddow (1945), Gillett, Haddow and Corbet (1959) on mosquitoes; Parker (1968), Foster (1967) on Muscidae; Caldwell and Dingle (1965, 1967) on *Oncopeltus*; Engelmann (1970) on cockroaches; Couturier and Robert (1958) on cockchafers; and de Wilde (1961) on Colorado beetle have stressed that there are temporal correlations between the insects' different activities and the abiotic factors of the environment. In *Oncopeltus* for example, feeding takes place late in the light period and is terminated by the onset of darkness. This period is especially important at the beginning of the insect's adult life since, on the 4th to the 10th days after adult ecdysis, the maximum frequency of copulation occurs simultaneously with the feeding maximum. Oviposition, on the other hand, which occurs later in life (8-15 days after adult ecdysis), reaches a peak about 8 hr after the beginning of the light period in the diurnal cycle and does not correspond in time with the peak of feeding and mating, and therefore does not conflict with these activities. Locomotion is the only other activity which occurs at the same time of day as the peak of oviposition behaviour and it seems likely that this is related to searching for oviposition sites.

However, biotic factors (population density, interactions between individuals) play an equally important role in determining the cyclical nature of the sequential activity of insects which depend on particular internal states or programmes. Deleurance's (1957) analysis and description of the "thème cyclique temporel fondamental" (fundamental temporal cyclical theme) of nest building of *Polistes*

THE EFFECTS OF THE PERFORMANCE OF ONE PIECE OF BEHAVIOUR ON SUBSEQUENT BEHAVIOUR

Influence of the feeding phase

Foster (1967) demonstrated that the courtship behaviour of *Scatophaga stercoraria* depends basically on the male having completed a phase of feeding activity. Males which have not fed adequately treat females as prey whereas males which have completed the feeding phase view females as objects for copulation. The amount of feeding required is not, however, fixed. The longer the period for which the males are deprived, the greater is the amount of food required. This increased requirement manifests itself as a greater intensity of predatory behaviour, and an increase in the amount of feeding and the amount of time spent doing so. It is not known exactly, in physiological terms, how the passage from the feeding behaviour phase to the sexual phase is brought about by the ingestion of certain quantities of food. It is known, however, that in this species, the development of the gonads of the males is dependent upon the number of prey eaten, starved individuals having very small testes.

Orr (1964a,b) using *Phormia regina*, and David, Fouillet and Arens (1971) using *Drosophila* have shown that, in addition to quantitative effects, there can be qualitative effects, this time in relation to the development of the ovaries in females. It is well known also that ovarian development occurs only after the intake of blood in haematophagous bugs such as *Rhodnius* and in anautogenous mosquitoes.

It seems improbable that feeding would have a direct metabolic effect on the gonads and the work of a number of authors indicates that an intermediary role is played by the products of different types of endocrine glands. This role is itself complex and the effects manifest themselves simultaneously (as shown in experiments with *Schistocerca* and *Sarcophaga*) on vitellogenesis and the metabolism of specific proteins, the function of which is still uncertain.

Many other factors modify the relationship between feeding and genital maturation and, among these, genetic factors are often mentioned. Church and Robertson (1966), and Cummings, Robin and Ganetzky (1971) have shown in *Drosophila* that genetic selection can lead to considerable individual differences in DNA and in the protein/DNA ratio. David (1971) has shown also in *Drosophila* that genetic factors influence the number of ovarioles, the rate of production of follicles by the ovarioles, the frequency of resorption of the follicles, ovulation, and the descent of ripe oocytes in the oviduct and oviposition.

Importance of secretory activity and locomotion before sexual behaviour

Production of sex pheromones

The almost universal role of pheromones in the sexual behaviour of insects is widely recognised and because of this I will not dwell on this aspect. I would point out only the cyclical character of the production of pheromones in Lepidoptera demonstrated by Shorey and Gaston (1965) and that, in the opinion of Roth and Barth (1964), the production of female pheromones, by all cockroaches which produce them (*Byrsotria fumigata* for example), is under the control of the corpora allata (CA). However, it must be said that neurosecretion by the pars intercerebralis (PI) can also play an important role and that we must remember to distinguish clearly between phenomena of pheromone production and the behavioural responses related to the partners' receptivity (Roth 1964).

Locomotor activity

I will, however, lay greater stress upon the relationship between locomotor activity and reproductive behaviour.

Michel (1972) has demonstrated that the neurosecretory cells of *Schistocerca* are empty after a period of flight. He was able to abolish the tendency to fly by destroying the PI and the corpora cardiaca (CC) but the precise controlling mechanisms are complex because the implanting of the PI taken from good fliers does not modify the flight activity of the recipient, whereas the transplanting of neurohaemal lobes of the CC causes flight in recipients that had previously been poor fliers. These results are compatible with those of Ozbas and Hodgson (1958) and of Milburn and Roeder (1962) obtained with the nerve cord of *Periplaneta*, or with those of Milburn, Weiant and Roeder (1960) and Haskell *et al.* (1965) concerning the activity of motor nerves of the metathoracic legs of cockroaches and locusts.

The CA, unlike other parts of the neurosecretory system appear to play no direct part in the regulation of flight, their only known effects being indirect as has been demonstrated by Nayar (1962), Highnam (1962a, b,c), Highnam and Haskell (1964), Thomsen and Lea (1968), Muller and Engelmann (1968), Lea and Thomsen (1969).

It is known that variations in locomotor activity, whether it be walking or flight, have an effect on oogenesis but it is not yet known how the effect is mediated. Highnam and Haskell (1964) and Hwang and Ma (1964) showed that forced flight hastened sexual maturation and shortened egg laying cycles in *Locusta* and *Schistocerca*. Michel (1972) confirmed that flying locusts mature more rapidly than ones kept confined in breeding cages and showed that the more intense the flight activity, the more rapid was the rate of development. Highnam and Haskell (1964), having shown that flight caused emptying of the CC, suggested that the neurosecretory material released activated the CA and that this, in turn, was responsible for the increase in the rate of development of the ovaries.

The results of Kennedy (1966), taken together with those just cited, make it clear that locomotor activity (walking and especially flying) can be regarded in many insects as being a physiological element involved in the regulation of reproductive behaviour. The motor programme is directly accessible to eco-ethological factors of the environment and must be included with the gonads and their controlling endocrine system in a retroactive cycle.

Importance of the courtship phase

Without going into details here about the various elements of courtship behaviour and their sequences, I wish to described how neurophysiological control mechanisms can interact with the performance of various pieces of behaviour to modify, further, certain aspects of the sequential regulation. Because of this, the performance of a discrete behavioural element can become a precise indicator of a particular internal state.

The studies on *Drosophila* spp. and related Diptera by Bastock and Manning (1955), Barton Browne (1957), Spieth (1952), Milani (1951a,b, 1956), Weidmann (1951), Maynard Smith (1956), Miller (1950), Ensign (1960), David (1971), David, Fouillet and Arens (1971), David and van Herrewege (1971), Petit (1958); on Acrididae by Perdeck (1957), Loher (1960, 1962), Loher and Huber (1964); and on cockroaches by Roth (1964), Roth and Barth (1964, 1967), Roth and Datea (1966), Roth and Stay (1959), Roth and Willis (1952a,b, 1954, 1957, 1960, 1961), Barth (1961a,b, 1962, 1963, 1964), Engelmann (1959, 1960a,b, 1962, 1970), Engelmann and Rau (1965), Grillou (1971, 1973), are some examples amongst many others.

In *Drosophila* the vigour of the courtship of the males and the receptivity of the females are dependent upon the temperature at which they have been raised (Cohet 1971, 1972). The intensity of the males' courtship is influenced by the period for which they have been mature (males 3 days old court more than ones one day old), by the time for which they have been isolated from females, by the state of receptivity of the females and by the number of previous copulations with receptive females. The variation in the intensity of courtship obviously represents changes in the balance between the tendency to copulate and the tendency to avoid females.

Each specific element of behaviour in males can be controlled either by internal components, independent of the female's behaviour (as in *D. pseudoobscura*), or its length and intensity can be controlled by the female's behaviour (as in *D. subobscura*). In each instance, however, the passage from one element to the next in a behavioural sequence is always controlled by feedback effects dependent upon the performance of each successive element.

There are two possible types of situation which can pertain to the development of courtship behaviour. The first is that in which the two sexes exchange specific stimuli throughout, and the second is that in which the male's motivational state is increased without any continuous specific stimuli coming from the female which he is courting. Amongst cockroaches, *Blatella germanica* is an example of the first situation. In this species the male must establish antennal contact with the female before beginning courtship behaviour, which depends throughout on the exchange of stimuli. In *Periplaneta americana*, on the other hand, the perception of the female sex pheromone causes the male to carry out its entire courtship behaviour and therefore provides an example of the second situation.

It is known that the various elements of courtship behaviour are primarily coordinated in the central nervous system. Huber (1960, 1963), for example, in his studies on the sexual songs of *Gomphocerus* and Gryllidae has shown that these are controlled by zones in the mushroom bodies and the central body. He demonstrated that small lesions in, or even total destruction of, one whole mushroom body did not reduce activity, but that recognition of the female, the assumption of the courtship posture, and song are suppressed after the making of large lesions to the calyces of both mushroom bodies or by saggital section of the brain. Furthermore, stridulation was suppressed after the sectioning of the cord at any level such that both sides of the cord were interrupted. Various minor lesions to the ventral nerve cord showed that there exists a central command pathway which probably does not divide before the mesothorax. It is certain, however, that these pathways cross over in the metathorax. The criterion for sexual receptivity in *Gomphocerus* females is the abandonment of the "first defence" and the passage from a passive unresponsive state to an active one in which they react to male stridulation. As soon as copulation is over the females show a "second defence" against attempts at copulation in which they prevent mating by kicking the males with their hind legs. This change is due to the presence of a spermatophore in the genital atrium for, if the spermatophore is removed, the female becomes receptive again after 30-120 min. Further evidence was provided by an experiment which showed that virgin females, in which the entrance of the spermathecal duct had been ligated, remained receptive after copulation. Additional proof was provided by electrophysiological monitoring of activity in the ventral nerve cord and by experiments in which it was shown that the sperm provide no chemical stimulation. In this insect, therefore, sensory input elicited by mechanical stimulation of the spermathecal duct by the sperm is conveyed to the last abdominal ganglion and produces a change in behaviour following mating. The pathway is purely neural.

In cockroaches, the available results give us considerable insight into the mechanisms controlling receptivity. Roth (1970) is of the opinion that mechanical stimulation, elicited by the introduction of the spermatophore, reaches the brain via a nervous pathway and causes the female to become unresponsive to male pheromone. This failure to respond means, not only that the female refuses attempts at mating, but also that she is inhibited from releasing, even though not from producing, female pheromone. Mated females are therefore less attractive to males than unmated ones.

Even though the experiments of Stay and Gelperin (1966) with *Pycnoscelus surinamensis* were perhaps not entirely convincing, they proved that the total or partial sectioning of nerve VN7 modified copulation behaviour and its consequences in bisexual cockroaches of this species. In a very interesting experiment, they demonstrated that the behavioural consequences of sectioning the nerve cord of cockroaches taken from a parthenogenetic stock were less important than in the bisexual stock. Thus the role of the stimulus due to copulation is closely related to the biology of the individual insect concerned.

It is important to point out here that, in discussing Roth's work, it is necessary to make a clear distinction between the production of pheromone by the females and the various elements of behaviour which make up courtship, the two kinds of response being independent in their expression. It is known that, when pheromone secretion is inhibited by, for example, allatectomy, the element of female courtship consisting of the licking of the male tergites persists for some time. It must be stated, however, that licking behaviour is not limited to male-female relationships but occurs also in male-male and even male-nymph relationships (Gautier and Morvan 1971) in which cases there is an increase in the level of excitation of both licked and licking animals.

There is no doubt that further detailed studies are required to establish more precisely which pieces of behaviour of cockroaches are characteristic of states linked specifically with mating behaviour. Grillou (1971) describes a particular reaction shown by females of *Blabera* just after copulation in response to antennal contact by males on the pronotum or the forewings, or to the passage of the antennae in their vicinity. Basically this behaviour consists of the partial opening of the plates at the tip of the abdomen. If stimulation is strong enough the mated female raises the wings to an angle of from 30° to 60° and brings the abdomen into a more horizontal position while, at the same time, opening the suranal and subgenital plates. The spermatophore is visible when the degree of opening is extreme but all degrees of opening or general posture occur. This behaviour can be displayed until the spermatophore produced by the male is finally rejected. This behavioural pattern is displayed by females from which the spermatophore has been removed and when mating is interrupted before insertion of the spermatophore or even only after the hooking of the subgenital plate by the phallomere.

I believe, as does Grillou, that the characteristic reaction is elicited by the stimulation of sensory organs in the edge of the subgenital plate (which is also innervated by nerve VN7) by the male phallomere and that it can to some extent be interpreted as a response indicating non-receptivity. The reasons for its persistence and for its fading are not yet clear but work on this aspect is now in progress. This reaction is reminiscent of that described by van den Assem (1970) in *Lariophagus* females which take up and maintain a non-receptive posture after taking up one indicating preparedness for copulation. This change of posture takes place after the time necessary for copulation, irrespective of whether or not it occurs, so long as the female has been exposed to stimuli from males (pseudo-virgin). It is reminiscent also of the posture of both Diptera and Orthoptera, in which the degree of protrusion of the genitalia is used as a signal to males.

It is apparent, however, that the afferent sensory information linked with copulation has effects other than the eliciting of the non-receptive behaviour of females. It is now proved that different sequences of the copulation phase are controlled by, and retroact upon, endocrine factors. Females of *Gomphocerus* which have been allatectomized at the end of nymphal development, or at the very beginning of imaginal development, remain unreceptive and fail to develop eggs. Allatectomy 24 or more hours after the adult ecdysis has no effect. In males allatectomy any time has no effect. Implantation of CA induces maturation in allatectomized females. The maturation occurs more rapidly if the implanted CA have been taken from a donor which is approaching maturity. Conversely, copulation elicits or increases the amount of CA activity and by so doing affects oocyte maturation (Roth 1964). Even if the increase in food intake by females during the first post-emergence phase does not seem to depend on the activity of the CA and ovaries, the maxima concerned with all other processes related to the reproduction of females are abolished by allatectomy and ovariectomy (Engelmann 1962). It seems that, before copulation, the increased food intake by females does not activate the CA sufficiently to cause normal vitellogenesis. The real stimulus for this activation would be copulation.

Importance of the ovipositional and parental behaviour phases

The female, having been prepared by copulation, by the completion of the development of the ovaries and by becoming non-receptive, is now ready to lay eggs, oviposition being regulated by numerous stimuli which affect, principally, sense organs in the genital region, the input from which is coordinated by cephalic and thoracic ganglia. Many papers on Hymenoptera (especially predators), Orthoptera and Diptera have been written about this problem. The experiments of David, Fouillet and Arens (1971) demonstrate how complex the interactions can be. These workers showed that a yeast which is very attractive for feeding in *Drosophila* can act as a deterrent for oviposition. However, the effect is influenced by genetic factors, by age and by the degree of sensitivity of the flies to the dirtiness of the cage!

Once again, the published works about ovoviviparous cockroahces allow a better analysis of the physiological relationships. During "gestation" in these cockroaches many different stimuli are said to be produced at the level of the incubation chamber: viz. nervous stimuli linked with the distension of the chamber (Roth 1968) which is sensed by specific mechanoreceptors (Grillou 1971; Brousse Gaury 1971, 1973), or the humoral stimuli arising either from the ootheca or from the chamber itself. These various stimuli could directly affect the 6th abdominal ganglion and, either by its acting as intermediary or by some other pathway, could affect the cephalic ganglia and the neuroendocrine system.

It is to the inhibition of the neuroendocrine system by these factors (the CA appear, histologically, to be inactive) that the blockage of the receptivity of females, the inactivation of ovaries and the inhibition of feeding cycles of "pregnant" females can be attributed. The implantation of nymphal CA into pregnant females restores their ovarian activity, as it does in other insects (Day 1943; Joly 1958; Scharrer and von Harnack 1958, 1961; Thomsen 1948; Wigglesworth 1948). Allatal implantation does not, however, reactivate the feeding cycles, which are normally associated with egg development, even though these females develop several series of eggs. The feeding cycle appears to be inhibited by the presence of eggs in the incubation chamber. Furthermore, the interdependency of inhibitory pathways to the CA is shown by the fact that the CA do not become active after removal of the ootheca until the insect eats more than 30 mg of food per day. Another physiological correlation concerns oxygen consumption which parallels feeding and therefore CA activity (Engelmann 1970; Sägesser 1960).

More detailed analyses of neural afferent and efferent messages to and from the 6th abdominal ganglion have given proof of the importance of neural activity in the passage of inhibitory information. Some experiments which are now in progress (Grillou, unpublished) will give, when finished, a better understanding of the roles of the different pathways. It has been shown already, however, that the state of receptivity displayed, when inhibition is removed by sectioning the ventral nerve cord of non-receptive cockroaches between the 5th and 6th abdominal ganglia, is characterized by a degree of alternation between non-receptivity and partial receptivity, and between these and other pieces of behaviour such as rejection of the ootheca or the laying of a new ootheca.

It is probable that the physiological state of an ovoviviparous female cockroach during pregnancy is comparable to that of female Dermaptera during the parental phase. Work by Vancassel (1967, 1968, 1971, 1972, and in press a,b) and Vancassel and Caussanel (1968) on *Labidura riparia* allows the analysis of the behavioural elements of this latter phase. Continuous observation from the time of emergence revealed that, 1 or 2 days after the imaginal moult, the female comes out of the moulting chamber she had previously prepared and eats. A steady increase in weight coincides with vitellogenesis. After 11 to 12 days, vitellogenesis is complete and feeding decreases. Throughout the feeding period the insect is seen to dig temporary burrows, to groom and to display elements of sexual behaviour. At this point of physiological development, the female will lay irrespective of whether she is virgin or mated. The behaviour of females differs markedly, however, according to their mating status.

Mated females shut themselves in closed burrows and lay all their eggs, a process which is completed in less than 12 hr. They lick their eggs, tend them until the time of hatching and then tend the nymphs for some time after that. Throughout this period they do not eat and vitellogenesis is suspended.

Virgin females usually lay only some of their eggs. Eggs are laid on the ground or in a temporary burrow and egg laying lasts 2-3 days. These females never lick or assemble their eggs and immediately start feeding voraciously. Vitellogenesis starts immediately and can be followed by further partial egg laying.

It seems, therefore, that stimuli supplied by copulation must have an effect on subsequent parental behaviour. If one rejects the hypothesis that a pheromone is responsible, the remaining possibility would seem to be that mechanical stimuli associated with copulation elicits the cycle of parental behaviour. It was shown that mating with castrated males induced females to carry out full parental behaviour including care of eggs for a period which corresponds to the time required for fertilised eggs to hatch. Copulation with castrated males as with normal males has a long term effect on behaviour, the parental behaviour often being repeated twice. Attempts at copulation by males from which the penis had been removed had less effect in that it induced parental behaviour only in some females and then only one parental cycle. These results are interpreted as indicating that the mechanical stimuli associated with the act of copulation can cause a short term effect lasting through only one parental cycle and that the transfer of seminal fluid to the female produces a long term effect which lasts throughout subsequent cycles. In *Labidura* copulation has no quantitative effect on vitellogenesis but plays a role in regulating the behaviour of the female, especially in bringing about the onset of cycles of parental behaviour. It must be stated also that sectioning of the abdominal nerve cord of virgin females may result in their displaying parental behaviour. Also the presence of eggs in the burrow plays a direct or indirect part in the maintenance of the parental cycle.

CONCLUSION

It would be interesting to analyse similarly the parental cycles in Sphegidae (Hymenoptera), the complexity of which has been revealed by the many descriptions of the sequence of behavioural elements consisting of burrow digging, the paralysing of prey and the stocking of the burrow. The publications of Steiner (1962) and of Truc (1972a,b) have shown that in these insects, the intensity of each act varies during the course of its unfolding and that this happening, in conjunction with the progressive "setting free" of each act by the one which preceded it, contributes to the sequential occurrence in a predetermined order of the acts which constitute nest building cycle.

A scheme of this type, which would be suitable for the analysis of its physiological correlations, occurs in social Hymenoptera in which the inter-individual relationships regulate particularly the level of egg laying activity (e.g. the studies with *Polistes* by Deleurance (1957) and Gervet (1962)). And, by passing from wasps to ants, we would be able, using the example of the *Eciton*, which has been so well studied by Schneirla (1957) and by Rettenmeyer (1963), to appreciate better the importance of reproductive behaviour, compared to that of other behaviour patterns, in the perpetually changing ecological framework in which the insect passes its life.

REFERENCES

van den ASSEM, J.: Courtship and mating in *Lariophagus distinguendus* (Först.) (Hymenoptera, Pteromalidae). Neth. J. Zool. 20, 329-352 (1970).

BAERENDS, G.P.: Fortpflanzungsverhalten und Orientierung der Grabwespe *Ammophila campestris*. Tijdschr. Ent. 84, 68-275 (1941).

BARTH, R.H.: Comparative and experimental studies on mating behavior in cockroaches. Ph.D. Thesis, Harvard University. (1961a).

BARTH, R.H.: Hormonal control of sex attractant production in the Cuban cockroach. Science, N.Y. 133, 1598-1599 (1961b).

BARTH, R.H.: The endocrine control of mating behavior in the cockroach *Byrsotria fumigata* (Guerin). Gen. comp. Endocr. 2, 53-69 (1962).

BARTH, R.H.: Endocrine-exocrine mediated behavior in insects. (Abstract). Proc. XVI Int. Congr. Zool. 3, 3-5 (1963).
BARTH, R.H.: The mating behavior of *Byrsotria fumigata*. Behaviour 23, 1-30 (1964).
BARTON BROWNE, L.: The effect of light on the mating behaviour of the Queensland fruit fly *Strumeta tryoni* (Frogg.). Aust. J. Zool. 5, 145-158 (1957).
BASTOCK, M., MANNING, A.: The courtship of *Drosophila melanogaster*. Behaviour 8, 85-111 (1955).
BROUSSE GAURY, P.: Soies mécanoréceptrices et poche incubatrice de blattes. Bull. biol. Fr. Belg. 105, 337-342 (1971).
BROUSSE GAURY, P.: Présence de mécanorécepteurs dans la bourse copulatoire de *Nauphoeta cinerea*, Dictyopteres Blaberidae, oxyhaloinae. C. r. hebd. Séanc. Acad. Sci., Paris (D) 276, 577-580 (1973).
CALDWELL, R.L., DINGLE, H.: Regulation of mating activity in the milkweed bug (*Oncopeltus*) by temperature and photoperiod. Am. Zoologist 5, 685 (1965).
CALDWELL, R.L., DINGLE, H.: The regulation of cyclic reproductive and feeding activity in the milkweed bug *Oncopeltus* by temperature and photoperiod. Biol. Bull. mar. biol. Lab., Woods Hole 133, 510-525 (1967).
CHURCH, R.B., ROBERTSON, F.W.: A biochemical study of the growth of *Drosophila melanogaster*. J. exp. Zool. 162, 337-352 (1966).
COHET, Y.: Mise en évidence de l'influence léthale de la copulation chez les Drosophiles femelles élevées à basse température. C. r. hebd. Séanc. Acad. Sci., Paris 273, 2542-2545 (1971).
COHET, Y.: Influence de l'environnement préimaginal sur l'activité sexuelle des adultes de *Drosophila melanogaster*. Effets d'une basse température de développement. C. r. hebd. Séanc. Acad. Sci., Paris 274, 3102-3105 (1972).
COUTURIER, A., ROBERT, P.: Recherches sur les migrations du hanneton commun (*Melolontha melolontha* L.). Annls Épiphyt. 9, 257-328 (1958).
CUMMINGS, M., ROBIN, M., GANETZKY, B.: Biochemical aspects of oogenesis in *Drosophila melanogaster*. J. Insect Physiol. 17, 2105-2118 (1971).
DAVID, J.: Influences génétiques sur les mécanismes contrôlant la fécondité des souches sauvages de *Drosophila melanogaster*. Annls Zool. Ecol. anim. 3, 493-500 (1971).
DAVID, J., FOUILLET, P., ARENS, M.F.: Influence répulsive de la levure vivante sur l'oviposition de la Drosophile : importance de la salissure progressive des cages et des différences innées entre les femelles. Rev. Comport. Anim. 5, 277-288 (1971).
DAVID, J., van HERREWEGE, J.: Fécondité et comportement de ponte chez *Drosophila melanogaster* : influence de diverses qualités de levure du commerce, du volume des cages et de la surface de la nourriture. Bull. biol. Fr. Belg. 105, 346-356 (1971).
DAY, M.F.: The function of the corpus allatum in muscoid Diptera. Biol. Bull. mar. biol. Lab., Woods Hole 84, 127-140 (1943).
DELEURANCE, E.P. : Contribution à l'étude biologique des *Polistes* (Hyménoptères Vespidés). I. L'activite de construction. Annls Sci. nat. (Zool.) 2e serie, 19, 91-222 (1957).
ENGELMANN, F.: The control of reproduction in *Diploptera punctata* (Blattaria). Biol. Bull. mar. biol. Lab., Woods Hole 116, 406-419 (1959).
ENGELMANN, F.: Hormonal control of mating behavior in an insect. Experientia 16, 69-70 (1960a).
ENGELMANN, F.: Mechanisms controlling reproduction in two viviparous cockroaches (Blattaria). Ann. N.Y. Acad. Sci. 89, 516-536 (1960b).
ENGELMANN, F.: Further experiments on the regulation of the sexual cycle in females of *Leucophaea maderae* (Blattaria). Gen. comp. Endocr. 2, 183-192 (1962).
ENGELMANN, F.: "The Physiology of Insect Reproduction". Oxford: Pergamon Press (1970).

ENGELMANN, F., RAU, I.: A correlation between feeding and the sexual cycle in *Leucophaea maderae* (Blattaria). J. Insect Physiol. 11, 53-64 (1965).
ENSIGN, S.E.: Reproductive isolation between *Drosophila tolteca* and related species. Evolution 14, 378-385 (1960).
FOSTER, W.: Hormone mediated nutritional control of sexual behavior in male dung flies. Science, N.Y. 158, 1596-1597 (1967).
GAUTIER, J.Y., MORVAN, R.: Contribution à l'étude des relations inter-individuelles chez les mâles de *Blabera craniifer* (Dictyoptères). Leurs variabilités dans le temps. Rev. Comport. Anim. 5, 45-51 (1971).
GERVET, J.: Etude de l'effet de groupe sur la ponte dans la société polygyne de *Polistes gallicus* L. (Hymén. Vesp.). Insectes soc. 9, 231-263 (1962).
GILLETT, J.D., HADDOW, A.J., CORBET, P.S.: Observations on the ovi-position cycle of *Aedes aegypti*. Ann. trop. Med. Parasit. 53, 35-41 (1959).
GRASSÉ, P.P.: L'essaimage des Termites : essai d'analyse causale d'un complexe instinctif. Bull. biol. Fr. Belg. 76, 347-382 (1942).
GRASSÉ, P.P., NOIRET, C.: La production des sexués néoténiques chez le Termite à cou jaune *Calotermes ruficollis* F.: inhibition germinale et inhibition somatique. C. r. hebd. Séanc. Acad. Sci., Paris 223, 569-571 (1946).
GRILLOU, H.: Etude de quelques uns des facteurs du controle de la réceptivité sexuelle chez *Blabera craniifer*. Thèse 3e Cycle, Rennes (1971).
GRILLOU, H.: A study of sexual receptivity in *Blabera craniifer* Burm. (Blattaria). J. Insect Physiol. 19, 173-193 (1973).
HADDOW, A.J.: The mosquitoes of Bwamba County, Uganda. II - Biting activity with special reference to the influence of micro-climate. Bull. ent. Res. 36, 33-73 (1945).
HASKELL, P.T., CARLISLE, D.B., ELLIS, P.E., MOORHOUSE, J.E.: Hormonal influences in locust marching behaviour. Proc. 12th Int. Congr. Ent. (London, 1964) 290-291 (1965).
HIGHNAM, K.C.: Neurosecretory control of ovarian development in *Schistocerca gregaria*. Q. Jl microsc. Sci. 103, 57-72 (1962a).
HIGHNAM, K.C.: Neurosecretory control of ovarian development in the desert locust. Mem. Soc. Endocr. 12, 379-390 (1962b).
HIGHNAM, K.C.: Neurosecretory control of ovarian development in *Schistocerca gregaria* and its relation to phase differences. Colloques int. Cent. natn. Rech. scient. 114, 107-121 (1962c).
HIGHNAM, K.C., HASKELL, P.T.: The endocrine systems of isolated and crowded *Locusta* and *Schistocerca* in relation to oocyte growth, and the effects of flying upon maturation. J. Insect Physiol. 10, 849-864 (1964).
HUBER, F.: Untersuchungen über die Funktion der Zentralnerven systems und ingbesondere der Gehirnes bei der Fortbewegung und der Lauterzengung der Grillen. Z. vergl. Physiol. 44, 60-132 (1960).
HUBER, F.: The role of the central nervous system in Orthoptera during the coordination and control of stridulation. *In* "Acoustic Behaviour of Animals" (Ed., R.G. Busnel). Amsterdam: Elsevier (1963).
HWANG, Guan-Hei, MA, Schin-Chun: Fat consumption and water loss of the Oriental migratory locust (*Locusta migratoria manilensis* Meyen) during flight and their relation to temperature and humidity. Acta zool. sin. 16, 372-380 (1964).
JOLY, P.: Les corrélations humorales chez les Acridiens. Annls Biol. 34, 97-118 (1958).
KENNEDY, J.S.: The balance between antagonistic induction and depression of flight activity in *Aphis fabae*. J. exp. Biol. 45, 215-228 (1966).
LARSEN, E., BRO, : The importance of master factors for the activity of noctuids. Studies on the activity of insects. Ent. Meddr 23. 352-374 (1943).

LEA, A.O., THOMSEN, E.: Size independent secretion by the corpus allatum of *Calliphora erythrocephala*. J. Insect Physiol. 15, 477-482 (1969).
LEBRUN, D.: La détermination des castes du termite à cou jaune. Thèse Doctorat, Paris. (1966).
LOHER, W.: The chemical acceleration of maturation process and its hormonal control in the male of the desert locust. Proc. R. Soc. (B) 153, 380-397 (1960).
LOHER, W.: Die Kontrolle des Weibchengesangs von *Gomphocerus rufus* L. (Acridinae) durch die Corpora allata. Naturwissenschaften 49, 406 (1962).
LOHER, W., HUBER, F.: Experimentelle Untersuchungen am Sexualverhalten des Weibchens der Heuschrecke *Gomphocerus rufus* L. (Acridinae). J. Insect Physiol. 10, 13-36 (1964).
LÜSCHER, M.: Hormonal control of caste differentiation in termites. Ann. N.Y. Acad. Sci. 89, 549-563 (1960).
MAYNARD SMITH, J.: Fertility, mating behaviour and sexual selection in *Drosophila subobscura*. J. Genet. 54, 261-279 (1956).
MICHEL, R.: Contribution à l'étude des facteurs de régulation de la tendance à voler chez le criquet pélerin *Schistocerca gregaria* (Forskal). Thèse Doctorat, Rennes. (1972).
MILANI, R.: Osservazioni sul corteggiamento di *Drosophila subobscura* Collin. Rc. Ist. lomb. Sci. Lett. 84, 3-14 (1951a).
MILANI, R.: Osservazioni comparative ed esperimenti sulle modalità del corteggiamento nelle cinque specie europee del gruppo *obscura*. Rc. Ist. lomb. Sci. Lett. 84, 48-58 (1951b).
MILANI, R.: Relations between courting and fighting in some *Drosophila* species. Sel. scient. Pap. Ist. sup. Sanità 1, 213-224 (1956).
MILBURN, N.S., ROEDER, K.D.: Control of efferent activity in the cockroach terminal abdominal ganglion by extracts of corpora cardiaca. Gen. comp. Endocr. 2, 70-76 (1962).
MILBURN, N., WEIANT, E.A., ROEDER, K.D.: The release of efferent nerve activity in the roach *Periplaneta americana* by extracts of the corpora cardiaca. Biol. Bull. mar. biol. Lab., Woods Hole 118, 111-119 (1960).
MILLER, D.D.: Mating behavior in *Drosophila affinis* and *Drosophila alonguin*. Evolution 4, 123-134 (1950).
MULLER, H.P., ENGELMANN, F.: Studies on the endocrine control of metabolism in *Leucophaea maderae* -(Blattaria). II. Effect of the corpora cardiaca on fat body respiration. Gen. comp. Endocr. 11, 43-50 (1968).
NAYAR, K.K.: Effects of injecting juvenile hormone extracts on the neurosecretory system of adult male cockroaches (*Periplaneta americana*). Mem. Soc. Endocr. 12, 371-378 (1962).
ORR, C.W.M.: The influence of nutritional and hormonal factors on egg development in the blowfly *Phormia regina* (Meig.). J. Insect Physiol. 10, 53-64 (1964a).
ORR, C.W.M.: The influence of nutritional and hormonal factors on the chemistry of the fat body, blood, and ovaries of the blowfly *Phormia regina* Meig. J. Insect Physiol. 10, 103-119 (1964b).
OZBAS, S., HODGSON, E.S.: Action of insect neurosecretion upon central nervous system in vitro and upon behavior. Proc. natn. Acad. Sci. U.S.A. 44, 825-830 (1958).
PAJUNEN, V.I.: Aggressive behaviour in *Leucorrhinia caudalis* Charp. (Odonata: Libellulidae). Ann. Zool. Fenn. 1, 357-369 (1964).
PARKER, G.A.: The sexual behaviour of the blowfly *Protophormia Terrae novae*. Behaviour 32, 291-308 (1968).
PERDECK, A.C.: The isolating value of specific song patterns in two sibling species of grasshoppers (*Chorthippus brunneus* Thunb. and *C. biggutulus* L.). Behaviour 12, 1-75 (1957).
PETIT, C.: Le déterminisme génétique et psycho-physiologique de la compétition sexuelle chez *Drosophila melanogaster*. Thèse, no 3181, Sr. A. (1958).

RETTENMEYER, C.W.: Behavioral studies of army ants. Kans. Univ. Sci. Bull. 44, 281-465 (1963).
ROTH, L.M.: Control of reproduction in female cockroaches with special reference to *Nauphoeta cinerea*. I. First pre oviposition period. J. Insect Physiol. 10, 915-945 (1964).
ROTH, L.M.: Reproduction of some poorly known species of Blattaria. Ann. ent. Soc. Am. 61, 571-579 (1968).
ROTH, L.M.: The stimuli regulating reproduction in cockroaches. Colloques int. Cent. natn. Rech. scient. (Sept. 1969) No. 189 (1970).
ROTH, L.M., BARTH, R.H.: The control of sexual receptivity in female cockroaches. J. Insect Physiol. 10, 965-975 (1964).
ROTH, L.M., BARTH, R.H.: The sense organs employed by cockroaches in mating behavior. Behaviour 28, 58-94 (1967).
ROTH, L.M., DATEA, G.P.: A sex pheromone produced by males of cockroach *Nauphoeta cinerea*. J. Insect Physiol. 12, 255-265 (1966).
ROTH, L.M., STAY, B.: Control of oocyte development in cockroaches. Science, N.Y. 130, 271-272 (1959).
ROTH, L.M., WILLIS, E.R.: A study of cockroach behavior. Am. Midl. Nat. 47, 66-129 (1952a).
ROTH, L.M., WILLIS, E.R.: Possible hygroreceptors in *Aedes aegypti* (L.) and *Blattella germanica* (L.). J. Morph. 91, 1-14 (1952b).
ROTH, L.M., WILLIS, E.R.: The reproduction of cockroaches. Smithson. misc. Collns 122, 1-49 (1954).
ROTH, L.M., WILLIS, E.R.: The biology of *Panchlora nivea* with observations on the eggs of other *Blattaria*. Trans. Am. ent. Soc. 83, 195-207 (1957).
ROTH, L.M., WILLIS, E.R.: The biotic associations of cockroaches. Smithson. misc. Collns 141, 1-470 (1960).
ROTH, L.M., WILLIS, E.R.: A study of bisexual and parthenogenetic strains of *Pycnoscelus surinamensis* (Blattaria : Epilamprinae). Ann. ent. Soc. Am. 54, 12-25 (1961).
SÄGESSER, H.: Über die Wirkung der corpora allata auf den Sauerstoffverbranch bei der Schabe *Leucophaea maderae* (F.). J. Insect Physiol. 5, 264-285 (1960).
SCHARRER, B., von HARNACK, M.: Histophysiological studies on the corpus allatum of *Leucophaea maderae*. I. Normal life cycle in male and female adults. Biol. Bull. mar. biol. Lab., Woods Hole 155, 508-520 (1958).
SCHARRER, B., von HARNACK, M.: Histophysiological studies on the corpus allatum of *Leucophaea maderae*. III. The effect of castration. Biol. Bull. mar. biol. Lab., Woods Hole 121, 193-208 (1961).
SCHNEIRLA, T.C.: Theoretical consideration of cyclic processes in Doryline ants. Proc. Am. phil. Soc. 101, 106-133 (1957).
SHOREY, H.H., GASTON, L.K.: Sex pheromones of noctuid moths. V. Circadian rhythms of pheromone-responsiveness in males of *Autographa californica, Heliothis virescens, Spectoptera exigua, Trichoplusia ni*. Ann. ent. Soc. Am. 58, 597-600 (1965).
SHOREY, H.H., GASTON, L.K., FUKUTO, T.R.: Sex pheromones of noctuid moths. I. A quantitative bioassay for the sex pheromone of *Trichoplusia ni* (Lepidoptera: Noctuidae). J. econ. Ent. 57, 252-254 (1964).
SPIETH, H.T.: Mating behavior within the genus *Drosophila* (Diptera). Bull. Am. Mus. nat. Hist. 99, 399-474 (1952).
STAY, B., GELPERIN, A.: Physiological basis of ovipositional behaviour in the false ovoviviparous cockroach, *Pycnoscelus surinamensis* (L.). J. Insect Physiol. 12, 1217-1226 (1966).
STEINER, A.: Etude du comportement prédateur d'un Hyménoptère sphégien : *Liris nigra*. Annls Sci. nat. (Zool.) 4, 1-126 (1962).
THOMSEN, E.: The gonadotropic hormones in Diptera. Bull. biol. Fr. Belg. (Suppl.) 33, 68-80 (1948).

THOMSEN, E., LEA, A.O.: Control of the medial neurosecretory cells by the corpus allatum in *Calliphora erythrocephala*. Gen. comp. Endocr. 12, 51-57 (1968).

TINBERGEN, N. MEEUSE, B.J.D., BOEREMA, L.K., VAROSSIEAU, W.W.: Die Balz des Samtfalters, *Eumenis (Satyrus) semele* (L.). Z. Tierpsychol. 5, 182-226 (1943).

TRUC, C.: Effet d'un retard apporté expérimentalement à l'exécution d'un acte spécifique sur le déroulement ultérieur du cycle nidificateur chez un sphégide chasseur de chenille : l'Ammophile *Podalinia hirsuta* Scapoli. C. r. hebd. Séanc. Acad. Sci., Paris 275, 429-432 (1972a).

TRUC, C.: Evolution du comportement au cours du cycle nidificateur chez un sphégide prédateur de chenilles : l'Ammophile *Podalinia hirsuta*. C. r. hebd. Séanc. Acad. Sci., Paris 275, 569-572 (1972b).

VANCASSEL, M.: Contribution à l'étude descriptive et expérimentale du cycle parental chez *Labidura riparia*. Thèse 3e Cycle, Fac. Sciences, Rennes (1967).

VANCASSEL, M.: Contribution à l'étude du rôle des yeux et des ocelles dans la prise et le maintien du rythme d'activité locomotrice circadien chez *Periplaneta americana*. Annls Épiphyt. 19, 159-164 (1968).

VANCASSEL, M.: Liaison entre comportement sexuel et parental chez *Labidura riparia*. Communication XII Conférence Intern. Ethologie (September 1971) (1971).

VANCASSEL, M.: Effet de la section de la chaîne nerveuse abdominale sur le cycle parental de *Labidura riparia* P. C. r. hebd. Séanc. Acad. Sci., Paris 274, 306-308 (1972).

VANCASSEL, M.: Eléments pour l'analyse du cycle parental chez *Labidura riparia* P. Rev. Comport. Anim. (In press a).

VANCASSEL, M.: Rapport entre les comportements sexuel et parental chez *Labidura riparia* P. Annls Soc. ent. Fr. (In press b).

VANCASSEL, M., CAUSSANEL, C.: Contribution à l'étude descriptive du comportement de *Labidura riparia* P. Rev. Comport. Anim. 2, 1-18 (1968).

WEIDMANN, V.: Uber den systematischen Wert von Balzhandlungen bei *Drosophila*. Revue suisse Zool. 58, 502-511 (1951).

WIGGLESWORTH, V.B.: The functions of the corpus allatum in *Rhodnius prolixus* (Hemiptera). J. exp. Biol. 25, 1-14 (1948).

de WILDE, J.: Extrinsic control of endocrine functions in insects. Bull. Res. Coun. Israel 10, 36-52 (1961).

WILLIAMS, C.B.: The times of activity of certain nocturnal insects, chiefly Lepidoptera, as indicated by a light trap. Trans. R. ent. Soc. 83, 523-555 (1935).

SENSORY INPUT AND THE INCONSTANT FLY

V.G. Dethier

Department of Biology, Princeton University, Princeton, N.J., U.S.A.

In his efforts to arrive at conclusions that have some universality the biologist customarily works with groups of animals rather than with individuals. He strives for reproducible results from which he can make predictions that have a high probability of being correct. To this end he carefully controls, in so far as it lies within his power, the characteristics of the experimental population and the ambient influences playing upon it. He is sensitive to inter-sample variability and variability from one individual to the next. He attempts to control for genetic variability by working with siblings or iso-genic strains. He seeks to control ontogenetic variation by striving to have all of his animals subjected to the same sets of developmental influences. He is attentive to sex, age, hormonal balance, nutritional state. He tries to ensure that the antecedant behavioral experience of each individual, its level of adaptation, its central excitatory and inhibitory states are the same.

When all of this is said and done, there remains the intra-individual variable. How constant is the individual even under the most rigorously controlled regime? Every experimenter has encountered the recalcitrant animal, the animal that simply will not respond as expected. All of us have longed at one time or another for a legitimate excuse to discard that particular animal whose response has provided a datum which sits in grand isolation far from the line in a graph. There is no gainsaying the fact that individuals misbehave (cf. Breland and Breland 1961). Throughout the many studies of the behavior of blowflies there had been a constant reminder of this idiosyncrasy.

One of the most intensively studied behavioral systems is the response of the proboscis of the blowfly to chemical stimulation. In our laboratory alone over 200,000 individual flies have been studied - as individuals. The usual variability in response was observed and was ascribed to uncontrollable differences among samples and among individuals. Naturally, every attempt was made to control the physiological state of the individual, but gradually the nagging thought recurred that there was an inherent instability in the individual and that the much sought after behavioral constancy under constant conditions was an illusion.

There is nothing new about this idea, but it provided the incentive for scrutinizing closely the detailed case histories of 17,000 individual flies for which complete records had been kept. In particular, comparisons were made of successive acceptance thresholds for sucrose. The conditions under which thresholds were measured were the following: All of the flies had been bred under "identical" conditions, even as to the amount of food available for *ad libitum* feeding. At the time of testing, their age, nutritional history, and the ambient influences, including temperature, relative humidity, light intensity, time of day, and season were identical. The tarsal acceptance threshold of each fly was measured by the ascending method of presentation (Dethier 1952). In this procedure the test solutions are made up as a series of ten doubling concentrations. Each fly (attached by the wings to a wax-tipped stick the afternoon before the experiment) is lowered gently to the surface of distilled water in a petri dish and held in position so that all tarsi touch the liquid for two seconds. If there is no extension of the proboscis, the fly is raised and then lowered as before to the surface of the least concentrated solution of the series. If still there is no response, the next higher solution is presented, and so on until the proboscis extends. At no time is the fly permitted to drink.

Fifteen minutes after the test, ample time for sensory disadaptation but not long enough for the state of nutrition or water balance to change, all of the flies were retested. A third test followed fifteen minutes later. During this period of experimentation each fly was maintained in water balance by being given several opportunities to drink between tests. In 88% of the cases the second and third response to sugar occurred at the same concentration as the first. For the remaining 12% it was one, two, or three doubling concentration steps above or below the first threshold. Although this amount of variability is tolerable, it is by no means insignificant especially considering the fact that the intervals between concentrations were doubling molarities. Had finer discrimination by the fly been required by making the intervals narrower, there is no doubt that the probability of a fly responding to the same concentration on successive trials would have been lower. A comparable study was made of rejection thresholds for forty-seven different unacceptable compounds, with similar results. Clearly, in so far as this particular pattern of behavior is concerned the fly is not the same from one moment to the next.

Assuming that all known variables have been controlled, it is customary in these circumstances to refer the inconstancy to some change in the internal state deep in the inner inaccessible reaches of the animal, change brought about by unknown antecedant events, or some oscillating endogenous system. Reference is not uncommonly made to mood, motivational state, or just plain "biological variability" (cf. Dethier 1966). An equally plausible explanation, also relating to central phenomena, is that sensory input from assumedly irrelevant sources varies and as a consequence alters central excitatory states, or, to be more explicit, inhibits or facilitates synapses relevant to the behavior under observation. A fly may, for example, have an itchy back, or an uncomfortable position; or one leg may have "gone to sleep". The fly's attention may be distracted. Just how plausible this last explanation may be is revealed by a recent study of "attentiveness" in the desert locust, *Schistocerca gregaria* (Camhi and Hinkle 1972). With this insect quantitative measurements were made of one particular way in which attention to sensory stimuli might affect central control. In tethered desert locusts changes in the angle of the wings evoke deflections of the abdomen which help control yaw - but only if the wings are beating in flight. In flight the central neuronal flight motor drives abdominal motor neurons in rhythmic bursts. Wing-angle inputs, which are adequate alone to drive these motor neurons, alter the number of spikes per burst, and the alterations are reciprocal in opposite nerves. In short, the switching of the abdomen's controlling neurons from "attentiveness" to "inattentiveness" is brought about by the central flight motor sensitizing interneurons to receive information about the angle of the wings.

An alternative to central change is peripheral change. Little attention has been directed to the periphery as a source of inconstancy. Anyone who has ever essayed electrophysiological monitoring of sensory output is acutely aware of the difficulty of obtaining reproducible results; however, the source of the variability is usually ascribed to unsatisfactory control of recording conditions. A different explanation, which is only infrequently explored, is that sense organs do indeed vary their output under constant conditions. This possibility has been explored in the chemoreceptive organs of the blowfly.

Consider first the perils of recording. Chemoreceptive systems are particularly vulnerable to artifactual influences, partly because of the difficulty of controlling the stimulus (as compared, for example, to visual systems) and partly because of the kind of electrophysiological techniques that must be employed. Two techniques are popularly used in the study of the contact chemoreceptors of insects: tip-recording and side-wall-recording. To record from the tip of a chemosensitive hair, an electrode is inserted into the body or extirpated mouthparts of the insect, and the recording electrode in the form of a micro glass pipette is lowered over the hair. The greatest disadvantage of tip-recording is that the pipette serves both as an electrode and as container for the stimulus; consequently, if the stimulus is not itself an electrolyte, it must be adulterated with one. Additionally there are the usual problems relating to changes in concentration and temperature resulting from evaporation, to maintenance of good electrical contact, and to other more subtle electrical problems (cf. Gillary 1966a,b). Furthermore, there is the ever present handicap of being unable to expose only one of the four receptors at a time to stimulation because all dendrites terminate in a common pore.

Although side-wall-recording is a superior technique, it does not provide unambiguous answers either. The recording electrode, a micro pipette filled with electrolyte, is placed in contact with a crack previously punched in the wall of the hair (Morita 1959). The tip of the hair is now free for the application of any stimulus, electrolyte or not. Except for the necessity of mixing the stimulating solution with an electrolyte, all of the hazards of tip-recording remain. There is an additional hazard. When the recording electrode is appressed to the side of the hair, a low level of "spontaneous" activity is recorded from some cells. The rate of this activity and the number of cells involved can be altered by the gentleness or roughness with which the electrode is applied; furthermore, the amount of activity generated by injury at the side can influence the amount of activity evoked by the stimulus at the tip (Dethier 1972).

When every precaution is taken to eliminate artifacts of recording as a source of variability, there still appears to be an inconstancy in the response of the contact chemoreceptors of the labellum. The study discussed in the following pages represents one attempt to measure variability within receptors and to assess its impact on behavior. The study consisted essentially of stimulating repeatedly the *largest* and *intermediate* hairs of the proboscis (nomenclature of Wilczek 1967) with water, sodium chloride, and sucrose. Responses to sodium chloride were monitored by tip- and side-wall recording; responses to water and to sugar, by side-wall-recording only. Each hair was stimulated no less than ten times and no more than twenty. Each stimulation lasted five seconds; the intervals between lasted two minutes. Thirty-seven hairs from fourteen different flies were examined. The flies were three- and four-day-old males. Some typical results are illustrated by records (Figs. 1-5) and presented graphically in Figures 6-11. Figures 1-5 relate to *largest* hair number 2 and depict stimulations number 1, 8, 11, 16, and 20. Figures 6 and 7 show the total number of impulses generated by the water receptor and by the salt receptor during the first second of continuous stimulation. Impulses were recorded at the tip. In the remaining graphs (Fig. 8-11) the lower sets of lines represent the number of impulses generated by each responding cell in the first 25 milliseconds of stimulation; the upper sets of lines represent the number generated in 1 second minus the number generated in the first 25 milliseconds. In other words, the phasic and tonic portions of the responses are plotted separately.

It is obvious that there is variation in all cases. Generally speaking, greater variability was observed with tip-recording than with

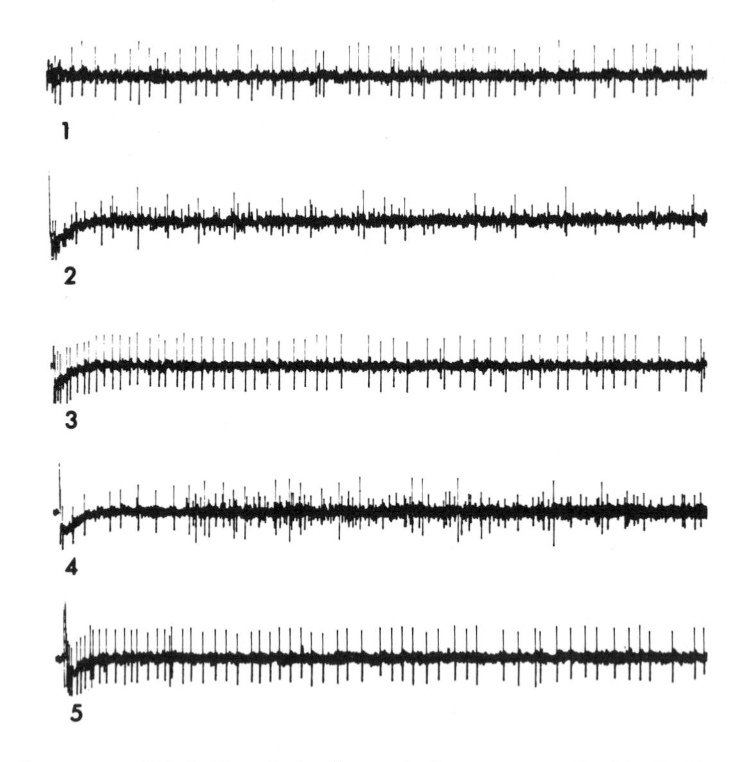

Figs. 1-5. Response of labellar hair *largest* 2 to repeated stimulation with 0.1 M NaCl. The records are from presentations 1, 8, 11, 16, and 20 respectively. Each stimulation lasted 5 sec of which the first 1.75 sec are shown. The inter-stimulation interval was 130 sec. Tip-recording was employed

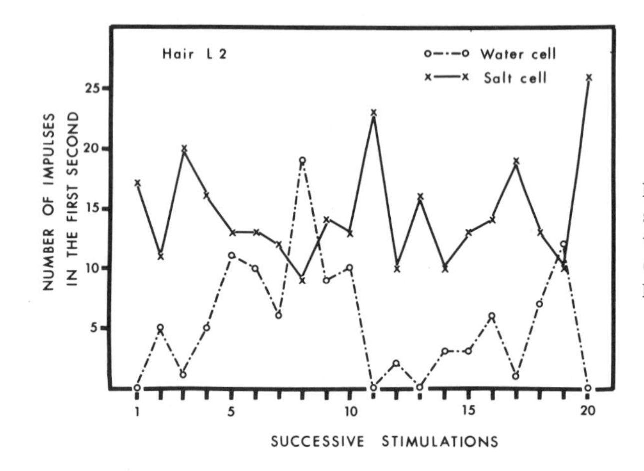

Fig. 6. Response of the water and salt cells of hair *largest* 2 to twenty repeated stimulations with 0.1 M NaCl. Conditions as in Figs. 1-5

side-wall-recording. There is a hint of cyclic fluctuations and of an inverse relationship between the number of impulses from the salt receptor and from the water receptor. Other workers who have studied the response of the salt receptor to different concentrations of sodium chloride have noticed that as the rate of responding of the salt receptor increases that of the water receptor decreases (Evans and Mellon 1962a;

Rees 1968). It is generally agreed that the effect can be attributed to chemical/physical differences in solutions of different concentrations and that it does not represent neural interaction (cf. Wolbarsht 1958). In the present study, however, all stimulations were designed to be at the same concentration. None the less, the results are suspicious, and the most obvious conclusion is, that despite all precautions, the concentration of the salt within the stimulating pipette varied over a range within which the receptor was able to distinguish differences.

It was as a check of this possibility that experiments were repeated and extended with side-wall-recording. Side-wall-recording permitted a close examination of the transient phase at the onset of stimulation (which was obscured before by the "on" artifact) and provided an opportunity for studying responses to water. No changes in concentration would be expected in water, and the temperature and relative humidity were constant from one presentation to the next.

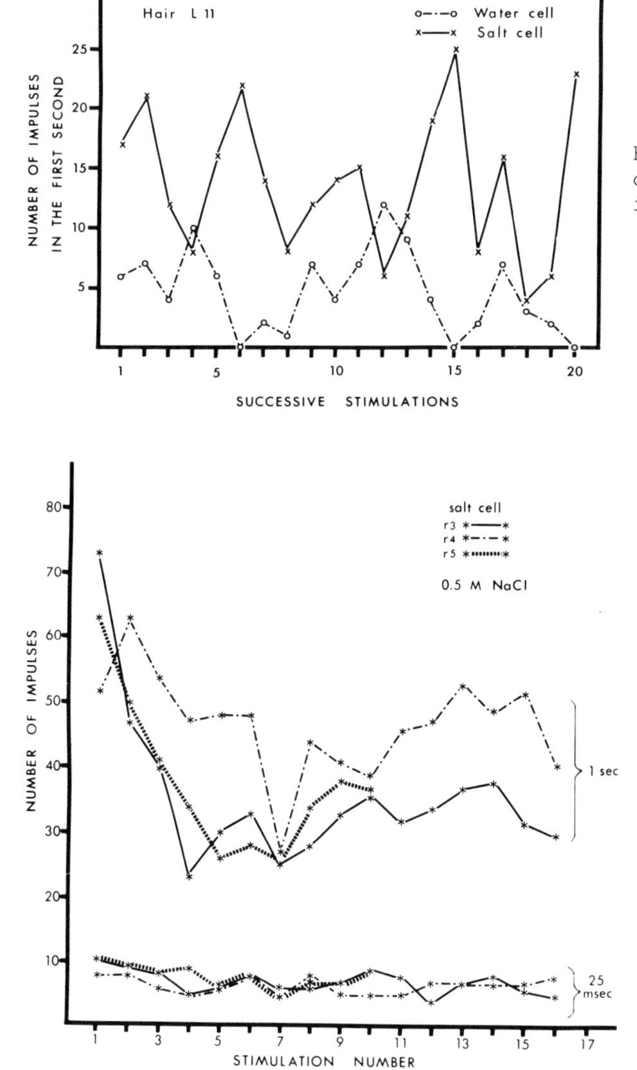

Fig. 7. Responses of water and cells of hair *largest* 11. Conditions as in Figs. 1-6

Fig. 8. Number of impulses generated by the salt cell in hairs *largest* r3, r4, and r5 to 16 repeated stimulations with 0.5 M NaCl. Each stimulation lasted 5 sec. The inter-stimulation interval was 130 sec. The bottom set of lines represents the impulses in the phasic part of the response (first 25 msec). The top set of lines represents the impulses in the first second of the tonic phase. Side-wall-recording was employed

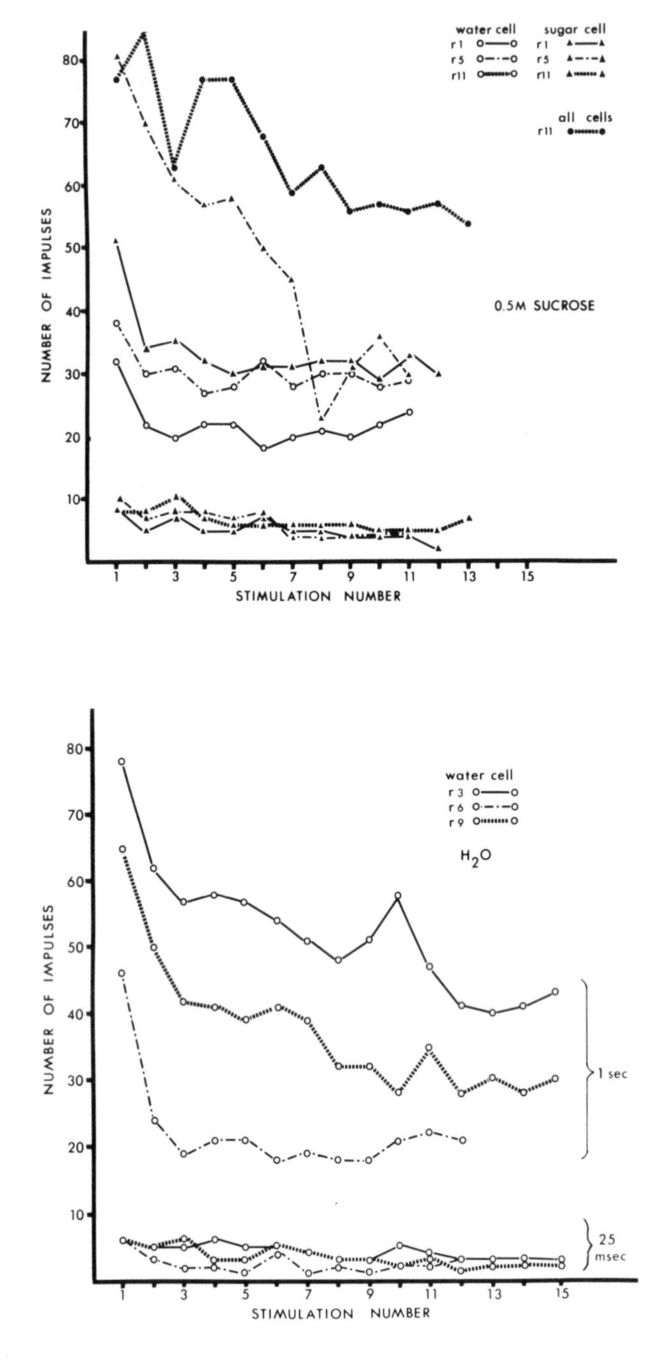

Fig. 9. Number of impulses generated by the water and sugar cells in hairs *largest* r1, r5, and r11 in response to 0.5 M sucrose. All conditions as in Fig. 8

Fig. 10. Number of impulses generated by the water cell in hairs *largest* r3, r6, and r9 in response to water. All conditions as in Fig. 8

An examination of Figures 8-11 reveals the following: (1) there are differences from one stimulus presentation to the next in the responses of *largest* and *intermediate* hairs to water, 0.5 M sodium chloride, and 0.5 M sucrose; (2) differences are non-cyclic and often of considerable magnitude; (3) differences occur both in the phasic and in the tonic parts of the response; (4) the salt receptor is slightly more irregular than the water receptor; (5) the sugar receptor is not characteristically more irregular upon repeated stimulations than are the others (cf. Hodgson and Roeder 1956; Hodgson 1957); (6) there is generally, but not always, a rapid decline in response during the first three presentations, suggesting that an interval of two minutes is insufficient for complete disadaptation (Gillary 1966b, Rees 1968, and others routinely allowed three minutes for recovery yet still noted adaptation within the first two presentations); (7) the *intermediate* hairs do not exhibit an initial decline and are more irregular than the *largest* hairs. It can be concluded that, apart from initial changes resulting from adaptation, there is variability in the response of labellar chemoreceptors upon repeated stimulation.

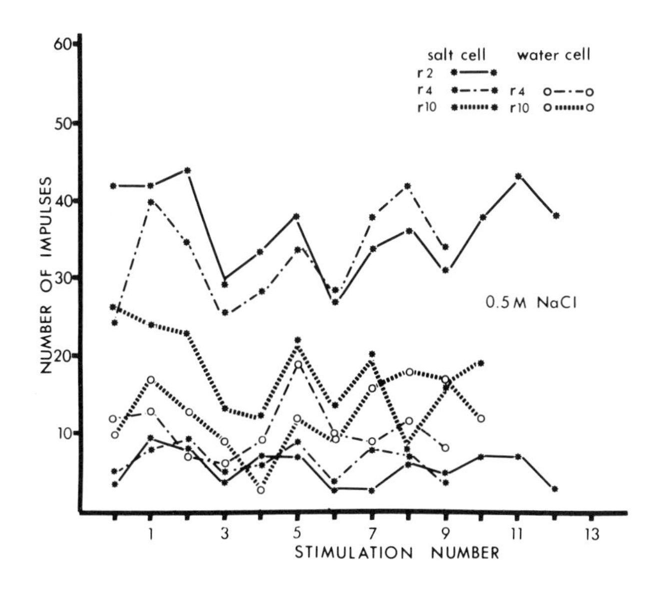

Fig. 11. Number of impulses generated by the water and salt cells in hairs *intermediate* r2, r4 and r10 in response to 0.5 M NaCl. All conditions as in Fig. 8

Detailed reports on the effect of repetitive stimulation are not numerous. Evans and Mellon (1962b) reported that brief tests with stimuli of constant intensity led to complete disappearance of the initial (phasic) portion after the fourth or fifth presentation and that "the effect was not a result of prior stimulation". The tonic phase was reported as "satisfactorily constant". Gillary (1966a), who carried out the most carefully controlled experiments up to that time, reported "...the stimulation procedure was capable of yielding responses within 10% of the mean better than 80% of the time for *normally* responding receptors, and quite often they were within 5%. In the current investigation a response was considered reproducible if it fell within 10% of the mean of all responses to the same applied stimulus. *Experiments yielding non-reproducible responses were considered unreliable and were discarded*". The italics are mine. Gillary's figure 2 depicts the results of twenty-five repeated stimulations with 0.4 M and 2.0 M sodium chloride of a duration of less than 1 sec every 10 sec. Because of the brief inter-stimulus interval the experiment is not strictly comparable to the present study; nevertheless, it shows a variation of about 10% of the mean. Rees (1968) reported that for the salt cell the 95% confidence limits are ± 5 impulses in the tonic phase "where the [rate] is constant enough to calculate them". In the unstimulated state the standard deviation varies from 30.2% to 53.8% of the mean.

Given the reality of inconstant sensory response one can ask two questions: what is the cause of the variation? - is it significant in so far as behavior is concerned? In answer to the first several possible causes can be considered. One is that changes occur in the accessory structures of the sensillum; a second, that there are changes in the responsiveness of the primary receptor itself. With respect to the first possibility either or both of two mechanisms could be involved: changes in the diameter of the pore at the tip of the hair; changes in the quantity of extracellular fluid extruded at the tip. It is quite possible that the tip of the hair opens and closes. This action has been shown to be a natural feature of some gustatory receptors of *Locusta migratoria* (L.) (Blaney, Chapman and Cook 1971; Bernays, Blaney and Chapman 1972; Bernays and Chapman 1972). Stürckow, Holbert and Adams (1967) have demonstrated that the tip of labellar hairs of *Phormia* may be single or bifurcate. High humidity and water cause the tip to "open" whereas treatment with such reagents as fixatives cause it to "close". Our electronmicrographs show, however, that it is not the pore in which the dendrites lie that opens and closes but rather that a line of weakness between the terminations of the two lumina allows the hair to "split". The severity of the treatment that is required to induce this action in *Phormia* argues against it being a natural phenomenon. In any case, it is difficult to imagine that successive applications of the same solution would cause reversible opening and closing.

The other possibility is more likely, namely, that there are changes in the nature or quantity of the extracellular fluid bathing the tips of the dendrites. Assuredly the dendrites are constantly bathed in a fluid. Under certain conditions, again not necessarily normal, an excess of this viscous fluid is extruded from the tip (Stürckow 1967). In a series of experiments in which a solution flowed constantly over the tip of the hair Stürckow (1967) observed in two cases a sudden increase in excitation without any change in stimulus. She suggested that there is a correlation between the frequency of action potentials generated by the salt receptor (and the water receptor) and the presence or absence of exudate.

An alternative to involvement of accessory structures is irregularity in the responsiveness of the primary receptor itself. Contrary to popular conceptions the responses of receptors to single stimuli are not very uniform. They are extremely ragged. This is even true of responses of such stable receptors as the visual cells of *Limulus* for which the stimulus can be impeccably controlled. The noisiness of this response is a reflection of random fluctuations in the membrane potential. "Noise" severely restricts the amount of information about the stimulus that the fiber can carry; however, it may be giving information about the state of the receptor that is useful to the organism. Preliminary analyses show that the labellar chemoreceptor is no exception in so far as noise is concerned. The train of impulses that it generates in response to a single continuing stimulus is a stochastic point process, stochastic because of random variation, point process because the spikes are essentially instantaneous and indistinguishable from one another. Processes of this kind can be stationary or non-stationary. In the first instance the underlying probability distribution does not depend on the time of observation, one sample of time being the same as another. Non-stationary processes are those in which there is a trend, as with adaptation. The response of the chemoreceptor was treated as stationary by splitting it into segments - the phasic portion and the tonic portion. Since counting and averaging spikes gives only crude information more detailed preliminary analyses were made. Spike interval measurements indicated that the response shows a Poisson distribution and is random. Higher order analyses showed no serial dependence when the stimuli were water, salt, or sugar.

Given the fact that the system is inherently noisy, it is hardly surprising that the response is not reproduced exactly upon successive stimulations. To what extent this variation reflects an inherent instability, that is independent of antecedant events, is difficult to surmise and even more difficult to test. An attempt was made to control one effect of previous stimulation, namely, adaptation. Whether or not there are longer lasting effects produced by repetitive stimulation is not known. Many compounds have inhibitory, synergistic, narcotic, and other effects of various durations. It is not inconceivable that excessive stimulation with such "normal" stimuli as water, sugar, and salt may also unstabilize the receptor; however, if the inter-stimulation time is increased in order to compensate for any of these hypothetical effects, other sources of variability (e.g., altered water balance, change in volume of the crop, etc.) are introduced.

If the variability is intrinsic, does it have any relevance to behavior or is it trivial? In seeking to answer this question it is well to consider the phasic and tonic portions of the response separately. Dethier (1968) and Getting (1971) have shown that the information which is significant for the *initiation* of feeding, that is, for extension of the proboscis, is contained in the transient phase - in the first few milliseconds. It is of interest that the rat also makes a clear gustatory discrimination within the first 250-600 msec of stimulation and must, therefore, be acting on information contained in the phasic as well as the tonic portion of the sensory response (Halpern and Tapper 1971). In the fly the number of impulses generated by a salt receptor within the first 80 msec can determine whether the response takes the form of acceptance or rejection (Dethier 1968). In the case of the sugar receptor extension of the proboscis (the haustellum only) is triggered by temporal summation of sensory activity during a 20 msec period after the first sensory impulse (Getting 1971). This is a very fine temporal discrimination, but it is not incompatible with the existence of variability in the generation of action potentials. If the variability is such that occasionally there are no intervals as short as 20 msec after the first impulse, there is quite simply no extension of the proboscis. In short, the invariance of response to a given stimulus would depend at the first level on how faithfully the receptor responds within the critically short interval.

Less is known about the contribution to behavior made by the tonic phase. In so far as extension of the haustellum is concerned, the duration and number of impulses in the motor neuron to the extensors of the haustellum are determined in part by the duration of sensory input; however, if the interspike interval falls below 20 msec, motor response cannot be maintained (Getting 1971). Thus, during a prolonged stimulation fluctuations in the tonic response which caused the interval to exceed the limit at any moment could terminate extension. If sensory responses to repetitive stimulation varied, as in this study, extension would be maintained in some cases and not in others even though the stimulus situation was identical in all. It is probable also that a continuous flow of information is necessary to maintain cibarial pumping, but this phase of feeding behavior has not been thoroughly investigated (cf. Rice 1970).

The foregoing remarks apply to laboratory situations where behavioral responses can be elicited by stimulation of a single receptor. That is a highly artificial situation. The fly's central nervous system is normally the recipient of information derived from a large population of chemoreceptors. If only a few were variable in their responses, their impact on behavior would be negligible; however, the fact that all thirty-seven of the hairs chosen in this study showed comparable variation suggests that they are representative of the population. If they varied in synchrony, one might expect to see this reflected in behavior of the fly when the whole labellum or all the tarsi were stimulated repeatedly for the reasons already given in the case of stimulation of single receptors. It is much more likely, however, that there is no synchrony. This likelihood could be tested experimentally only by simultaneous recording from many receptors or by monitoring integrated responses in the whole labellar nerve. Neither of these approaches is currently possible.

One might well ask why sensory variability is tolerated. It could be that the extent of variability is trivial as far as the requirements of the animal are concerned. Possibly it is energetically uneconomical to build and operate high fidelity systems particularly when random variability can be smoothed out by interneurons at higher levels. In most sensory systems (e.g., the olfactory, visual, and auditory systems of vertebrates and the visual and auditory systems of insects) there are successive levels of convergence giving rise to neural responses at successive levels of abstraction. As Knight (1972) has pointed out "unusually precise over-all results arise from the functioning of components which have very modest precision in their individual structure and behavior".

It is obvious that profound fluctuations in the responsiveness of an animal have their genesis in changing internal status or "irrelevant" sensory input affecting the receptivity of the central nervous system to specific incoming sensory information. In this respect all animals are inconstant. It is also clear that sensory systems themselves are inconstant. Some of the changes that occur in them are irreversible as, for example, those caused by age. In *Phormia* both the sensitivity of labellar receptors and the percent of the population that is operational decrease with age (Stoffolano, this volume). Other changes are induced by antecedant stimulation. The most transient of these is the level of adaptation. Longer lasting but still reversible are narcosis and various kinds of inhibition. Any one of these peripheral effects can affect behavioral responsiveness when all other variables are controlled. There is also an intrinsic variability in sensory systems existing independently of outside influences. In laboratory situations, especially where one is attempting to correlate behavior with activity in a single receptor, this inconstancy can influence behavior. In nature it is swamped by the domination of factors acting on the central nervous system.

REFERENCES

BERNAYS, E.A., BLANEY, W.M., CHAPMAN, R.F.: Changes in chemoreceptor sensilla on the maxillary palps of *Locusta migratoria* in relation to feeding. J. exp. Biol. 57, 745-753 (1972).

BERNAYS, E.A., CHAPMAN, R.F.: The control of changes in peripheral sensilla associated with feeding in *Locusta migratoria* L. J. exp. Biol. 57, 755-763 (1972).

BLANEY, W.M., CHAPMAN, R.F., COOK, A.G.: The structure of the terminal sensilla on the maxillary palps of *Locusta migratoria* (L.), and changes associated with moulting. Z. Zellforsch. mikrosk. Anat. 121, 48-68 (1971).

BRELAND, K, BRELAND, M.: The misbehavior of organisms. Am. Psychol. 16, 681-684 (1961).

CAMHI, J.M., HINKLE, M.: Attentiveness to sensory stimuli: central control in locusts. Science, N.Y. 175, 550-553 (1972).

DETHIER, V.G.: Adaptation to chemical stimulation of the tarsal receptors of the blowfly. Biol. Bull. mar. biol. Lab., Woods Hole 103, 178-189 (1952).

DETHIER, V.G.: Insects and the concept of motivation. *In* "Nebraska Symposium on Motivation" (Ed. D. Levine), pp. 105-136. Lincoln,

Nebraska: Univ. Nebraska Press (1966).
DETHIER, V.G.: Chemosensory input and taste discrimination in the blowfly. Science, N.Y. 161, 389-391 (1968).
DETHIER, V.G.: Sensitivity of the contact chemoreceptors of the blowfly to vapors. Proc. natn. Acad. Sci. U.S.A. 69, 2189-2192 (1972).
EVANS, D.R., MELLON, DeF.: Electrophysiological studies of a water receptor associated with the taste sensilla of the blowfly. J. gen. Physiol. 45, 487-500 (1962a).
EVANS, D.R., MELLON, DeF.: Stimulation of a primary taste receptor by salt. J. gen. Physiol. 45, 651-661 (1962b).
GETTING, P.A.: The sensory control of motor output in fly proboscis extension. Z. vergl. Physiol. 74, 103-120 (1971).
GILLARY, H.L.: Quantitative electrophysiological studies on the mechanism of stimulation of the salt receptor of the blowfly. Doctoral Dissertation, The Johns Hopkins Univ., Baltimore, Maryland. (1966a).
GILLARY, H.L.: Stimulation of the salt receptor of the blowfly. I. NaCl. J. gen. Physiol. 50, 337-350 (1966b).
HALPERN, B.P., TAPPER, D.N.: Taste stimuli: quality coding time. Science, N.Y. 171, 1256-1258 (1971).
HODGSON, E.S.: Electrophysiological studies of arthropod chemoreception. II. Responses of labellar chemoreceptors of the blowfly to stimulation by carbohydrates. J. Insect Physiol. 1, 240-247 (1957).
HODGSON, E.S., ROEDER, K.D.: Electrophysiological studies of arthropod chemoreception. I. General properties of the labellar chemoreceptors of Diptera. J. cell. comp. Physiol. 48, 51-76 (1956).
KNIGHT, B.W.: Dynamics of encoding in a population of neurons. J. gen. Physiol. 59, 734-766 (1972).
MORITA, H.: Initiation of spike potentials in contact chemosensory hairs of insects. III. D.C. stimulation of generator potential of labellar chemoreceptor of *Calliphora*. J. cell. comp. Physiol. 54, 189-204 (1959).
REES, C.J.C.: The effect of aqueous solutions of some 1:1 electrolytes on the electrical response of the type 1 ("salt") chemoreceptor cell in the labella of *Phormia*. J. Insect Physiol. 14, 1331-1364 (1968).
RICE, M.J.: Cibarial stretch receptors in the tsetse fly (*Glossina austeni*) and the blowfly (*Calliphora erythrocephala*). J. Insect Physiol. 16, 277-289 (1970).
STÜRCKOW, B.: Occurrence of a viscous substance at the tip of the labellar taste hair of the blowfly. *In* "Olfaction and Taste. II" (Wenner-Gren Centre International Symposium Series, Vol. 8) (Ed. T. Hayashi), pp. 707-720. Oxford: Pergamon Press. (1967).
STÜRCKOW, B., HOLBERT, P.E., ADAMS, J.R.: Fine structure of the tip of chemosensitive hairs in two blowflies and the stable fly. Experientia 23(9), 780-782 (1967).
WILCZEK, M.: The distribution and neuroanatomy of the labellar sense organs of the blowfly *Phormia regina* Meigen. J. Morph. 122, 175-201 (1967).
WOLBARSHT, M.L.: Electrical activity in the chemoreceptors of the blowfly. II. Responses to electrical stimulation. J. gen. Physiol. 42, 413-428 (1958).

CONTROL OF FEEDING AND DRINKING IN DIAPAUSING INSECTS

J.G. Stoffolano, Jr.

Department of Entomology, University of Massachusetts, Amherst, Massachusetts, U.S.A.

The insects provide an excellent example of the diverse morphological, physiological, and behavioral adaptations involved in obtaining food and water. In addition to coping with the problem of obtaining a given food and water supply, insects frequently face periods completely devoid of either of these commodities. These periods of stress have placed selective pressure on insects forcing them to develop specific strategies to counter the natural pressures imposed by the adverse environmental conditions.

In addition to the lack of food during the winter months, insects in the temperate region are confronted with freezing temperatures. Adult insects have solved these problems by developing various migratory strategies that either place them in a new favorable environment or else put them at their overwintering site where they survive the winter in a state of arrested development (dormancy). The Monarch Butterfly, *Danaus plexippus* (L.), in North America is an example of a long distance migrant that flies to a favorable environment. In the fall, large numbers migrate in a south-easterly direction from their northern breeding grounds to southern retreats. The Colorado potato beetle, *Leptinotarsa decemlineata* (Say), however, has solved the overwintering problems in another way. In the fall, large numbers of adults fly to their overwintering grounds where they remain throughout the winter in a dormant state.

Both migration and dormancy have similar physiological, ecological, and behavioral components that are under hormonal control (Dingle 1972; Johnson 1969). It is evident that these strategies provide for escape from stress conditions, the one in space and the other in time, and that both probably evolved from the same syndrome (Dingle 1972).

Diapause, a type of dormancy which occurs in response to extreme and long adversities, is a strategy evolved by insects to survive periods of low or high temperatures, and lack of water or food. Diapause can occur at any stage in the life of insects but usually only occurs during one stage in the development of a given species. Recently, Mansingh (1971) schematically outlined the interplay of environmental stimuli with the neuro-endocrine systems involved in insect diapause. He also outlined the three basic periods of diapause (pre-diapause, diapause, and post-diapause) and the various phases (preparatory, induction, refractory, activated and termination) (see Fig. 1) associated with it.

Since diapause is an adaptation of environmental stress, focus should be placed on the various physiological, ecological and behavioral mechanisms developed by the organism to survive these unfavorable periods. Some of the physiological mechanisms developed by diapausing adult insects are a reduced oxygen uptake, a reduced body water content, increased lipids, fat body hypertrophy and a reproductive arrest. An ecological mechanism present in almost all diapausing adult insects is some type of migratory strategy (Dingle 1972; Johnson 1969). Few experimental reports have been given on the behavioral mechanisms developed by diapausing adult insects. In general, diapausing adults usually become photonegative, form aggregations in refuges, refuse to mate, are not attracted to baits and usually cease feeding.

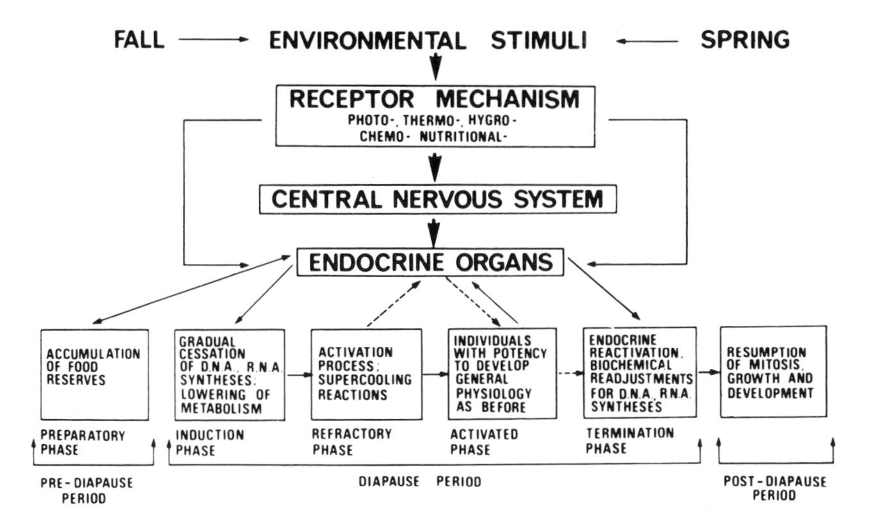

Fig. 1. Schematic representation of the relationships between environmental factors and the neuro-endocrine systems involved in insect diapause. (From Mansingh 1971.)

As with migratory insects, diapausing adults are usually non-reproductive and usually fail to feed. The significance of the failure of a migratory strategist to feed and reproduce are being investigated by Dingle and his co-workers (Dingle, Caldwell, this volume). The reproductive arrest associated with adult diapause has been adequately studied (Beck 1968; Danilevskii 1965; Lees 1955), but the role of feeding prior to, during, or after diapause has not been experimentally investigated. The bulk of information available on feeding during diapause is in the form of casual observations and represents little experimental work. This paper is designed to explore some of these casual observations, to relate our knowledge of the behavior of diapausing adults to what we already know about feeding and drinking under non-diapause inducing conditions and to present some recent work demonstrating the role of the central nervous system in regulating ingestion in the diapausing adult blowfly, *Phormia regina* (Meigen).

FEEDING AND DRINKING BEFORE AND DURING DIAPAUSE

Do diapausing adults eat and drink?

Lees (1955) stated that "The alternate patterns of behaviour that are typical of many insects with facultative diapause must be considered". Later de Wilde (1970) reported that diapause represents not only extrinsic control over growth, differentiation and reproduction, but also over behavior. This hypothesis, that diapause represents extrinsic control over behavior, is well founded but needs experimental support. Since feeding and drinking are such vital and important behaviors for individual survival, their measurement prior to, during, and after diapause becomes important.

An important question to answer is whether insects entering or in a winter diapause take water. The factors regulating water intake in insects have been aptly discussed and reviewed by several workers (Barton Browne 1964; Dethier and Evans 1961; Edney 1957; Evans 1961; Barton Browne 1968; Barton Browne and Dudziński 1968). From the work of Salt (1953, 1961) and Asahina (1969), it seems evident that an insect that eats or drinks during winter diapause faces the danger of freezing since both of these activities reduce the insect's cold-hardiness. We know very little about the factor(s) regulating water intake in overwintering insects and such studies should provide us with a more complete understanding of the control of water intake in insects in general.

Reports have been made that certain diapausing insects cease feeding (Guerra and Bishop 1962; Hodek 1967; McMullen 1967; Wallis 1959) whereas several investigators have reported that, while feeding may continue during diapause, it is reduced or intermittent (Ankersmit 1964; Burges 1960; Davey 1956; Siew 1966). Recently, Tauber and Tauber (1973) demonstrated that under long days the presence or absence of prey alone could induce, avert, or terminate the imaginal reproductive diapause in *Chrysopa mohave* Banks. They also noted that winter diapause is under photoperiodic control only until about December, after which time the absence of prey probably contributes to maintenance of the diapause. Adults exposed to short days entered diapause even when fed prey. Whether adults in the field during the fall would consume prey was not discussed. It is evident that a more complete investigation is needed to understand the mechanism(s) controlling feeding prior to, during, and after diapause.

Selective feeding prior to diapause

The ability of organisms to increase consumption of specific dietary components in response to periods of metabolic stress, represents a common homeostatic behavior among animals (Denton 1967; Leshner, Collier and Squibb 1971; Richter and Barelare 1938; Rozin 1967; Soulairac 1967); however, the mechanisms regulating such behaviors are poorly understood. The concept of behavioral homeostasis has been discussed by McFarland (1970) and seems to apply to insects during the fall in temperate regions. Dethier (1969) reported that some larval insects are seized by a hyperphagia that he likened to the hyperphagia that seizes certain species of birds prior to migration and also certain mammals prior to hibernation. Whether such a hyperphagia is exhibited by adult insects destined to diapause in the fall is unknown. From Mansingh's scheme (see Fig. 1), it appears evident that such a period should exist during the preparatory phase when large accumulations of food reserves are set aside for the winter. Leshner, Collier and Squibb (1971) recently demonstrated that rats housed at 2°C and offered diets containing different concentrations of carbohydrates and proteins selected a higher percentage of the diet containing more carbohydrate. Does such selection occur in insects, and if so, what are the mechanisms underlying such behavior?

Dethier (1961), and Strangeways-Dixon (1959, 1961) both demonstrated that flies actively selected a carbohydrate or protein diet depending on their physiological state. The mechanisms underlying selective feeding in non-diapausing insects have been studied but no attempt has been made to examine the factors operative in diapausing insects. No environmental stress was placed on the organisms in either Dethier's or Strangeways-Dixon's work. In the desert locust, *Schistocerca*, two major growth phases have been recognized, somatic and reproductive growth (Mordue *et al.* 1970). The amount of protein produced during somatic and reproductive growth was shown to be approximately the same, but twelve times as much carbohydrate was taken during the somatic as in the reproductive phase (Mordue *et al.* 1970). Hill, Luntz and Steele (1968) have shown that during this somatic growth phase the locusts select bran in preference to lettuce. Luntz (1968) has suggested that this selection of bran over lettuce may be because the insects can readily obtain the required amount of the carbohydrate from bran without encountering the problem of water loading. It is evident then that insects can and do actively select the type of diet according to their internal physiological state. No work has been done, however, on the ability of insects, during the preparatory phase of diapause to actively select a carbohydrate diet over a protein one. It would also be worth-while to determine whether or not the insect, prior to diapause, increases its carbohydrate intake over that when it is not stressed, i.e., under non-diapausing conditions.

What can the study of adult mosquitoes tell us about feeding prior to and/or during diapause?

Mosquitoes, as other Diptera, can either produce one batch of eggs without a previous protein meal (autogenous) or else require a protein meal prior to ovarian development (anautogenous). Some species have both autogenous and anautogenous strains. Autogenous forms, however, require a blood meal for development of the second and subsequent batches of eggs.

Adult female mosquitoes fail to take a bloodmeal prior to or during diapause. Since we will be looking in detail at the role of feeding during diapause in the adult blowfly, *Phormia regina*, which also requires protein for its egg development, a brief look at the role of feeding and ovarian development in the mosquito may provide us with some information that may apply to our study of the control of feeding in our model system for *Phormia*. Exactly why does the female mosquito need a bloodmeal anyway, and what is the relationship between blood feeding and ovarian development? In a series of excellent experiments, Lea (1970) provided some answers to these questions.

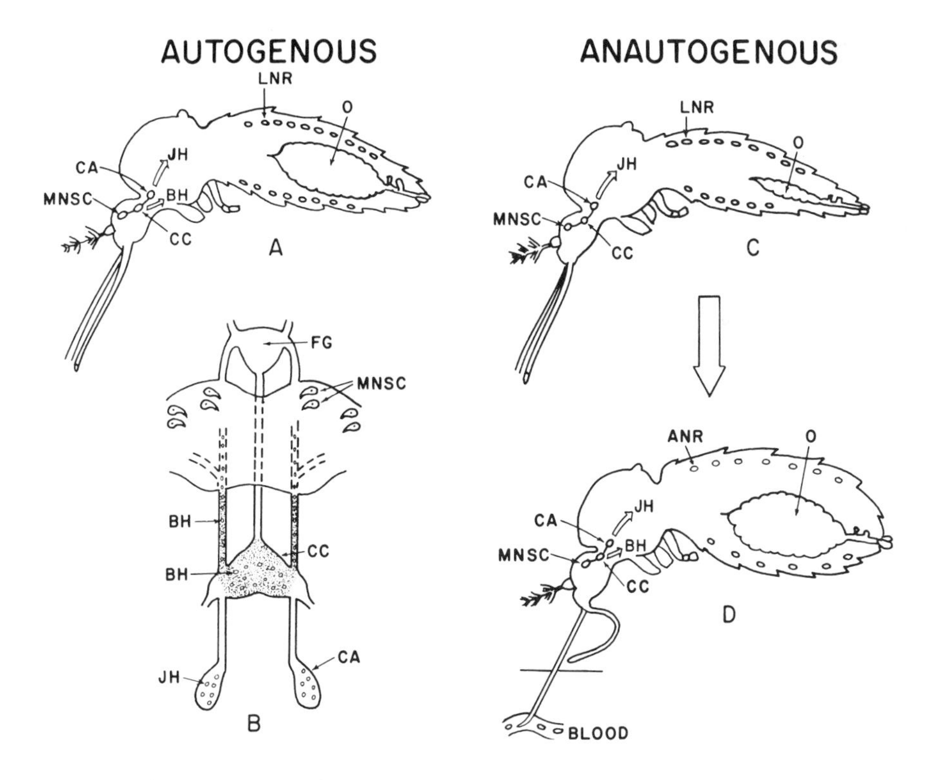

Fig. 2. Schematic representation of the endocrine glands and their secretions that are involved in the control of ovarian development in autogenous (A and B) and anautogenous (C and D) mosquitoes

Both juvenile hormone (JH) and brain hormone (BH) are required for ovarian development (O) in most Diptera. The JH is believed to be involved in protein synthesis (Orr 1964a; Wilkens 1969), while BH appears to be necessary for the uptake of yolk protein (vitellogen) (Clift 1971; Pratt and Davey 1972; Telfer 1965). In an autogenous mosquito both JH and BH are present in the hemolymph at emergence; consequently, the females are able to produce one batch of eggs without a bloodmeal (see Fig. 2A). The protein reserves for vitellogenesis are obtained from the larval nutrient reserve (LNR), at this time in the form of pupal fat body. This fat body, unlike adult fat body that contains only lipid and glycogen, also contains protein. The BH from the medial neurosecretory cells (MNSC) passes down the axon to the corpora cardiaca (CC) where it is stored (see Fig. 2B). This material is stored in the CC until it is released into the hemolymph. The corpora allata (CA) produces and secretes JH, necessary for vitellogenesis, into the hemolymph. In an anautogenous form (Fig. 2C), Lea (1970) showed that the ovaries failed to develop because of failure of the CC to release the BH. He found that JH was already in the hemolymph shortly after adult emergence. One function of the bloodmeal is to provide the stimulus causing the CC to release the BH. Once the female has fed on blood (Fig. 2D), both JH and BH are present in the hemolymph and ovarian development can proceed. The question that emerges from this scheme is, how does blood feeding cause the CC to release the BH? Larsen and Bodenstein (1959) hypothesized from their work that the sustained stretching of the gut by a bloodmeal triggered a series of events leading to maturation of the ovaries. Recently, Bellamy and Bracken (1971) suggested that the Larsen and Bodenstein hypothesis was incorrect and suggested that the factor(s) causing egg development to proceed were "nutritional" and/or were due to a "metabolite" in the hemolymph that resulted from protein feeding. It is evident, especially in anautogenous forms, that the female must obtain a bloodmeal prior to ovarian development.

It is a well established fact that females entering diapause fail to bite (de Buck and Swellengrebel 1934; Schaefer, Miura and Washino 1971; Tate and Vincent 1932, 1936; Washino 1970; Washino and Bailey 1970; Washino, Gieke and Schaefer 1971). The only experimental work designed to study the effect of photophase on feeding behavior was done by Danilevskii and Glinyanaya (1958), Eldridge (1963), Hosoi (1954), and Tate and Vincent (1932). These investigations demonstrated that, at short photophases, adult mosquitoes destined to diapause failed to bite, but no effort was made to elucidate the underlying mechanism(s). Stoffolano (1968), working with diapausing and non-diapausing face flies, *Musca autumnalis* DeGeer, showed that diapausing adult flies had an elevated tarsal acceptance threshold to sugars and proteins. This was attributed to stretching of the abdomen by fat hypertrophy and also to the slow emptying of the crop. Both of these factors can activate stretch receptors, thus sending negative feedback to the central nervous system that counters the peripheral input (Dethier 1969). This hypothesis was tested by Chen (1969) who, using mosquitoes, reported that these factors did not influence their biting behavior. Her work suggests that some other factor, possibly hormonal influence on the CNS, resulted in failure of mosquitoes to bite in the fall. It seems likely that such a hormonal influence is the primary cause for failure to bite or feed in the fall since fat hypertrophy does not occur until several days after feeding has begun. It is clear, in any case, that some factor has to act on the central or peripheral systems, from the time of emergence, to prevent feeding on protein. It is possible that the negative feedback provided by distension of the abdomen and slow emptying of the crop in overwintering mosquitoes may result in the maintenance of the winter inhibition, but apparently some other initial factor prevents feeding on a host prior to diapause induction.

What would happen if a mosquito does take a bloodmeal in the fall? It has been reported that cold tolerance and survival are reduced in female *Culex restuans* (Wallis 1959), and *C. pipiens* (Hall 1967) that take a bloodmeal prior to overwintering. Clements (1963) also reported that females emerging in spring or summer, and which then pass through one or more gonadotrophic cycles, fail to enter diapause in the fall. It is suggested that a bloodmeal would switch (via the hormonal system) the female's physiology from a diapausing to a non-diapausing state. Because of this it is evolutionarily adaptive for a mosquito destined to overwinter not to bite.

It is evident that feeding on blood plays an extremely important role in the ultimate physiological state of the adult female mosquito. The mechanisms resulting in the failure of females in the fall to exhibit the "biting-drive" described by Clements (1963) still remains a mystery. What is the underlying physiological mechanism that results in this diapause behavior? Is the control over biting regulated by the olfactory or gustatory receptors and then is the control central and/or peripheral?

Phormia regina - a model system

The report by Stoffolano (1968) raised the question as to whether the elevated tarsal acceptance threshold of diapausing face flies was due to peripheral or central inhibition or both. Experiments using face fly were designed to answer this question, but after consideration, it was decided that face fly was not a suitable candidate for an investigation of this problem, as we lacked information about the factors regulating its feeding and also about the electrophysiological characteristics of its receptors. Since there is a large body of relevant information about *Phormia regina* (cf., Dethier 1969) it was decided that this species, if it were found to have an adult diapause, would be the best insect for such a study. An examination of the literature showed that an adult diapause had not been reported for *Phormia*, but Wallis (1962) and Dondero and Shaw (1971) provided evidence that *Phormia regina* overwintered in the temperate region as an adult. These reports suggested an adult diapause. Experiments at Princeton University (Stoffolano, unpublished data) and the University of Massachusetts (Stoffolano, Calabrese and Greenberg, in press), showed that *Phormia* has an imaginal diapause that is induced by short photophase and low temperature.

Before proceeding to use *Phormia* as our model, it is necessary to know whether the tarsal acceptance thresholds for non-diapausing and diapausing *Phormia* adults differ. Then, in order to establish the site of control it must be determined whether there is a distinct difference between the input and the output. An attempt at this has been made by comparing, behaviorally, the tarsal acceptance thresholds and intakes of diapausing and non-diapausing flies and also by comparing, electrophysiologically, the mean impulse frequency of their labellar chemoreceptors. A decision as to the center for control can be made according to whether or not behavioral differences are reflected by electrophysiological differences in the peripheral chemoreceptors.

Tarsal acceptance thresholds

Experiments at Princeton University on *Phormia regina* have shown that, when tested on various concentrations of sucrose, diapausing adults had elevated tarsal acceptance thresholds when compared to non-diapausing adults (Stoffolano, in press) (see Fig. 3). Before assessing this effect, however, it is necessary to examine in detail the changes with age in non-diapausing flies. Fig. 3 shows that non-diapausing adults, 3½ days old were highly responsive. Percentages of flies responding to various concentrations being 1M sucrose, 90%; 0.5M, 85%; 0.25M, 81%; and 0.125M, 69%. These data are in agreement with previous results of Dethier and

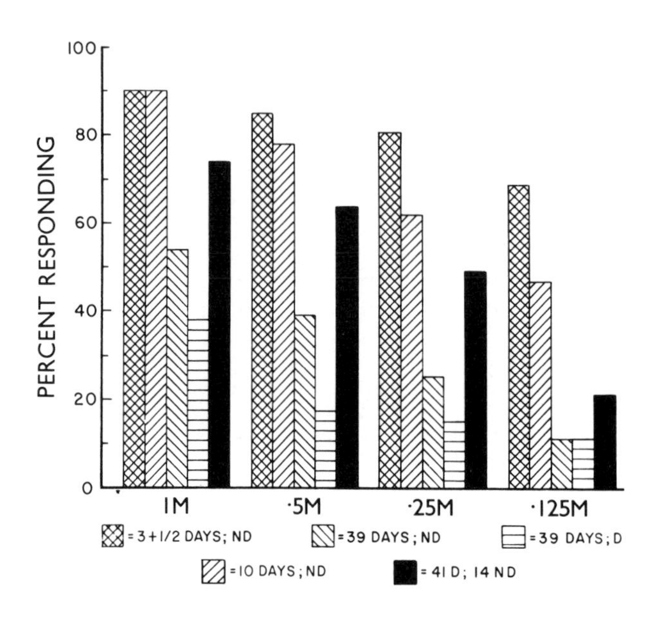

Fig. 3. Effects of age and diapause on the tarsal acceptance thresholds of *Phormia regina* to various concentrations of sucrose

his co-workers on non-diapausing flies. The results reveal, however, that as the non-diapausing fly ages, its tarsal acceptance threshold increases. This finding is consistent with the work of Dethier and Chadwick (1948) who reported that age was one of the factors that influenced gustatory thresholds. At 3½ days, 90% of the non-diapausing flies responded to 1M sucrose, at 10 days, 90% responded whereas only 54% responded at 39 days. The influence of age on the percentage responding is seen even better at the lower concentrations of sucrose tested (0.5M, 0.25M and 0.125M) (see Fig. 3).

How does one explain this aging influence on the gustatory theshold? Rees (1970), and recently Stoffolano (1973), both demonstrated that as flies age the percentage of sugar sensilla that are operative decreases. Since Dethier (1969) stated that the input from these receptors is additive, one can see how a decrease in the number of operative sensilla may result in an increase in the threshold with age.

Comparison of the tarsal acceptance threshold of non-diapausing (39 days old, exposed to 18°C and a 16 hr photophase) to that of diapausing flies (39 days old, exposed to 18°C and a 9 hr photophase) reveals that a smaller proportion of diapausing flies of comparable age responded to 1M, 0.5M and 0.25M sucrose solution. Fig. 3 shows that 54% of the non-diapausing flies, 39 days old, responded to 1M sucrose compared with only 38% of diapausing flies. A similar trend exists for flies tested at 0.5M and 0.25M sucrose. The question that now emerges is whether the elevated tarsal acceptance thresholds of these diapausing and non-diapausing flies, 39 days old, as compared with that of young non-diapausing flies, is due to the same or diffent mechanisms.

The above results suggest that the factor or factors causing the elevated tarsal acceptance threshold of non-diapausing and diapausing flies of comparable age may be the same. In order to test this, flies 41 days old and in diapause were placed at non-diapause inducing conditions (27°C and a 16 hr photophase) for 14 days and were tested when they were 55 days old. These flies showed a much lower threshold than did diapausing flies 39 days old. Results shown in Fig. 3 reveal that diapausing flies, 41 days old then removed to non-diapause inducing conditions for 14 days had 74% responding to 1M sucrose compared to 38% responding for diapausing flies 39 days old, and 54% for non-diapausing flies 39 days old. In other words, something happened to the diapausing flies during the exposure to the non-diapause inducing conditions that resulted in a release of the factor(s) that produced the elevated threshold. From these results, one would like to hypothesize that the elevated tarsal acceptance threshold in non-diapausing flies was due to aging whereas the elevated threshold in diapausing flies was due, at least in part, to a diapause factor which was removed when the flies were placed at non-diapause inducing conditions. This view is supported by the observation that when non-diapausing flies (39 days old) were removed from a chamber maintained at 18°C with a 16 hr photophase and placed at 27°C with a 16 hr photophase for the scheduled 3 day starvation period prior to testing, they all failed to survive the starvation period. Diapausing flies of the same age, however, survived a 3 day starvation period.

Table 1. Mean intake and duration of drink of diapausing and non-diapausing *Phormia regina* fed a 0.125M sucrose solution.

Physiological condition and age when tested	Intake (μl)	Seconds
N-D*	24.7	140.6
D*	6.0	37.5
D-41 days; N-D, 8 days	22.0	168.1

N-D = non-diapause inducing conditions
D = diapause inducing conditions
* = 40 days old

Intakes

In order to show another kind of difference between diapausing and non-diapausing flies, the amount of 0.125M sucrose solution a fly would imbibe during one continuous, uninterrupted drink was measured. All flies were offered granulated sucrose, fresh beef liver and water daily. Prior to testing, all the food, except water, was removed and the flies were starved for 3 days. Non-diapausing, 40 day old flies had a greater mean intake (24.7 μl) and a mean duration of sucking (140.6 sec) when compared to diapausing flies 41 days old (6.0 μl and 37.5 sec) (Table 1). The intake data support the threshold data previously discussed for these two groups. One would not expect flies in diapause to imbibe much sucrose solution because of their elevated tarsal acceptance thresholds; but, what would be the intake of diapausing flies when returned to non-diapause inducing conditions? Earlier experiments showed that the tarsal acceptance thresholds of diapausing flies decreased when returned to non-diapause inducing conditions, thus it seems logical that the intake would increase. This is exactly what happened. Flies in diapause for 41 days and then returned to non-diapause inducing conditions for 8 days, showed an increase in the amount of 0.125M sucrose solution imbibed, 22.0 μl, compared to 6.0 μl for flies still in diapause. They also drank for a period, 168 sec, which is comparable to the period for non-diapausing controls (140 sec). It is interesting to note that diapausing flies, when returned to non-diapause inducing conditions for 8 days, had an intake (22 μl) comparable to the non-diapausing group (24.7 μl). These data suggest that some factor inhibitory to feeding is removed when diapausing flies are returned to non-diapause inducing conditions. So far, the tarsal threshold and intake data compliment one another. These results do not tell us whether the inhibition removed, when diapausing flies are returned to non-diapause inducing conditions, is central and/or peripheral. They do, however, suggest central control.

Electrical recordings

What information can we obtain by comparing recordings from the labellar sensilla of diapausing and non-diapausing flies? A comparison between the receptor responses of non-diapausing and diapausing flies of comparable age (60 days) when the labellar receptors were stimulated with a 0.5M sucrose solution showed that the frequency (imp/sec) did not differ significantly between the two groups (see Table 2). The impulse frequency for the non-diapausing flies was 34.0 imp/sec for the sugar sensilla while the frequency for the diapausing group was 27.3 imp/sec.

Table 2. Effect of diapause on the mean impulse frequency and percentage of inoperative sugar labellar chemoreceptor sensilla using 0.5M sucrose as testing solution.

Age (days)	Frequency (imp/sec)	Inoperative sensilla	No. flies tested	No. sensilla recorded from
ND, 60	34.0	27.1	6	70
D, 60	27.3	60.2	11	70

ND = non-diapause; D = diapause

An interesting and unexpected result was that the number of inoperative sugar sensilla, for the diapausing flies, was nearly double that of the non-diapausing group (see Table 2). This suggests that diapausing flies emerge from their overwintering sites with part of their peripheral nervous system non-functional. One could then ask whether the neural homeostasis, involved in feeding behavior of flies emerging from winter sites, breaks down? The intake and tarsal threshold data, however, demonstrate that when diapausing flies were returned to non-diapause inducing conditions their intakes and tarsal thresholds were normal. In other words, somehow the flies maintained homeostasis. A similar situation is discussed by Frolkis (1966) who demonstrated that, in aging mammals, there was a decrease of neural influences upon effector organs but, that as the organisms aged, the sensitivity to hormones increased. Consequently, the aging mammal maintains homeostasis by developing a different neuronal strategy. A similar mechanism may exist for aging, diapausing flies. It is suggested that as flies age the sensitivity of the labellar, and probably tarsal, sensilla increases to compensate for the increased number of inoperative sensilla. This hypothesis is in part supported by the work of Gillary (1966) who showed that the labellar receptors of 16 day old, non-diapausing *Phormia* were more sensitive to salt than were those of younger flies. It is highly improbable that the tarsal receptors of diapausing flies are in one case (during diapause) insensitive but then later (during non-diapause inducing conditions) return to normal functioning. A more likely explanation, and one that is supported by my work, is that inhibition is central and not peripheral at the receptor level in diapausing flies and that under non-diapause inducing conditions this inhibition is removed. Knowledge of the factor or factors causing this inhibition and how it is integrated to influence the peripheral receptors remains unknown.

So far, we have demonstrated that failure of diapausing adult *Phormia regina* to imbibe much sucrose solution can be attributed to an elevated tarsal acceptance threshold. The mechanism producing this change in threshold still remains obscure. It appears that the inhibitions are in the central nervous system and not at the peripheral chemoreceptors. In an attempt to better understand the mechanisms regulating metabolic homeostasis throughout the life of the blowfly, a brief discussion of the factors already known to regulate feeding in non-diapausing flies will be given.

The excellent work of Dethier and his co-workers has provided us with an understanding of the factors involved in the feeding behavior of the blowfly under non-diapause inducing conditions (Dethier 1969). In general, feeding results from an interplay of excitation due to peripheral chemoreceptors which is countered by inhibitions from stretch receptors in the foregut that monitor crop emptying (Gelperin 1967) and also from stretch receptors in the abdomen that monitor abdominal distension (Dethier and Gelperin 1967; Gelperin 1971a). It is the inhibitory information from these sources that results in elevation of the tarsal acceptance threshold. Evidence for this was provided electrophysiologically, and also behaviorally, by sectioning the recurrent and abdominal ventral nerves, both of which result in hyperphagia. Recently, however, Getting and Steinhardt (1972) demonstrated that the labellar threshold is not influenced by these sources of negative feedback. Gelperin (1971b, 1972) has reviewed the literature on the regulation of feeding and has also produced a model to account for metabolic homeostasis of carbohydrates in the blowfly. He did not intend this model to account for protein uptake nor did he consider the diapause state. The latter is understandable, since diapause in *Phormia* had not yet been reported. However, in the light of new information, some provision should now be made in this model to account for protein feeding and also the control mechanisms involved in feeding prior to, during, and after diapause. Another fact that was evident from re-examination of this model was that the work on the blowfly has been concentrated mainly on the gustatory chemoreceptors located on the tarsi and labellum. Recently laboratory evidence (Dethier, personal communcation), together with reports from the literature, suggest that the antennae of the blowfly may be more important in feeding than previously thought.

At present, no electrophysiological responses of *Phormia*'s antennae to odors have been recorded and the identification of these receptors rests on indirect evidence (Dethier 1969). Topical application and ablation experiments have pointed to the antennae as the olfactory receptors of the fly; however, behavior after the removal of all known sources of olfactory reception still suggested another sensitive site. It has recently been established by Dethier (1972) that the contact chemoreceptors of the blowfly are sensitive to vapors. In the light of this new information, the role of olfaction in the feeding behavior of the adult blowfly will have to be investigated and an attempt will have to be made to incorporate and integrate the information obtained with our present knowledge of the gustatory receptors. Only then can we obtain a more thorough understanding of feeding behavior patterns in adult flies. The model proposed in Fig. 4 is an extension of Gelperin's model and provides for the incorporation of the olfactory receptors.

It is suggested that, during the summer months, the long photophase and high temperatures induce a non-diapause syndrome in the adult blowfly. In order to account for this influence of photophase and temperature on the whole system, heat and light have been added to the model as sources of energy (Fig. 4). It is believed that the odor of a dead

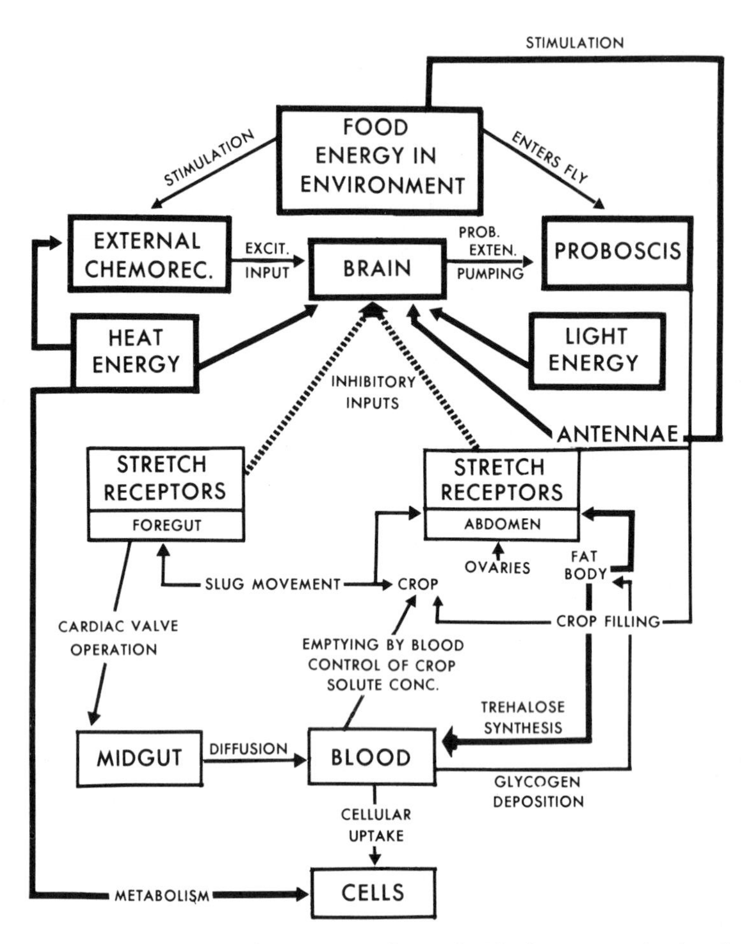

Fig. 4. Model system for the maintenance of metabolic homeostasis in the blowfly, *Phormia regina*

animal attracts females prior to the maturation of their ovaries. Contact with the carcass results in stimulation of tarsal chemoreceptors, proboscis extension occurs, and protein feeding commences. It has previously been shown that protein is necessary in the diet of female blowflies (Orr 1964a,b) for egg development and that this protein feeding somehow triggers the endocrines needed in egg development. Bennettová-Řežábová (1972) has recently shown that sectioning of the ventral nerve cord prevents ovarian development in *Phormia regina*. She suggests that, somehow, information from the central nervous system, possibly via the MNSC from the brain, is necessary for normal fat body metabolism and egg development.

When the ovaries are completely developed, further feeding on protein may be limited by inhibitory feedback, that either elevates the tarsal acceptance threshold or somehow influences the response of the female to odors. Pospíšil (1958) demonstrated a switch in odor preference of adult female flies that was dependent on the stage of the gonadotrophic cycle. Virgin females with undeveloped ovaries responded positively to odors from associated decaying animal tissue, whereas during the gravid stage, these females failed to respond. Stoffolano (1968) demonstrated that gravid females of *M. autumnalis* had elevated tarsal acceptance thresholds to sucrose solutions, suggesting that the ovaries acted on the stretch receptors of the abdomen, in a manner similar to a full crop, and thus were able to provide the necessary feedback to prevent feeding (see Fig. 4). Whether this applies to protein feeding remains to be established. It is suggested that, in the fall, the shortening of the photophase accompanied by lower temperatures, results in an alteration of olfactory response which is probably correlated with changes in olfactory receptor thresholds. How this is brought about, and whether it is centrally or peripherally controlled, remains unknown. The fact that adult blowflies fail to frequent protein baits in the fall (Schoof and Savage 1955) is evidence for the occurrence of this change in olfactory threshold. Instead, adult flies during the fall, are probably feeding on carbohydrates (mainly plant nectars). The environmental factors have thus produced a shut-down of the endocrines involved in ovarian development. Similar endocrine control of behavior, as influenced by changes in the physical environment, have been demonstrated for other insects (Mordue *et al.* 1970 and de Wilde 1970). If protein feeding did occur, it would force the female into a non-diapause state; consequently, as reported for mosquitoes earlier, their cold hardiness and ability to survive low temperatures might be decreased. It is possible that the presumed carbohydrate feeding in the fall would result in the formation of large stores of fat which in turn would produce a pressure on the abdominal stretch receptors. Such a state may then result in elevation of the tarsal acceptance threshold and thus failure to feed. It is also believed that, during the winter, many adult flies possess a full crop (cf. Stoffolano 1968). As spring approaches, the changing photophase and rise in temperature probably causes an increase of locomotor activity which is followed by emptying of the crop and gradual depletion of the fat reserves. This removal of negative feedback results in lowering of the tarsal acceptance threshold and somehow a reduction in olfactory threshold. Now the fly seeks a protein source. Protein feeding and a reversal in the endocrine strategy of the fly under non-diapause inducing conditions results in egg development.

SUMMARY AND CONCLUDING REMARKS

In order to study the factors controlling feeding and drinking in diapausing insects, the adult blowfly, *Phormia regina*, has been chosen as the model system. The literature on feeding and drinking in diapausing insects is scanty and provides little information with which to generate new ideas. Since very little work has been reported on diapause in adult blowflies, it was felt that a review of imaginal diapause in mosquitoes could provide enough information to generate some new hypotheses as to the underlying mechanisms that control feeding in *Phormia*. This information can be integrated with the case study of feeding behavior on non-diapausing *Phormia regina* to develop a hypothetical model, thus permitting the investigator to re-examine the role of olfactory and the gustatory receptors in the control of feeding during non-diapause inducing conditions.

The statement of Chen, that mosquitoes refuse a bloodmeal which is a reflection of changes occurring in the neuroendocrine system in response to low temperature and short photophase, raises several new ideas. She suggests that there is some neuroendocrine influence on the nervous system that alters the biting behavior of mosquitoes in the fall (Chen 1969). This work does not agree with the hypothesis originally proposed by Stoffolano (1968). While discussing the concurrent changes associated with the fly's reproductive cycle, and its response to protein and carbohydrate, Dethier (1969) noted that these changes could occur either at the receptor level or in the central nervous system. He reported that such changes could be brought about by one or more hormones or by changes in the protein titer and that this could be accomplished by the hormones increasing or decreasing the sensitivity of the fly to protein. Preliminary experiments discounted hormonal control; however, Dethier concluded that the idea of hormonal control still warrants further investigations.

It is believed that the central nervous system of insects will be found to be as important as that of higher animals in regulating feeding and drinking and that the study of the mechanisms controlling these vital acts in *Phormia regina* will provide much new information as to the role of environmental factors and hormones on this aspect of the insect's overt behavior.

ACKNOWLEDGEMENTS

The author would like to extend his appreciation to Professor V.G. Dethier and Princeton University for providing the necessary facilities to complete portions of this study. Appreciation is also extended to M. Tauber and T.M. Peters for reading the manuscript and providing helpful suggestions.

This work was supported by National Science Foundation Postdoctoral Research Grant No. 40055 and was conducted in the laboratory of Professor V.G. Dethier at Princeton University.

REFERENCES

ANKERSMIT, G.W.: Voltinism and its determination in some beetles of cruciferous crops. Meded. LandHoogesch. Wageningen. 64, 1-60 (1964).

ASAHINA, E.: Frost resistance in insects. *In* "Advances in Insect Physiology" (Eds., J.W.L. Beament, J.E. Treherne and V.G. Wigglesworth), pp. 1-49. New York: Academic Press (1969).

BARTON BROWNE, L.: Water regulation in insects. A. Rev. Ent. 9, 63-82 (1964).

BARTON BROWNE, L.: Effects of altering the composition and volume of the hemolymph on water ingestion of the blowfly, *Lucilia cuprina*. J. Insect Physiol. 14, 1603-1620 (1968).

BARTON BROWNE, L., DUDZIŃSKI, A.: Some changes resulting from water deprivation in the blowfly, *Lucilia cuprina*. J. Insect Physiol. 14, 1423-1434 (1968).

BECK, S.D.: "Insect Photoperiodism", 288 pp. New York: Academic Press (1968).

BELLAMY, R.E., BRACKEN, G.K.: Quantitative aspects of ovarian development in mosquitoes. Can. Ent. 103, 763-773 (1971).

BENNETTOVÁ-ŘEŽÁBOVÁ, B.: The regulation of vitellogenesis by the central nervous system in the blow fly, *Phormia regina* (Meigen). Acta ent. Bohemoslav. 69, 78-88 (1972).

de BUCK, A., SWELLENGREBEL, N.H.: Behaviour of Dutch *Anopheles atroparvus* and *messeae* in winter under artificial conditions. Riv. Malar. 13, 404-416 (1934).

BURGES, H.D.: Studies on the dermestid beetle *Trogoderma granarium* Everts. IV. Feeding, growth, and respiration with particular reference to diapause larvae. J. Insect Physiol. 5, 317-334 (1960).

CHEN, S.S.: The neuroendocrine regulation of gut protease activity and ovarian development in mosquitoes of the *Culex pipiens* complex and the induction of imaginal diapause. Diss. Abstr. B 30(9), 4187-B (1969).

CLEMENTS, A.N.: "The Physiology of Mosquitoes". Oxford: Pergamon Press 393 pp. (1963).

CLIFT, A.D.: Control of germarial activity and yolk deposition in non-terminal oocytes of *Lucilia cuprina*. J. Insect Physiol. 17, 601-

606 (1971).
DANILEVSKII, A.S.: "Photoperiodism and Seasonal Development of Insects", 283 pp. Edinburgh and London: Oliver and Boyd (1965).
DANILEVSKII, A.S., GLINYANAYA, E.I.: Dependence of the gonadotrophic cycle and imaginal diapause on changes in day-length. Uchen. Zap. liningr. gos. Univ. (Ser. Biol. Nauk) 46, 34-52 (1958).
DAVEY, K.G.: The physiology of dormancy in the sweetclover weevil. Can. J. Zool. 34, 86-98 (1956).
DENTON, D.A.: Salt appetite. *In* "Handbook of Physiology", Vol. 1 (Ed., C.F. Code), pp. 433-459. Washington, D.C.: Am. Physiol. Soc. (1967).
DETHIER, V.G.: Behavioral aspects of protein ingestion by the blowfly *Phormia regina* Meigen. Biol. Bull. mar. biol. Lab., Woods Hole 121, 456-470 (1961).
DETHIER, V.G.: Feeding behavior of the blowfly. *In* "Advances in the Study of Behavior", Vol. 2 (Eds., D.S. Lehrman, R.A. Hinde and E. Shaw), pp. 111-266. New York: Academic Press (1969).
DETHIER, V.G.: Sensitivity of the contact chemoreceptors of the blowfly to vapors. Proc. natn. Acad. Sci. U.S.A. 69, 2189-2191 (1972).
DETHIER, V.G., CHADWICK, L.E.: Chemoreception in insects. Physiol. Rev. 28, 220-254 (1948).
DETHIER, V.G., EVANS, D.R.: Physiological control of water ingestion in the blowfly. Biol. Bull. mar. biol. Lab., Woods Hole 121, 108-116 (1961).
DETHIER, V.G., GELPERIN, A.: Hyperphagia in the blowfly. J. exp. Biol. 47, 191-200 (1967).
DINGLE, H.: Migration strategies of insects. Science, N.Y. 175, 1327-1335 (1972).
DONDERO, L., SHAW, F.R.: The overwintering of some muscoidean Diptera in the Amherst area of Massachusetts. Proc. ent. Soc. Wash. 73, 52-53 (1971).
EDNEY, E.B.: "The Water Relations of Terrestrial Arthropods", 109 pp. London: Cambridge University Press. (1957).
ELDRIDGE, B.F.: The influence of daily photoperiod on blood-feeding activity of *Culex tritaeniorhynchus* Giles. Am. J. Hyg. 77, 49-53 (1963).
EVANS, D.R.: Control of the responsiveness of the blowfly to water. Nature, Lond. 190, 1132-1133 (1961).
FROLKIS, V.V.: Neuro-humoral regulations in the aging organism. J. Geront. 21, 161-167 (1966).
GELPERIN, A.: Stretch receptors in the foregut of the blowfly. Science, N.Y. 157, 208-210 (1967).
GELPERIN, A.: Abdominal sensory neurons providing negative feedback to the feeding behavior of the blowfly. Z. vergl. Physiol. 72, 17-31 (1971a).
GELPERIN, A.: Regulation of feeding. A. Rev. Ent. 16, 365-378 (1971b).
GELPERIN, A.: Neural control systems underlying insect feeding behavior. Am. Zoologist 12, 489-496 (1972).
GETTING, P.A., STEINHARDT, R.A.: The interaction of external and internal receptors on the feeding behaviour of the blowfly, *Phormia regina*. J. Insect Physiol. 18, 1673-1681 (1972).
GILLARY, H.L.: Stimulation of the salt receptor of the blowfly. J. gen. Physiol. 50, 337-350 (1966).
GUERRA, A.A., BISHOP, J.L.: The effect of aestivation on sexual maturation in the female alfalfa weevil (*Hypera postica*). J. econ. Ent. 55, 747-749 (1962).
HALL, D.W.: Factors associated with hibernation of *Culex pipiens* Linnaeus in Central Indiana. M.S. Thesis, Purdue University (1967).
HILL, L., LUNTZ, A.J., STEELE, P.A.: The relationships between somatic growth, ovarian growth, and feeding activity in the adult desert locust. J. Insect Physiol. 14, 1-20 (1968).
HODEK, I.: Bionomics and ecology of predaceous Coccinellidae. A. Rev.

Ent. 12, 79-104 (1967).
HOSOI, T.: Egg production in *Culex pipiens pallens* Coquillett - II. Influence of light and temperature on activity of females. Jap. J. med. Sci. Biol. 7, 75-81 (1954).
JOHNSON, C.G.: "Migration and Dispersal of Insects by Flight", 763 pp. London: Methuen & Co. Ltd. (1969).
LARSEN, J.R., BODENSTEIN, D.: The humoral control of egg maturation in the mosquito. J. exp. Zool. 140, 343-378 (1959).
LEA, A.O.: Endocrinology of egg maturation in autogenous and anautogenous *Aedes taeniorhynchus*. J. Insect Physiol. 16, 1689-1696 (1970).
LEES, A.D.: "The Physiology of Diapause in Arthropods", 151 pp. London: Cambridge University Press (1955).
LESHNER, A.I., COLLIER, G.H., SQUIBB, R.L.: Dietary self-selection at cold temperatures. Physiol. Behav. 6, 1-3 (1971).
LUNTZ, A.J.: Neurosecretory activity and growth during reproductive development in *Schistocerca gregaria* Forsk. M.Sc. Thesis, University of Sheffield. (1968).
MANSINGH, A.: Physiological classification of dormancies in insects. Can. Ent. 103, 983-1009 (1971).
McFARLAND, D.J.: Behavioral aspects of homeostasis. *In* "Advances in the Study of Behavior", pp. 1-26. (Eds., D.S. Lehrman, R.A. Hinde and E. Shaw). New York: Academic Press (1970).
McMULLEN, R.D.: A field study of diapause in *Coccinella novemnotata* (Coleoptera: Coccinellidae). Can. Ent. 99, 42-49 (1967).
MORDUE, W., HIGHNAM, K.C., HILL, L., LUNTZ, A.J.: Environmental effects upon endocrine-mediated processes in locusts. *In* "Hormones and the Environment" (Eds., G.K. Benson and J.G. Phillips), pp. 111-136. London: Cambridge University Press. [Mem. Soc. Endocr. 18] (1970).
ORR, C.W.M.: The influence of nutritional and hormonal factors on egg development in the blowfly, *Phormia regina* (Meigen). J. Insect Physiol. 10, 53-64 (1964a).
ORR, C.W.M.: The influence of nutritional and hormonal factors on the chemistry of the fat body, blood, and ovaries of *Phormia regina* (Meigen). J. Insect Physiol. 10, 103-120 (1964b).
POSPÍŠIL, J.: Some problems of the smell of saprophilic flies. Čas. čsl. Spol. ent. 55, 316-334 (1958).
PRATT, G.E., DAVEY, K.G.: The corpus allatum and oogenesis in *Rhodnius prolixus* (Stdl.). I. The effects of allatectomy. J. exp. Biol. 56, 201-214 (1972).
REES, C.J.C.: Age dependency of response in an insect chemoreceptor sensillum. Nature, Lond. 227, 740-742 (1970).
RICHTER, C.P., BARELARE, B.,Jr.: Nutritional requirements of pregnant and lactating rats studied by the self-selection method. Endocrinology 23, 15-24 (1938).
ROZIN, P.: Thiamine specific hunger. *In* "Handbook of Physiology", Vol. 1 (Ed., C.F. Code), pp. 411-431. Washington, D.C.: Am. Physiol. Soc. (1967).
SALT, R.W.: The influence of food on the cold-hardiness of insects. Can. Ent. 85, 261-269 (1953).
SALT, R.W.: Principles of insect cold-hardiness. A. Rev. Ent. 6, 55-74 (1961).
SCHAEFER, C.H., MIURA, T., WASHINO, R.K.: Studies on the overwintering biology of natural populations of *Anopheles freeborni* and *Culex tarsalis* in California. Mosquito News 31, 153-157 (1971).
SCHOOF, H.F., SAVAGE, E.P.: Comparative studies of urban fly populations in Arizona, Kansas, Michigan, New York, and West Virginia. Ann. Ent. Soc. Am. 48, 1-12 (1955).
SIEW, Y.C.: Some physiological aspects of adult reproductive diapause in *Galeruca tanaceti* (L.) (Coleoptera: Chrysomelidae). Trans. Ent. Soc. Lond. 118, 359-374 (1966).
SOULAIRAC, A.: Control of carbohydrate intake. *In* "Handbook of Physiology", Vol. 1 (Ed., C.F. Code), pp. 387-398. Washington, D.C.: Am.

Physiol. Soc. (1967).
STOFFOLANO, J.G.,Jr.: The effect of diapause and age on the tarsal acceptance threshold of the fly, *Musca autumnalis*. J. Insect Physiol. 14, 1205-1214 (1968).
STOFFOLANO, J.G.,Jr.: Effect of age and diapause on the mean impulse frequency and failure to generate impulses in labellar chemoreceptor sensilla of *Phormia regina*. J. Geront. 28, 35-39 (1973).
STOFFOLANO, J.G.,Jr.: Central control of feeding in diapausing *Phormia regina* (Meigen). J. exp. Biol. (In press).
STOFFOLANO, J.G.,Jr., CALABRESE, E.J., GREENBERG, S.: Imaginal facultative diapause in the black blowfly, *Phormia regina* (Meigen). Ann. Ent. Soc. Am. (In press).
STRANGEWAYS-DIXON, J.: Hormonal control of selective feeding in female *Calliphora erythrocephala* Meig. Nature, Lond. 184, 2040-2041 (1959).
STRANGEWAYS-DIXON, J.: The relationship between nutrition, hormones, and reproduction in the blowfly, *Calliphora erythrocephala* (Meig.). I. Selective feeding in relation to the reproductive cycle, the corpus allatum volume, and fertilisation. J. exp. Biol. 38, 225-235 (1961).
TATE, P., VINCENT, M.: Influence of light on the gorging of *Culex pipiens* L. Nature, Lond. 130, 366-367 (1932).
TATE, P., VINCENT, M.: The biology of autogenous and anautogenous races of *Culex pipiens* L. (Diptera: Culicidae). Parasitology 28, 115-145 (1936).
TAUBER, M.J., TAUBER, C.A.: Nutritional and photoperiodic control of the seasonal reproductive cycle in *Chrysopa mohave* (Neuroptera). J. Insect Physiol. 19, 729-736 (1973).
TELFER, W.H.: The mechanism and control of yolk formation. A. Rev. Ent. 10, 161-184 (1965).
WALLIS, R.C.: Diapause and fat body formation by *Culex restuans* Theobald. Proc. Ent. Soc. Wash. 61, 219-222 (1959).
WALLIS, R.C.: Overwintering activity of the blowfly, *Phormia regina*. Ent. News 73, 1-5 (1962).
WASHINO, R.K.: Physiological condition of overwintering female *Anopheles freeborni* in California (Diptera: Culicidae). Ann. ent. Soc. Am. 63, 210-216 (1970).
WASHINO, R.K., BAILEY, S.F.: Overwintering of *Anopheles punctipennis* (Diptera: Culicidae) in California. J. Med. Ent. 7, 95-98 (1970).
WASHINO, R.K., GIEKE, P.A., SCHAEFER, C.H.: Physiological changes in the overwintering females of *Anopheles freeborni* (Diptera: Culicidae) in California. J. Med. Ent. 8, 279-282 (1971).
de WILDE, J.: Hormones and insect diapause. *In* "Hormones and the Environment" (Eds., G.K. Benson and J.G. Phillips), pp. 487-514. London: Cambridge University Press. [Mem. Soc. Endocr. 18] (1970).
WILKENS, J.L.: The endocrine control of protein metabolism as related to reproduction in the fleshfly, *Sarcophaga bullata*. J. Insect Physiol. 15, 1015-1024 (1969).

THE REGULATION OF FOOD INTAKE BY ACRIDIDS

E.A. Bernays and R.F. Chapman

Centre for Overseas Pest Research, College House, Wrights Lane, London, England

Grasshoppers, like many other insects, tend to take their food in discrete meals of several minutes duration separated by more or less extended intervals without feeding (Blaney, Chapman and Wilson, in press). This pattern of activity demands some physiological control, both to initiate and to stop feeding, and in this paper we discuss the possible mechanisms involved in the control of a single meal eaten by an actively feeding nymphal insect. We are not, in general, concerned with the frequency of feeding or the amounts consumed over long periods. Most of our own work has been concerned with *Locusta migratoria* L. We believe that the regulatory mechanisms which occur in this species are found throughout the Acrididae, but there are certainly differences in detail between species such that the relative importance of the different mechanisms varies. In this paper we try to put our work into perspective; it is not intended as a full review of feeding behaviour in Acrididae.

Feeding is an element in a continuum of behaviour, any aspect of which may be influenced by what has gone before. Because of the inter-relationships of the regulatory mechanisms it is convenient to consider first the end of feeding.

THE END OF FEEDING

If a grasshopper is eating certain foods it eats until the foregut is full so that meal size depends largely on how empty the foregut is to start with (Bernays and Chapman 1972a). The implication in such cases is that distension of the foregut is of major importance in bringing feeding to an end. On other foods, however, feeding stops before the foregut is filled and in these cases adaptation of the mouthpart sensilla and chemical inhibition may be relatively more important.

Distension of the foregut

The foregut fills in a regular manner during a feed. In *Locusta* nymphs deprived of food for eight hours, no backward movement of food from the foregut to the midgut occurs until some minutes after the end of feeding (Fig. 1), so that during the course of a meal there is a progressive filling of the foregut. The last part to fill is the extreme anterior end, suggesting that stretch receptors in this region may be concerned with regulating meal size. This possibility has been investigated by cutting various nerves of the stomatogastric system and determining the effects of the operation on meal size. All of these operations are likely to have multiple effects because all the nerves contain large numbers of axons (Cook, personal communication), but only when the posterior pharyngeal nerves are severed is a larger than normal meal taken. Rowell (1963) observed that cutting the recurrent nerve of *Schistocerca* resulted in hyperphagia. His cuts were made close to the frontal ganglion and thus almost certainly involved cutting the posterior pharyngeal nerves as well. From these experiments, it appears certain that a major factor regulating meal size is inhibitory input from the stretch receptors in the wall of the foregut, posterior to those described by Clarke and Langley (1963) which are associated with the anterior pharyngeal nerve or the frontal connectives (Allum, personal communication).

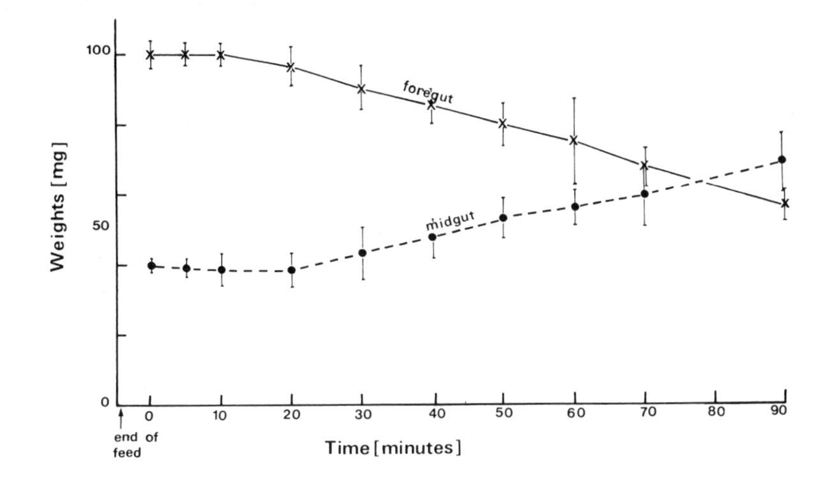

Fig. 1. The weights of the foregut and midgut of *Locusta* following a feed. Notice that the midgut does not start to increase in weight, indicating the backward passage of food from the foregut, until 30 min after feeding. The earlier drop in foregut contents is presumably due to digestion beginning while the food is still in the crop

Even on palatable foods, the action of the foregut stretch receptors is influenced by other inputs to the central nervous system. For instance, although the quantity of *Poa* eaten is less than that of *Agropyron*, it too is controlled by the degree of foregut distension. Neither grass appears to contain chemical inhibitors and it seems that the difference must lie in their phagostimulatory properties. There is also evidence of habituation to meal size since this was influenced by the size of previous meals (Goodhue, 1962; Bernays and Chapman 1972a). This suggests that the stretch receptors may undergo long term changes in sensitivity.

In fluid feeding insects distension of the gut also produces an increase in the size of the abdomen, but in *Locusta* there is no change in volume of insects as a result of feeding, the volume of the meal being accommodated by a collapse of the air sacs. Further, information concerning changes in body form would probably be relayed to the brain via the ventral nerve cord, but cutting the cord immediately posterior to either the pro- or the meta-thoracic ganglion has no effect on meal size (Bernays and Chapman, in press).

Adaptation of the mouthpart sensilla

The chemoreceptors of the mouthparts of *Locusta* adapt to a very low level of input within a short time of stimulation (Blaney, personal communication), and Barton Browne, Moorhouse and van Gerwen (in prep. a) have shown that adaptation regulates meal size in *Chortoicetes* when it is fed on solutions of single chemicals. In this case each meal lasts for only two or three minutes. Adaptation could also account for the small amounts of food eaten in one meal in experiments in which phagostimulants are presented singly. The amount eaten in this case increases with the concentration of the phagostimulant (Cook, personal communication), and appears to reflect the increased input from the sensilla (Blaney, personal communication), but even the largest amounts eaten fall far short of a meal on grass (Sinoir 1968b). Analysis of time lapse films of *Locusta* feeding on pith discs impregnated with sucrose shows that only 7% of the feeds lasted more than 5 min and the feeds were not aggregated into discrete meals (Cook, personal communication). In view of the results of Barton Browne, Moorhouse and van Gerwen (in prep. a), the failure to eat more of such a highly stimulating food could well result from adaptation.

In an insect with continuous access to food, over 20% of the feeds last for more than 5 min and the feeds are aggregated into meals (Blaney, Chapman and Wilson, in press), whereas insects deprived of food feed continuously for 15-20 min. In such cases it appears that adaptation cannot be important and this is also suggested by experiments in which the foregut is filled with an appropriate amount of agar through a cannula. After this treatment no further feeding occurs despite the fact that the mouthpart chemoreceptors have not been stimulated chemically. Clearly, adaptation of the chemoreceptors is not an essential feature of meal size regulation. This is further emphasised by experiments in which an aqueous extract of grass is dripped on to the mouthparts for 20 min before the insect is offered grass. The initial feeding behaviour appears unaffected and the meal size is not reduced.

It seems likely that in normal feeding complete adaptation of all the mouthpart receptors does not occur. In the case of the palps, the intermittent stimulation of the terminal sensilla as a result of palpation offsets adaptation (Blaney, personal communication) and they are used only intermittently throughout the meal so that recovery from adaptation is quite possible. But the sensilla within the cibarial cavity are continually bathed with fluid during the meal and so presumably are continuously stimulated. Perhaps the sensilla do become fully adapted, but nevertheless feeding is driven by the continued input from the palps, or perhaps the chemical complexity of the normal food is such that the nature of the stimulants is continually varying and complete adaptation does not occur.

Chemical inhibition

On less palatable foods, appreciably smaller meals are taken and Sinoir (1970) has suggested that this is the result of adaptation. However, in a number of cases feeding is terminated by chemical inhibition rather than lack of stimulation. For instance, on seedling grasses the average meal size of male fifth instar *Locusta* nymphs is always less than 50% of the meal size of mature grasses, but removing materials from the seedlings with chloroform makes them much more palatable, while the extract can be applied to acceptable food to make it less acceptable (Bernays, Chapman and Horsey, in prep.). If the dried extract, dispersed in water, is applied to the mouthparts for some minutes before feeding it has no effect on the amount eaten if the food which is offered is entirely palatable, but it completely prevents feeding if the food also contains the inhibitory substance. The insect has reached the stage where further inhibitory input is not tolerated.

Many of the chemoreceptors on the mouthparts of *Schistocerca* respond to azadirachtin (Haskell and Schoonhoven 1969) and perhaps also to other inhibitory substances. Haskell and Mordue (1969) have shown that the A3 group of sensilla on the labrum of *Schistocerca* are particularly important in the rejection of food containing azadirachtin. On the other hand Sinoir (1970) found that the A3 sensilla were not important in the rejection by *Locusta* of food containing 'scillarène', and Le Berre and Louveaux (1969) suggest that sensilla on the mandibles are important in detecting feeding inhibitors. Williams (1954) and Abushama (1968) found that removal of the palps enhanced the amount of feeding on distasteful plants. From these examples it appears that inhibition involves a variety of sensilla on different elements of the mouthparts.

We infer that on normal food the inhibition associated with certain chemical and mechanical aspects of food intake is integrated, within the central nervous system, with the positive tendency to feed. The end of feeding occurs when the inhibitory effects become dominant. On chemically simple foods, however, feeding may stop as a result of the failure of the positive inputs resulting from adaptation of the sensilla.

THE CONSEQUENCES OF FEEDING

Feeding affects many aspects of the insect's physiology and development, but here we are concerned only with those consequences of feeding which might affect subsequent feeding behaviour. The act of distending the foregut has a number of short term effects which are indicated in Fig. 2. First the information from the stretch receptors is relayed to the brain so that feeding behaviour is switched off. A slightly more long term effect, also promoted via the brain, is to release one or more hormones from the corpora cardiaca (CC). One effect of the hormones is to cause the pores on the terminal sensilla of the palps to close so that the sensilla are no longer functional. This effect, as measured by changes in electrical resistance across the tips of the palps, persists for an hour or more, but by two hours after feeding the sensilla are again fully functional (Bernays, Blaney and Chapman 1972; Bernays and Chapman 1972b). The closure of the sensilla may be concerned in determining the end of the meal, and also effectively prevents further feeding for a period after the meal by cutting off the sensory input. The release of a diuretic hormone is also promoted by feeding, presumably with the effect of offsetting any increase in blood volume as a result of the absorption of water (Mordue 1969; Bernays and Chapman 1972b).

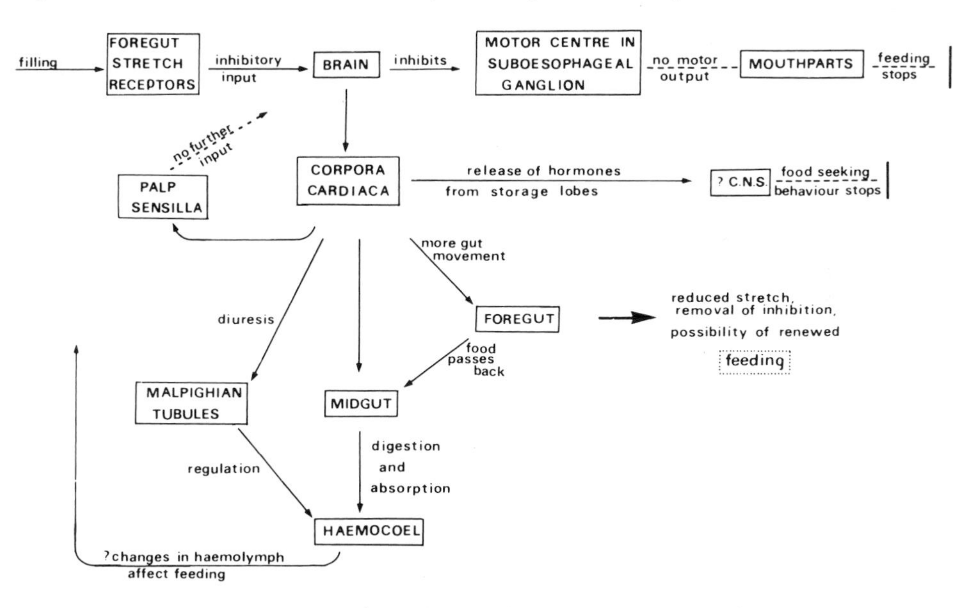

Fig. 2. Diagrammatic representation of the changes which occur in the insect as a result of filling the foregut

There are other changes following feeding which probably result from hormone release. For instance, locomotor activity is reduced after feeding and the same effect is produced in insects deprived of food for 4 hr by filling the foregut with agar or by injecting homogenates of the storage lobes of the CC into the haemocoel (Table 1). It is also found that the beginning of the next meal is delayed by injection of blood from recently fed insects although the size of the meal is not affected (Table 2).

Table 1. The effects on activity of artificially filling the foregut with agar or injecting corpus cardiacum (CC) homogenates into the haemocoel of unfed insects.

Treatment	Control		Experimental	
	Number of insects	% of time active	Number of insects	% of time active
Crop agar-filled	25	38	25	18
Injection of CC homogenate	5	50	5	15

Table 2. The effects on timing of the next meal, and on meal size, of injecting blood from recently fed or unfed insects into the haemocoel of unfed insects.

Donor blood	fed	unfed
Number of recipients	27	27
Time to next feed (min)	68	41
Meal size (mg)	81	81

Other authors have obtained very variable responses following injection of CC homogenates, but Haskell *et al.* (1965) and Mitchell (1971) observed reduced marching activity of *Locusta* nymphs following the injection of relatively high titres. The effect appeared to be dependent on the degree of food deprivation of the recipients and was totally ineffective in insects deprived for some time, as Moorhouse (1968) also found in *Schistocerca*. Finally, Cazal (1969) has shown that homogenates of the storage lobes of the CC of *Locusta* enhance the active movements of the foregut. If it is true that the hormone responsible for this activity is released as a result of feeding, this might be the factor responsible for the beginning of foregut emptying some 15 min after feeding.

There are no direct data concerning the changes in the levels of CC hormone which occur after feeding, but the electrical resistance across the tips of the palps, which is dependent on the level of hormone present (Bernays and Mordue 1973), decreases rapidly (Fig. 3) (Bernays, Blaney and Chapman 1972). W. Mordue (1969) also suggests that release of the diuretic hormone stops at the end of feeding.

Changes may occur in the composition of the haemolymph after a meal as a result of rapid absorption from the food, but these are not adequately documented. Moorhouse (1968) has shown that the concentration of potassium in the blood of *Schistocerca* increases slightly in the 30 min after a meal, but then decreases again, and it seems likely that, in general, other substances absorbed from the food will also increase during this period. Subsequently the concentrations of potassium (Moorhouse 1968), and of amino acids in the blood, decrease (Hill and Goldsworthy 1970) although the volume of haemolymph decreases at the same time (Fig. 3). The concentration of trehalose, on the other hand, remains fairly constant for some hours (Mitchell 1971).

The contents of the foregut begin to decrease in volume from the time the feed ends. The decrease is slight at first and due to digestion alone, but subsequently the backward passage of food is also involved.

At 30°C the foregut of *Locusta* fifth instar hoppers is completely empty of food in about 5 hr.

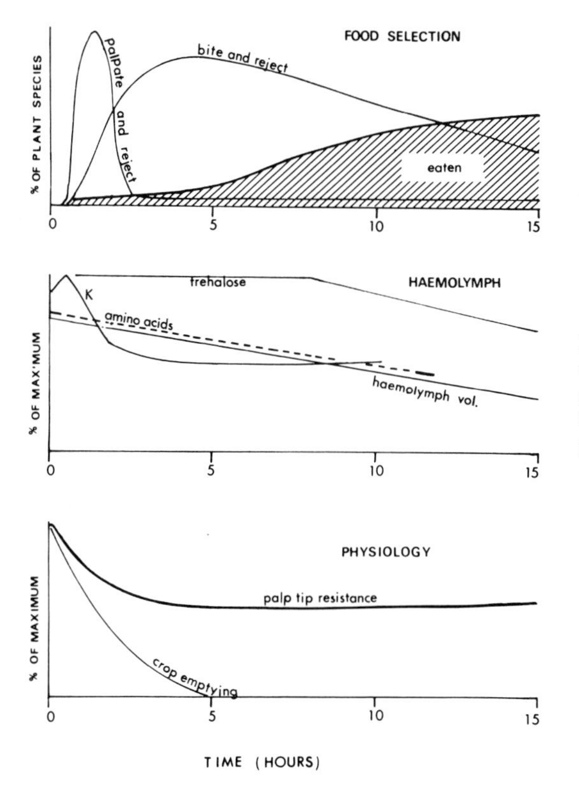

Fig. 3. Diagrammatic representation of the changes in feeding behaviour and related phenomena occurring after a meal by *Locusta*. The vertical axes represent approximate percentages of maximum values

THE BEGINNING OF FEEDING

The state of readiness to feed

If the fall in hormone levels suggested by the reduction in palp resistance in Fig. 3 can be taken as an indication that parallel falls occur in the levels of other fractions released from the CC by feeding, the physiological and behavioural changes induced by feeding will soon be reversed. Within an hour the insect is potentially active with fully functional palp sensilla and at the same time the emptying of the foregut will reduce the inhibition on feeding behaviour.

Hence the insect returns to a state of readiness to feed and may be induced to begin the feeding sequence by any stimulus, visual, tactile or chemical which promotes activity, whether the stimulus is specific to food or not. At first relatively few of the population respond, but as the period of food deprivation increases the proportion of nymphs responding increases (Moorhouse 1968, 1971; Bernays and Chapman 1970). In these instances, the newly imposed stimuli have a general arousal effect (Kennedy and Moorhouse 1969) which leads to the insect moving and responding to the food. In the absence of stimuli promoting activity, the beginning of feeding is delayed, while factors which reduce locomotor activity also reduce the likelihood of feeding. For instance, interfeed periods are longer in darkness than in the light (Blaney, Chapman and Wilson, in press). In these instances, where a stimulus promoting movement is lacking, activation presumably results from the expression of some central nervous endogenous activity.

While feeding commonly depends on preceding locomotor activity, the two are not directly correlated. The most clear cut feeding sequences are observed in the laboratory in generally inactive insects. Such an insect after a period of total inactivity will suddenly become active and move directly towards the food, palpating as it goes. Having 'tested' the food with its palps it then begins to feed and continues to do so for some minutes.

In other individuals, however, locomotion is the dominant activity and feeding appears to occur simply because they blunder into the food, rather than by a directed response. In these circumstances feeds are taken at frequent intervals, but they are only very short (Blaney, Chapman and Wilson, in press).

Since any of the three activities, feeding, locomotion and resting, may occur in the unchanging environment of a small experimental cage it seems likely that the dominant activity is determined by phenomena within the insect. Further, the three aspects of activity are virtually mutually exclusive, only one of them occurring at any one time, although the change from one to the other may be rapid. This singleness of action demands some complex central nervous integration, not only of stimuli arising from the external and internal environments, but also of the central nervous effects of successive activities (Kennedy 1967).

Locating and recognising the food

Attraction of the activated insect to the food from a distance involves visual and olfactory cues (Moorhouse 1968, 1971; Mulkern 1969), and whether or not feeding then begins depends on the receipt of favourable stimuli from the plant.

There is no evidence that the antennae have a role in the acceptance of suitable food (Williams 1954), although on some aromatic plants stimulation of the antennae alone may lead to rejection (Goodhue 1962). The sensilla of the antennae respond to moisture (Waldow 1970) and the response of *Schistocerca* to moist surfaces varies greatly depending on its degree of desiccation (Kendall 1971), but there is no clear evidence that this affects the amount of food eaten (but see Sinoir 1966).

Kendall (1971) has shown that the tarsal receptors of *Schistocerca* are sensitive to leaf-surface waxes, contact with a surface impregnated with such waxes causing the insect to pause in its forward movement and to touch the surface with its palps. The latter have a key role in food selection and in their absence feeding is delayed (A.J. Mordue, in prep.). The sensilla at the tips of the palps enable the insect to distinguish the leaves of acceptable and unacceptable plants simply by touching the leaf surface (Bernays and Chapman 1970; Blaney and Chapman 1970), and suitable chemical stimulation leads to a continuation of the feeding sequence. The simultaneous mechanical stimulation of the palps causes the insect to lower its head and move across the leaf surface until it reaches an edge. Probably visual and a variety of tactile stimuli are important in the recognition of an edge, and Sinoir (1969) has shown that stimulation of the mechanoreceptors on the lateral parts of the labrum leads to biting.

Biting releases chemicals within the leaf and these spread over the mouthparts together with saliva. Here there is an extensive array of sensilla (Thomas 1966) and on the inside of the cibarial cavity of the grasshopper *Zonocerus variegatus* there are over 2,000 sensilla, many of which are chemoreceptors with a number of neurones (Cook 1972). Various studies have been made which indicate that the sensilla on the mouthparts are sensitive to a wide range of phagostimulants: sugars (Dadd 1960; Thorsteinson 1958; Sinoir 1968a; Cook, pers. comm.), lipids (Dadd 1960; Chauvin 1951; Thorsteinson and Nayar 1963; Mehrotra and Rao 1972), amino acids (Thorsteinson 1960; Cook, pers. comm.), vitamins (Thorsteinson 1958) and organic acids (Thorsteinson 1960). Synergism or inhibition between the various substances may also occur (Thorsteinson 1960).

After biting has occurred, continued feeding depends on the presence of phagostimulants since Goodhue (1962) has shown that grass extracted with acetone and water, and hence providing a virtually inert substance, is no longer eaten by *Schistocerca* unless a suitable phagostimulant is added. Whether or not feeding occurs on normal leaves depends on the balance of these phagostimulants with possible inhibitors. Inorganic salts may have inhibitory effects in high concentrations (Dadd 1960; Thorsteinson 1960), but since salts are universally present in terrestrial plants, generally in low concentration, they are unlikely to be generally important. Much more important are organic inhibitors; steroids and alkaloids have been shown to be important in this respect by Dadd (1960) and Harley and Thorsteinson (1967). If these inhibitors can be removed by soaking the plant in a suitable solvent the plants become acceptable (Bernays, Chapman and Horsey, in prep.).

Changes in selectivity due to food deprivation

If food is not available to the insect its activity increases and feeding responses are made to generalised stimuli. Thus, whereas in a normal feeding cycle biting only follows after palpation on a chemically suitable substrate, with food deprivation the insect will bite at objects presenting the correct visual (Williams 1954) or mechanical (Sinoir 1969; Dadd 1960) characteristics even though these may be chemically inert. In these cases the palps have a purely mechanical role or none at all (Sinoir 1969). As a result of this behaviour the insects bite plants which are normally rejected at palpation (Blaney and Chapman 1970; Bernays and Chapman 1970). Because, for practical reasons, most experimental work on feeding behaviour of acridids has been carried out on insects deprived of food, the concept has arisen that they are random biters, selecting their food largely on the basis of its internal chemical characteristics (Pfadt 1949; Dadd 1963; Mulkern 1969; Sinoir 1970).

If the period of food deprivation is prolonged, some inhibitory chemicals become less important and plants are eaten which, after shorter periods of deprivation, are rejected (Fig. 3) (Bernays and Chapman 1970). This reduction in selectivity, and the reduction in importance of the palps in making the selection, indicates some changes within the insect as the period of food deprivation gets longer.

A number of changes in haemolymph composition occur which coincide with the changes in food selection. For instance, in 4-day-old nymphs of *Locusta*, the blood volume and the concentration of amino acids falls over the first 12 hr of food deprivation (Hill and Goldsworthy 1970). That such changes could affect selection is suggested by experiments of Sinoir (1966, 1968a) who shows that desiccation and the nutritional state of the insect do affect the amount of food eaten over a period. It is, however, possible that the reduction in selectivity results from increased endogenous activity within the central nervous system, the positive effects of which are to override to an increasing extent the inhibitory inputs of various plant chemicals. The fact that the marching speed of hoppers increases rapidly over the first two and a half hours of starvation, and then increases slowly for at least the next five hours (Ellis 1951), suggests that the changes resulting from food deprivation have a generalised effect and are not specific to feeding.

THE CONTINUATION OF FEEDING

Continuous feeding may depend on continuous favourable inputs from the peripheral sensilla. Alternatively, it may be supposed that, once it has started, feeding will continue as a result of endogenous activity until it is switched off by inhibitory inputs. Various experiments indicate the importance of endogenous activity and suggest that one effect of the sensory input from the peripheral sensilla is to modify this activity. For instance if, during the early part of a meal, the insect loses contact with the food, its food searching activity is greatly enhanced, the insect casting about from side to side and palpating much more persistently than in the period before feeding: the nearer the beginning of the meal, the more persistent the palpation (Fig. 4). This seems to suggest a positive feedback from the food itself, and this is also suggested by experiments in which whole grass extract is dripped onto the mouthparts before feeding; the subsequent size of the meal is on average 17% greater than that of normal insects. Barton Browne, Moorhouse and van Gerwen (in prep. b) also suggest that the act of feeding on highly stimulating material may enhance subsequent feeding activity.

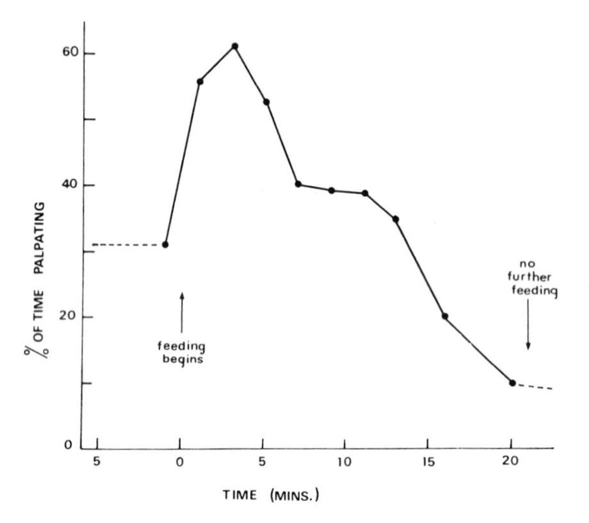

Fig. 4. The percentage of time spent palpating following loss of contact with the food at different stages in the meal

A general dynamic function in various activities has been demonstrated for the antennae by Bayramoglu-Ergene (1966), and Sinoir (1969, 1970) suggests that both the palps and the antennae are important in this respect in feeding. However, the lengths of meals taken by insects with palps amputated are very little less than those of intact insects (A.J. Mordue, in prep.).

Hence there is evidence that the peripheral sensilla promote continuous feeding through their effect on a general endogenous central nervous activity which is switched onto the feeding pathway. This does not, however, preclude the possibility of a more direct connection between sensory input and feeding activity, although no such connection has been demonstrated.

While it is possible to account for many of the observed features of feeding behaviour in terms of responses to stimulation or inhibition of the sensilla, there are, nevertheless, certain features which strongly suggest that feeding behaviour as a whole is driven by some endogenous central nervous activity. The role of the peripheral systems appears then to modulate the form and intensity with which this activity is expressed. It is this endogenous activity which, initially, leads to movement and location of the food and we suggest that, having arrived at the food, it is switched so as to initiate and subsequently to maintain the feeding response.

ACKNOWLEDGEMENTS

Much of the work which we describe in this paper derives from a group of colleagues working at the Centre for Overseas Pest Research and in the University of London. They are Mr. W.M. Blaney, Miss A.G. Cook, Dr. M.D. Kendall, Dr. G. Mitchell, Mrs. A.J. Mordue and Mr. R. Yeadon. To them we are indebted for their cooperation and for permission to use unpublished data.

REFERENCES

ABUSHAMA, F.T.: Food-plant selection by *Poecilocerus hieroglyphicus* (Klug) (Acrididae: Pyrgomorphidae) and some of the receptors involved Proc. R. ent. Soc. Lond. (A) 43, 96-104 (1968).

BARTON BROWNE, L., MOORHOUSE, J.E., van GERWEN, A.C.M.: An investigation of mechanisms bringing about the termination of meals of water or sucrose solutions by the Australian plague locust, *Chortoicetes terminifera*. (In prep. a)

BARTON BROWNE, L., MOORHOUSE, J.E., van GERWEN, A.C.M.: Evidence for the development of an excitatory state in the nervous system of the locust, *Chortoicetes terminifera*, during the ingestion of sucrose solutions. (In prep. b)

BAYRAMOGLU-ERGENE, S.B.: Untersuchungen uber die dynamogene Funktion der Antennen bei *Anacridium aegyptium*. Z. vergl. Physiol. 52, 362-369 (1966).

BERNAYS, E.A., BLANEY, W.M., CHAPMAN, R.F.: Changes in chemoreceptor sensilla on the maxillary palps of *Locusta migratoria* in relation to feeding. J. exp. Biol. 57, 745-753 (1972).

BERNAYS, E.A., CHAPMAN, R.F.: Experiments to determine the basis of food selection by *Chorthippus parallelus* (Zetterstedt) (Orthoptera: Acrididae) in the field. J. Anim. Ecol. 39, 761-776 (1970).

BERNAYS, E.A., CHAPMAN, R.F.: Meal size in nymphs of *Locusta migratoria*. Entomologia exp. appl. 15, 399-410 (1972a).

BERNAYS, E.A., CHAPMAN, R.F.: The control of changes in peripheral sensilla associated with feeding in *Locusta migratoria*. J. exp. Biol. 57, 755-763 (1972b).

BERNAYS, E.A., CHAPMAN, R.F.: The regulation of feeding in *Locusta migratoria*. Internal inhibitory mechanisms. Entomologia exp. appl. (In press).

BERNAYS, E.A., CHAPMAN, R.F., HORSEY, J.: The distastefulness of seedling grasses to insects. (In prep.)

BERNAYS, E.A., MORDUE, A.J.: Changes in the palp tip sensilla of *Locusta migratoria* in relation to feeding: the effects of different levels of hormone. Comp. Biochem. Physiol. 45A, 451-454 (1973).

BLANEY, E.M., CHAPMAN, R.F.: The functions of the maxillary palps of Acrididae (Orthoptera). Entomologia exp. appl. 13, 363-376 (1970).

BLANEY, W.M., CHAPMAN, R.F., WILSON, A.: The pattern of feeding of *Locusta migratoria* (Orthoptera, Acrididae). Acrida (In press)

CAZAL, M.: Actions d'extraits de corpora cardiaca sur le peristaltisme intestinal de *Locusta migratoria*. Archs Zool. exp. gén. 110, 83-89 (1969).

CHAUVIN, R.: Sur les facteurs responsables de l'attraction que manifestant les acridiens pour le son. Bull. Off. natn. anti-acrid. Algér. 1, 15-18 (1951).
CLARKE, K.U., LANGLEY, P.A.: Studies on the initiation of growth and moulting in *Locusta migratoria migratorioides*. III. The role of the frontal ganglion. J. Insect Physiol. 9, 411-421 (1963).
COOK, A.G.: The ultrastructure of the A1 sensilla on the posterior surface of the clypeo-labrum of *Locusta migratoria migratorioides* (R. & F.). Z. Zellforsch. mikrosk. Anat. 134, 539-554 (1972).
DADD, R.H.: Observations on the palatability and utilisation of food by locusts, with particular reference to the interpretation of performances in growth trials using synthetic diets. Entomologia exp. appl. 3, 283-304 (1960).
DADD, R.H.: Feeding behaviour and nutrition in grasshoppers and locusts. Adv. Insect Physiol. 1, 47-109 (1963).
ELLIS, P.E.: The marching behaviour of hoppers of the African migratory locust (*Locusta migratoria migratorioides* R. & F.) in the laboratory. Anti-Locust Bull. No. 7, 46 pp. (1951).
GOODHUE, R.D.: The effects of stomach poisons on the desert locust. Ph.D. Thesis, University of London. (1962).
HARLEY, K.L.S., THORSTEINSON, A.J.: The influence of plant chemicals on the feeding behaviour, development and survival of the two-striped grasshopper, *Melanoplus bivittatus* (Say), Acrididae: Orthoptera. Can. J. Zool. 45, 305-319 (1967).
HASKELL, P.T., CARLISLE, D.B., ELLIS, P.E., MOORHOUSE, J.E.: Hormonal influences in locust marching behaviour. 12th Int. Congr. Ent., London, 1964, 290-291 (1965).
HASKELL, P.T., MORDUE (Luntz), A.J.: The role of mouthpart receptors in the feeding behaviour of *Schistocerca gregaria*. Entomologia exp. appl. 12, 561-610 (1969).
HASKELL, P.T., SCHOONHOVEN, L.M.: The function of certain mouthpart receptors in relation to feeding in *Schistocerca gregaria* and *Locusta migratoria migratorioides*. Entomologia exp. appl. 12, 423-440 (1969).
HILL, L., GOLDSWORTHY, G.J.: The utilization of reserves during starvation of larvae of the migratory locust. Comp. Biochem. Physiol. 36, 61-70 (1970).
KENDALL, M.D.: Studies on the tarsi of *Schistocerca gregaria* Forskål. Ph.D. Thesis, University of London (1971).
KENNEDY, J.S.: Behaviour as physiology. *In* "Insects and Physiology" (Eds. J.S.L. Beament and J.E. Treherne). Edinburgh: Oliver & Boyd (1967).
KENNEDY, J.S., MOORHOUSE, J.E.: Laboratory observations on locust responses to windborne grass odour. Entomologia exp. appl. 12, 489-503 (1969).
LE BERRE, J.E., LOUVEAUX, A.: Equipement sensoriel des mandibules de la larve du premier stade de *Locusta migratoria* L. C. r. hebd. Séanc. Acad. Sci., Paris (D) 268, 2907-2910 (1969).
MEHROTRA, K.N., RAO, P.J.: Phagostimulants for locusts: studies with edible oils. Entomologia exp. appl. 15, 208-213 (1972).
MITCHELL, G.A.: Studies on the control of locust haemolymph carbohydrate levels and their influence on behaviour. Ph.D. Thesis, University of Wales (1971).
MOORHOUSE, J.E.: Locomotor activity and orientation in locusts. Ph.D. Thesis, University of London (1968).
MOORHOUSE, J.E.: Experimental analysis of the locomotor behaviour of *Schistocerca gregaria* induced by odour. J. Insect Physiol. 17, 913-920 (1971).
MORDUE, A.J.: Function of the palps of *Schistocerca gregaria*. (In prep.)
MORDUE, W.: Hormonal control of Malpighian tube and rectal function in the desert locust, *Schistocerca gregaria*. J. Insect Physiol.

15, 273-285 (1969).
MULKERN, G.B.: Behavioural influences on food selection in grasshoppers (Orthoptera: Acrididae). Entomologia exp. appl. 12, 509-523 (1969).
PFADT, R.E.: Food plants as factors in the ecology of the lesser migratory grasshopper, *Melanoplus mexicanus* (Sauss.). Bull. Wyo. agric. Exp. Stn 290, 50 pp. (1949).
ROWELL, C.H.F.: A method of chronically implanting stimulating electrodes into the brains of locusts, and some results of stimulation. J. exp. Biol. 40, 271-284 (1963).
SINOIR, Y.: Interaction du deficit hydrique de l'insecte et de la teneur en eau de l'aliment dans la prise de nourriture chez le criquet migrateur, *Locusta migratoria migratorioides*. C. r. hebd. Séanc. Acad. Sci., Paris (D) 262, 2480-2483 (1966).
SINOIR, Y.: Etudes de quelques facteurs conditionnant la prise de nourriture chez les larves du criquet migrateur, *Locusta migratoria migratorioides* (Orthoptera: Acrididae). I. Facteurs externes. Entomologia exp. appl. 11, 195-210 (1968a).
SINOIR, Y.: Etudes de quelques facteurs conditionnant la prise de nourriture chez les larves du criquet migrateur, *Locusta migratoria migratorioides* (Orthoptera: Acrididae). II. Facteurs internes. Entomologia exp. appl. 11, 443-449 (1968b).
SINOIR, Y.: Le role des palpes et du labre dans le comportement de prise de nourriture chez la larve du criquet migrateur. Annls Nutr. Aliment. 23, 167-194 (1969).
SINOIR, Y.: Quelques aspects du comportement de prise de nourriture chez la larve de *Locusta migratoria migratorioides* R. & F. Annls Soc. ent. Fr. (N.S.) 6, 391-405 (1970).
THOMAS, J.G.: The sense organs of the mouthparts of the desert locust (*Schistocerca gregaria*). J. Zool., Lond. 148, 420-448 (1966).
THORSTEINSON, A.J.: Acceptability of plants for phytophagous insects. 10th Int. Congr. Ent., Montreal, 1956. 2, 599-602 (1958).
THORSTEINSON, A.J.: Host selection in phytophagous insects. A. Rev. Ent. 5, 193-218 (1960).
THORSTEINSON, A.J., NAYAR, J.K.: Plant phospholipids as feeding stimulants for grasshoppers. Can. J. Zool. 41, 931-935 (1963).
WALDOW, U.: Elektrophysiologische Untersuchungen an Feuchte-, Trocken- und Kalter-rezeptoren auf der Antenne der Wanderheuschrecke *Locusta*. Z. vergl. Physiol. 69, 249-283 (1970).
WILLIAMS, L.H.: The feeding habits and food preferences of Acrididae and the factors which determine them. Trans. R. ent. Soc. Lond. 105, 423-454 (1954).

NEURAL MECHANISMS OF RESPONSE MODIFICATION IN INSECTS

J.M. Camhi

Section of Neurobiology and Behavior, Cornell University, Ithaca, N.Y., U.S.A.

The last few decades have seen dramatic changes in our concepts of the neural control of invertebrate behavior. The extent of these changes becomes clear on reading the following statement, written by Sherrington twenty-five years ago:

> The behaviour of the spider is reported to be entirely reflexive, but reflex action ... would go little way toward meeting the life of external relation of a horse or cat or dog, still less of ourselves. As life develops it would seem that ... conscious behaviour tends to replace [the] reflex ...
>
> (Sherrington 1947).

While no one would argue with the basic theme of this paragraph, two of its implications have recently been seriously challenged. On the one hand, we now know that invertebrates, far from being entirely reflexive, possess numerous central neuronal networks which are controlled only slightly, or not at all, by sensory inputs (for review see DeLong 1971). Secondly, the implied hiatus between the machine-like spider and the much less stereotyped mammal has been partially spanned by numerous findings suggesting previously unrecognized complexities to invertebrate behavior and central integration (see, for instance, Hoyle 1964, 1970).

The recently discovered central mechanisms which, along with reflexes form our presently available conceptual currency, include neuronal oscillators consisting of single neurons (Alving 1968; Chen, von Baumgarten and Takeda 1971; Mendelson 1971; Strumwasser 1967) and those known or presumed to consist of networks (for instance Hagiwara 1961; Maynard 1966, 1972; Pearson 1972; Wilson 1961; Page and Wilson 1970). Controlling the frequency of some of these are certain central units, the temporal command fibers (Dando and Selverston 1972; Davis and Kennedy 1972a,b,c; Bowerman and Larimer, in press b; Mendelson 1971), and synchronizing the timing between oscillators, are distinct coordinating fibers (Stein 1971). Also present are postural command fibers (Bowerman and Larimer, in press a; Evoy and Kennedy 1967; Kennedy 1968; Kennedy, Evoy and Hanawalt 1966; Kennedy *et al.* 1967; Larimer and Eggleston 1971; Larimer and Kennedy 1969), units which, through their spatially extensive synapses, individually can drive coordinated behaviors. In summary, then, insect and other invertebrate central nervous systems (CNS) no longer appear to be the relatively passive switchboards that they were believed to be a few years ago, and, indeed, many behaviors appear to operate by the turning on of complex central "motor tapes" (Hoyle 1964). Actually, the often-mentioned economy of neuron numbers in insects may be partially offset by increased capabilities of certain central cells, much as protozoa substitute cellular complexity for division of labor. These new views of central organization will be important later in this paper, where I suggest that response modification in insects in some cases implies cellular complexities not yet directly observed within the CNS.

Paralleling these developments has been a growing awareness of the intricacy of insect behavior. Although ethologists and other field workers have long recognized highly complex insect sensory-perceptual capabilities (e.g., Lindauer 1961; von Frisch 1967), there is a great conceptual distance between work of this kind and physiologically oriented studies which more often focus on the movements of specific body segments or muscles. But even at this behaviorally more restricted level, important new forms of variability have appeared. Habituation has been studied in arthropods at the cellular level within neural pathways mediating specific movements (Zucker, Kennedy and Selverston 1971) and has come to be recognized as a general property of behavioral or neural responses (Horn and Hinde 1970). Furthermore, complex interactions among different sensory inputs and centrally derived patterns in insects have been recognized (Bentley 1969a; Camhi and Hinkle 1972; Hinkle and Camhi 1972; Kutsch 1969; Page and Wilson 1970; Roeder 1970; Camhi, unpublished; Sherman, Novotny and Camhi, unpublished). In general, even seemingly simple motor patterns are turning out to be surprisingly complicated, examples being the escape behaviors of cockroaches (Dagan and Parnas 1970; Harris and Smyth 1971; Milburn and Bentley 1971; Parnas *et al.* 1969; Spira, Parnas and Bergmann 1969a,b), crayfish (Larimer *et al.* 1971; Schramek 1970; Wine and Krasne 1972) and other invertebrates (Bullock and Horridge 1965). Aside from suggesting closer ties to complex vertebrate behaviors, these findings have further increased the likely usefulness of such invertebrates as insects, crustaceans and molluscs as model systems (Hoyle 1970) for understanding response modification in phyletically more advanced animals.

One consequence of central neuronal complexity and independence is to limit the conditions under which sensory inputs drive motor outputs. The study of these conditions is largely a new development in the field, and my purpose in this paper is to define and explore them.

The minimum condition required for a sensory input to drive a motor output is that the two be connected by a neuronal pathway having excitatory synapses. Though obvious, the actual study of pathways at the unit level has been presented with such obstacles that at present only a few are fairly well understood (Zucker, Kennedy and Selverston 1971; Nichols and Purves 1970; Kupfermann and Kandel 1969; Castellucci *et al.* 1970). A crucial step in this type of analysis has been the discovery of identified neurons; that is, cells which can be individually recognized anatomically and/or physiologically in each member of a species. This is now possible not only for those relatively scarce neurons with giant axons or cell bodies (Bullock and Horridge 1965; Selverston and Remler 1972) but also for large and moderately sized insect motoneurons (Bentley 1970; Camhi and Hinkle 1972, in prep.; Cohen and Jacklet 1967) and even for medium size crustacean interneurons (Wiersma and Hughes 1961; Zucker, Kennedy and Selverston 1971). The use of procion yellow (Remler, Selverston and Kennedy 1968; Stretton and Kravitz 1968) and cobalt (Pitman, Tweedle and Cohen 1972a) intracellular staining techniques together with their application to whole peripheral nerves (Iles and Mulloney 1971) has recently given great impetus to this pursuit. Also new is the increased number of cases where spikes and synaptic potentials can be recorded intracellularly in insects (Bentley 1969a; Pitman, Tweedle and Cohen 1972b; Crossman, Kerkut and Walker 1972). Of possible great future value is the establishment of electrical excitability in insect somata following axotomy or colchicine treatment (Pitman, Tweedle and Cohen 1972b). Another technique whose contribution will probably prove immeasurably valuable is computer assisted analysis of EM serial sections (Macagno, Lopresti and Leventhal 1973).

But even where a pathway is present, activating it does not always culminate in a motor act, or not always the same motor act. We can distinguish two types of mechanism which can control whether, and in what way, stimulating a given pathway will result in a response. I shall call these *homo-pathway* and *hetero-pathway* control, or modification of the response, which I define as follows. The neural activity driving a given behavior can begin in either sensory receptors or a central program generator, and concludes with the muscle response. I will label the nerve circuit connecting this starting and end point the *effective pathway* (in some cases I discuss pathways which actually have no known motor output). Alteration in the response resulting from changes within one or more neuron of the effective pathway I shall call *homo-pathway* modification. Examples are habituation, homo-synaptic facilitation and defacilitation, temporal summation, refractoriness, and output strength as controlled by variations in spatio-temporal input patterns, or by varying input strength.

By contrast, effective pathway modification can result from inflow of neural signals originating outside this pathway; notably in other receptors or other central pattern generators. (This in-current pathway need not produce its own motor output.) This is *hetero-pathway* modification, and the signal which causes it I shall call the *modifier signal**. This formalization is similar to that devised by Rowell (1971c), who distinguished between "stimulus related" and "stimulus unrelated" behavioral increments and decrements. In this paper, I shall stress hetero-pathway modification because it is the area of my current work and interest.

HOMO-PATHWAY MODIFICATION

One puzzling example of homo-pathway modification is the blanking of visual responsiveness by exposure to blue light, and its return by red exposure in mutant fruitflies (Cosens and Briscoe 1972). This strange case is not unlike habituation, which is a decremental process observed at the behavioral or neurophysiological level in numerous responses (Horn and Hinde 1970), and which appears to be a bona fide homo-pathway process, the decrement occurring at synapses within the effective pathway (Zucker, Kennedy and Selverston 1971; Pinsker *et al.* 1970; Kupfermann *et al.* 1970; Castellucci *et al.* 1970; Rowell 1971c). I shall not consider habituation further, since it has recently been extensively reviewed (Horn and Hinde 1970).

In another form of homo-pathway modification, the intensity of a given sensory input may determine the degree to which a pathway's activity transcends the *local sign* properties of a reflex and extends through *irradiation* to more distant motor outputs (Sherrington 1906). In his studies of reflexes, Sherrington found that weak, noxious stimuli applied to the foot of a spinal dog or monkey evoked a weak flexion response which was local (*local sign*), but that the response strength and its spatial extent increased as the stimulus intensity was elevated (*irradiation*). If the stimulus position was changed, the general properties of the reflex remained the same, but the output geometry conformed to the new locus. The entire constellation of related reflexes was called a *type reflex*. Local sign and irradiation are now widely known properties of reflexes in vertebrates and invertebrates (Bullock and Horridge 1965). For instance, one case of irradiation is the spread of the ventilatory rhythm at times of stress to muscles which as a rule do not show this rhythm (Miller 1960, 1971c). Intracellular recordings of events related to irradiation have recently been reported (Pinsker *et al.*1970; Prior 1972).

These reflex parameters are particularly well demonstrated by the behavior of the giant Madagascar cockroach, *Gromphadorhina portentosa* (Camhi, unpublished). Figs. 1A and 1B show posterior dorsal views of

---

* Modifier signals can be hormonal, but only neural signals will be considered here.

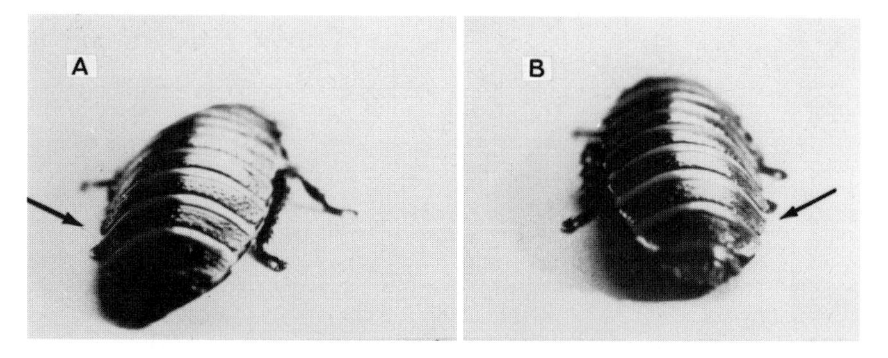

Fig. 1. A. and B. *Gromphadorhina portentosa* photographed from above and behind, just after a single brush stroke had been delivered to the sites indicated by the arrows

this insect. Just before the photograph shown in Fig. 1A was taken the insect was stroked along its left side with a camel hair brush, and a similar stroke to the right side just preceded the taking of Fig. 1B. The cockroach responds by moving the stimulated side ventrad, a maneuver accomplished by three separate, coordinated motor outputs. One of these, "segmental yaw", is discussed here, while the other two motor outputs are described in the next section.

Segmental yaw is a movement of a given segment, with respect to the anterior adjacent segment, in the yawing plane such as to close the soft intersegmental membrane ipsilaterally and expose it contralaterally. By waxing a pin to each segment in the yawing plane and filming from above, accurate measures of the component angles of the response can be attained. Fig. 2A shows a graph of the segmental yaw angles assumed, following tactile stimulation of the left margin of the sixth abdominal segment. The greatest response is local (local sign) and a stronger stimulus evokes both a greater local response and further spread (irradiation). In the example shown, the joint between abdominal segments 2 and 3 gives no discernible response for a weak, but does for a strong stimulus. Controls in which stimuli were delivered to a hardened wax droplet adhering to the cuticle surface, which resulted in almost no response, show that the signal which spreads on stronger stimulation is neuronal, not mechanical. Fig. 2B shows the local response to a stimulus on the left fourth abdominal segment of the same animal. The two responses belong to the same Sherringtonian type reflex.

A different form of response spread occurs when the effector normal for a response is lost under natural or experimental conditions. Among arthropods, leg autotomy is a natural event, and various neuronal adaptations for this contingency have been developed. Luco and Aranda (1964) report that amputating a forelimb of the cockroach *Blatta orientalis*, which normally guides the labial grooming of the contralateral antenna, results in an immediate switch to the use of the remaining foreleg. The same occurs in *Periplaneta americana* (personal observation). Luco and Aranda further report that the removal of both forelegs results in a learned (requiring about one week) ability to use the contralateral middle leg. Unlearned changes in the abdominal grooming response of *Periplaneta* have been demonstrated by Moran, Camhi and Eisner (unpublished). They showed that topically irritated loci are scratch by the ipsilateral hind leg in intact animals but that after autotomy or amputation of this limb the scratching is performed by the ipsilateral middle leg, the movements being well directed and coordinated. Loci too far posterior for the middle leg to reach are groomed instead by the contralateral hind leg. Thus the grooming receptive field for the hind leg is fractionated when this leg is lost.

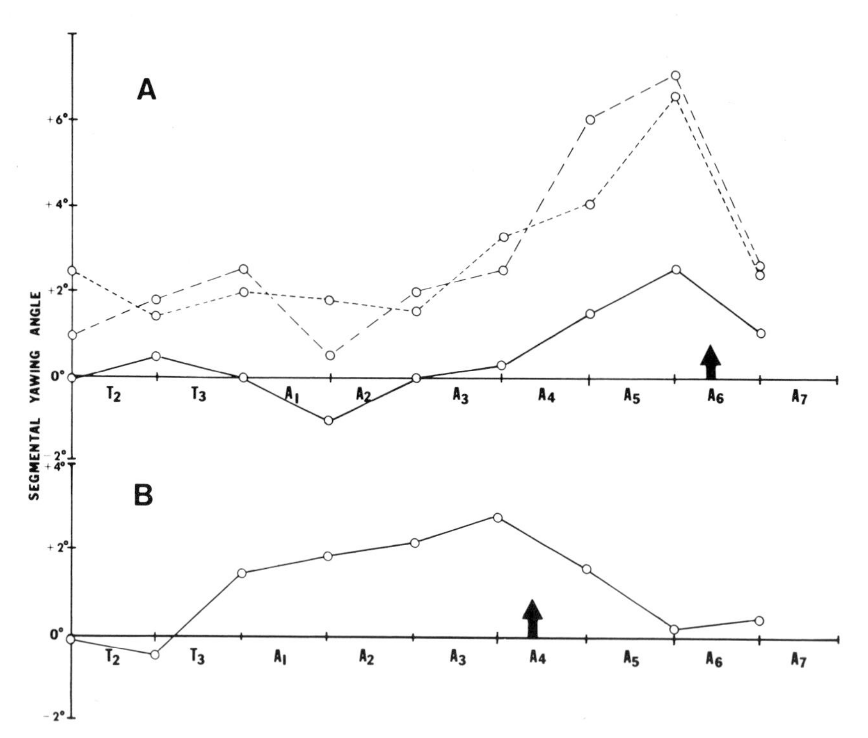

Fig. 2. "Segmental yaw" response of *Gromphadorhina portentosa* to tactile stimulation. The amount of intersegmental contraction following local tergite stimulation is plotted for each joint of the thorax and abdomen. Positive angles represent contraction ipsilateral to the stimulus; negative angles represent expansion ipsilateral to the stimulus. T_2 and T_3 - thoracic segments 2 and 3. A_1-A_7 - abdominal segments 1-7.

A. Stimulus to left margin of sixth abdominal tergite. Solid line - weak stimulus. Dashed line - stronger stimulus. Dotted line - strong stimulus presented five minutes later.

B. Single stimulus to left margin of fourth abdominal tergite

A related series of observations is that concerning the changes in an insect's gait following selective leg amputation. This work has been reviewed by Wilson (1966), who argued effectively against the concept of "plasticität" which had implied a switch to a novel neuronal pathway when a limb is lost. Instead, Wilson claimed that the main effect of amputation is to cause a subtle shift, along a continuum, in the interval between the onset of successive peristaltic waves of stepping, beginning in the metathoracic leg and proceeding anteriorly. This interval shift parallels that which occurs in the intact animal when stepping frequency is varied, and thus presumably the amputee uses the same, not a novel, neuronal mechanism. The theory is elegant and the fit with various unexpected observations is remarkably good, in spite of minor problems raised by Delcomyn's (1971b) observations, and the lack of fit to the data on *Carausius* (Wendler 1967).

A further example of phase changes among different motor units driven by a neural walking oscillator, described in detail in the next section, supports Wilson's view that this type of response modification can result from subtle changes within pathways, rather than the use of different pathways. The argument is in line with the joint requirements of central neuronal economy and complexity in insects. On the other hand, the observations on grooming reflexes involving newly employed outputs and sensory fractionation suggest much more profound response modification. Although it is premature to generalize, a reasonable hypothesis is that insects employ the same central mechanisms for as many different functions as possible, resorting to new pathways or mechanisms only where necessary, as in the case of a response switching to a different limb after one limb is lost.

It is not really clear whether these effects of limb loss should be thought of as homo- or hetero-pathway modifications, since in no case is it known what neural input is responsible for the change in gait (for instance, see Wendler 1967). Likely possibilities include sensory units sensing directly the leg's lesion, loss of reafference during motor outputs to the amputated limb, inability to remove the grooming stimulus, or to balance properly during walking (excluded by Wendler 1967) or efference copy information that a leg has been autotomized (in cases of autotomy). The last seems unlikely, since animals with autotomized hind legs, and those whose hind legs were amputated at the same or certain different sites, were essentially equivalent in their recruitment of the middle leg for abdominal grooming (Moran, Camhi and Eisner, unpublished).

HETERO-PATHWAY MODIFICATION

Depression and Facilitation

In other cases, however, there is clear indication that an effective pathway is depressed or facilitated by hetero-pathway modifier signals. For instance, Gelperin (1966, 1967) and Dethier and Gelperin (1967) have shown that pre-drinking proboscis extension in blowflies, which is excited by tarsal or labellar sugar stimulation, is inhibited by foregut and body wall stretch receptors, as well as by tarsal or labellar salt receptors. In fireflies, Case and Buck (1963), Case and Trinkle (1968), and Carlson and Copeland (1972) have shown that visual detection of a light flash inhibits spontaneous or evoked flashes for a brief duration, and these authors suggest that this inhibition plays a role in courtship flash synchronization. In crayfish, the population of ommatidia contributing to receptive fields of specific visual interneurons (space constancy fibers) changes as statocyst inputs record tilts of the animal (Wiersma and Yamaguchi 1967). A similar situation obtains for cortical (Horn and Hill 1969) and collicular (Bisti, Maffei and Piccolino 1972) visual units of cats subjected to body tilt.

While these processes grade on the one side into more commonplace convergence of excitatory and inhibitory synaptic influences upon postsynaptic elements, they also grade into more complex interactions. For instance, Dethier, Solomon and Turner (1965) showed that responsiveness of a blowfly's proboscis extension system to stimulation of a single labellar water receptor is increased by prior sugar stimulation of a single sugar receptor. The elevated responsiveness outlasts the sugar cell's spiking activity by about one minute. Sugar stimulation does not affect the spike activity of the water receptor, recorded peripherally, so the interaction appears to be central. Related examples of general tonic sensory input acting as a modifier of an effective pathway include the elevation of locomotory oscillator frequency, caused normally by those proprioceptors excited during locust flight (Wilson and Gettrup 1963) and crab walking (Cohen 1965). Other insect responses are tonically excited by ocellar (Goodman 1970) or inhibited by optic (Godden and Goldsmith 1972) inputs.

In other cases, two stimuli evoking conflicting complex motor acts have been experimentally presented simultaneously. For example, Moran Camhi and Eisner (unpublished) observe in the cockroach *Periplaneta* that the scratching reflex of a hind leg to an ipsilateral abdominal irritant may be blocked by simultaneous placement of an irritant on the other side. The stimulus threshold is elevated, and the two opposite legs alternate in bouts of scratching their own sides - a striking parallel to scratch reflex conflicts described by Sherrington (1906) in spinal dogs. Also relevant here is Rowell's (1971b) finding that visual interneurons of a locust are less responsive to repeated movement stimuli while the insect is grooming its antennae, than otherwise. In the slug *Pleurobranchaea*, Davis and Mpitsos (1971) describe a stereotyped righting reflex evoked by turning the animal over. However, if food is offered the slug while inverted, righting is deferred while stereotyped eating sequences ensue, following which a completely normal righting response occurs. Moreover, righting does not occur when the animal locomotes upside down on the water surface. It is presumed that righting behavior is triggered by gravitational cues sensed by statocysts (Wood and von Baumgarten 1972), and these findings suggest that the input from the statocysts to the righting mechanism, or this mechanism itself, can be inhibited under certain circumstances.

A further intriguing example of response modification is that reported by Roeder (1970) in acoustic neurons of the noctuid moth brain. Whereas these neurons usually faithfully follow ultrasonic inputs which drive bat evasion behavior, they often switch abruptly out of, and back into, this responsive mode. They can switch also into a bursting mode in which the burst rhythm can be phase locked to the wingbeat, though this can also occur in the absence of detectable flight muscle activity. No consistent modifier signal for these complex changes has been found.

Another type of hetero-pathway modification is the dishabituation of a habituated response, which often can be brought about by exciting some neurons outside the habituated pathway (Horn and Hinde 1970). In the abdominal ganglion of *Aplysia*, dishabituation of the habituated gill-withdrawal reflex results from heterosynaptic facilitation (Castallucci *et al.* 1970; Kandel and Tauc 1965a,b) and appears to consist of a process different from the mere restoration of the habituated mechanism to its original condition. In a locust visual interneuron, dishabituation results from sensory cues of a variety of modalities, or from struggling or other movements of the animal (Rowell 1971a). In this case dishabituation restores the visual neuron's responsiveness only to about its pre-stimulated level. Whereas habituation is specific to the particular ommatidia excited, dishabituation extends to the cell's entire receptive field.

Rowell (1971b) argues for a rather direct role of multi-modal afferent inputs in setting the level of a general arousal system, reflected in the visual interneuron's responsiveness, but his experiments seem to me not to rule out the possibility of central networks operating in this fashion. In fact, his experiments showing that this cell is generally rendered less responsive during antennal grooming (Rowell 1971b) appear to argue more strongly for central than for peripheral control of visual responsiveness, since reafferent inputs very much like those which attend grooming cause dishabituation at times other than during the grooming program. One can readily conceive of a central grooming program generator, or "motor tape", which while exciting grooming outputs, inhibits visual inputs.

As Rowell (1971b) points out, it is often difficult to distinguish between afferent, reafferent, and efference copy signals as modifiers of responsiveness. Attempts to deal with this question through transection experiments have been instructive only in a general sense. Experiments by Rowell (1961, 1964, 1965, 1969), for instance, suggest, through lesions and mass recording experiments, that the inhibition of a prothoracic grooming reflex which is released by lesions of specific connectives, may result largely from a non-specific neural influence of combined afferent and central origin. Findings by Roeder, Tozian and Weiant (1960) established the existence of widespread inhibitory control of endogenous activity patterns by the suboesophageal ganglion, but the mechanisms could not be explored. A similar case is that of the decerebrate cat as studied by Wall (1967). Here reversible spinalization by cold block leads to several changes in single motoneuron responsiveness to mechanical stimuli including increased responsiveness and increased receptive field size. In some units spinalization causes a switch from responsiveness mainly to deep receptors, to that mainly to cutaneous receptors. Numerous other examples are presented in a careful review by Rowell (1971c). These, on the basis of results with decapitated insects, generally support the role of somewhat deep central modifier signals, perhaps in addition to ones of sensory origin. However, in at least one insect behavior, the righting response of the cockroach *Gromphadorhina*, a highly complex sequence of stereotyped movements occurs completely normally in decapitated individuals (Camhi, unpublished).

Along different lines, Bentley and Hoy (1970) show that although the neural mechanisms for adult cricket stridulation were present in the nymph, these were inhibited until adulthood by descending inhibition. It has long been known that mushroom body lesions in the adult cricket brain can lead to continuous stridulatory calling (for instance, see Bentley 1969a), mediated by a central oscillator now known to be located in the pro- and/or mesothoracic ganglia (Kutsch and Otto 1972). Another adult behavior for which the wiring is present in the nymph, namely flight, apparently has no such inhibition, but would generally not be called forth in nymphs because of tarsal inhibition (Bentley and Hoy 1970). Reingold and Camhi (unpublished) have similarly shown that in cockroaches certain abdominal grooming reflexes change as nymphs molt into adults. In a minority of late instar nymphs, the adult pattern occurs, but only after decapitation.

The specific pathways of these descending inhibitions are unknown in all cases studied, but it is instructive that Bentley (1969b) found cricket motoneurons to be inhibited by inhibitory postsynaptic potentials (IPSPs) (studied in a largely deafferented individual with mushroom body lesions) except during motor acts when the inhibition was lifted. Crossman, Kerkut and Walker (1972) similarly record a predominance of IPSPs over excitatory postsynaptic potentials (EPSPs) in defined efferent locust and cockroach neurons. Atwood and Wiersma (1967) have found descending fibers in the crayfish circumoesophageal ganglion which inhibits either swimmeret or walking leg beating. Also of possible importance here are common inhibitory efferent neurons which in Crustacea (Wiersma 1961) may depress muscle activity during molting, although in insects they appear not to function in this way (Pearson and Bergman 1969). Crossman, Kerkut and Walker (1972) have recently described common efferent fibers in cockroaches and locusts which, while possibly inhibitory, are probably different from those studied by Pearson and Bergman, and have an unknown function.

The most powerful approach to the study of hetero-pathway modification would be to find an animal which exhibits this property and has very large nerve cells. There has been some success along these lines with molluscs (Castellucci *et al.* 1970; Davis and Mpitsos 1971), and increasing success in crayfish (e.g., Zucker, Kennedy and Selverston 1971). Secondly, one can take advantage of the massive amount of information on pathways becoming available for a few animals with smaller neurons - notably crayfish. Thus, for instance, Page and Sokolove (1971) showed that during slow abdominal flexion, "accessory nerve" activity inhibits the segmental abdominal stretch receptors and thus turns off resistance reflexes. Enough of the circuitry is known to suggest that the source of the inhibition may be either stretch receptors of other segments (Eckert 1961a) or perhaps the command fibers which presumably drive the flexion (Evoy and Kennedy 1967; Kennedy *et al.* 1967). Finally, one can try to find within the output some signature which allows identification of the modifier signal. Camhi and Hinkle (1972, in prep.) have been able to employ this approach in studying hetero-pathway control of flight stabilizing reflexes in locusts. The following description is abstracted mainly from these two papers.

The locust *Schistocerca gregaria* in tethered flight responds to rotations of the wind about the head in a horizontal arc, with lateral, rudder-like abdominal movements. The same wind rotations evoke no abdomen response if the insect is not flying (Camhi 1970a,b; Camhi and Hinkle 1972, in prep.). (The stimuli elicited by wind rotation mimic those produced by the aerodynamic consequences of flight yaw, and the response is appropriate for yaw correction.) The effective pathway here begins with wind receptor cells on the head (Camhi 1969; Weis-Fogh 1949, 1956) and terminates in the motoneurons of the abdomen's dorsal longitudinal muscles. During flight only, these motoneurons show rhythmic spike bursts phase-locked to the wingbeat. At such times, pivoting of the wind enhances the burst strength of ipsilateral motoneurons (increases the number of spikes per burst and brings in larger diameter axons) and depresses the burst strength of contralateral motoneurons. The phase with respect to the wingbeat is unchanged. In flightless periods, both the bursts and motoneuron responsiveness to wind pivots are absent. Thus this burst rhythm is a reflection of a bursting modifier signal. The modifier signal clearly must originate either in the central flight oscillator (Wilson 1961) (or other oscillators coupled to it) or in rhythmic reafferent sources. The latter can be excluded as essential, since abdominal motor axons in a central nervous system completely isolated from the periphery (except in the head) show the same bursting behavior and burst changes in response to wind pivots, only when thoracic motor nerve stump recordings show flight frequency bursts. Again, the thoracic and abdominal rhythms are phase locked. Other experiments show that wind velocity changes lead to abdominal posturing in the vertical plane, a response which is possibly useful in stall avoidance at low flight speeds (Camhi 1970b). This response involves the same sensory hairs (operating now as wind velocity detectors as inputs, and the same abdominal muscles as outputs, but in this case working with dorsal and ventral longitudinal muscles as synergists pitted against the ventral longitudinal muscles as antagonists. The central flight oscillator plays essentially the same role here in setting the responsiveness of the pathway.

Recently Litte and Camhi (unpublished) have shown that visually mediated roll correcting reflexes of locusts also occur only during flight. Goodman (1965) and Wilson (1968a) had shown that roll stabilization is activated visually and results in the locust rotating first the head and then the body in alignment with an artificial horizon line or a dorsal light. Fig. 3 shows our recordings of head movements made by a tethered locust in a wind tunnel in response to repeat rotations of a line representing an artificial horizon. If the locust stops flying, whether or not the wind persists unchanged, the head rotation response ceases. This is not due to general fatigue, since if the insect flies again moments later, the response is restored in full. While we do not yet know whether the central flight oscillator provides the modifier signal in this case, the parallels with the abdomen's responsiveness to wind changes suggest this as a likely possibility. Another possible, but as yet unexplored case, in which the flight oscillator may provide a modifier signal is that of the maintenance in the open position, during flight, of the second and third thoracic spiracles. This occurs in spite of ongoing activity of the central ventilatory oscillator which periodic-

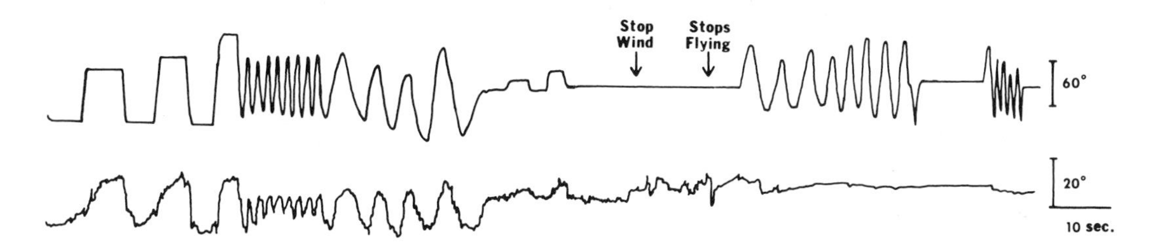

Fig. 3. Locust head rotations in response to rotations of an artificial horizon line stimulus. Tracings from chart recorder printout. Locust is flying in a wind tunnel, looking through the open end (wind drawn over body by fan pulling air from behind). Top trace - angle of artificial horizon line; bottom trace - angle of head. Turning off wind stops flight, following which visual input no longer drives head rotations

ally closes all other spiracles, and normally closes these as well (Miller 1960). Similar coupling and decoupling of spiracular and other ventilatory outputs from central respiratory oscillators have been observed in mantids, beetles, locusts and cockroaches (Hoyle 1964; Miller 1971a,c, and personal communication).

The results of our studies on locusts show a remarkable parallel with recent findings on dogfish locomotion. Roberts and Russell (1972) have shown that efferent neurons known to inhibit synaptically lateral line sensory hair cells (Roberts and Ryan 1971; Russell and Roberts 1972) are active almost exclusively during locomotory movements. At these times the efferent axons show bursts of spikes phase locked to the locomotory rhythm. This rhythm is generated by a central oscillator (Roberts 1969) although afferent inputs interact strongly with the oscillator (Lissmann 1946a,b). The obvious parallel to the locust is that, in both cases, a central locomotory oscillator serves as a modifier signal to deal with the conditions created by its own activity. Whereas in the locust the modifier is excitatory, in the dogfish it is inhibitory, and presumably functions to prevent lateral line overload during swimming (Roberts 1972; though see Russell and Roberts 1972). Interestingly, the lateral line efferents are also sometimes excited in rhythm with the gill movements.

A similar situation obtains in *Xenopus* (Russell 1971) where voluntary movements, even as minute as an eye-blink, induce spikes in the inhibitory efferent fibers to the lateral line. Interestingly, in *Xenopus* motor activity to small tonic muscles does not evoke lateral line efferent activity, a situation which parallels the crayfish's separation of fast and slow motor and sensory pathways (Kennedy and Takeda 1965a,b). Other examples, in vertebrates, of movement-related inhibition of sensory inputs have been described in visual systems (for instance, Johnstone and Mark 1971; Bizzi and Evarts 1971).

Invertebrate examples which parallel these include the sensing of water vibrations in crayfish by defined interneurons, the activity of which are inhibited during walking movements (Taylor 1968). This inhibition can precede and outlast visible leg movements by as much as 20 seconds and one minute respectively, suggesting that the modifier signal has a central rather than reafferent origin. Also in crayfish, Wiersma and Yamaguchi (1967) showed that visual responses of interneurons called "jittery movement fibers" are suppressed during eye-stalk movements. Again, Wiersma and coworkers (Arechiga and Wiersma 1969; Wiersma 1970; Wiersma and Yamaguchi 1967) showed that visual interneurons have greater responsiveness to light when the animal is moving than when it is quiescent. Just preceding the elevated responsiveness, they were often able to record spike activity in efferent fibers in the optic nerve, called "activity fibers" which may convey the modifier signal. In crabs, voluntary movement also increased spike frequencies in eye-stalk motor axons but inhibited those which cause protective eye-stalk retractions (Wiersma and Fiore 1971). Again, the source of the modifier signal may be central, since forced movements of the appendages were ineffective, whereas voluntary movements were effective in producing these changes. Moreover, the spiking of activity fibers waned only gradually following voluntary movements. Other related observations in crayfish are that lateral giant fiber outputs to fast flexor muscles are inhibited during swimming behavior (Schramek 1970); that similar synaptic inhibition of lateral giant fibers follows spikes in single medial or lateral giant fibers (Roberts 1968a,b); and that there is inhibition following single medial giant spikes upon the abdominal stretch receptor cells (Eckert 1961b; Roberts 1968b). The inhibitory interactions among these three separate tail flick pathways are particularly interesting since these are now known to mediate three different forms of tail flick behavior and have different output connections (Larimer *et al.* 1971; Mittenthal and Wine 1973; Schramek 1970; Wine and Krasne 1972). This finding directly parallels those of Diamond and Yasargil (1966) on teleost fish, that exciting one Mauthner neuron, which evokes an evasive tail flip, synaptically inhibits the contralateral Mauthner cell, which mediates an oppositely directed tail flip. These findings also relate to others, not covered here, which show that there is central regulation of stretch receptor sensitivity by controlling receptor muscle tension in lepidopterous insects (Finlayson and Lowenstein 1958) and in crayfish (Kennedy, Evoy and Fields 1966).

In the cockroach *Periplaneta*, Dagan and Parnas (1970) have made the startling observation that giant fibers appear to be unable (or perhaps less able than smaller fibers) to drive leg motoneurons, a finding which contradicts the time honored view that the giants mediate evasive behavior (see Roeder 1963). In view of the lability of giant fiber systems in dissected preparations (see, for instance, Wine and Krasne 1972), it would be of greatest interest to try to replicate some of their findings in unrestrained insects. Also of great interest from Dagan and Parnas is the observation that giant fibers excite antennal muscles which whip the antennae into a forward "startle position", and possibly ready them for tactile performance. The authors thus suggest a general alerting role for the giant fibers of the cockroach.

A number of examples of hetero-pathway modification emerge from my current studies of tactile reflexes and righting behavior in the cockroach *Gromphadorhina*. The following description is taken from papers in preparation by Camhi and by Sherman, Novotny and Camhi. I shall first describe tactile reflexes as observed in the stationary, quiescent insect, and then contrast these with responses to identical stimuli presented while the insect is engaged in righting behavior. I described in the previous section part of the reflex response of this insect to tactile stimulation of the lateral margin of the tergites - the segmental yaw response (Figs. 1 and 2). The total reflex is more complex, however, since it includes two additional, simultaneous outputs. In the output I call "segmental roll" a responding segment moves in the rolling plane with respect to the anterior adjacent segment, the side stimulated moving ventrad. The other output is leg flexion ipsilateral, and leg extension contralateral to the stimulus. The last two responses work in concert to move the stimulated side ventrad, thus presenting the heavy segmental tergites to any predator causing the stimulus. (Segmental yaw covers, and thus protects, the softer, intersegmental membranes ipsilaterally.) All three of these reflexes show conjoint local sign and irradiation properties. For instance, a soft, posterior abdomen stimulus generally evokes segmental yaw and roll in the posterior part of the abdomen, but no leg response, whereas leg responses plus anterior segmental yaw and roll result from thoracic stimulation.

Experimental Analysis of Insect Behaviour

Edited by L. Barton Browne

Erratum

Figure 5 on page 71 has been printed upside down.
See below the figure correctly composed.

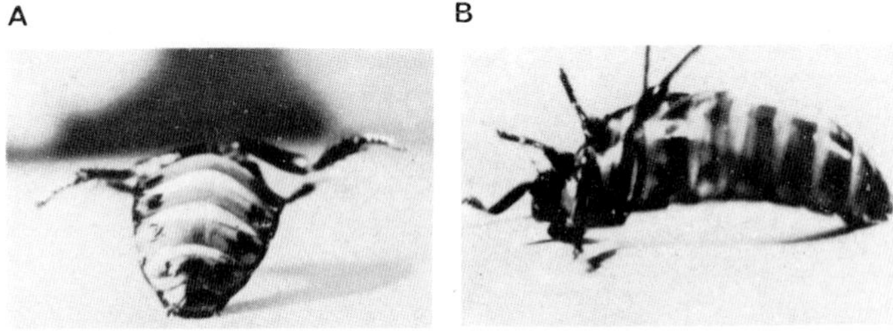

Springer-Verlag New York Heidelberg Berlin 1974

Fig. 4. *Gromphadorhina portentosa*. Top - normal posture, plus labial grooming of right front tarsus. Bottom - just following medial tactile stimulation of the protergite. Head moves to a central position and is retracted. Body segments bend ventrad, especially within the thorax. Lines are drawn tangential to the dorsal surface of these segments

There are three aspects of these responses whose special importance will become clear momentarily. The first is that stimulation of the midline at the anterior and posterior body extremes (prothorax and last few abdominal segments) evokes bilaterally symmetrical ventrad movements mainly of the nearby segments (Fig. 4). (This can be viewed as an extension of the local sign property to a medial stimulus site for which neither left nor right segmental yaw or roll is appropriate.) The second point is that stimulation of the lateral margins of the abdominal or thoracic tergites evokes no movement of the head or the antennae away from their normal orientation. (In fact tactile stimuli sometimes *return* the head to a straight position (Fig. 4.).) And thirdly, the direction of all the movements I have described is preserved relative to the body, independent of any change in body orientation relative to gravity.

When the insect is turned on its back, a highly stereotyped, though complex, series of righting movements occurs. Immediately, one observes a profound dorsad flexion of the entire thorax and abdomen, such that the body forms an arch supported on the substrate by its front and rear extremes (Fig. 5A). Soon the unstable arch topples to a metastable position with one lateral margin of all the tergites in contact with the ground (just beginning in Fig. 5B). Presumably in response to this lateral tactile stimulus, the three-part tactile reflex outlined above

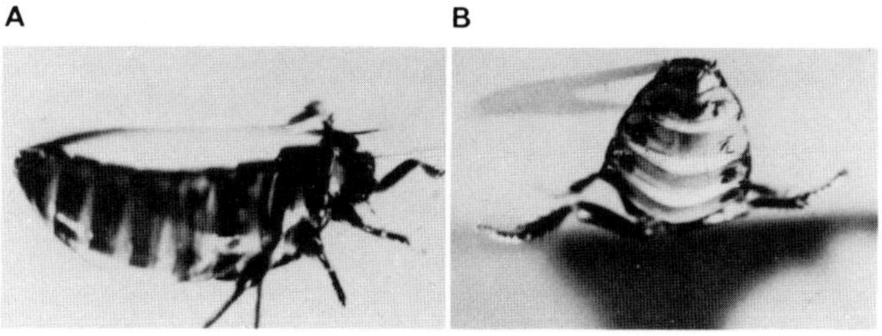

A B

Fig. 5. *Gromphadorhina* righting response.

A. Dorsad arching, first phase of the response.

B. Insect just beginning to topple to a metastable position on its right side which, when contacting the ground, will reflexively induce several motor acts results. The three components, namely segmental yaw and roll and ipsilateral (down) leg flexion and contralateral (up) leg extension, are all little modified from their appearance in stationary individuals. Experiments on tethered individuals engaged in this righting behavior (righting can be induced by removing tarsal contact, as described below), in which I use controlled tactile stimuli, demonstrate that stimuli applied alternately to the left and right tergal margins evoke alternately complete three-part left and right behaviors, and in the absence of such stimulation neither behavior occurs.

There are striking differences, however, between responses evoked by tactile stimulation of the tergal margin while the insect is stationary and while it is righting. In this and the next subsection I describe four such differences. The first difference is that during righting, in either a free or a tethered insect, lateral stimuli evoke head turning, the ventral aspect being turned toward the stimulated side, and a backward flexion of the ipsilateral antenna. These motions bring this antenna in contact with the ground which it then rapidly palpates. As I mentioned above, this response is never driven by even strong tactile stimuli in stationary insects.

The second difference between stationary and righting individuals is in a sense the most dramatic change possible, namely a reversal of reflex sign. As I have mentioned, when *Gromphadorhina* is standing still it responds to tactile stimuli of the mid-anterior ridge of the protergite, or of the posterior few segments, with local ventrad movements (Fig. 4). These result from contractions of ventral longitudinal muscles. Fig. 6A shows the response of a tethered stationary insect (holding a balsa block) to anterior protergite stimulation, and Fig. 7A shows ventral longitudinal muscle spike responses to this stimulus.

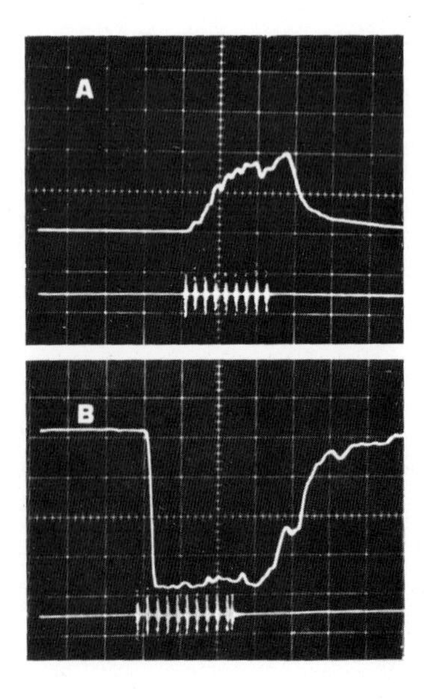

Fig. 6. Response of whole body of tethered *Gromphadorhina* to tactile stimulation of medial protergite. Upper trace of each photo - body movement; ventrad upward, dorsad downward. Lower trace of each photo - monitor of repeated tactile stimuli.

A. Insect holding balsa block.

B. Just after balsa block removed. Each division = 0.5 sec

By contrast to these ventrad movements, righting behavior begins with a *dorsad* arch (Fig. 5A) which can be evoked by turning the insect on its back, or for a tethered individual, by removing "ground" contact (contact with balsa block). (The insect's orientation with respect to gravity is incidental to the direction of the response, which is always dorsad.) Moreover, in this arched position a free righting insect's protergite and posterior abdominal segments are receiving constant tactile stimulation from the substrate. In tethered individuals showing this dorsad arch, protergite or posterior abdominal tactile stimulation reflexively evokes further *dorsad* arching, as seen in Fig. 6B - a movement opposite to that seen in stationary, non-righting insects. Fig. 7B shows that during righting, protergite stimuli evoke no spikes in the ventral longitudinal muscles, but do excite activity in the dorsal longitudinals. Often, as shown in this figure, the muscle shows background activity induced by release of tarsal contact, and this is enhanced by protergite stimuli.

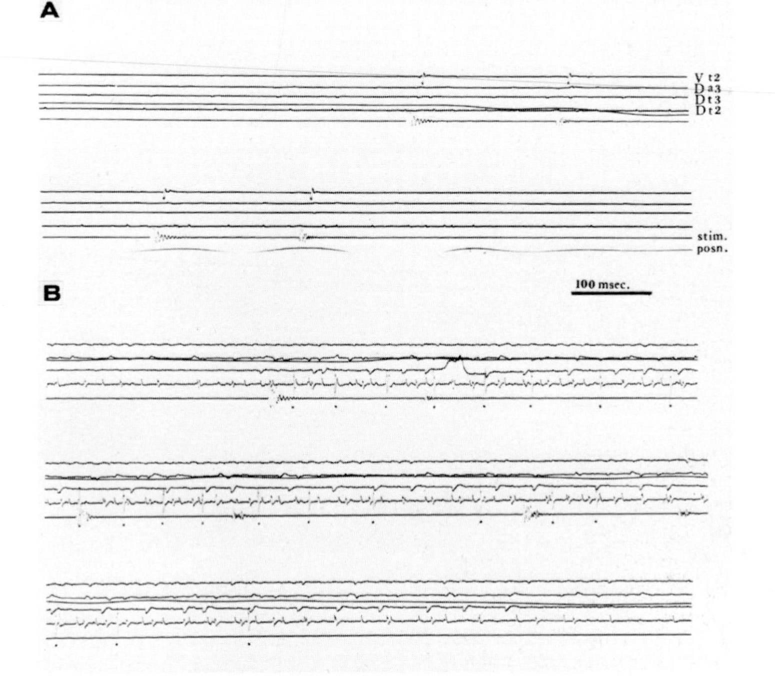

Fig. 7. Muscle action potentials and body movements recorded from tethered *Gromphadorhina* in response to controlled tactile stimuli. Vt2 - ventral longitudinal muscles of the second thoracic segment. Da3,Dt3,Dt2 - dorsal longitudinal muscles of third abdominal, third thoracic, and second thoracic segments respectively. stim. - monitor of electrically controlled pokes to protergite. posn. - record of body position; downward is ventrad, upward is dorsad.

A. Insect holding a balsa block, ventral longitudinal muscle spikes and ventrad movement are evoked.

B. Balsa block has just been removed, resulting in some dorsad arching and ongoing activity in dorsal longitudinal muscles. Stimuli then evoke further dorsad arching, and enhanced dorsal longitudinal spiking, with *no* ventral longitudinal spikes. The dots indicate the timing of a single large unit whose spike frequency is enhanced at least five times by the repeated stimuli

Remarkably, this reflex reversal, as well as the entire normal righting sequence including restoration of proper orientation, can be evoked just as easily in decapitated as in intact cockroaches (except, of course, for the head and antennal responses). Therefore, not only the effective pathways, but also the modifier signals are essentially independent of the head ganglia.

On the basis of the evidence presented, if would appear that the modifier signal which switches the protergite response from the ventrad to the dorsad mode originates in leg receptors sensing ground contact. However, it appears that tarsal contact release is only one sensory cue which can transform the behavior by affecting a *central* mechanism. For instance, Fig. 7B shows that breaking tarsal contact activates a continuous train of spikes, and we find the dorsad response continues only as long as this train continues. In some cases the train has the appearance of a bursting oscillator at a reasonable flight frequency - 20/sec - though *Gromphadorhina* is totally wingless. A second line of evidence, supporting central control relates to the observation that following repeated righting behavior of free or tethered individuals, most ultimately habituate and will now remain inverted and perfectly motionless for a variable period which can be as long as thirty minutes. Light tactile pokes given anywhere on the body during this period evoke responses characteristic of stationary - not righting individuals. If left alone, however, such an insect almost always performs, after a variable period, a completely normal righting response beginning with a dorsad arch and concluding with restored orientation. The sharp, abrupt onset of righting is not set off by loss of contact of the legs to any substrate or to each other; rather it appears as though some central trigger has just switched the animal into a righting mode.

The third difference between responses of stationary and righting *Gromphadorhina* is that, during righting, the legs on the upper side will respond with grabbing motions at any tactilely stimulated point within their reach. Even very gentle pokes to the coxae, sternites or other surfaces evoke immediate movements by the nearest leg. The stationary cockroach, free or tethered, with or without tarsal contact (e.g. inverted and habituated to righting) will not show this response even to strong tactile stimuli.

Oscillatory Phase Changes

This final category which I treat concerns phase changes in the outputs of central oscillators, and has been studied almost exclusively in insects. Wilson (1968b) has reviewed aspects of this topic, and the present treatment is largely supplementary to his paper. Kammer (1968, 1970, 1971) has extensively studied phase changes during flight warm-up and flight turns in Lepidoptera. During warm-up, some normally antagonistic motor units are excited synchronously. The result is that there is almost no wing movement, but instead only exothermic muscular contractions. When flight temperature is reached, the phases of these motor units either (depending upon species) slide gradually apart, or fire in both phases for a brief period and then only in antagonism. The occurrence of gradual sliding, with constant frequency, seems to be a compelling criterion for the existence of a single oscillator with variable output phases, i.e., of a commonality of some or most of the underlying central circuitry during warm-up and flight. The modifier signal in one species appears to originate in thermoreceptors within the thoracic ganglia themselves (Hanegan and Heath 1970). Similar changes from warm-up to stridulatory calling occur in tettigoniids (Heath and Josephson 1970). Related to phase changes is Miller's (1971b) finding that characteristic changes in burst and interburst activity occur in mantid spiracular motor rhythms following various stimuli, while the phasing with respect to the abdominal ventilatory pumping is maintained. This parallels the more detailed studies of Dando and Selverston (1972) on command fiber control of the decapod stomatogastric ganglion. Also relevant, though not actually a phase change, is a switch from walking controlled by extensor bursts to that controlled by flexor bursts in locusts (Hoyle 1964).

Though oscillatory phase changes occur commonly in dipteran flight in the form of phase multi-stability (Wyman 1966, 1969a,b; Levine 1973; Mulloney 1970a), some of these cases may result merely from random perturbations operating on a mutual inhibitory network, whereas others, which are accompanied by transient frequency increases, may involve what could be called a modifier signal (Wyman 1970). Occasionally, dramatic changes in coupling have been seen (Mulloney 1970b). But since in these and other fibrillar flight systems, muscle contraction is independent of motoneuron and muscle spike times, phase changes have no known behavioral effect (Wyman 1969a,b; Levine 1973). Consequently, any wingbeat modification must result from either a general frequency increase to all muscles (Wyman 1970) or to specific cuticular tensing muscles (Nachtigal and Wilson 1967), or from changes from a beating to a bursting pattern. Such bursting patterns may be employed in the courtship display of *Drosophila*, where individual movements of the wing closely resemble those during flight (Bennet-Clark and Ewing 1968) though, interestingly, only one wing is operated at a time. Species differences in courtship pattern may have undergone strong selection, as reproductive isolating mechanisms, since *Drosophila* species with similar flight frequencies show very different courtship wing frequencies (Waldron 1964). In the fibrillar flight muscles of bumble bees, single spikes within continuous trains show no phase preference between muscles during flight, but synchronous bursts, and synchronous spikes within these bursts, can occur among all flight muscles during pre-flight warm-up (Kammer and Heinrich 1972; Mulloney 1970b).

In the cockroach *Gromphadorhina* we have studied (Sherman, Novotny and Camhi, unpublished) what is perhaps a more complex oscillatory phase shift within the muscles of *individual* legs as the insect engages in either walking or righting behavior. This is the fourth and last of the differences between stationary and righting cockroaches mentioned above. Evidence that cockroach and other insect walking rhythms derive from central oscillators, though not studied directly in this species, is presented by Pearson (1972), Pearson and Iles (1970), and Hoy and Wilson (1969), and is strongly suggested by Wilson (1966) and Delcomyn (1971a,b). Moreover, the walking gait of *Gromphadorhina* is very much like that of other cockroaches (Delcomyn 1971a,b) in that a tripod rhythm is employed at almost all stepping frequencies (maximum frequency 6 Hz); bilateral mesothoracic amputees show frequency dependent phases between ipsilateral legs; and the ratio of protraction time to retraction time for each leg increases with increased stepping frequencies. I shall present evidence that, as in the flight system of Lepidoptera, there appears to be some commonality in the oscillatory networks and in the motoneurons employed in the two behaviors, though they differ in bursting phase. Moreover, I shall show that this phase shift within one leg can influence the activity of a contralateral leg, a finding which relates this work to that described in an earlier section on phase changes between the oscillators of different legs.

As I have mentioned, when this insect is placed on its back, it first arches dorsally and then falls to a meta-stable position with one side touching the ground (down side), with down legs flexed and the up legs extended. Superimposed upon this leg posture, however, the down legs immediately begin kicking rhythmically outward against the substrate, these movements being the agents of restoring the insect to an upright orientation. On a slick surface, or for a tethered insect in air, this rhythmic kicking may continue for many seconds or even minutes.

The leg kicking rhythm has several features in common with walking: (1) It generally occurs at 5-6 Hz, the maximum walking frequency, though frequencies covering almost the whole range of walking have been observed. (2) The legs are phased with respect to each other as in walking.

Usually all three down legs kick, and when this occurs they adopt the tripod pattern; by selective tactile stimulation in tethered insects, different combinations of legs can be made to kick (e.g. L3,L2; L2,L1; L1,R1; etc.) and these almost always move in the pattern expected on the basis of the tripod rhythm. (3) Bilateral mesothoracic leg amputees use patterns like those employed by similarly operated insects in walking. These factors suggest that the same oscillator or oscillators may underlie both walking and righting leg rhythms, and stronger evidence for this point is presented below.

There is a subtle but important difference, however, between the walking and righting leg oscillations. Whereas in walking any given leg moves anteriorly and posteriorly, in righting it moves primarily outward against the ground. This change comes about largely by observable phase alterations in the movements of specific leg segments with respect to each other. Underlying this is a phase change within identified muscles, as demonstrated in Fig. 8A and B. Fig. 8A shows simultaneous electromyograms from the main coxal remotor and the main femoral extensor of the right middle leg during walking. Fig. 8B is taken

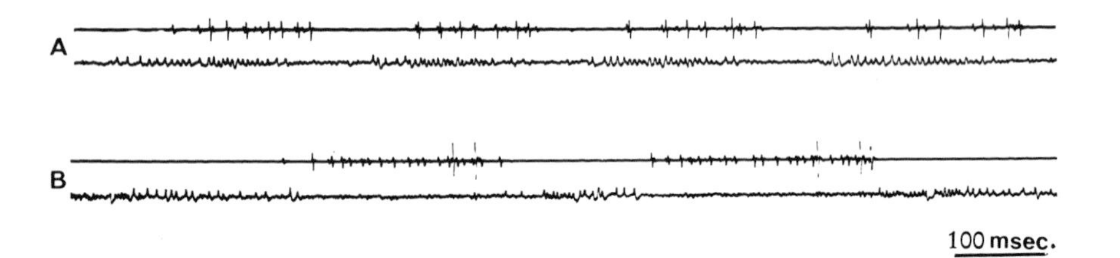

Fig. 8. Electromyograms from leg muscles of *Gromphadorhina* during walking and righting. Top trace in each pair is from the main coxal remotor, bottom trace from the main femoral extensor.

A. Tethered walking.

B. Tethered righting. Time mark same for botn records. Amplification decreased for top trace, compared with A

moments later during righting in which the right side is down. Notice that the phases of the rhythmic bursts from the two muscles are drastically changed. Nevertheless, the burst form is similar, and it is highly likely, from spike height analyses, that the same motor units are involved in the two behaviors. Such phase changes are seen in every individual and are reproducible under widely varying conditions of load. Specifically, patterns characteristic of righting behavior are shown by righting insects tethered in air or on surfaces of different degrees of slickness, and those characteristic of walking by insects walking under various loads and at different angles with respect to gravity. Fig. 9, a phase histogram from one typical tethered individual, demonstrates that the phases form two distinct populations.

Often while the down legs are kicking, the up legs show no discernible rhythm, but are held stiffly out, and if one records from the homologous muscles in these up legs, they show continuous, non-bursting spike activity. It thus appears that the oscillators of the down legs can be turned on independently, and as I have said, one can to some extent select which legs are turned on by appropriately placed lateral tergite stimulation. On many occasions during righting, however, one can see rhythmic movements in one or more of the up legs which appear to occur in anti-phase with its contralateral homolog. We have recorded during such periods simultaneously from the homologous muscles in both meso-

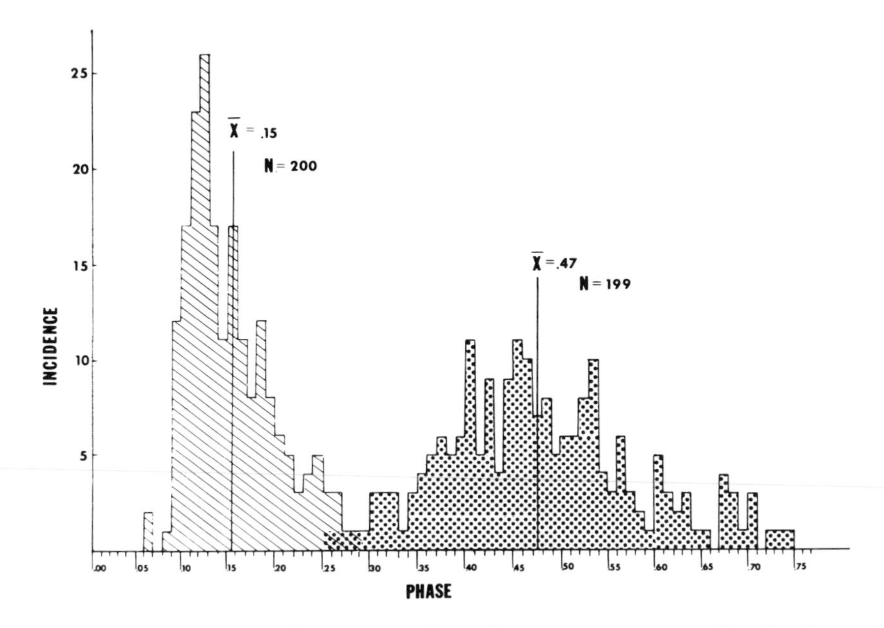

Fig. 9. Phase histogram for spike bursts in main coxal remotor and main femoral extensor of a middle leg during walking and righting in *Gromphadorhina*. Stippled bars are for bursts recorded during walking, crosshatched during righting. Phases were computed as the interval from the start of one femoral muscle burst to the start of the next coxal muscle burst, divided by the interval from the start of the same femoral burst to the start of the next femoral burst. Bins equal 0.01 phase increments

C_d
F_d
C_u
F_u
100 msec

Fig. 10. Electromyograms from coxal and femoral muscles of *both* middle legs during righting in *Gromphadorhina*. C_d and F_d - main coxal remotor and femoral extensor, respectively, of the down side. C_u and F_u - main coxal remotor and femoral extensor, respectively, of the up side. Phases of C_d in F_d = 0.32, 0.42 (fairly typical righting phases). Phases of C_u in F_u = 0.18, 0.20 (typical walking phases). Phases of C_d in C_u = 0.80, 0.66 (typically 0.5 during walking). Phases of F_d in F_u = 0.47, 0.43 (typically 0.5 during walking)

thoracic legs. As shown in Fig. 10, the rhythms of the two legs are phase locked, as was generally true when both were bursting. However, while the down leg was employing righting phases, the up leg was using walking phases. Also, the bursts of either muscle in the up leg are not quite in anti-phase with their contralateral homolog, but rather appear to adopt a phase determined both by the timing of the contralateral homolog burst, and by the burst of the other ipsilateral muscle.

While it is possible that the rhythmic movements of the up legs have a function in righting, none is apparent from observing the behavior. In fact, in free individuals engaged in righting, the movements of the up legs are very slight or absent, as contrasted with the profound rhythmic kicking of the down legs. It may be, therefore, that the up leg rhythm is a reflexion of neuronal couplings which are employed during walking, the walking oscillator being employed by the down legs during righting. The phase locking of contralateral legs and the predictable shift from perfect antagonism because of the phase shift of the down legs both support this argument.

This concept of a single oscillator used in both behaviors is in line with the hypothesis I proposed earlier, namely that insects employ the same neural circuit for as many different operations as possible, making whatever subtle changes are required. In this light, rhythmic grooming behavior and swimming movements are further possible uses of leg oscillators, and should be investigated with this in mind.

The modifier signal used to shift phases in *Gromphadorhina* is unidentified, but is probably the same signal used to switch other outputs during this insect's righting response. The modifier appears to operate by permitting a lateral tactile input to turn on a walking oscillator, and to do so with phase alternation in the ipsilateral legs. Since the burst structure is essentially unchanged in the same motoneurons it would seem most likely that the modifier signal would act by changing the couplings between antagonistic motor or pre-motor fibers, rather than affecting these units directly. Similar arguments have been advanced concerning other systems (Wilson 1968b; Hanegan and Heath 1970).

GENERAL DISCUSSION

Much of the body of this paper supports the opening argument, that insects and other invertebrates show a degree of behavioral and central neuronal complexity not generally suspected until quite recently. It seems likely that this trend towards the demonstration of complexity will continue, and that many more examples of response modification will be discovered. Insects are likely to play an increasingly prominent role in this area, because of their numerous behavioral and neuro-anatomical advantages. Consequently, an extensive effort on one or a few favorable systems seems to be an important present aim.

On the main subject of this paper, hetero-pathway modification, the findings are so fragmentary that it is difficult to generalize. One speculation I will make, however, is that modifier signals will increasingly be found to derive from central neuronal sources. For instance, complex "decision making" processes are likely to involve deep central integration. Also, large numbers of central locomotory oscillators probably use an efference copy of their rhythm as a modifier signal (witness the striking parallel between locusts and dogfish sharks), and the same should be true for non-rhythmic motor outputs. Moreover, couplings among these oscillators probably are quite commonly varied, especially in cases where several limbs are involved. Both these arrangements of central oscillators show neuronal economy.

Secondly, I shall comment upon the notion that response modification may operate by non-specific, or diffuse connections. In one case, the "total amount of irrelevant input" diffusely entering a ganglion through small diameter axons has been implicated as a modifier signal (Rowell 1965). Moreover, Hoyle (1964) argues for considerable indeterminacy of central connections, although Miller (1971c) points out that many of these connections can become specifically meaningful under certain conditions.

There seems to be some confusion about the significance, in terms of neuronal connectivity, of terms such as "non-specific", "indeterminate", and "diffuse". A diffusely connected neuron can be highly specific, as is true for giant fibers and command fibers. The presence of non-specific, or indeterminate connections would imply that the system containing them either operates by virtue of, or can operate adequately in spite of, non-specificity. Though this remains a possibility, in almost no case has the techniques used been precise enough to determine whether this is so. In one case where the resolution needed to study small fiber connectivity was achieved, a fairly high degree of anatomical precision and cell-to-cell specificity was found (Macagno, Lopresti and Levinthal 1973).

As more and more subtle questions are asked about nervous systems, one is forced to pay ever-increasing attention to behavioral and neuro-anatomical considerations in choosing a research subject. For instance, the righting behavior of *Gromphadorhina* is stereotyped, fast and abrupt, features which suggest the use of relatively few and large neurons. Since simultaneous and coordinated response modification occurs in all regions of the body during righting, a spatially distributed cellular substrate carrying the modifier signal throughout the ventral cord is implicated. A distributed system in insects offers advantages over a strictly intra-ganglionic system, in that it permits the possibility of recording the results of intermediate integrative steps in the inter-ganglionic connectives. A continuing drawback is the relative difficulty of recording synaptic potentials.

Command fibers are distributed interneurons with specific contacts. These fibers have been discovered as common elements in invertebrate central nervous systems, but have not been reported in insects. The output connections of such command fibers can be both excitatory and inhibitory, and their integrative effects can be very complex (Dando and Selverston 1972; Atwood and Wiersma 1967). It is not unlikely that there exist distributed interneurons like command fibers, which alter connections between, or the responsiveness of, other neurons, while not necessarily commanding any action themselves. Atwood and Wiersma (1967) suggest that some crayfish cells may act this way. Our findings on *Gromphadorhina* suggest the presence of such neurons, and our current work is focussed upon finding and describing them.

How might such hypothetical neurons increase a pathway's responsiveness? One possible mechanism is for them to excite a neuron of the effective pathway at a low level, depolarizing its spike trigger zone toward, or above, threshold, thus rendering it more responsive to small synaptic inputs from its driving neurons within the effective pathway. Such an explanation would fit with my observations on ventrad-to-dorsad switching in the protergite reflex in *Gromphadorhina*. Fig. 7 shows a case in which the dorsal-longitudinal muscles fire when righting occurs, but increase their firing upon protergite stimulation.

Another possible mechanism is for the hypothetical modifier neuron to disinhibit an effective pathway (Kandel, Frazier and Wachtel 1969), i.e., to synaptically inhibit cells which tonically inhibit effective pathway neurons. In this regard, one is reminded of Bentley's (1969a) observation that ongoing inhibition occurs in insect motoneurons and is lifted when these engage in motor acts. Though the interactions predicted here have not yet been directly observed, they are not out of line with the developing picture of complex central neuronal interactions, as outlined in the introduction.

One caution which it will be important to heed is that of Wine and Krasne (1972), that relatively intact animals perform very differently from highly dissected preparations. In studying subtle events like response modification, which may increasingly involve complex higher central integration, it may be particularly important to bear this in mind. This difficulty, pointed out clearly years ago by Roeder (1963), may in some measure limit how much we can learn about complex central functioning, by operating as a negative feedback on our experimental performance. The minimum requirement is to proceed with increasing caution, and in a step-by-step fashion, in progressing from observations of whole animals in relatively natural surroundings to single cell neurophysiological experiments. The challenge of the coming years will be to combine such caution with ingenious new techniques for peering deeper into the brain.

ACKNOWLEDGEMENTS

I acknowledge the helpful criticism of this manuscript by George Casaday, and I thank Dr. Louis Roth, U.S. Army Laboratories, Natick, Mass., for providing the starting culture of cockroaches.

The original work presented here was supported by NIH Grant # NS 09083.

REFERENCES

ALVING, B.: Spontaneous activity in isolated somata of *Aplysia* pacemaker neurons. J. gen. Physiol. 51, 29-45 (1968).
ARECHIGA, H., WIERSMA, C.A.G.: The effect of motor activity on the reactivity of single visual units in the crayfish. J. Neurobiol. 1, 53-69 (1969).
ATWOOD, H., WIERSMA, C.A.G.: Command interneurons in the crayfish central nervous system. J. exp. Biol. 46, 249-261 (1967).
BENNET-CLARK, H., EWING, A.: The wing mechanism involved in the courtship of *Drosophila*. J. exp. Biol. 49, 117-128 (1968).
BENTLEY, D.: Intracellular activity in cricket neurons during generation of song patterns. Z. vergl. Physiol. 62, 267-283 (1969a).
BENTLEY, D.: Intracellular activity in cricket neurons during the generation of behavior patterns. J. Insect Physiol. 15, 677-699 (1969b).
BENTLEY, D.: A topographical map of the locust flight system motor neurons. J. Insect Physiol. 16, 905-918 (1970).
BENTLEY, D., HOY, R.: Postembryonic development of adult motor patterns in crickets: a neural analysis. Science, N.Y. 170, 1409-1411 (1970).
BISTI, S., MAFFEI, L., PICCOLINO, M.: Variations of the visual responses of the superior colliculus in relation to body roll. Science, N.Y. 175, 456-457 (1972).
BIZZI, E., EVARTS, E.: Translational mechanisms between input and output. Neurosci. Res. Prog. Bull. 9(1), 31-59 (1971).
BOWERMAN, R., LARIMER, J.: Command fibers in the circumesophageal connectives of crayfish. I. Tonic fibers. (In press a).
BOWERMAN, R., LARIMER, J.: Command fibers in the circumesophageal connectives of crayfish. II. Phasic fibers. (In press b).
BULLOCK, T., HORRIDGE, G.: "Structure and Function in the Nervous System of Invertebrates". San Francisco: Freeman (1965).
CAMHI, J.: Locust wind receptors: I. Transducer mechanics and sensory response. J. exp. Biol. 50, 335-348 (1969).
CAMHI, J.: Yaw-correcting postural changes in locusts. J. exp. Biol. 52, 519-531 (1970a).
CAMHI, J.: Sensory control of abdomen posture in flying locusts. J. exp. Biol. 52, 533-537 (1970b).
CAMHI, J., HINKLE, M.: Attentiveness to sensory stimuli: central control in locusts. Science, N.Y. 175, 550-553 (1972).
CAMHI, J., HINKLE, M.: Response modification by the central flight oscillator of locusts. (In preparation).
CARLSON, A., COPELAND, J.: Photic inhibition of brain stimulated firefly flashes. Am. Zoologist 12, 479-487 (1972).
CASE, J., BUCK, J.: Control of flashing in fireflies. II. Role of central nervous system. Biol. Bull. mar. biol. Lab., Woods Hole

125, 234-250 (1963).
CASE, J., TRINKLE, M.: Light inhibition of flashing in the firefly *Photuris missouriensis*. Biol. Bull. mar. biol. Lab., Woods Hole 135, 476-485 (1968).
CASTELLUCCI, V., PINSKER, H., KUPFERMANN, I., KANDEL, E.R.: Neuronal mechanisms of habituation and dishabituation of the gill-withdrawal reflex in *Aplysia*. Science, N.Y. 167, 1745-1748 (1970).
CHEN, C., von BAUMGARTEN, R., TAKEDA, R.: Pacemaker properties of completely isolated neurons in *Aplysia californica*. Nature New Biol. 233, 27-29 (1971).
COHEN, M.: The dual role of sensory systems: detection and setting central excitability. Cold Spring Harb. Symp. quant. Biol. 30, 587-600 (1965).
COHEN, M., JACKLET, J.: The functional organization of motor neurons in an insect ganglion. Phil. Trans. R. Soc. (B) 252, 561-569 (1967).
COSENS, D., BRISCOE, D.: A switch phenomenon in the compound eye of the white-eyed mutant of *Drosophila melanogaster*. J. Insect Physiol. 18, 627-632 (1972).
CROSSMAN, A., KERKUT, G., WALKER, R.: Electrophysiological studies on the axon pathways of specified nerve cells in the central ganglia of two insect species, *Periplaneta americana* and *Schistocerca gregaria*. Comp. Biochem. Physiol. 43A, 393-415 (1972).
DAGAN, D., PARNAS, I.: Giant fibre and small fibre pathways involved in the evasive response of the cockroach, *Periplaneta americana*. J. exp. Biol. 52, 313-324 (1970).
DANDO, M., SELVERSTON, A.: Command fibers from the supra-oesophageal ganglion to the stomatogastric ganglion in *Panularus argus*. J. comp. Physiol. 78, 138-175 (1972).
DAVIS, W., KENNEDY, D.: Command interneurons controlling swimmeret movements in the lobster. I. Types of effects on motorneurons. J. Neurophysiol. 35, 1-12 (1972a).
DAVIS, W., KENNEDY, D.: Command interneurons controlling swimmeret movements in the lobster. II. Interaction of effects on motorneurons. J. Neurophysiol. 35, 13-19 (1972b).
DAVIS, W., KENNEDY, D.: Command interneurons controlling swimmeret movements in the lobster. III. Temporal relationships among bursts in different motoneurons. J. Neurophysiol. 35, 20-29 (1972c).
DAVIS, W., MPITSOS, G.: Behavioral choice and habituation in the marine mollusk *Pleurobranchia californica* MacFarland. Z. vergl. Physiol. 75, 207-232 (1971).
DELCOMYN, F.: The locomotion of the cockroach *Periplaneta americana*. J. exp. Biol. 54, 443-452 (1971a).
DELCOMYN, F.: The effect of limb amputation on locomotion in the cockroach *Periplaneta americana*. J. exp. Biol. 54, 453-469 (1971b).
DeLONG, M.: Central patterns of movement. Neurosci. Res. Prog. Bull. 9(1), 10-30 (1971).
DETHIER, V., GELPERIN, A.: Hyperphagia in the blowfly. J. exp. Biol. 47, 191-200 (1967).
DETHIER, V., SOLOMON, R., TURNER, L.: Sensory input and central excitation in the blowfly. J. comp. physiol. Psychol. 60, 303-313 (1965).
DIAMOND, J., YASARGIL, G.: A mutual crossed inhibition of the effects of Mauthner-axon excitation in goldfish spinal cords. J. Physiol., Lond. 185, 11P (1966).
ECKERT, R.: Reflex relationships of the abdominal stretch receptors of the crayfish. I. Feedback inhibition of the receptors. J. cell. comp. Physiol. 57, 149-162 (1961a).
ECKERT, R.: Reflex relationships of the abdominal stretch receptors of the crayfish. II. Stretch receptor involvement during the swimming reflex. J. cell. comp. Physiol. 57, 163-174 (1961b).
EVOY, W., KENNEDY, D.: The central nervous organization underlying control of antagonistic muscles in the crayfish. I. Types of

command fibers. J. exp. Zool. 165, 223-238 (1967).
FINLAYSON, L., LOWENSTEIN, O.: The structure and function of abdominal stretch receptors in insects. Proc. R. Soc. (B) 148, 433-449 (1958).
von FRISCH, K.: "The Dance Language and Orientation of Bees". Cambridge, Mass.: Harvard University Press (1967).
GELPERIN, A.: Investigations of a foregut receptor essential to taste threshold regulation in the blowfly. J. Insect Physiol. 12, 829-841 (1966).
GELPERIN, A.: Stretch receptors in the foregut of the blowfly. Science, N.Y. 157, 208-210 (1967).
GODDEN, D., GOLDSMITH, T.: Photoinhibition of arousal in the stick insect *Carausius*. Z. vergl. Physiol. 76, 135-145 (1972).
GOODMAN, L.: The role of certain optomotor reactions in regulating stability in the rolling plane during flight in the desert locust, *Schistocerca gregaria*. J. exp. Biol. 42, 385-407 (1965).
GOODMAN, L.: The structure and function of the insect dorsal ocellus. Adv. Insect Physiol. 7. 97-195 (1970).
HAGIWARA, S.: Nervous activity in the heart of crustacea. Ergebn. Biol. 24, 288-311 (1961).
HANEGAN, J., HEATH, J.: Temperature dependence of the neural control of the moth flight system. J. exp. Biol. 53, 629-639 (1970).
HARRIS, C., SMYTH, T.: Delayed firing of giant axons in the American cockroach. J. Insect Physiol. 17, 1565-1577 (1971).
HEATH, J., JOSEPHSON, R.: Body temperature and singing in the katydid, *Neoconcephalus robustus* (Orthoptera, Tettigoniidae). Biol. Bull. mar. biol. Lab., Woods Hole 138, 272-285 (1970).
HINKLE, M., CAMHI, J.: Locust motoneurons: bursting activity correlated with axon diameter. Science, N.Y. 175, 553-556 (1972).
HORN, G., HILL, R.: Modifications of receptive fields of cells in the visual cortex occurring spontaneously and associated with body tilt. Nature, Lond. 221, 186-188 (1969).
HORN, G., HINDE, R. (Eds.): "Short Term Changes in Neural Activity and Behavior". London: Cambridge University Press (1970).
HOY, R., WILSON, D.: Rhythmic motor output in the leg motorneurons of the milkweed bug, *Oncopeltus*. Fedn Proc. Fedn Am. Socs exp. Biol. 28, 588 (1969).
HOYLE, G.: Exploration of neuronal mechanisms underlying behavior in insects. *In* "Neural Theory and Modeling" (Ed., R. Reiss). Stanford, Calif.: Stanford University Press, pp. 346-376 (1964).
HOYLE, G.: Cellular mechanisms underlying behavior - neuroethology. Adv. Insect Physiol. 7, 349-444 (1970).
ILES, J., MULLONEY, B.: Procion yellow staining of cockroach motor-neurons without the use of microelectrodes. Brain Res. 30, 397-400 (1971).
JOHNSTONE, J., MARK, R.: The efference copy neurone. J. exp. Biol. 54, 403-414 (1971).
KAMMER, A.: Motor patterns during flight and warm-up in Lepidoptera. J. exp. Biol. 48, 89-109 (1968).
KAMMER, A.: A comparative study of motor pattern during pre-flight warm-up in hawkmoths. Z. vergl. Physiol. 70, 45-56 (1970).
KAMMER, A.: The motor output during turning flight in a hawkmoth, *Manduca sexta*. J. Insect Physiol. 17, 1073-1086 (1971).
KAMMER, A., HEINRICH, B.: Neural control of bumblebee fibrillar muscles during shivering. J. comp. Physiol. 78, 337-345 (1972).
KANDEL, E., FRAZIER, W., WACHTEL, H.: Organization of inhibition in abdominal ganglion of *Aplysia*. I. role of inhibition and disin-hibition in transforming neural activity. J. Neurophysiol. 32, 496-508 (1969).
KANDEL, E., TAUC, L.: Heterosynaptic facilitation in neurones of the abdominal ganglion of *Aplysia delipans*. J. Physiol., Lond. 181, 1-27 (1965a).
KANDEL, E., TAUC, L.: Mechanism of heterosynaptic facilitation in the

giant cell of the abdominal ganglion of *Aplysia delipans*. J. Physiol., Lond. 181, 28-47 (1965b).
KENNEDY, D.: Input and output connections of single arthropod neurons. *In* "Physiological and Biochemical Aspects of Nervous Integration" (Ed., F.D. Carlson). Englewood Cliffs, N.J.: Prentice-Hall. (1968).
KENNEDY, D., EVOY, W., DANE, B., HANAWALT, J.: The central nervous organization underlying control of antagonistic muscles in the crayfish. II. Coding of position by command fibers. J. exp. Zool. 165, 239-248 (1967).
KENNEDY, D., EVOY, W., FIELDS, H.: The unit basis of some crustacean reflexes. *In* "Nervous and Hormonal Mechanisms of Integration". Symp. Soc. exp. Biol. 20, 75-109 (1966).
KENNEDY, D., EVOY, W., HANAWALT, J.: Release of coordinated behavior in crayfish by single central neurons. Science, N.Y. 154, 917-919 (1966).
KENNEDY, D., TAKEDA, K.: Reflex control of abdominal flexor muscle in the crayfish. I. The twitch system. J. exp. Biol. 43, 211-227 (1965a).
KENNEDY, D., TAKEDA, K.: Reflex control of abdominal flexor muscles in the crayfish. II. The tonic system. J. exp. Biol. 43, 229-246 (1965b).
KUPFERMANN, I., CASTELLUCCI, V., PINSKER, H., KANDEL, E.: Neuronal correlates of habituation and dishabituation of the gill-withdrawal reflex in *Aplysia*. Science, N.Y. 167, 1743-1745 (1970).
KUPFERMANN, I., KANDEL, E.R.: Neuronal controls of a behavioral response mediated by the abdominal ganglion of *Aplysia*. Science, N.Y. 164, 847-850 (1969).
KUTSCH, W.: Neuromaskuläre aktivität bei verschiedenen Verhaltensweisen von drei Grillenarten. Z. vergl. Physiol. 63, 335-378 (1969).
KUTSCH, W., OTTO, D.: Evidence for spontaneous song production independent of head ganglia in *Gryllus campestris* L. J. comp. Physiol. 81, 115-119 (1972).
LARIMER, J., EGGLESTON, A.: Motor programs for abdominal positioning in crayfish. Z. vergl. Physiol. 74, 388-402 (1971).
LARIMER, J., EGGLESTON, A., MASUKAWA, L., KENNEDY, D.: The different connections and motor outputs of lateral and medial giant fibers in the crayfish. J. exp. Biol. 54, 391-402 (1971).
LARIMER, J., KENNEDY, D.: The central nervous control of complex movements in the uropods of crayfish. J. exp. Biol. 51, 135-150 (1969).
LEVINE, J.: Neural control of flight in wild type and mutant *Drosophila melanogaster*. Ph.D. Thesis, Yale University. (1973).
LINDAUER, M.: "Communication Among Social Bees". Cambridge, Mass.: Harvard University Press (1961).
LISSMAN, H.: The neurological basis of the locomotory rhythm in the spinal dogfish (*Scyllium caniciela, Acanthias vulgaris*). I. Reflex behavior. J. exp. Biol. 23, 143-161 (1946a).
LISSMAN, H.: The neurological basis of the locomotory rhythm in the spinal dogfish (*Scyllium caniciela, Acanthias vulgaris*). II. The effect of deafferentation. J. exp. Biol. 23, 162-176 (1946b).
LUCO, J., ARANDA, L.: An electrical correlate to the process of learning. Acta physiol. latinoam. 14, 274-288 (1964).
MACAGNO, E., LOPRESTI, V., LEVINTHAL, C.: Structure and development of neuronal connections in isogenic organisms: Variations and similarities in the optic system of *Daphnia magna*. Proc. natn. Acad. Sci. U.S.A. 70, 57-61 (1973).
MAYNARD, D.: Integration of crustacean ganglia. Symp. Soc. exp. Biol. 20, 111-150 (1966).
MAYNARD, D.: Simpler networks. Ann. N.Y. Acad. Sci. 193, 59-72 (1972).
MENDELSON, M.: Oscillator neurons in crustacean ganglia. Science, N.Y. 171, 1170-1173 (1971).
MILBURN, N., BENTLEY, D.: On the dendritic topology and activation of cockroach giant interneurons. J. Insect Physiol. 17, 607-623 (1971).

MILLER, P.: Respiration in the desert locust. III. Ventilation and the spiracles during flight. J. exp. Biol. 37, 264-278 (1960).
MILLER, P.: Rhythmic activity in the insect nervous system. I. Ventilatory coupling of a mantid spiracle. J. exp. Biol. 54, 587-597 (1971a).
MILLER, P.: Rhythmic activity in the insect nervous system. II. Sensory and electrical stimulation of ventilation in a mantid. J. exp. Biol. 54, 599-607 (1971b).
MILLER, P.: Rhythmic activity in the insect nervous system: Thoracic ventilation in nonflying beetles. J. Insect Physiol. 17, 395-405 (1971c).
MITTENTHAL, J., WINE, J.: Connectivity patterns of crayfish giant interneurons: visualization of synaptic regions with cobalt dye. Science, N.Y. 179, 182-184 (1973).
MULLONEY, B.: Organization of flight motor of Diptera. J. Neurophysiol. 33, 86-95 (1970a).
MULLONEY, B.: Impulse patterns in the flight motor neurons of *Bombus californicus* and *Oncopeltus fasciatus*. J. exp. Biol. 52, 59-77 (1970b).
NACHTIGAL, W., WILSON, D.: Neuro-muscular control of dipteran flight. J. exp. Biol. 47, 77-97 (1967).
NICHOLLS, J., PURVES, D.: Monosynaptic chemical and electrical connexions between sensory and motor cells in the central nervous system of the leech. J. Physiol., Lond. 209, 647-667 (1970).
PAGE, C., SOKOLOVE, P.: Crayfish muscle receptor organ: role in regulation of postural flexion. Am. Zoologist 11, 669 (1971).
PAGE, C., WILSON, D.: Unit responses in the metathoracic ganglion of the flying locust. Fedn Proc. Fedn Am. Socs exp. Biol. 29, 590 (1970).
PARNAS, I., SPIRA, M., WERMAN, R., BERGMANN, F.: Non-homogeneous conduction in giant axons of the nerve cord of *Periplaneta americana*. J. exp. Biol. 50, 635-649 (1969).
PEARSON, K.: Central programming and reflex control of walking in a cockroach. J. exp. Biol. 56, 173-193 (1972).
PEARSON, K., BERGMAN, S.: Common inhibitory motoneurons in insects. J. exp. Biol. 50, 445-471 (1969).
PEARSON, K., ILES, J.: Discharge patterns of coxal levator and depressor motoneurons of the cockroach *Periplaneta americana*. J. exp. Biol. 52, 139-165 (1970).
PINSKER, H., KUPFERMANN, I., CASTELLUCCI, V., KANDEL, E.: Habituation and dishabituation of the gill-withdrawal reflex in *Aplysia*. Science, N.Y. 167, 1740-1742 (1970).
PITMAN, R., TWEEDLE, C., COHEN, M.: Branching of central interneurons: intracellular cobalt injection for light and electron microscopy. Science, N.Y. 176, 412-414 (1972a).
PITMAN, R., TWEEDLE, C., COHEN, M.: Electrical responses of insect neurons: augmentation by nerve section or colchicine. Science, N.Y. 178, 507-509 (1972b).
PRIOR, D.: A neural correlate of behavioral stimulus intensity discrimination in a mollusc. J. exp. Biol. 57, 147-160 (1972).
REMLER, M., SELVERSTON, A., KENNEDY, D.: Lateral giant fibers of crayfish: location of somata by dye injection. Science, N.Y. 162, 281-283 (1968).
ROBERTS, A.: Recurrent inhibition in the giant-fiber system of the crayfish and its effect on the excitability of the escape response. J. exp. Biol. 48, 545-567 (1968a).
ROBERTS, A.: Some features of the central co-ordination of a fast movement in the crayfish. J. exp. Biol. 49, 645-656 (1968b).
ROBERTS, B.: Spontaneous rhythms in the motoneurons of spinal dogfish (*Scyliorhinus canicula*). J. mar. biol. Ass. U.K. 49, 357-378 (1969).
ROBERTS, B.: Activity of lateral-line sense organs in swimming dogfish. J. exp. Biol. 56, 105-118 (1972).

ROBERTS, B., RUSSELL, I.: The activity of lateral-line efferent neurones in stationary and swimming dogfish, *Scyliorhinus canicula* L. J. exp. Biol. 57, 435-448 (1972).
ROBERTS, B., RYAN, K.: The fine structure of the lateral-line sense organs of dogfish. Proc. R. Soc. (B) 179, 157-169 (1971).
ROEDER, K.: "Nerve Cells and Insect Behavior". Cambridge, Mass.: Harvard University Press. (1963).
ROEDER, K.: Episodes in insect brains. Am. Scient. 58, 378-389 (1970).
ROEDER, K., TOZIAN, L., WEIANT, E.: Endogenous nerve activity and behavior in the mantis and cockroach. J. Insect Physiol. 4, 45-62 (1960).
ROWELL, C.H.F.: The structure and function of the prothoracic spine of the desert locust, *Schistocerca gregaria* Forskal. J. exp. Biol. 38, 457-469 (1961).
ROWELL, C.: Central control of an insect segmental reflex. I. Inhibition by different parts of the central nervous system. J. exp. Biol. 41, 559-572 (1964).
ROWELL, C.: The control of reflex responsiveness and the integration of behavior. *In* "The Physiology of the Insect Central Nervous System" (Eds., J. Treherne and J. Beament). London: Academic Press. (1965).
ROWELL, C.: Central control of an insect segmental reflex. II. Analysis of the inhibitory input from the metathoracic ganglion. J. exp. Biol. 50, 191-201 (1969).
ROWELL, C.: Variable responsiveness of a visual interneurone in the free-moving locust, and its relation to behavior and arousal. J. exp. Biol. 55, 727-747 (1971a).
ROWELL, C.: Antennal cleaning, arousal and visual interneurone responsiveness in a locust. J. exp. Biol. 55, 749-761 (1971b).
ROWELL, C.: Incremental and decremental processes in the insect central nervous system. *In* "Short Term Changes in Neural Activity and Behavior" (Eds., G. Horn and R. Hinde). London: Cambridge University Press, pp. 237-280 (1971c).
RUSSELL, I.: The role of the lateral line efferent system in *Xenopus laevis*. J. exp. Biol. 54, 621-641 (1971).
RUSSELL, I., ROBERTS, B.: Inhibition of spontaneous lateral-line activity by efferent nerve stimulation. J. exp. Biol. 57, 77-82 (1972).
SCHRAMEK, J.: Crayfish swimming: alternating motor output and giant fiber activity. Science, N.Y. 169, 698-700 (1970).
SELVERSTON, A., REMLER, M.: Neuronal geometry and activation of crayfish fast flexor motoneurons. J. Neurophysiol. 35, 797-814 (1972).
SHERRINGTON, C.: "The Integrative Action of the Nervous System". New Haven: Yale University Press. (1906).
SHERRINGTON, C.: "The Integrative Action of the Nervous System". Forward to the second edition. New Haven: Yale University Press. (1947).
SPIRA, M., PARNAS, I., BERGMANN, F.: Organization of the giant axons of the cockroach *Periplaneta americana*. J. exp. Biol. 50, 615-627 (1969a).
SPIRA, M., PARNAS, I., BERGMANN, F.: Histological and electrophysiological studies on the giant axons of the cockroach *Periplaneta americana*. J. exp. Biol. 50, 629-634 (1969b).
STEIN, P.: Intersegmental coordination of swimmeret motoneuron activity in crayfish. J. Neurophysiol. 34, 310-318 (1971).
STRETTON, A., KRAVITZ, E.: Neuronal geometry: determination with a technique of intracellular dye injection. Science, N.Y. 162, 132-134 (1968).
STRUMWASSER, F.: Types of information stored in single neurons. *In* "Invertebrate Nervous Systems", (Ed., C.A.G. Wiersma). Chicago and London: Chicago University Press. pp. 291-319 (1967).
TAYLOR, R.: Water-vibration reception: A neurophysiological study in unrestrained crayfish. Comp. Biochem. Physiol. 27, 795-805 (1968).

WALDRON, I.: Courtship sound production in two sympatric sibling *Drosophila* species. Science, N.Y. 144, 191-193 (1964).
WALL, P.: The laminar organization of dorsal horn and effects of descending impulses. J. Physiol., Lond. 188, 403-423 (1967).
WEIS-FOGH, T.: An aerodynamic sense organ stimulating and regulating flight in locusts. Nature, Lond. 163, 873-874 (1949).
WEIS-FOGH, T.: Biology and physics of locust flight. IV. Notes on sensory mechanisms in locust flight. Phil. Trans. R. Soc. (B) 239, 553-584 (1956).
WENDLER, G.: The co-ordination of walking movements in arthropods. *In* "Nervous and Hormonal Mechanisms of Integration". Symp. Soc. exp. Biol. 20, 229-249 (1967).
WIERSMA, C.A.G.: The neuromuscular system. *In* "The Physiology of Crustacea", (Ed., T. Waterman). New York: Academic Press. pp. 191-240 (1961).
WIERSMA, C.A.G.: Reactivity changes in crustacean neural systems. *In* "Short Term Changes in Neural Activity and Behavior", (Eds., G. Horn and R. Hinde). London: Cambridge University Press. pp. 211-235 (1970).
WIERSMA, C.A.G., FIORE, L.: Factors regulating the discharge frequency in optimotor fibers of *Carcinus maenas*. J. exp. Biol. 54, 497-505 (1971).
WIERSMA, C.A.G., HUGHES, G.: On the functional anatomy of neuronal units in the abdominal cord of the crayfish, *Procambarus clarkii* (Girard). J. comp. Neurol. 116, 209-261 (1961).
WIERSMA, C.A.G., YAMAGUCHI, T.: Integration of visual stimuli by the crayfish central nervous system. J. exp. Biol. 47, 409-431 (1967).
WILSON, D.: The central nervous control of flight in a locust. J. exp. Biol. 38, 471-490 (1961).
WILSON, D.: Insect walking. A. Rev. Ent. 11, 103-122 (1966).
WILSON, D.: Inherent asymmetry and reflex modulation of the locust flight motor pattern. J. exp. Biol. 48, 631-641 (1968a).
WILSON, D.: The nervous control of insect flight and related behavior. Adv. Insect Physiol. 5, 289-338 (1968b).
WILSON, D., GETTRUP, E.: A stretch reflex controlling wingbeat frequency in grasshoppers. J. exp. Biol. 40, 171-185 (1963).
WINE, J., KRASNE, F.: The organization of escape behavior in the crayfish. J. exp. Biol. 56, 1-18 (1972).
WOOD, J., von BAUMGARTEN, R.: Activity recorded from the statocyst nerve of *Pleurobranchaea californica* during rotation and at different tilts. Comp. Biochem. Physiol. 43A, 495-502 (1972).
WYMAN, R.: Multistable firing patterns among several neurons. J. Neurophysiol. 29, 807-833 (1966).
WYMAN, R.: Lateral inhibition in a motor output system. I. Reciprocal inhibition in dipteran flight motor system. J. Neurophysiol. 32, 297-306 (1969a).
WYMAN, R.: Lateral inhibition in a motor output system. II. Diverse forms of patterning. J. Neurophysiol. 32, 307-314 (1969b).
WYMAN, R.: Patterns of frequency variation in dipteran flight motor units. Comp. Biochem. Physiol. 35, 1-16 (1970).
ZUCKER, R., KENNEDY, D., SELVERSTON, A.: Neuronal circuit mediating escape response in crayfish. Science, N.Y. 173, 645-650 (1971).

BOREDOM AND ATTENTION IN A CELL IN THE LOCUST VISUAL SYSTEM

C.H.F. Rowell

Department of Zoology, University of California at Berkeley, Berkeley, California, U.S.A.

In the past 5 years or so, several workers on crustaceans, molluscs and insects have turned their attention to the lability of the behaviour of individually identified interneurones. I first became involved in this approach around 1966, having become dissatisfied with the sort of results that can be obtained from an analysis of larger subsets of the nervous system. Since then I have worked on a variety of interneurones in the auditory and visual systems of locusts and other Orthoptera, along with several colleagues, especially G. Horn, J. Mackay, and M. O'Shea. I should like to acknowledge my intellectual debt, in the early days, to the work of Horridge and co-workers (Horridge *et al.* 1965), who first showed the rich variety of units which could be characterized by micro-electrode recording from the insect brain, and who were not afraid to say that the properties of some units seemed to be labile and affected by antecedent events.

The neurone which we know most about is the descending contralateral movement detector (DCMD) of the locust. Its extracellular properties, and those of its ipsilateral homologue, the DIMD, have been exhaustively studied in grasshoppers and crickets by Burtt and Catton in the '50s (1954, 1956, 1959, 1960), by Palka (1967a,b, 1969, 1972) and by myself (1971a,b,c; Rowell and Horn 1967, 1968; Horn and Rowell 1968) as they are among the largest axons in the ventral nerve cord, and the spikes are conspicuous in extracellular records. Only this year we have elucidated its anatomy. We succeeded in this by first determining its position in the nerve cord by splitting techniques and the use of fine surface electrodes, and by correlating our results with the distribution of large axons as seen in cross section of the cord under low power EM magnification. We then probed the area so defined with microelectrodes until we penetrated the axon, which we recognised from its response characteristics, previously determined extracellularly. The axon was marked by electrophoretic dye injection. Repeating this process in several different connectives, we found that the position of the axon was constant in the cord throughout its length and between different individuals, and that it could be identified visually when the connective was transilluminated. We then elucidated its anatomy in the brain by axonal iontophoresis (after Iles and Mulloney 1971) of dye from the connectives, which filled the integrating segment, dendrites, and the cell body. Measurement of the position of the cell body in the brain allowed us to position microelectrodes stereotactically and to penetrate the (invisible) cell through the sheath, and to confirm our identification by injecting dye into the soma and thus into the rest of the cell. We commenced this work with the dye procion yellow, but progress was much more rapid after we adopted the cobalt sulphide technique of Pitman, Tweedle and Cohen (1972); Fig. 1A shows a cobalt preparation. The cell body is situated in dorsal posterior protocerebrum, behind the calyx of the corpus pedunculatus, a little way lateral to the midline. It sends its neurite to a thickened integrating segment in the protocerebrum, bearing dendrites; one fine branch has been traced from the integrating segment to near the base of the optic lobe, and it is possible that it may run to the lobula. A thin axon crosses the midline, thickens abruptly to 10-15 microns, and descends through the tritocerebrum to the connectives. Similar techniques elucidated the structure of the

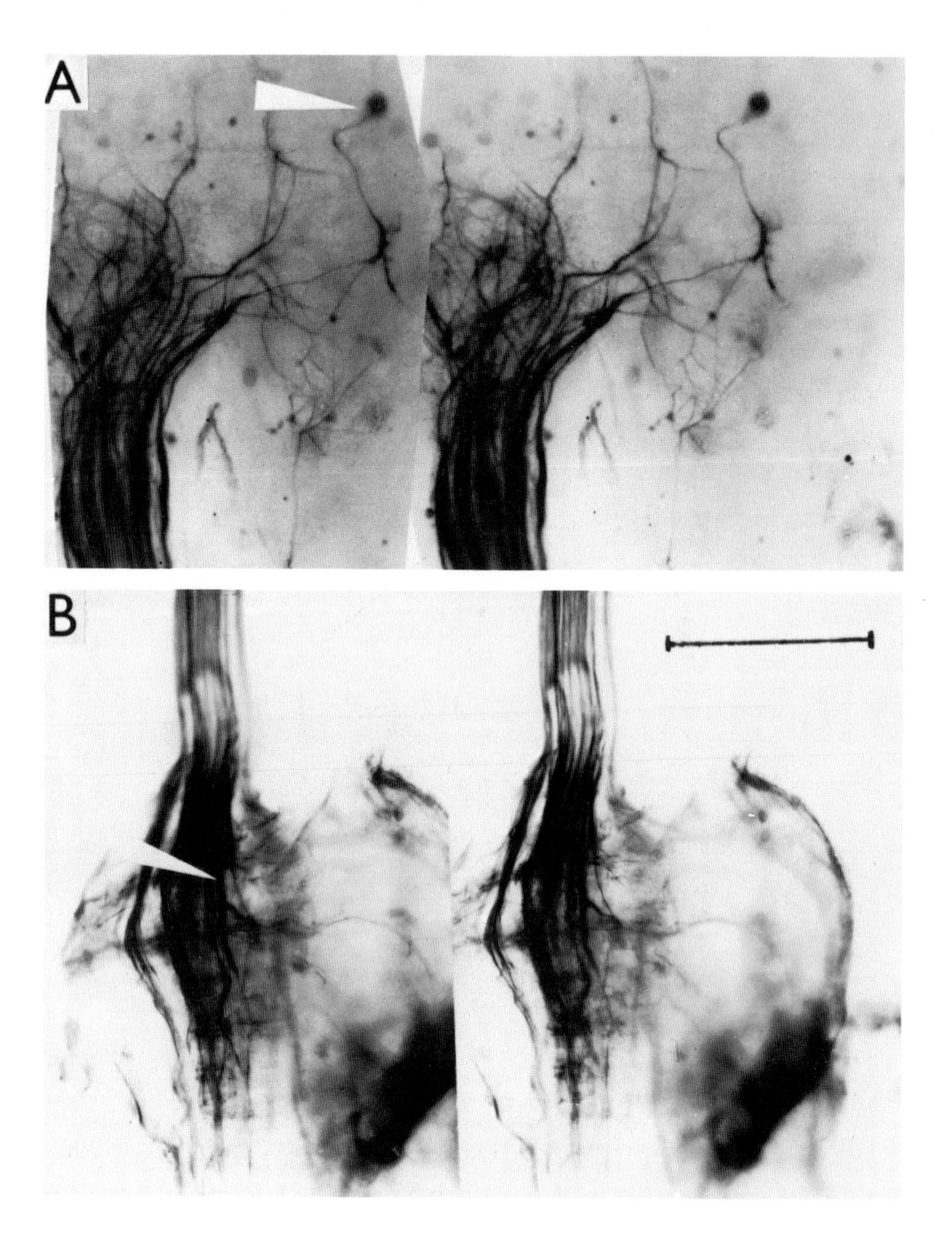

Fig. 1. Whole mount preparations of brain (A) and metathoracic ganglion (B) of *Schistocerca vaga* in which the DCMD neurone, along with others, has been filled with cobalt by iontophoresis along the left hand side connective, and subsequent precipitation as the sulphide. Scale mark in B applies to both figures, and equals 500 microns. In A the brain is viewed from the back; the DCMD soma is situated in the lower protocerebrum (arrow); the looping and descending neurite, integrating segment, and axon crossing the deuto- and tritocerebrum are clearly seen. Note the marked thickening of the axon on entering the connective. At least one fine branch, not visible in this photograph, runs from the integrating segment towards the optic lobe. In B the metathoracic ganglion is viewed from above. The DCMD axon (arrow) branches extensively, but only some of the ipsilateral ones are visible here. Both figures are stereo pairs, and should be viewed with an appropriate optical device

neurone in the thoracic ganglia. It passes through the suboesophageal ganglion without branching, branches slightly in the pro- and mesothoracic ganglia, and ends with extensive branching in the metathoracic ganglion (Fig. 1B). There is one such axon in each connective.

The response characteristics of the DCMD are well known, and the mechanisms by which they are produced modelled by Palka (1967a) and Horn and Rowell (1968), though these models have not yet been tested by recording from antecedent cells in the visual pathway. The basic response seems to be to a rapid decrease of illumination at perhaps a single ommatidium, which produces both excitatory and weaker, but longer lasting, inhibitory effects. When these effects are summated over the whole eye, the result is phasic, but extreme, sensitivity (less than 1°) to novel abrupt movement of small contrasting objects in the visual field. A consequence of the inhibition is that the unit is not excited much by any stimulus which affects the whole of the monocular visual field simultaneously, and is thus stabilised against apparent movement caused by the animal's own locomotion, or overall changes in light intensity.

As yet we are unable to state with any surety the main function of this neurone in the thoracic ganglia. We know a great deal about what it does *not* do, but so far we have been unable to demonstrate any effect of its activity on a thoracic motor pattern. Its anatomy suggests strongly that it might relate to jumping, which is almost the only exclusively metathoracic motor pattern, and this would correlate well with its response characteristics, which seem to fit it for giving visual warnings. As driving this unit alone does not elicit jumping, the hypothesis assumes that it is a synergist requiring other input as well to cause tne jump. An analogous situation has recently been described in the crayfish, where several prominent sensory interneurones, singly ineffective, have been found to be capable of activating, syngergistically, the giant fibre which mediates the tail-flip (Zucker, Kennedy and Selverston 1971). Other functions have been suggested (e.g. Rowell 1971a), but await experimental verification.

The response of the DCMD to adequate stimulation is highly labile, and two distinct processes are separable. These are habituation (i.e. a decremental response to a repeated stimulus) and incrementation of response as a result of behavioural arousal. I will discuss each of these processes in turn.

HABITUATION

The DCMD habituates to repeated stimuli. In order for this to be apparent, three conditions are necessary. First, the stimuli must be applied to the same areas of eye; secondly, the interval between stimuli must be shorter than a minimum which suffices for recovery, and which varies with the intensity of the stimulus; and lastly, the animal should not undergo an increase in arousal between the presentations. When these conditions are met, the DCMD response, as measured by the number of spikes per stimulus presentation, decreases exponentially to a plateau level characteristic of the stimulus, the interstimulus interval, and the degree of arousal. Under suitable conditions, habituation can be detected at stimulus intervals of 10 min, and recovery after a number of series may take days.

If the stimulus is moved to another area of the eye, even without a rest interval of any sort, the response is immediately regained (Fig. 2). The smallest effective distance of movement needed for the regaining of

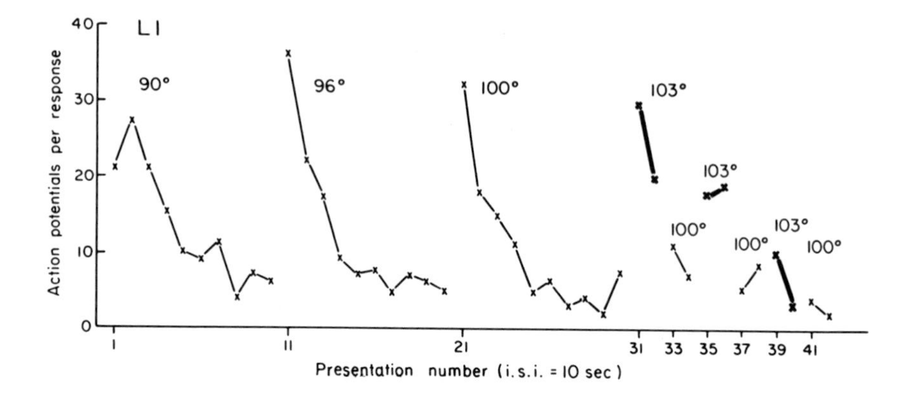

Fig. 2. DCMD response recovery due to passive movement of the retina relative to the axis of movement of the visual stimulus (a 5° black disc moved back and forwards horizontally, lateral to the animal's head). The animal is mounted on a shaft which can be rotated through a few degrees during an interstimulus interval. Displacement of the axis by progressively 6, 4 and 3° causes on each occasion a complete recovery. (From Rowell 1971c.)

the response is approximately the same as the ommatidial separation. This result shows that the decremental process responsible for habituation of the response is localised at a level of the optic pathway prior to any appreciable convergence. Spatial projection of the retinal image is continued to the level of the lobula or third optic ganglion in insects. The later elements of the chain would seem the most suitable site for habituation. As most visual units recorded from the optic lobes (e.g., Horridge *et al.* 1965; Northrup and Guignon 1970) do not show decrement to repeated stimuli, to place the site of habituation early in the visual process would mean carrying two spatial projections to higher centres, one decremental and one non-decremental. Similarly, the DCMD shows preferences for some directions of movement, which are interpreted as implying convergence of many directionally sensitive, small-field, cells. It would seem uneconomical to derive this information more than once and as non-habituating directionally sensitive cells are also found, the habituation process is likely to be localised subsequent to the analysis of movement, which probably takes place in the medulla. There is no evidence for a retinal projection extending as far as the protocerebrum, so the site is probably at a synaptic layer in the medulla or lobula.

The actual cellular mechanism responsible for the decrement is not known, though *a priori*, several are possible. Experimental evidence so far has eliminated both a progressive build-up of inhibition (e.g., by feed-forward or recurrent collateral inhibition) and also the hypothesis of active gating by a neural network entrained to the stimulus repetition frequency. The most economical hypothesis is that decrement represents progressively less release of transmitter substance from a presynaptic terminal with finite stores and resynthesis capabilities (Horn and Rowell 1968).

We have recently been able to obtain intracellular recordings from the soma of the DCMD which show at least some of the post-synaptic activity which underlies the pattern of spike activity detected by extracellular recording (Fig. 3). The position of the soma varies little from preparation to preparation, and it is possible to locate it stereotactically, or sometimes by reference to the external tracheation of the brain. Good penetrations last for more than an hour, and the preparation allows simultaneous free movement of the animal's legs and wings. The results of intracellular recording confirm the deductions made above from extra-

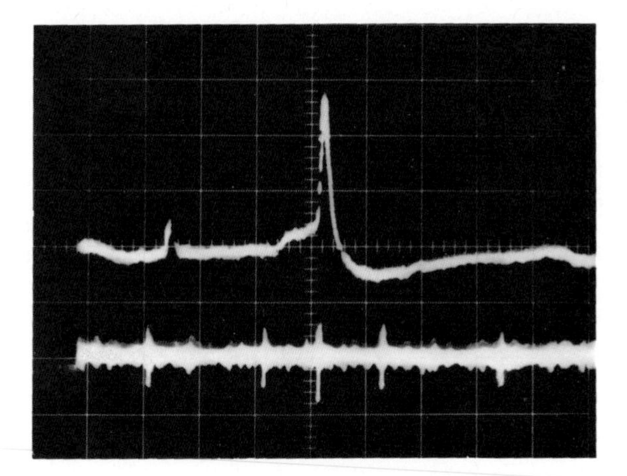

Fig. 3. Intracellular record from the soma of the DCMD during visual stimulation (top trace) and simultaneous extracellular record from the surface of tritocerebrum. Calibration, 10 ms and 6.25 mv per division. Note EPSP giving rise to an attenuated propagated action potential (PAP) (intracellular recordings from the conducting region of the axon of the cell show overshooting PAPs of 80 mv or more), followed by a pronounced negative going after-potential. The small deflection of the upper trace after 16 ms is caused by the passage of a constant current calibration pulse through the recording electrode

cellular work. During habituation, the number of propagated action potentials (PAPs) seen in the cell, in response to fixed visual stimulation, decreases; the decrement is not associated with inhibitory postsynaptic potentials (IPSPs) or with tonic conductance changes of the cell membrane. The decremental process appears to be taking place at least one neurone peripheral to the DCMD. There is no change in the frequency of "spontaneous" EPSPs in the DCMD during habituation of the response to any one retinal site; presumably these arise from the remainder of the retinal projection.

In the absence of knowledge of the function of the DCMD output, it is speculative to assess the function of its habituation properties. Certainly it makes the unit very sensitive to novel, as opposed to familiar, stimuli. This would be an appropriate property for a neurone conveying warning information. The subject is made considerably more complex by a consideration of the effects of active and passive movement of the animal or of the head. During active movement, response decrement is always less marked and may be absent (see below). At low levels of activity, however, decrements can be seen, and it is then noticeable that voluntary movements of the animal, which change the position of the stimulus on the retina, do not produce the same dramatic recovery that would be the case were the displacement caused experimentally by passive movement of the head. A specially obvious instance is seen during antennal cleaning. This behaviour involves moving the head around all three major axes by up to 25°. There is no recovery of the DCMD response during this process (Fig. 4A). It can easily be shown that the DCMD response is not inhibited during antennal cleaning, as a new visual stimulus presented to the eye during cleaning elicits a normal full scale response (Fig. 4B); it is only "familiar" stimuli which have already produced habituation that are unaffected by the change in retinal position. This is an example of a facility possessed by all mobile visual animals, that of being able to distinguish between exogenously and endogenously produced movement on the retina, but as in other animals the neural

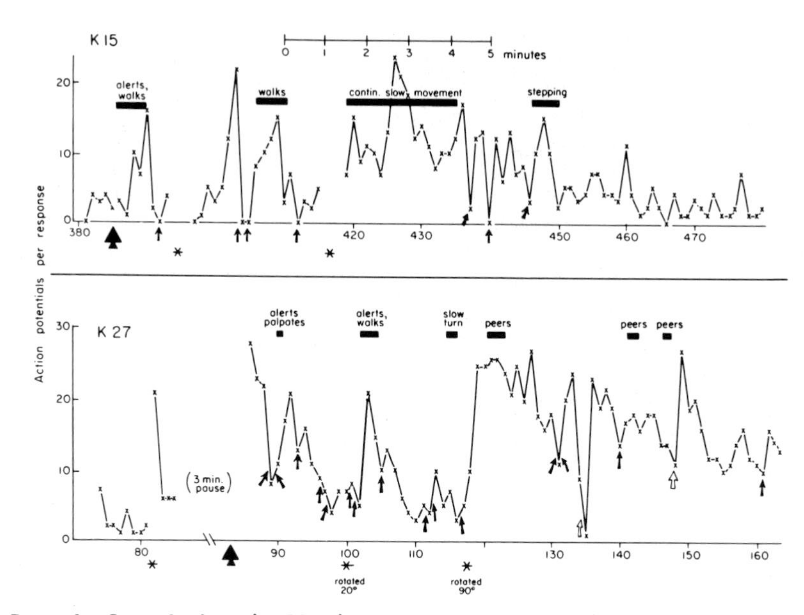

Fig. 4A. Records from 2 chronically implanted, freely moving animals. The DCMD is responding to a visual stimulus (target movement) once every 10 sec; the X axis gives the presentation number. After a weak chemical stimulus is applied to the antenna (large arrow) the animals clean the antenna several times (small solid arrows) and K 27 also spontaneously cleaned the unstimulated antenna (small open arrows). Although cleaning is associated with head movement, there was no recovery of response of the DCMD corresponding to movement of the target axis on the retina (compare with Fig. 2). Instead, cleaning takes place at times of significantly low DCMD responsiveness

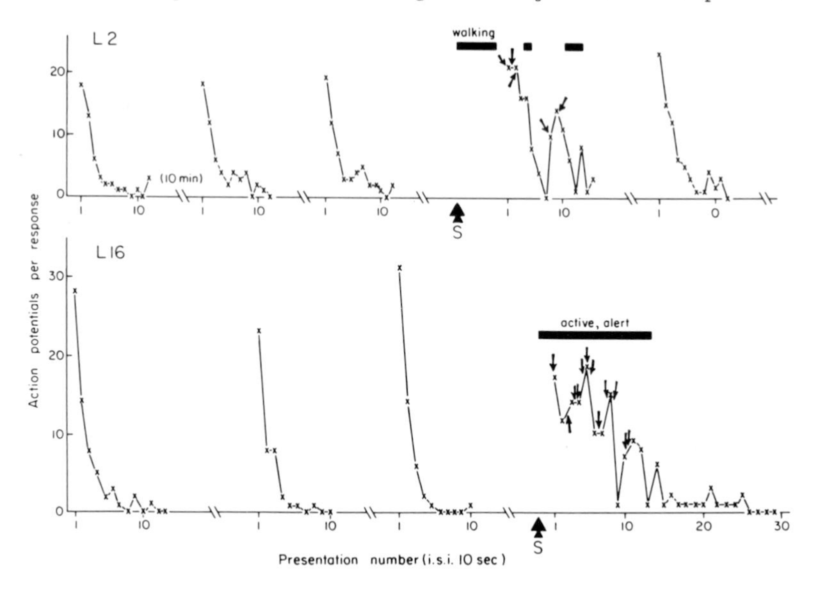

Fig. 4B. A similar experiment to those shown in 4A. The first 3 curves show control decremental responses of the DCMD separated by intervening 10 min rest intervals. After a further rest interval, a chemical stimulus is applied to the antenna (large arrow) in the absence of visual stimulation. As the animal starts to clean the antenna (small dark arrows) the visual stimulation is commenced. Under these circumstances there is no reduction in DCMD response during antennal cleaning. The effect seen in 4A is therefore not due to automatic inhibition of the visual response during cleaning. (Modified from Rowell 1971c.)

mechanism responsible is not yet understood. Presumably the animal must compute the required correction from a knowledge of the movement of the head. As passive movement of the head does not produce the same effect, this knowledge is not simply a question of proprioception, and it is likely that it derives from the motor command to the neck muscles. The correction also demands a short-term memory to differentiate between familiar and unfamiliar stimuli. This could be achieved by some such system as that seen in the optokinetic system of the crab (e.g. Horridge 1968). Perhaps, more economically, it could also be derived from the information contained in the differentially habituated retinal projection which feeds the DCMD. This might be an additional function of the habituation process, though the read-out mechanism required by such a system is not obvious.

INCREMENTATION AND AROUSAL

The first indication of arousal effects in the DCMD was a chance finding that the preparation showed changes in responsiveness which seemed to correlate with changes in sleep and wakefulness, but not with circadian intervals. We then found that a habituated preparation could often be dishabituated (i.e., response recovery without a rest interval) by applying electric shocks to the CNS, though the efficacy of this treatment soon waned (Rowell and Horn 1968).

During our early experiments, we had noted occasional aberrant sequences in which the habituation of response of an otherwise "good" preparation showed sudden incrementation, apparently unrelated to our paradigm. In these experiments, the head was rigidly clamped and the remainder of the animal was restrained in a loosely fitting brass tube behind the clamp, which incidentally prevented us from seeing its limbs. Later (Rowell 1971b,c) recording was via chronically implanted electrodes in the freely moving or ball-running animal (Fig. 5), in which the full range of behaviour could be observed. It then became apparent that both spontaneous and evoked dishabituation was associated with what can be called behavioural arousal, as measured by small changes in a variety of motor activities, including posture, respiration, movement of antennae and mouthparts, defaecation, locomotion, peering, and similar behaviour patterns.

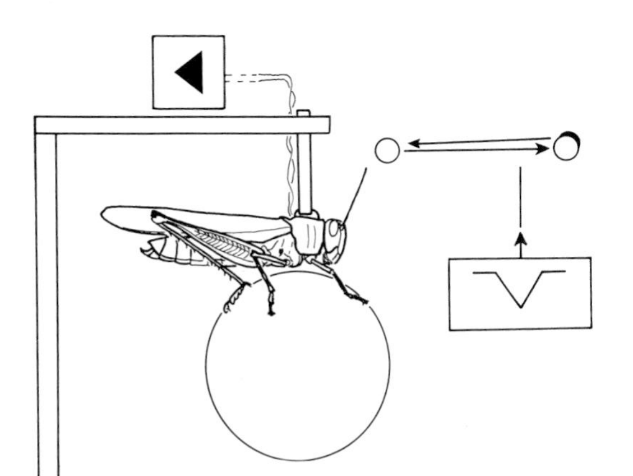

Fig. 5. Experimental arrangement for studying association of DCMD response and motor behaviour. The animal is waxed to a support by the pronotum, and carries with its legs a light foam plastic sphere. Chronically implanted leads record the activity of the DCMD axon in the thoracic connective. A visual target, moved by a servo motor driven by a waveform generator, is placed opposite the appropriate eye

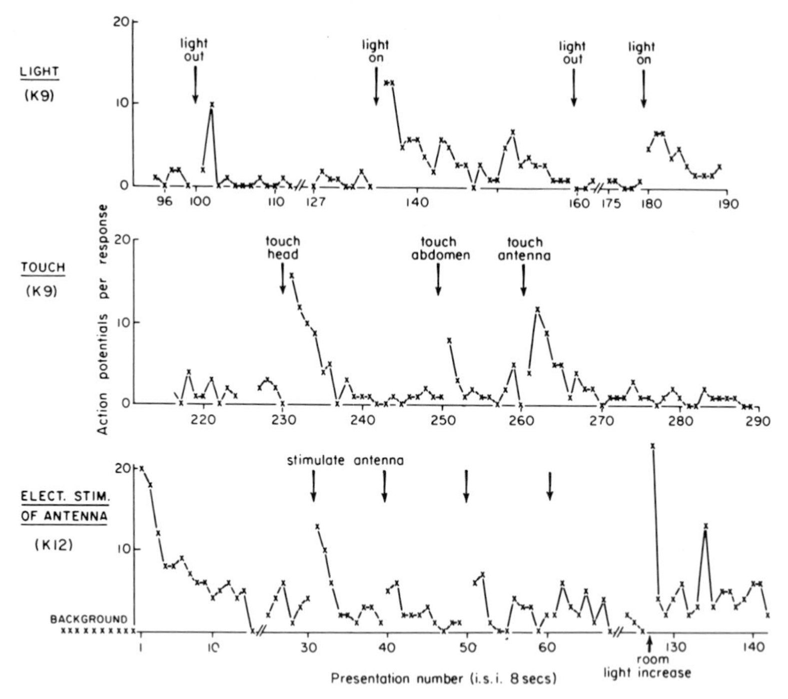

Fig. 6. Examples of dishabituation of DCMD response by a variety of sensory stimuli. Repetition of the same dishabituating stimulus, as in lines 1 (light ON and OFF) and 3 (constant current stimulation of the antenna) lead to a rapid waning of its effectiveness, but do not affect the efficacy of other modalities. These records were made from restrained and dissected animals. (Further discussion and examples can be found in Rowell, 1971b.)

Almost any sensory stimulus strong enough to "catch the animal's attention" would dishabituate the visual unit's response - once. On repetition, the effect always waned, but a new modality would again be effective (Fig. 6). Stimuli which induced movements of the limbs were often especially effective. It appears to be the sensory reafference from the movement which is effective, rather than a collateral signal from the efferent motor system, as sensory input remains to some extent effective in preparations in which all motor output from the brain has been abolished by cutting. The increments in responsiveness of the DCMD often shows a close correlation, though with an appreciable latency, with the activity of proprioceptive interneurones in the cord. Sensory inputs of other modalities can also cause dishabituation after all proprioception or motor output has been abolished by lesion. Dishabituation appears then to be a true sensory dishabituation, often associated with increase in motor activity, and which can follow any of a wide variety of sensory inputs of differing modality, provided only that they are sufficiently novel and potent.

Sensory input not only has sudden incrementing effects on the DCMD (as in dishabituation), but also exerts a tonic influence on the response level of the DCMD and on its decrement during habituation. Progressive reduction in input, brought about by restraint and lesion, causes progressive increase in the rate of habituation, and a lowering of the plateau response level which succeeds the decremental phase. In other words, minimally restrained or freely moving animals habituate more slowly and less than do restrained or deafferented preparations (Fig. 7). Animals which are highly aroused, and especially those which are actively walking or flying, show no regular decrement in their visual response; the number of spikes per response fluctuates markedly, but shows no overall trend. As soon as activity or arousal lessens, however, a regular decrement is observed (Fig. 8).

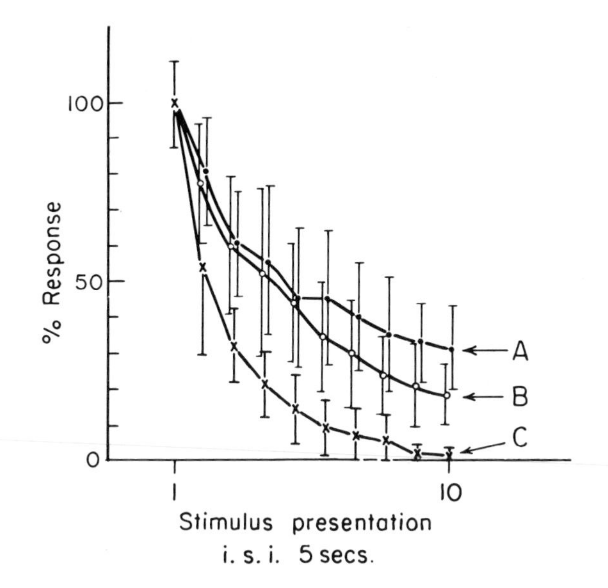

Fig. 7. Dependence of rate of habituation and of plateau level of response when habituation is complete, on level of general sensory input. The three curves each derive from several individuals, each tested a number of times, and the results are normalised as percentages of individual first responses. The vertical bars indicate two standard deviations, and the curves follow the mean response. Curve A, chronically implanted animal, intact, walking on a ball (see Fig. 5); curve B, restrained and dissected animal, nervous system intact; curve C, restrained and dissected animal, record made from isolated brain, with one eye and one antenna still connected to it

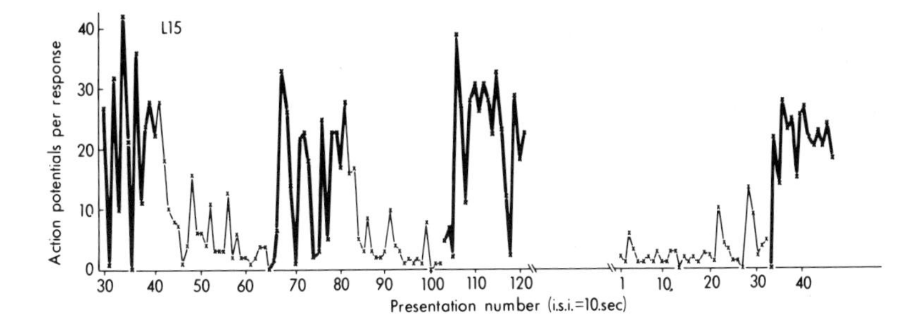

Fig. 8. Effect of locomotor activity on DCMD responsiveness. The record is obtained from a chronically implanted ball-running animal, to the head of which a cardboard shield has been waxed so that it cannot see its own legs. The animal alternates between struggling to remove this shield, and resting (shown as thick and thin lines on the curve, respectively). While the animal is aroused and struggling, DCMD responses fluctuate very much, but show no regular decremental trend and the mean remains high. During periods of quiescence a regular decrement supervenes, and the response drops to a low plateau level until arousal takes place again. In this animal, the eye is stabilised with respect to the target axis, and its struggles do not move the retina at all. (Modified from Rowell 1971b.)

In the mammalian nervous systems, very similar effects, in either neural or whole animal behaviour, are ascribed to the activity of efferent pathways, which may be either specific to the sensory modality in question, as in the various corticofugal tracts, or more generalised, as in the effects of the midbrain reticular formation. The physiological mechanisms by which they influence the ascending sensory tracts are not well understood. At least some cases of dishabituation in spinal interneurones are thought to represent compensatory enhancement elsewhere in the pathway, which has the effect of cancelling the decrement brought about during habituation by a process of algebraic summation in transmission properties of the pathway as a whole. What can we say about the mechanisms of arousal in the case of insects?

First, the indications are that the DCMD is influenced by a non-specific, rather than a specifically visual or yet more narrowly defined, arousal system. The evidence for this is that the same sensory lesions which bring about tonic effects on the responsiveness of the DCMD also influence, but inversely, the responsiveness of,for example, grooming reflexes (Rowell 1964). During normal spontaneous behaviour, chronically implanted recording electrodes show the same inverse relationship between the two activities. One cannot demonstrate, however, any direct inhibition between the two which might account for this inverse probability; during, for example, antennal cleaning, it is possible to elicit large DCMD responses to new visual stimuli, and the cleaning continues in spite of the DCMD response. There is also some suggestion, not yet properly investigated, that certain auditory neurones show parallel changes in responsiveness with the DCMD in response to sensory stimuli leading to arousal.

As to mechanism, the indications are that there are large differences from what has been described in the spinal cord. A major feature of the DCMD is that the first responses of the unit to a given stimulus are always the highest ones. Arousal or transient dishabituation can lead to *a restoration* of a habituated response up to the initial level, but one never sees *an enhancement* of response significantly beyond that obtained during the first presentations. (See, e.g., Figs. 4B and 6). The simplest explanation of these findings is that the dishabituation process "resets" the synapses which are responsible for the initial decrement during habituation. This could be achieved, for example, by episynaptic input to the terminals, giving temporary presynaptic inhibition. The appreciable latency of the dishabituation (often several seconds) is consistent with this hypothesis. Dishabituation can be shown not to be site-specific on the retinal projection, as is the case with the habituation process, but to affect simultaneously all of the field. I have suggested elsewhere that a candidate for this function would be one of the tangential cells of the medulla or lobula, which appear to be cells arising in the central protocerebrum and having an output to the entire retinal projection of the various optic ganglia. Such a cell could also act to summate diverse sensory input, in the simplest model; but it is more likely, in view of the evidence presented above, that this sensory accumulation function is accomplished by a separate antecedent network which feeds several different outputs, and with differing sign.

If the argument, that the arousal input acts to modulate the same synapses which are responsible for the decrement, is dismissed, then the most probable site of such an input would be at the DCMD itself, after convergence is complete; the effect is, after all, simultaneous over all of the field. In such a case, one might hope to see postsynaptic events corresponding to such an input during intracellular recording from the DCMD. This does not seem to happen. Our experiments to date have not shown any correlate of dishabituation in the DCMD, other than the enhanced response. (There is, however, variation in the amount of slow potential activity which can be "seen" down the neurite from preparation to preparation, and it is possible that we have not yet seen the right cell to demonstrate the effect.) In a negative way, this evidence supports the original hypothesis as to a more peripheral site of action of dishabituation and tonic arousal effects.

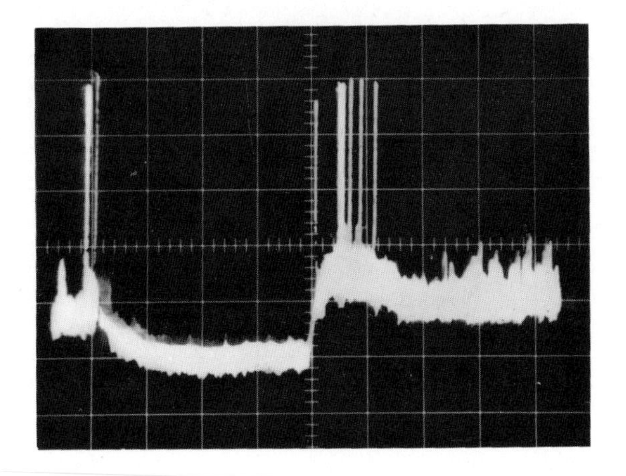

Fig. 9. Intracellular electrode from the soma of an unidentified protocerebral visual interneurone located near to that of the DCMD. Light OFF (a step function decrease in intensity, but not to darkness) produces first a brief depolarisation and a few spikes, then a maintained hyperpolarisation of which at least a part is not an artifact of the photochemistry of the electrode array, and during which the frequency of incoming EPSPs decreased markedly. At light ON there is a larger depolarisation, with extensive spiking, followed by a brief silent period, and then a resumption of a normal high level of EPSPs. Note that the maintained hyperpolarisation is shared with the DCMD under similar circumstances, but in all other respects the cells are almost exact opposites (see Fig. 10 for comparison)

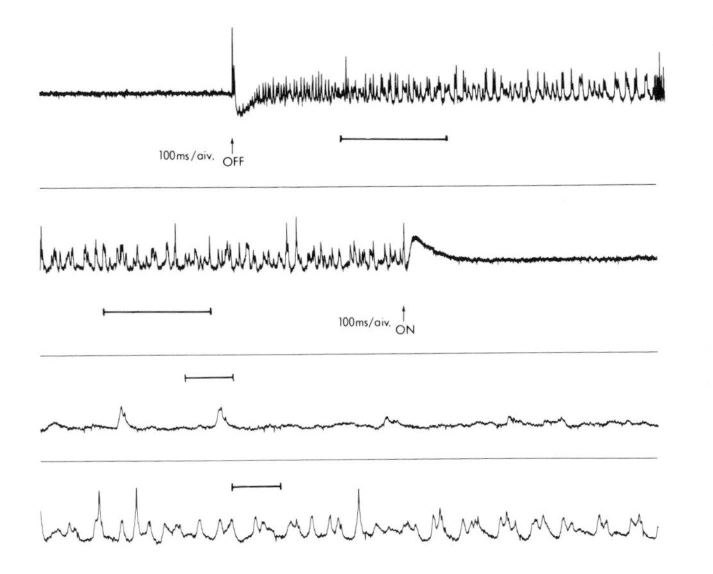

Fig. 10. Intracellular record from the soma of the DCMD. Top line, response of light OFF. The cell briefly depolarises, gives a single spike, and then hyperpolarises. The preparation is AC coupled to the recorder, and the maintained hyperpolarisation which accompanies the period of decreased illumination is reduced by differentiation to a slow inflection at the start of the dark period (compare with the DC record of Fig. 9). Note that shortly after light OFF, there is a steady increase in EPSPs seen in the cell, and that these eventually synchronise into a roughly rhythmic input at approximately 20/sec frequency. Second line, light ON after several seconds of reduced illumination. The hyperpolarisation is lost, there is no depolarisation or spiking from the cell, and the EPSP level is immediately reduced. Third and fourth lines; excerpts from the same records at a higher chart speed to show comparative levels of post-synaptic activity in the cell while normally light adapted (i.e., immediately before light OFF) and soon after light OFF. Note unitary EPSPs in the former (line 3) and compound summating and synchronising EPSPs in the latter (line 4). Time calibrations: line 1 and 2 - 1 sec; lines 3 and 4, 200 msec

So far, the intracellular recording has been used only to present negative evidence. I would like to close by mentioning some interesting positive results obtained in this way, which may suggest what at least some of the inputs to the extensive dendritic branches in protocerebrum accomplish. During recording under normal conditions of room illumination, one sees a certain level of EPSP bombardment in the DCMD, which rarely (<1/sec) leads to spiking. If the general illumination level is suddenly reduced, the cell spikes once or twice, and is then subject to several millivolts maintained hyperpolarisation (relative to the illuminated state). At light ON, the potential is restored to its former level and there may be again the odd spike at the transition. We find these same characteristics in a number of other visual interneurones in the protocerebrum (Fig. 9). In the DCMD, however (Fig. 10) the low-illumination period is also characterised by a greatly increased EPSP barrage, in terms of both amplitude and frequency, far greater than could be explained by the amplitude enhancement associated with hyperpolarisation. Sometimes this activity will settle down to regularly cycling, compound EPSPs at a frequency of 10-20/sec. At light ON, this activity is abruptly curtailed. If the low level illumination is maintained, activity slowly decreases and falls to normal levels over a period of minutes, which presumably corresponds to a dark adaptation process. The dark period EPSPs do not in themselves initiate spiking, and there is so far no evidence that, by summation with other inputs, they can modify the threshold of the cell towards normally effective visual stimuli, though it is true that no one has so far looked for this effect. If it turned out to be so, this EPSP barrage might temporarily compensate the DCMD for the loss of input associated with the reduction of signal from the light-adapted eye which presumably occurs when the intensity of ambient light is lowered. It may be that intracellular recording from the DCMD is going to tell us more about central nervous mechanisms of dark adaptation in insects than about mechanisms of habituation and arousal!

REFERENCES

BURTT, E.T., CATTON, W.T.: Visual perception of movement in the locust. J. Physiol., Lond. 125, 566-580 (1954).

BURTT, E.T., CATTON, W.T.: Electrical responses to visual stimulation in the optic lobes of the locust and certain other insects. J. Physiol., Lond. 133, 68-88 (1956).

BURTT, E.T., CATTON, W.T.: Transmission of visual responses in the nervous system of the locust. J. Physiol., Lond. 146, 492-514 (1959).

BURTT, E.T., CATTON, W.T.: The properties of single unit discharges in the optic lobes of the locust. J. Physiol., Lond. 154, 479-490 (1960).

HORN, G., ROWELL, C.H.F.: Medium and long-term changes in the behaviour of visual neurones in the tritocerebrum of locusts. J. exp. Biol. 49, 143-169 (1968).

HORRIDGE, G.A.: Five types of memory in crab eye responses. *In* "Physiological and Biochemical Aspects of Nervous Integration" (Ed., F.D. Carlson). Englewood Cliffs, N.J.: Prentice Hall (1968).

HORRIDGE, G.A.: SCHOLES, J.H., SHAW, S., TUNSTALL, J.: Extracellular recordings from single neurones in the optic lobe and brain of the locust. *In* "The Physiology of the Insect Central Nervous System" (Ed., J.E. Treherne and J.W.L. Beament). London: Academic Press (1965).

ILES, J.F., MULLONEY, B.: Procion yellow staining of cockroach motor neurones without use of microelectrodes. Brain Res. 30, 397-400 (1971).

NORTHRUP, R.B., GUIGNON, E.F.: Information processing in the optic lobes of the lubber grasshopper. J. Insect Physiol. 16, 691-713 (1970).
PALKA, J.: An inhibitory process influencing visual responses in a fibre of the ventral nerve cord of locusts. J. Insect Physiol. 13, 235-248 (1967a).
PALKA, J.: Head mcvement inhibits locust visual unit's response to target movement. Am. Zoologist 7, 728 (1967b).
PALKA, J.: Discrimination between movements of eye and object by visual interneurones of crickets. J. exp. Biol. 50, 723-732 (1969).
PALKA, J.: Moving movement detectors. Am. Zoologist 12, 497-505 (1972).
PITMAN, R.M., TWEEDLE, C.D., COHEN, M.J.: Branching of central neurons: intracellular cobalt injection for light and electron microscopy. Science, N.Y. 176, 412-414 (1972).
ROWELL, C.H.F.: Central control of an insect segmental reflex. I. Inhibition by different parts of the central nervous system. J. exp. Biol. 41, 559-572 (1964).
ROWELL, C.H.F.: The Orthopteran descending movement detector (DMD) neurones: a characterisation and review. Z. vergl. Physiol. 73, 167-194 (1971a).
ROWELL, C.H.F.: Variable responsiveness of the visual interneurone in the free-moving locust, and its relation to behaviour and arousal. J. exp. Biol. 55, 727-747 (1971b).
ROWELL, C.H.F.: Antennal cleaning, arousal, and visual interneurone responsiveness in the locust. J. exp. Biol. 55, 749-761 (1971c).
ROWELL, C.H.F., HORN, G.: Response characteristics of neurones in an insect brain. Nature, Lond. 216, 702-703 (1967).
ROWELL, C.H.F., HORN, G.: Dishabituation and arousal in the response of single nerve cells in an insect brain. J. exp. Biol. 49, 171-183 (1968).
ZUCKER, R.S., KENNEDY, D., SELVERSTON, A.I.: Neuronal circuit mediating escape responses in the crayfish. Science, N.Y. 173, 645-650 (1971).

THE ANTENNAE OF INSECTS AS AIR-CURRENT SENSE ORGANS AND THEIR RELATIONSHIP TO THE CONTROL OF FLIGHT

M. Gewecke

Institut für Zoologie, Universität Düsseldorf, Germany

Air-current sense organs are organs that can be stimulated by movements of the surrounding air and which because of this are able to control positional behaviour. In insects, the cuticular sensilla are used for the perception of air currents. Frequently they are true hair sensilla which are arranged in patches and, therefore, together form a complex sense organ (Boyd and Ewer 1949; Weis-Fogh 1949, 1950, 1956b; Camhi 1969; Smola 1970). Other types of mechanoreceptors such as campaniform sensilla or scolopideal sensilla which lie within or under the cuticle cannot be stimulated by air currents directly. A sensillum of this type can perceive air currents only if it has auxiliary structures capable of receiving the mechanical stimulus and then of conducting it to the mechanoreceptor itself. As with the hair patches, no single receptor of this type is, alone, an air-current sense organ but only one component of the more complex organ to which it belongs, for example, the antenna.

The definition of an air-current sense organ given above requires not only that the single receptors are excitable by movement to the stimulus receiving structures, but also that the input elicited is evaluated by the central nervous system as being relevant to the control of positional behaviour. Information about this latter aspect is generally obtained by behavioural investigation and, indeed, by experiments of this type, it has already been shown that hair patches and antennae function as air-current sense organs in insects.

In this paper I will describe behavioural work which refers especially to the antennae of locusts and flies, which, although different in structure, have been shown to have similar functions in flight control.

MOVEMENT MECHANISM OF THE ANTENNAE

In pterygota, the antenna consists of the basal scapus, the pedicellus and the segmented flagellum.

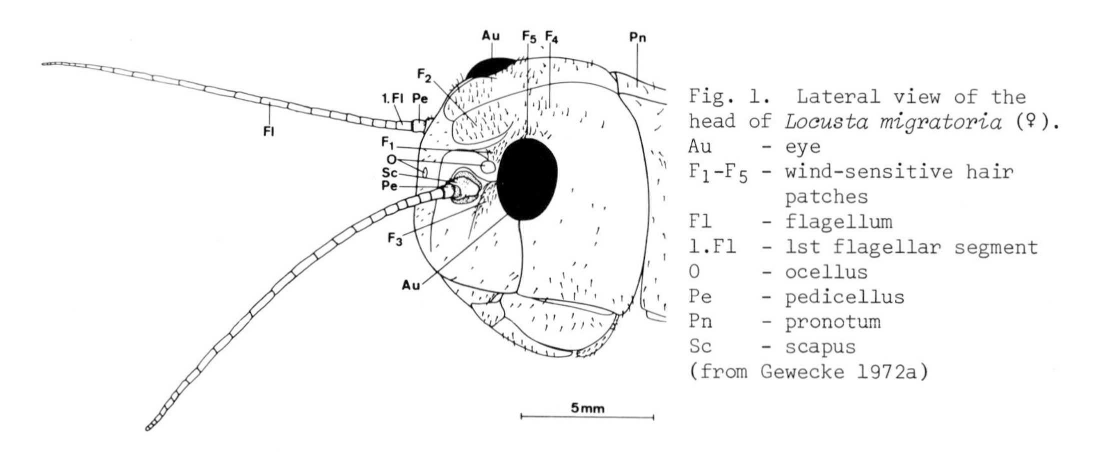

Fig. 1. Lateral view of the head of *Locusta migratoria* (♀).
Au - eye
F_1-F_5 - wind-sensitive hair patches
Fl - flagellum
1.Fl - 1st flagellar segment
O - ocellus
Pe - pedicellus
Pn - pronotum
Sc - scapus
(from Gewecke 1972a)

In locusts the form of the antenna is somewhat primitive (Fig. 1). Only the first two joints of the antenna of, for example, *Locusta migratoria*, can be moved actively (Fig. 2): the antenna can be raised or lowered about a horizontal axis (G_1-G_2) at the joint between the head capsule and scapus by two muscles of the head capsule. The distal part of the antenna can be turned laterally or medially about a vertical axis (G_3-G_4) at the joint between scapus and pedicellus by the two muscles of the scapus. All the other joints between the pedicellus and the flagellum and between the flagellar segments are either only passively moved by exterior forces or not at all. The passive movements of the flagellum about the vertical axis (G_5-G_6) of the pedicellus-flagellum joint are the adequate stimuli for the mechanoreceptors of the pedicellus: Johnston's organs, the campaniform sensilla, and a chordotonal organ (Gewecke 1972a).

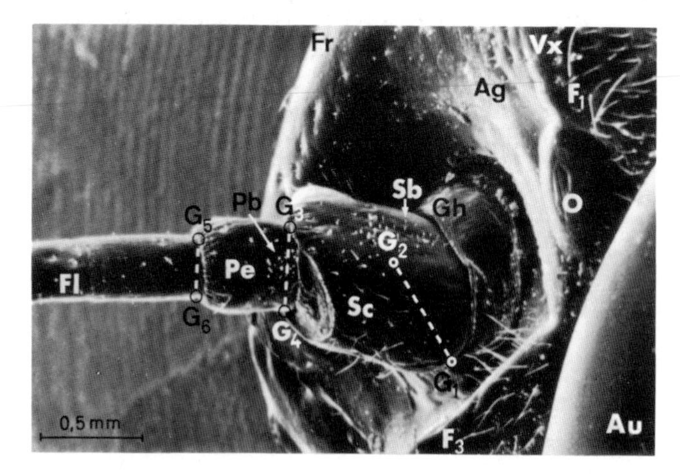

Fig. 2. Lateral view of the proximal part of the antenna of *Locusta*.
Ag - antennal cavity
Au - eye
F_1(F_3) - wind-sensitive hair patch
Fl - flagellum
Fr - frons
G_1-G_6 - joint articulations
Gh - joint membrane
O - ocellus
Pb - hairs on the pedicellus
Pe - pedicellus
Sb - hairs on the scapus
Sc - scapus
Vx - vertex
(from Gewecke 1972b)

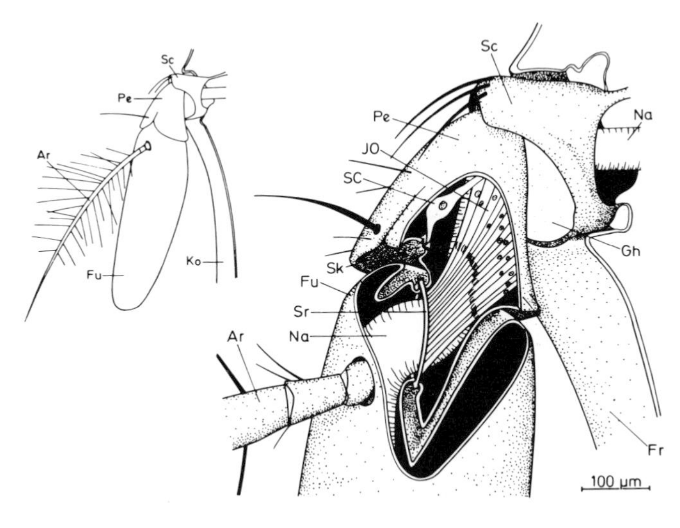

Fig. 3. On the left: Lateral view of the antenna of *Calliphora erythrocephala* (♀). On the right: Lateral view of the proximal part of the antenna; pedicellus (Pe) and funiculus (Fu) partly opened.
Ar - arista
Fr - frons
Gh - joint membrane
JO - Johnston's organ
Ko - head capsule
Na - nervus antennalis
SC - campaniform sensillum
Sc - scapus
Sk - cuticular dome
Sr - joint connection
(from Gewecke and Schlegel 1969)

The antenna of the blowfly, *Calliphora erythrocephala* (Fig. 3), differs greatly from that of the locust. In the blowflies the joint between head capsule and scapus does not participate in the movements of the antenna. All active movements are performed in the scapus-pedicellus joint about a horizontal and a vertical axis by the four muscles of the scapus. The first flagellar segment has developed into a large funiculus on the side of which the rest of the flagellar segments are situated. Together these form the feathered arista. The mechanoreceptors of the pedicellus, Johnston's organ (which is composed of several hundred scolopideal sensilla) and a single campaniform sensillum, are stimulated by passive movements of the funiculus about the vertical axis of the pedicellus-flagellum joint (Gewecke 1967a).

STIMULATION OF THE ANTENNAL MECHANORECEPTORS BY AIR CURRENTS

Flagellar steady deviation

Locusts and flies hold their antennae stretched forward during flight (Fig. 1). The flagella in locusts and the aristae in blowflies are stimulus-receiving structures, and are pushed backwards by pressure from frontal air currents. A torque (M) is, thereby, generated in the pedicellus-flagellum joint causing a rotation of the flagellum that results in a flagellar steady deviation measured as an angle (β). In the antenna of the blowfly this deviation is directly proportional to the speed of the frontal air currents (Gewecke 1967b). The torque (M), however, increases with the square of the flagellar deviation (β; Fig. 4). Therefore, only within the range of small flagellar deviations does the pedicellus-flagellum joint work as a weak spring with little restoring torque ($D^* = \Delta M/\Delta\beta$). An increase in β causes the spring to become stiffer.

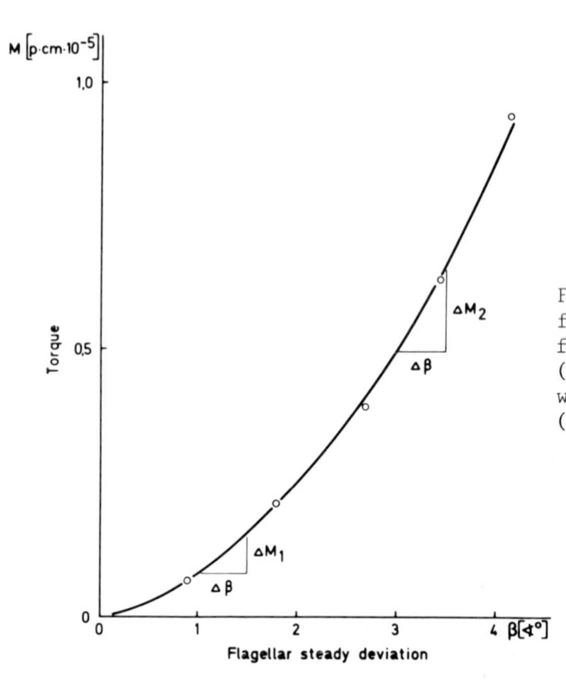

Fig. 4. The torque (M) in the pedicellus-flagellum joint of *Calliphora* (♀) as a function of the flagellar steady deviation (β). The stiffness of the joint increases with the restoring torque ($D^* = \Delta M/\Delta\beta$). (After Gewecke 1967b)

In *Calliphora* the lateral rotation of the flagellum, together with the flagellar steady deviation, stimulate the single campaniform sensillum of the pedicellus, which is a phasic-tonic receptor. The initial high impulse frequency, the phasic part of the discharge of this receptor, increases with increasing angular speed of the lateral flagellar rotation until it reaches a maximum. On the other hand, the plateau frequency of discharge increases linearly as a function of β up to β = 10° (Schlegel 1970). In this way, the campaniform sensillum is able to measure the lateral steady deviation (β) caused by frontal air currents during flight.

Flagellar vibration

Superimposed on the flagellar steady deviation (β) in the pedicellus-flagellum joint of *Calliphora* and, in fact, generated by the flight sound itself are passive vibrations of the flagellum (arista and funiculus) with respect to the pedicellus (Gewecke and Schlegel 1970). The resonance frequency (f_o) of these flagellar vibrations, when the pedicellus-flagellum joint is in resting position, is higher (f_o = 280 Hz; Fig. 5: V_A = 0 m/sec) than the wing-beat frequency (f_w = 140 to 200 Hz). With increasing speed of the air current both the steady deviation (β) and the restoring torque (D*) of the pedicellus-flagellum joint increases. In addition, the resonance frequency (f_o) also rises (Fig. 5) since it depends on the restoring torque (D*):

$$f_o = \frac{1}{2\pi} \cdot \sqrt{\frac{D^*}{\Theta}}$$

(Θ is the momentum of inertia). At the same time the amplitude of the flagellar vibrations (σ) decreases within the range of normal wing-beat frequencies (i.e., f_w = 140 to 200 Hz; Fig. 5).

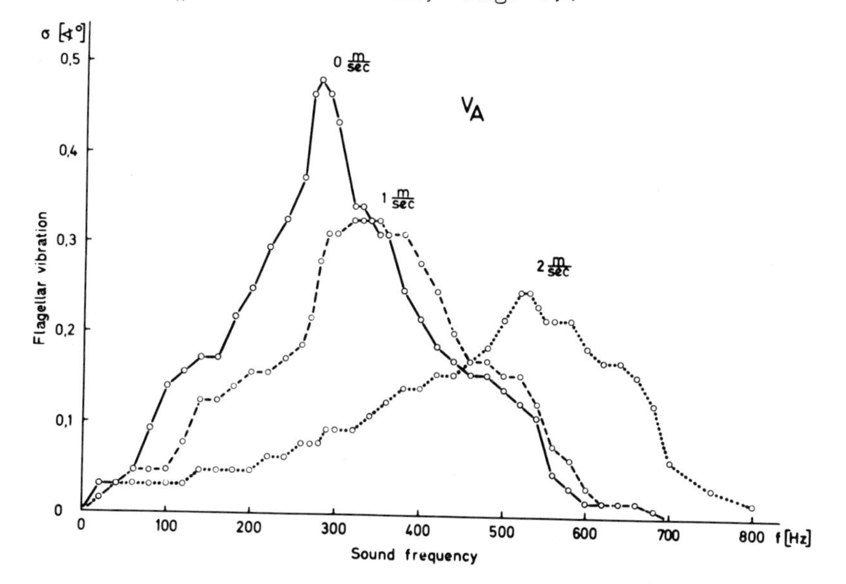

Fig. 5. Mechanical resonance curves of the antenna of *Calliphora* (♀) at different air speeds (V_A), i.e., at different flagellar steady deviations (β = 0°; 1.1°; 2.2°). The effective sound speed is 2.3 cm/sec. (After Gewecke and Schlegel 1970)

Johnston's organ of *Calliphora*, which is stimulated by these flagellar vibrations, is a phasic receptor (Burkhardt and Schneider 1957; Westecker 1957; Burkhardt 1960a,b; Burkhardt and Gewecke 1965). The majority of its sense cells (or perhaps all of them) respond only to movements, each phase of which causes a single spike. One group of sense cells responds to lateral rotations of the flagellum and another one to medial rotations. Therefore, during rhythmic stimulation, the frequency of the compound potential of Johnston's organ is twice that of the stimulus frequency (Fig. 7).

The amplitude of the compound potential of Johnston's organ depends on both the amplitude of the flagellar vibration and its frequency (Burkhardt 1960b; Gewecke and Schlegel 1970). The organ is most sensitive within the range of normal wing-beat frequencies (Fig. 6). The maximal

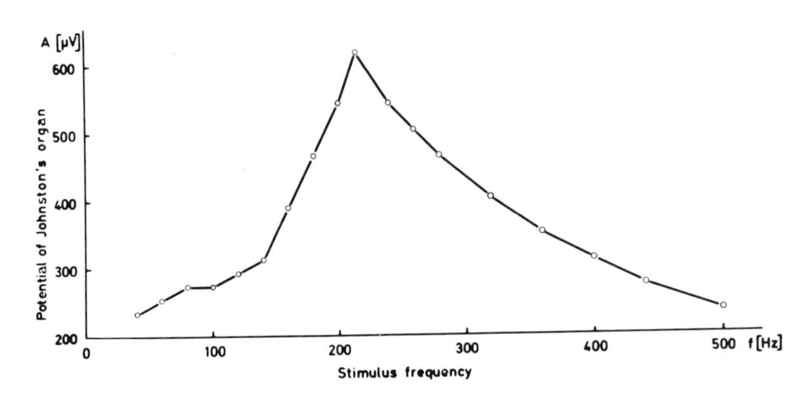

Fig. 6. Physiological resonance curve of the Johnston's organ of *Calliphora* (♀). The amplitude of the flagellar vibration caused by the direct mechanical stimulation was kept constant (σ = 0.3°). (After Gewecke and Schlegel 1970)

response of the sense cells at frequencies of about 200 Hz is not caused by the mechanical resonance which occurs at much higher frequencies (Fig. 5) but depends on properties of the stimulus conducting system or the sense cells themselves. These properties are distinct from the mechanical resonance and can be attributed to some kind of physiological resonance.

The change of the flagellar vibrations caused by frontal air currents influences the form of the biphasic potential of Johnston's organ. In particular, positive or negative acceleration of the air currents (which rotates the vibrating flagellum laterally or medially with respect to the pedicellus) cause characteristic changes in the pattern of excitation (Fig. 7). Specifically, during acceleration of the air currents from 0 to 1 m/sec one group of sense cells is maximally excited and the other group is nearly silent. During deceleration from 1 m/sec to 0 the excitation is nearly equal in both groups of sense cells but higher than when there are no air currents. Consequently, the antennae of blowflies are specialized acceleration sense organs that indicate changes in the flight speed.

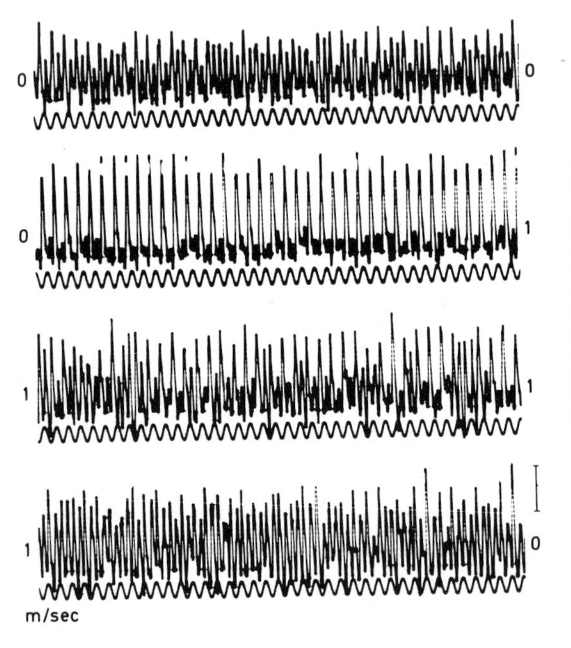

Fig. 7. The influence of the speed (m/sec) of air currents and their positive (0 → 1) and negative (1 → 0) acceleration (i.e., lateral or medial rotations of the vibrating flagellum with respect to the pedicellus) on the compound potential of the Johnston's organ (upper beam; vertical bar indicates 200 μV) of *Calliphora* (♀). The lower beam indicates the voltage of the stimulating loudspeaker with a frequency of 180 Hz. (After Gewecke and Schlegel 1970)

The information about the flight speed, itself, might be coded in the amplitude of the antennal vibration (Fig. 5). To test this hypothesis antennae of blowflies in tethered flight were stimulated by different vibrations which, themselves, were superimposed on different steady deviations (β) of the flagellum (Gewecke and Schlegel 1970). The speed of the air currents generated by the flapping wings was evaluated as a behavioural output parameter. When $\beta = 0$ the flies produce an air current with a speed of about 1.3 m/sec.

If the flagellar vibrations generated by the flight sound are diminished by pushing the aristae medially, the air current speed decreases only slightly. On the contrary, if the vibrations are diminished by the same degree by pushing the aristae laterally, thereby simulating normal flight conditions, the air current decreases significantly. The air current remains at this low level if the vibrations of the flagellum then are increased to the original value. This result shows, as suggested by Burkhardt (1960a,b), that it is the ultimate position of the pedicellus-flagellum joint of the vibrating antenna (and not the amplitude of the vibrations) which provides the final information for the perception of the flight speed. Therefore, it seems that the measurement of air speed by the antennae depends more on which of the scolopideal sensilla of the complex Johnston's organ are stimulated by mechanical vibration than on the intensity of their stimulation.

It is curious that the steady deviation of the flagellum as a tonic component is measured by Johnston's organ, a phasic receptor. For the fly, however, this method of measurement is advantageous since in this way both flight speed and its accelerations can be simultaneously adapted to altered flight conditions.

The results demonstrate, then, that there are two mechanoreceptors in the antenna of flies which are stimulated by the speed and acceleration of air currents: a single campaniform sensillum and the Johnston's organ. Their relationship to the control of flight has been clarified with the help of behavioural investigations.

CONTROL OF FLIGHT BEHAVIOUR

Antennal movements

Insects at rest can move their antennae independently of one another in different directions. In tethered flight in front of a wind tunnel, however, the antennae are symmetrically stretched forward in flight position (Fig. 8, on the right). The flagellae in *Locusta* and the aristae in *Calliphora* lie approximately horizontal thereby forming a definite angle in the horizontal plane (the antennal angle γ) with the body axis. Without air current the antennal angle (γ) is about 50° in both locusts and blowflies (Gewecke 1967b, 1972b). With increasing speed (V_A) of frontal air currents of the wind tunnel, this antennal angle (γ) decreases within the range of normal flight speeds (Fig. 8: L_n, C_n). In other words the antennae are moved forward.

After the elimination or blocking of all the mechanoreceptors of the pedicellus the antennal angle (γ) is independent of the air speed (Fig. 8: L_i, C_i). This finding shows that these receptors, which are stimulated by the flagellar steady deviation (β), control the active forward movement of the antenna about the vertical axis of the scapus-pedicellus joint (Figs. 2, 3). Further experiments suggest that in *Calliphora* the single campaniform sensillum of the pedicellus (Fig. 3), which is stimulated by β, controls the active antennal movements during flight.

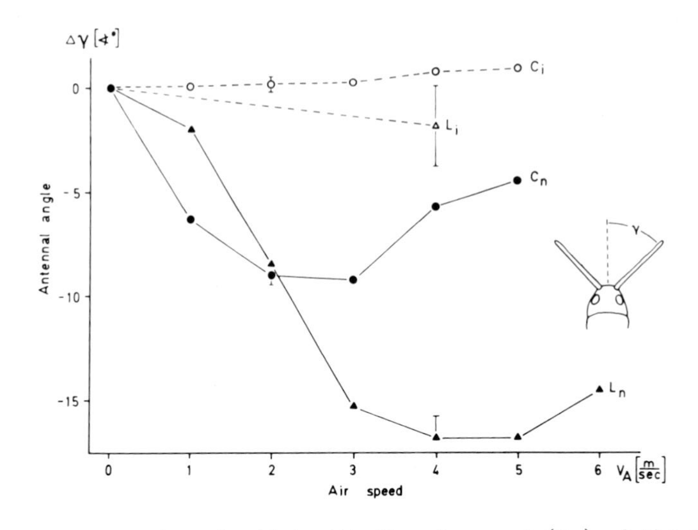

Fig. 8. Change of antennal angle ($\Delta\gamma$) with the air speed (V_A) of the wind tunnel.
C_n - normal *Calliphora*
C_i - *Calliphora* with pedicellus-flagellum joint immobilized
L_n - normal *Locusta*
L_i - *Locusta* with pedicellus-flagellum joint immobilized.
Vertical bars indicate the standard error (± S.E.). (After Gewecke 1967b, 1972b)

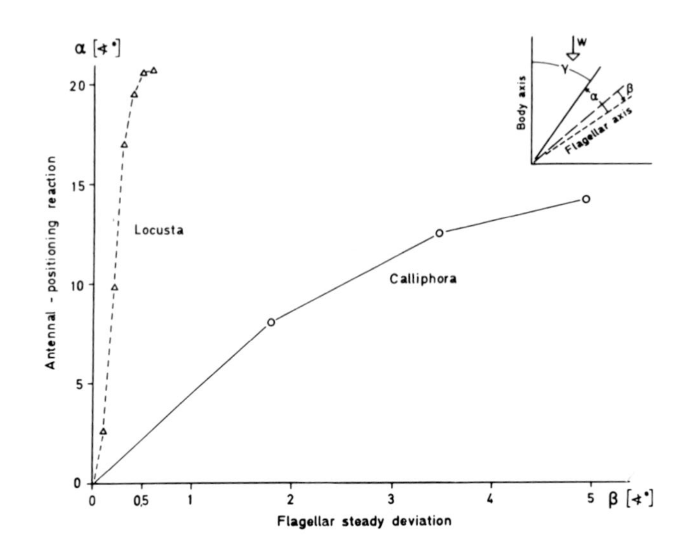

Fig. 9. Dependence of the active change of pedicellar position (the antennal-positioning reaction, α) on the passive flagellar steady deviation (β) in the pedicellus-flagellum joint.
W - wind direction
γ - antennal angle.
(After Gewecke 1967b, 1972b)

These active movements seem to function as an antennal-positioning reaction which actively rotates the pedicellus and the flagellum medially about an angle (α). This active change of pedicellar position (α) is greater than the opposite passive deviation (β) in the pedicellus-flagellum joint on which it depends (Fig. 9), and, therefore, the antennal angle (γ) decreases in normal animals during flight (Fig. 8).

The various angles which describe the antennal movements are shown in Fig. 9 (right upper corner): a frontal wind pushes the flagellum backwards and laterally and results in a change of flagellar position (β). This passive movement causes the active change of pedicellar position (α) in the opposite direction (the antennal-positioning reaction) and in this way determines the complex antennal angle (γ).

The antennal-positioning reaction affects, as a negative feedback mechanism, both the torque (M) in the pedicellus-flagellum joint (which during frontal air currents depends on the antennal angle, γ) and, consequently, the flagellar steady deviation (β; Fig. 4). As a result of the antennal-positioning reaction, then, the pedicellus-flagellum joint works as a relatively weak spring at high as well as at low flight speeds of the insect. It is this reaction which keeps the mechanoreceptors of the pedicellus, which ultimately control flight, within their operating range.

Flight movements

Wing-stroke angles, wing-beat frequency and flight speed have also been evaluated as part of the investigation of flight control by the antennae.

Wing-stroke angles are only crude indicators of the very complicated wing movements of insects (Weis-Fogh 1956a; Jensen 1956; Nachtigall 1966; Zarnack 1972). Variations in this parameter show that flight is somewhat changed, but they do not indicate exactly what particular changes

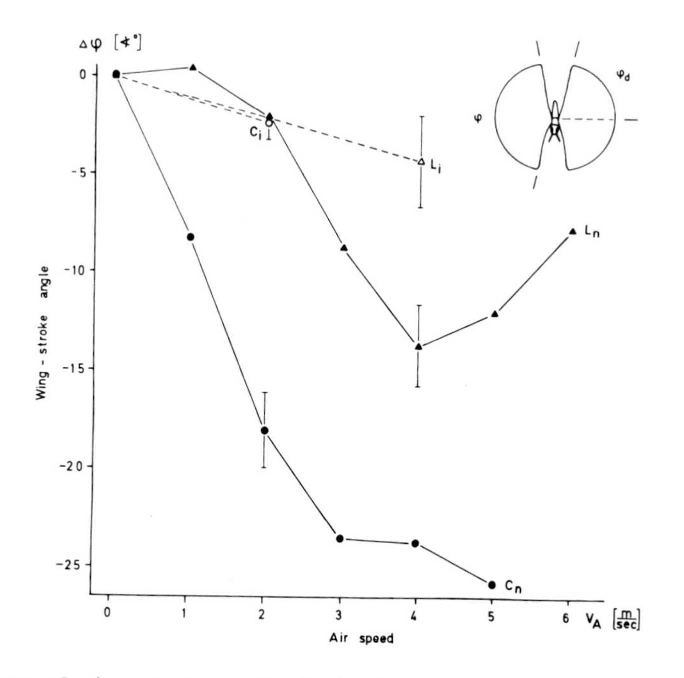

Fig. 10. Change of wing-stroke angle (Δφ) with the air speed (V_A) of the wind tunnel. In *Locusta* only the change of the dorsal part of the wing-stroke angle ($Δφ_d$) of the forewing is evaluated. See Fig. 8.
(After Gewecke 1967b, 1972b)

are occurring. Furthermore, it seems possible that there might be changes in the flight movements which are not expressed in wing-stroke angles. Changes may occur, for example, in the three-dimensional path or the angle of attack of the wings. In spite of these disadvantages, the method of measuring wing-stroke angle has proved to be useful, for it is simple, and gives some valuable results.

In tethered flight without an air current from the wind tunnel, the wing-stroke angle (φ) is about 140° in both the fore- and hindwings of *Locusta* (Gewecke 1972b) and about 145° in the wings of *Calliphora* (Gewecke 1967b). With increasing air speed (V_A) the wing-stroke angle (φ) decreases in the fore- and hindwings of *Locusta* and also in the wings of *Calliphora* within the range of normal flight speeds (Fig. 10: L_n, C_n).

This reduction in the wing-stroke angles is, however, less pronounced in both organisms when the pedicellus-flagellum joints have been immobilized by wax colophony (Fig. 10: L_i, C_i). This observation suggests that the large decrease in wing-stroke angles observed in normal animals is, at least partly, induced by the mechanoreceptors of the pedicellus. In *Calliphora* it is Johnston's organ which controls the wing-stroke angle (Gewecke 1967b).

Flight direction

The results of experiments in which the antennae were asymmetrically stimulated (i.e., experiments in which one antenna was amputated or immobilized during stimulation by frontal air currents, or in which the air currents came from a lateral direction) show that in locusts (Gewecke 1972b), as well as in flies (Gewecke 1967b), the antennal-positioning reaction of an antenna is controlled entirely by its own mechanoreceptors. However, there are differences between locusts and flies with respect to the neural connection of the antennal mechanoreceptors with the flight motor neurones.

For example, in *Calliphora* the neural connections are unilateral and, therefore, the antennae can help to control the direction of flight in the horizontal plane. If the animal makes a sudden unintentional turn, the impact pressure acting on the leading antenna is less than that acting on the other. This will induce a decrease of the wing-stroke angle on the side of the more strongly stimulated antenna and will thus cause the turn to be counteracted (Schneider 1953; Gewecke 1967b).

In *Locusta*, on the contrary, the antennae influence the wing-stroke angles on both sides of the body equally. Accordingly, then, the control of wing-stroke angles by the antennae cannot take part in the control of flight direction.

Flight speed

The role of the antennae as air-current sense organs for the flight control of insects in free flight has also been established by means of behavioural experiments. Blowflies from which the antennae have been removed have been shown to fly faster than they did before the removal (Burkhardt and Schneider 1957; Schneider 1965). The conclusion to be drawn from this observation is that in normal animals the antennae, stimulated by the frontal air currents, reduce the flight speed via a negative feedback mechanism.

The function of the antennae for the control of flight speed relative to the surrounding air in *Locusta* has been shown by means of tethered flight experiments (Gewecke 1972c). At a flight balance (Weis-Fogh 1956a: Pendulum) in front of a wind tunnel, the air speed of which is regulated indirectly by the animal itself, both high wing-beat frequency and high flight speed are observed during the first minutes of flight (Fig. 11). The level of both parameters then decreases. The height correlation of these parameters shows that under these circumstances the flight speed depends on the wing-beat frequency.

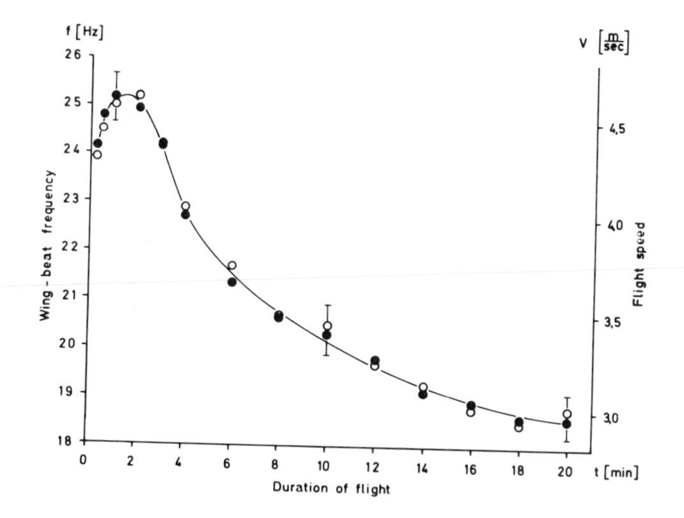

Fig. 11. The identical dependence of the wing-beat frequency (f: o) and the flight speed (V: •) on the duration of flight (t) at the flight balance in *Locusta migratoria*. (After Gewecke 1972c)

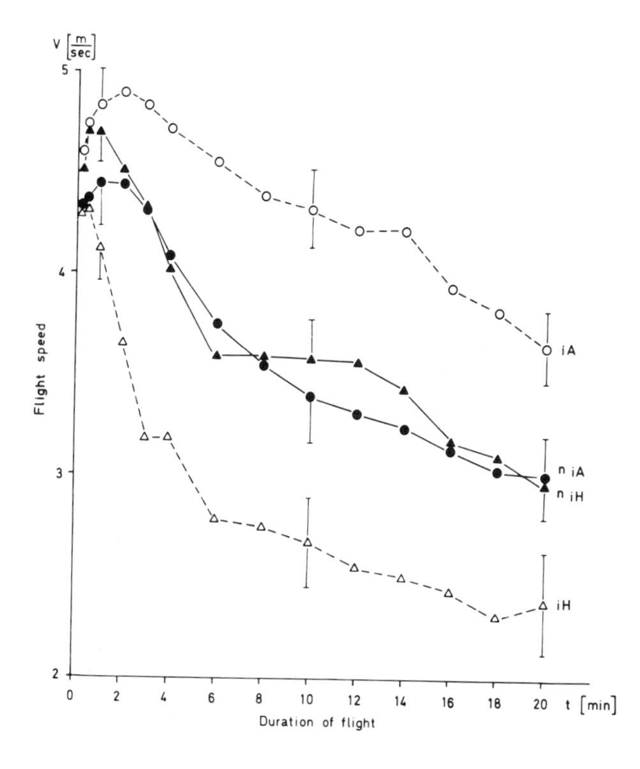

Fig. 12. Flight speed (V) of *Locusta* at different durations of flight (t).
n_{iA} (iA) - animals before (after) immobilizing the proximal joints of the antennae
n_{iH} (iH) - animals before (after) immobilizing the wind-sensitive hair patches
(After Gewecke 1972c)

Locusts in which the mechanoreceptors of the pedicelli have been blocked by the immobilization of the proximal joints of the antennae (including the pedicellus-flagellum joints) fly faster (Fig. 12: iA) than they did before (Fig. 12: n_{iA}). On the other hand, when the wind-sensitive hairs on the frons and vertex (Figs. 1, 2) are covered with wax colophony, the speed of flight is considerably reduced (Fig. 12: iH). On the basis of these results, then, it seems clear that the two types of air-current sense organs of locusts play different roles in the control of the flight speed. Specifically, the hairs on the frons and the vertex stimulate flight. This function is especially well demonstrated by locusts with immobilized antennae which fly at maximal speed (Fig. 12: iA). However, as in flies, the antennae of locusts throttle flight, and after immobilization of the stimulating hairs they fly very slowly (Fig. 12: iH).

DISCUSSION

Comparative consideration of the antennae of insects as air-current sense organs

Behavioural investigations have demonstrated that the antennae are used for the perception of air currents by insects representative of a number of orders. This has been shown not only for groups composed of insects which fly but also for groups composed of flightless insects.

The antennae of bugs (Heran 1962) and beetles (Birukow 1958) control anemotaxis during odour orientation, and some beetles use their antennae in their anemomenotactic orientation (Linsenmair 1969, 1970).

In locusts (Gewecke 1970, 1972b,c), aphids (Johnson 1956), bees (Heran 1959), mosquitoes (Bässler 1958), and flies (Hollick 1941; Burkhardt and Schneider 1957; Burkhardt and Gewecke 1965; Schneider 1965; Gewecke 1967b) the antennae control flight. The antennae of these insects are very different in form but similar in function. In all of these cases it is the flagellum which is the stimulus-receiving structure which conducts the aerodynamic forces, generated by the air currents during flight, to the mechanoreceptors of the pedicellus. These receptors, Johnston's organ, other chordotonal organs and the campaniform sensilla, perceive the passive movements of the flagellum with respect to the pedicellus.

In locusts, bees and flies, the antennae are held in the typical flight position during flight: they are stretched forward so that the flagellae lie in a horizontal plane. If the flagellae are passively pushed backwards by frontal air currents the animal actively reduces the angles between the flagellae and the direction of the air currents. This antennal movement is considered as an antennal-positioning reaction which adapts the operating range of the pedicellar mechanoreceptors to the particular flight speed. The antennal-positioning reaction is similar in these three orders of insects even though they are not closely related.

In flight control, the antennae of these insects also have a similar function even if their flight mechanism is quite different (Pringle 1957, 1965; Wilson 1968). Not only are the wing-stroke angles regulated similarly but also the flight speed. The flight speed, relative to the surrounding air, is controlled by a negative feedback mechanism with the antennae working as sensory units. Into this primary system a secondary feedback mechanism is inserted which acts to control the operating range of the pedicellar mechanoreceptors by the antennal-positioning reaction.

Biological importance of flight control by the antennae

The most important role of the antennae as air-current sense organs is, probably, to enable the flying insect to maintain an economical flight speed, an ability which must be of particular importance to insects making long distance migrations. Locusts, for example, usually live in large open areas and it is more useful for them to be able to make long flights than to be able to make the complicated flight manoeuvres characteristic of bees and flies.

There are somewhat conflicting reports about the orientation of migrating locusts. Rainey (1959) states that *Schistocerca gregaria*, in migrating swarms, fly generally with random orientation to the direction of swarm displacement, and that only the individuals flying at the periphery of the swarm show any particular orientation and that this is directed towards the main body of the swarm. It is postulated (Rainey 1951) that the swarm is transported passively by monsoonal winds to zones of convergence where rain storms develop and thus create favourable conditions for the development of eggs and larvae. More recently Waloff (1972) has provided evidence that the direction of flight in swarming *S. gregaria* is orientated, at least for a substantial proportion of their time aloft, in the direction of swarm displacement. In either event, however, it is of advantage to the locusts to be able to control their speed of flight so as to achieve greatest economy, but it is apparent that the economical flight speed would be different in the two circumstances.

1. In the situation described by Rainey, in which orientation is random, it is clear that a maximum rate of displacement would be achieved if the insects were to remain in the air as long as possible and that it would be of no advantage for them to have a high speed in relation to the air. The economical flight speed for maximum displacement is, therefore, that which keeps the insect airborne with the expenditure of a minimum of energy. The finding of Weis-Fogh (1952) that the metabolic rate of flying locusts increases exponentially with flight speed leads to the conclusion that the locusts should fly as slowly as possible (Haskell 1960).
2. If, however, migrating locusts are orientated in the direction of swarm displacement, i.e. if their flight is directed towards the destination of their migration, the energy output per actual distance flown should be minimized. This is similar to the control of flight speed relative to the surrounding air reported for other insects, for example in bees when collecting food (Heran 1959).

During the first minutes after take-off locusts fly with high wing-beat frequency and high flight speed (Fig. 11) so as to produce sufficient lift (Weis-Fogh 1956a). After reaching a definite altitude the flight speed is reduced because it would be uneconomical to continue to fly so fast. This reduction of flight speed and the resulting reduction in fuel consumption is controlled by the antennae (Fig. 12), which probably function as sensory units of a negative feedback mechanism.

It seems clear that, because they act in this way, the antennae play a very important role in the control of flight speed, especially during extensive long range migrations of insects, by acting as air-current sense organs.

ACKNOWLEDGEMENT

I thank Dr. Gregory Fowler very much for critically reading the manuscript.

REFERENCES

BÄSSLER, U.: Versuche zur Orientierung der Stechmücken: Die Schwarmbildung und die Bedeutung des Johnstonschen Organs. Z. vergl. Physiol. 41, 300-330 (1958).
BIRUKOW, G.: Zur Funktion der Antennen beim Mistkäfer. Z. Tierpsychol. 15, 265-276 (1958).
BOYD, K., EWER, D.W.: Flight responses in grasshoppers. S. Afr. Sci. 2, 168-169 (1949).
BURKHARDT, D.: Untersuchungen zur Gehörphysiologie der Insekten (Aktionspotentiale des Johnstonschen Organs von *Calliphora erythrocephala*). Verh. dt. zool. Ges. 1959, 215-220 (1960a).
BURKHARDT, D.: Action potentials in the antennae of the blowfly (*Calliphora erythrocephala*) during mechanical stimulation. J. Insect Physiol. 4, 138-145 (1960b).
BURKHARDT, D., GEWECKE, M.: Mechanoreception in arthropoda: the chain from stimulus to behavioral pattern. Cold Spring Harb. Symp. quant. Biol. 30, 601-614 (1965).
BURKHARDT, D., SCHNEIDER, G.: Die Antennen von *Calliphora* als Anzeiger der Fluggeschwindigkeit. Z. Naturf. (B) 12, 139-143 (1957).
CAMHI, J.M.: Locust wind receptors. I. Transducer mechanics and sensory response. J. exp. Biol. 50, 335-348 (1969).
GEWECKE, M.: Der Bewegungsapparat der Antennen von *Calliphora erythrocephala*. Z. Morph. Ökol. Tiere 59, 95-113 (1967a).
GEWECKE, M.: Die Wirkung von Luftströmung auf die Antennen und das Flugverhalten der Blauen Schmeißfliege (*Calliphora erythrocephala*). Z. vergl. Physiol. 54, 121-164 (1967b).
GEWECKE, M.: Antennae: another wind-sensitive receptor in locusts. Nature, Lond. 225, 1263-1264 (1970).
GEWECKE, M.: Bewegungsmechanismus und Gelenkrezeptoren der Antennen von *Locusta migratoria* L. (Insecta, Orthoptera). Z. Morph. Tiere 71, 128-149 (1972a).
GEWECKE, M.: Antennen und Stirn-Scheitelhaare von *Locusta migratoria* L. als Luftströmungs-Sinnesorgane bei der Flugsteuerung. J. comp. Physiol. 80, 57-94 (1972b).
GEWECKE, M.: Die Regelung der Fluggeschwindigkeit bei Heuschrecken und ihre Bedeutung für die Wanderflüge. Verh. dt. zool. Ges. 1971, 247-250 (1972c).
GEWECKE, M., SCHLEGEL, P.: Der Mechanismus der Antennenschwingungen von *Calliphora*. Verh. dt. zool. Ges. 1968, 399-404 (1969).
GEWECKE, M., SCHLEGEL, P.: Die Schwingungen der Antenne und ihre Bedeutung für die Flugsteuerung bei *Calliphora erythrocephala*. Z. vergl. Physiol. 67, 325-362 (1970).
HASKELL, P.T.: The sensory equipment of the migratory locust. Symp. zool. soc. Lond. 3, 1-23 (1960).
HERAN, H.: Wahrnehmung und Regelung der Flugeigengeschwindigkeit bei *Apis mellifica* L. Z. vergl. Physiol. 42, 103-163 (1959).
HERAN, H.: Anemotaxis und Fluchtorientierung des Bachläufers Velia caprai Tam. (= V. currens F.). Z. vergl. Physiol. 46, 129-149 (1962).
HOLLICK, F.S.J.: The flight of the dipterous fly *Muscina stabulans* Fallén. Phil. Trans. R. Soc. (B) 230, 357-390 (1941).
JENSEN, M.: Biology and physics of locust flight. III. The aerodynamics of locust flight. Phil. Trans. R. Soc. (B) 239, 511-552 (1956).
JOHNSON, B.: Function of the antennae of aphids during flight. Aust. J. Sci. 18, 199-200 (1956).
LINSENMAIR, K.E.: Anemomenotaktische Orientierung bei Tenebrioniden und Mistkäfern (Insecta, Coleoptera). Z. vergl. Physiol. 64, 154-211 (1969).
LINSENMAIR, K.E.: Die Interaktion der paarigen antennalen Sinnesorgane bei der Windorientierung laufender Mist-und Schwarzkäfer (Insecta, Coleoptera). Z. vergl. Physiol. 70, 247-277 (1970).

NACHTIGALL, W.: Die Kinematik der Schlagflügelbewegungen von Dipteren. Methodische und analytische Grundlagen zur Biophysik des Insektenflugs. Z. vergl. Physiol. 52, 155-211 (1966).
PRINGLE, J.W.S.: Insect flight. Cambridge monographes exp. Biol. 9, (1957).
PRINGLE, J.W.S.: Locomotion: Flight. *In* "The physiology of Insecta" (Ed., M. Rockstein), Vol. 2, pp. 283-329. New York, London: Academic Press (1965).
RAINEY, R.C.: Weather and the movements of locust swarms: a new hypothesis. Nature, Lond. 168, 1057-1060 (1951).
RAINEY, R.C.: Some new methods for the study of flight and migration. Proc. XV Int. Congr. Zool. (London, 1958), 866-870 (1959).
SCHLEGEL, P.: Die Leistungen eines Gelenkrezeptors der Antenne von *Calliphora* für die Perzeption von Luftströmungen. Elektrophysiologische Untersuchungen. Z. vergl. Physiol. 66, 45-77 (1970).
SCHNEIDER, G.: Die Halteren der Schmeißfliege (*Calliphora*) als Sinnesorgane und als mechanische Flugstabilisatoren. Z. vergl. Physiol. 35, 416-458 (1953).
SCHNEIDER, P.: Vergleichende Untersuchungen zur Steuerung der Fluggeschwindigkeit bei *Calliphora vicina* Rob.-Desvoidy (Diptera). Z. wiss. Zool. 173, 114-173 (1965).
SMOLA, U.: Untersuchung zur Topographie, Mechanik und Strömungsmechanik der Sinneshaare auf dem Kopf der Wanderheuschrecke *Locusta migratoria*. Z. vergl. Physiol. 67, 382-402 (1970).
WALOFF, Z.: Orientation of flying locusts, *Schistocerca gregaria* (Forsk.), in migrating swarms. Bull. ent. Res. 62, 1-72 (1972).
WEIS-FOGH, T.: An aerodynamic sense organ stimulating and regulating flight in locusts. Nature, Lond. 164, 873-874 (1949).
WEIS-FOGH, T.: An aerodynamic sense organ in locusts. Proc. 8th Int. Congr. Ent. (Stockholm, 1948), 584-588 (1950).
WEIS-FOGH, T.: Fat combustion and metabolic rate of flying locusts (*Schistocerca gregaria* Forskål). Phil. Trans. R. Soc. (B) 237, 1-36 (1952).
WEIS-FOGH, T.: Biology and physics of locust flight. II. Flight performance of the desert locust (*Schistocerca gregaria*). Phil. Trans. R. Soc. (B) 239, 459-510 (1956a).
WEIS-FOGH, T.: Biology and physics of locust flight. IV. Notes on sensory mechanisms in locust flight. Phil. Trans. R. Soc. (B) 239, 553-584 (1956b).
WESTECKER, M.: Elektrophysiologische Untersuchung der Antennenreaktion von *Calliphora erythrocephala* bei Luftschall. Inauguraldissertation, Würzburg (1957).
WILSON, D.M.: The nervous control of insect flight and related behavior. Adv. Insect Physiol. 5, 289-338 (1968).
ZARNACK, W.: Flugbiophysik der Wanderheuschrecke (*Locusta migratoria* L.). I. Die Bewegungen der Vorderflügel. J. comp. Physiol. 78, 356-395 (1972).

RHYTHMIC ACTIVITIES AND THE INSECT NERVOUS SYSTEM

P.L. Miller

Department of Zoology, Oxford, England

The rhythmic activities of arthropods have received a lion's share of the attention of neurophysiologists during the last two decades. For example, the swimmeret beat of crayfish and lobsters, walking in insects and crabs, ventilation in *Limulus*, crabs and insects, and insect flight, stridulation, grooming and swimming have all been the subjects of a number of studies: many more aspects remain to be explored as much insect behaviour is characterised by rhythmic swaying, rocking, twitching, wagging, waving and beating. In no case is the neural machinery completely understood, but properties of the motor output common to some examples suggest that similar types of neural interaction may underlie different rhythmical activities. Moreover some similarities to vertebrate locomotory patterns begin to appear. Curarized spinal dogfish for example are like some insects in that they may continue to produce locomotory patterns of motor impulses in the absence of phasic input (Roberts 1969). The stepping of cats is said to depend on the "cooperative activity of a relatively simple internal pattern-generating network" affected by proprioceptive input and set in motion by a system of descending neurones which act like the command neurones of Crustacea (Evarts 1971). Such suggested similarities encourage the further study of arthropod rhythmical systems and the search for common mechanisms among them.

In this article, I shall try to draw together some common features of several insect rhythmical systems, while always remaining aware of the likelihood of blurring or glossing over differences between them in a too zealous search for *the* common rhythmic mechanism. Even when it becomes possible to account in neural terms for a rhythmic activity, we shall still be a long way from understanding its initiation and termination, and the means whereby it meshes into the total behaviour of an insect. The account which follows must therefore at many points be qualitative and imprecise.

CENTRAL PROGRAMMES

Central programmes are thought to underlie many forms of rhythmical activity. Such programmes are inherited and when triggered they produce rather stereotyped motor activity which may interact with, but is not entirely dependent upon, consequent input from sense organs. They constitute "instinctive" acts and resemble tapes which are played through in a linear fashion. Since those we are concerned with are repetitive, the tapes may be thought of as looped (Hoyle 1970). A single loop may last a few msec (flight), or seconds (courtship behaviour), or over an hour (pre-eclosion behaviour). One of the most complex and long-lasting central programmes known is that which controls the pre-eclosion behaviour of saturniid pupae, and presumably of other insects. In this, a sequence of complex cycles of motor activity lasting about 72 min is initiated by a hormone from the median neurosecretory cells of the brain and can be carried through without reference to the periphery: it can take place, for example, in insects with the pupal cuticle already peeled off, and even in an isolated abdominal nerve cord (Truman 1971; Truman and Sokolove 1972; Riddiford and Truman, this volume). The complex courtship sequence of *Gomphocerippus rufus*, comprising three sub-units of activity and lasting 8-9 sec at 28°C, can be carried through after considerable peripheral disruption (Loher and Huber 1966; Elsner and Huber 1969). Cleaning movements may have some of the characteristics of central programmes, for example antennal cleaning of locusts (O'Shea 1972), or in diopsid flies which attempt to clean an excised eye for long periods (Seibt 1972), or in cockroaches which similarly attempt to clean a non-existent antenna (Laudien 1970). Less complex programmes may pattern the output for flight (Wilson 1966b) and for walking (Iles and Pearson 1969; Pearson and Iles 1970), but in the latter proprioception plays a more important role.

All motor outputs, even the most simple 'reflex' responses to stimuli, contain centrally derived elements of temporal and spatial patterning not present in the original signal, but the concept of a central programme is most usefully reserved for complex outputs which can produce the basic pattern without phasic re-stimulation, often cyclically. Central programmes are normally unmodified by individual experience. Thus, Bentley (1971) found that stridulatory patterns in the cricket *Teleogryllus* were inherited through the action of several genes on more than one chromosome, and that even the number of action potentials in a motor burst might be genetically determined: individual differences in diet, temperature, light cycle, time of year and population density had no effect on the song pattern.

OSCILLATORY SYSTEMS

Attempts have been made in several systems to identify the components of an oscillatory system, and in particular to recognise a cell or group of cells which acts as the oscillator and produces the basic patterning expressed in an appendage or segment through the action of intermediate cells. Such a system may be thought of as divisible into the following components, and evidence will be discussed below which goes some way towards justifying this division in a few systems:

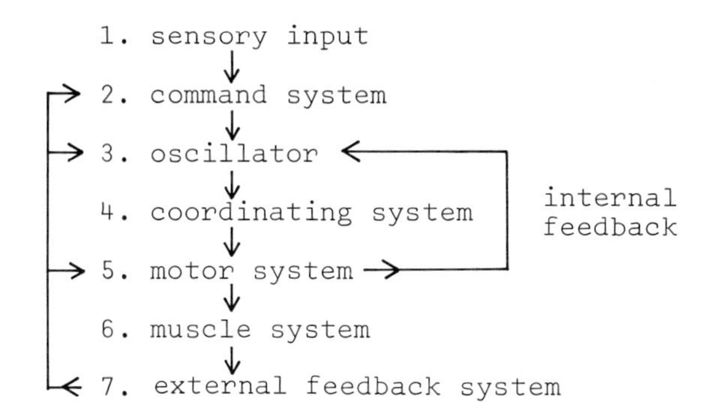

Each level in the above scheme may represent a separate layer of cells and a hierarchy of control can be envisaged. Alternatively, several components may co-exist in one cell layer; for example, the oscillator, coordinator and motor systems may reside in parts of a single motor neurone, or group of interconnecting motor neurones. Although recent evidence shows that insect motor neurones have considerable anatomical complexity (Burrows 1973c), the physiological data suggest that some of the functions at least are performed by separate cell layers.

CHARACTERISTICS OF OSCILLATORS

The oscillator is thought of as a cell or group of cells whose activity establishes the basic motor pattern. In no arthropod system has such a central oscillator been identified with certainty, but peripherally situated oscillators are known, for example in the crustacean cardiac and stomatogastric ganglia. Single oscillatory cells are known also in the central nervous system (CNS) of *Aplysia*, and some may continue their characteristic burst formation after complete isolation from the ganglion (Chen, von Baumgarten and Takeda 1971), but their function is unknown. Similar burst-formers have been thought to underlie several types of rhythmic activity in insects (Miller 1966; Bentley 1969b; Pearson and Iles 1970; Delcomyn 1971b). Identification of a central oscillator cell has recently been claimed by Mendelson (1971) in the crab in which it is believed to control the scaphognathite beat which ventilates the gill cavity. It produces undulating potential changes without spikes which in turn drive antagonistic sets of motor neurones supplying the scaphognathite muscles. The mechanism producing the undulation may be similar to that responsible for burst-formation in other oscillators, but it acts at sub-threshold levels.

Another type of burst-forming mechanism was postulated by Wilson (1964, 1966b) for the locust flight pattern generator. In this a symmetrical pair of cells, receiving a common drive, is joined by reciprocal inhibitory connections and each rapidly accumulates refractoriness, or has an auto-inhibitory mechanism, to turn off a burst. Thus, one fires a short burst, then stops, thereby lifting inhibition from the other, which in turn fires a short burst. The alternating bursts produced can then drive elevator and depressor motor neurones whose high thresholds account for the spacing of the antagonistic motor spikes.

Endogenous activity in a neurone may take the simpler form of continuous trains of spikes. This type of activity may be found in some insect motor and interneurones and it may play a part in more complex pattern formation. For example, cells with this activity may be rhythmically inhibited by burst-forming cells, as was postulated for the inspiratory motor neurones in the locust abdomen (Miller 1966).

Although attempts to identify the oscillators responsible for the flight motor pattern of insects have so far failed (Page 1970), there is much indirect information about their activity from the work of Wilson and his colleagues, and from a more recent study by Burrows (1973b) on the locust, *Chortoicetes*. Burrows studied one pair of motor neurones to the tergosternal muscles (113: elevators) and one pair to the first basalars (127: depressors), by recording intracellularly from the neuron somata. Anti-dromic stimulation of a tergosternal unit caused a burst of excitatory post-synaptic potentials (EPSPs) to be produced in the contralateral homologous unit after an interval of 25 msec, with a spike following after 40-45 msec, the period for one wing beat cycle in this species; there was also a weaker ipsilateral effect after the same interval. He could detect no direct interactions between the pair of units and the action of one on the other probably took place through interneurones which had a built-in delay mechanism, perhaps resulting from a reverberating circuit. Subsequent oscillations, each separated by the same interval, were sometimes recorded. The postulated interneurones may form a part of the flight oscillator system which receives feedback from, and in turn feeds onto, both the ipsi- and the contralateral motor neurones. In addition, tergosternal-unit spikes caused responses in the basalar units after an interval equal to about half a wing beat cycle, but there was no action in the reverse direction. In Burrows's model (Fig. 1) each tergosternal neurone is supplied with a separate interneuronal delay circuit, and a further delay circuit is shown between the elevator and the depressor neurone. Burrows's records are from four of the total of about eighty flight motor neurones and

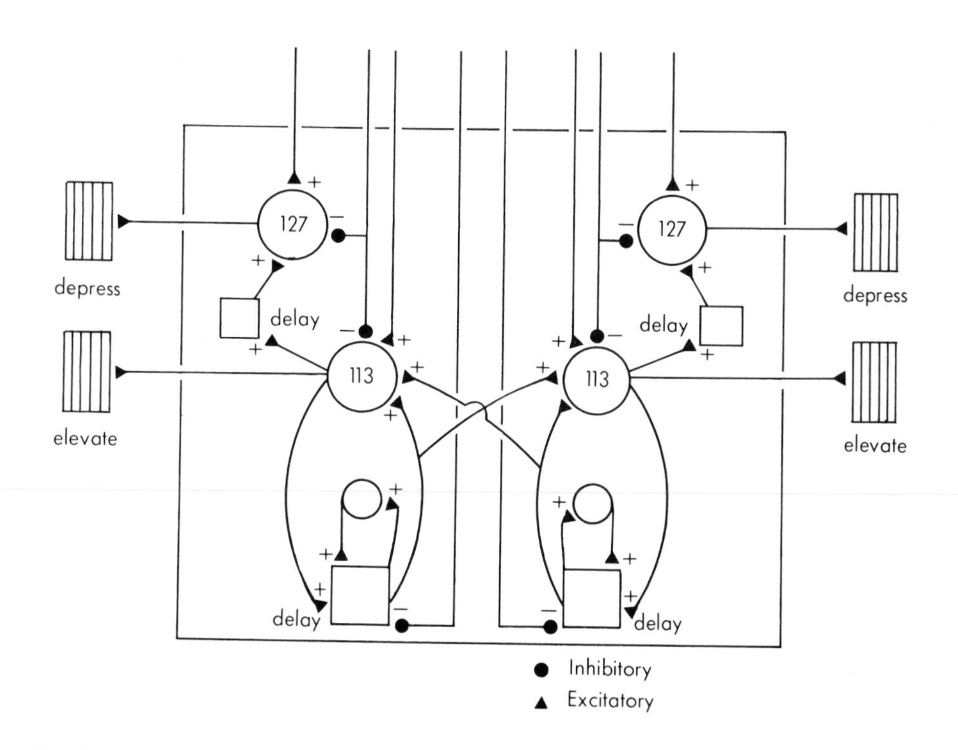

Fig. 1. A summary of some of the synaptic pathways shown to exist by intracellular recording and anti-dromic stimulation between two pairs of homologous motor neurones which contribute to the flight motor output in the metathoracic ganglion of a locust, *Chortoicetes*. One pair (113) supplies the tergosternal muscles which act as wing elevators: the other pair (127) supplies the first subalars which act as depressors. Anti-dromic stimulation of one 113 unit produces an excitatory response in the contralateral 113 and a weaker ipsilateral response after a delay approximately equal to the wing beat cycle time. A delay circuit is shown as a square representing an unknown number of intermediate synapses: it feeds onto ipsi- and contralateral 113 units. Excitation of 113 also excites the antagonist 127 after a delay equal to about half a wing beat cycle: a smaller square represents the delay circuit between the antagonists. Cell 127 has no action on 113. Further inputs are shown which have excitatory or inhibitory effects on 113 and 127. Cells 113 may be symmetrically driven by bursts of EPSPs at the flight frequency without spiking at times when no motor spikes can be detected. This may represent the activity of an interneuronal flight oscillator. (From Burrows 1973b)

there may be many interacting oscillators with similar properties which contribute to the regularity of the output. Kendig (1968) showed that in degenerating preparations of the locust, different flight oscillators might uncouple from each other and show independent activity - an observation which agrees with the above. Roeder (1970) found, in the brain of a moth, evidence for another flight oscillator whose action might be important for the initiation and maintenance of flight but not for the patterning of the output (see also Svidersky 1967).

Some forms of rhythmical activity have an asymmetrical output, one of the two active phases being relatively constant in duration while the other varies with the cycle time. For example, in the crab, *Emerita*, abdominal flexion which is the powerstroke in swimming is relatively fixed in duration compared to the recovery stroke (Paul 1971), and the powerstroke in the beat of lobster swimmerets is similarly less variable than the recovery stroke (Davis 1969a). In the calling song of *Gryllus campestris*, the interchirp interval is constant in duration and independent of the chirp length (Kutsch 1969). In locust ventilation, the motor burst in the inspiratory nerves shows less variation of duration than that in the expiratory nerves (Lewis, Miller and Mills 1973). In cockroach walking, leg protraction is less variable than retraction, the latter being more affected by sensory input, but both decrease in duration to some extent as the stepping frequency rises (Pearson and Iles 1970; Delcomyn 1971a).

Some rhythmical activities showing this property may be driven by a relaxation oscillator whose relaxation, controlled intrinsically, determines the length of the phase of constant duration. Its rate of recharge may be dependent on the inflow of energy from elsewhere and therefore be more variable (cf. Bentley 1969b; Graham 1972). Ventilation frequency for example may be controlled partly by the activity of interneurones which respond to gas tensions in anterior regions of the CNS. However, for cockroach walking, Delcomyn (1971b) favours a model involving a sinusoidally oscillating element whose output is chopped by a second element to produce a series of bursts. Burst length can be determined by a change of oscillator frequency and by alteration of the threshold of the chopper.

Even in flight the output may not be symmetrical. In locusts, the upstroke is faster than the downstroke and elevators are more affected by sensory input than are depressors (Waldron 1967a,b). Likewise, in dragonflies the first upstroke is more variable in duration than the downstroke (Pond, unpublished). Temporal asymmetries may be common in many rhythmic systems; spatial (left-right) asymmetries also occur and will be discussed later.

The cycle in some forms of rhythmic activity is highly predictable in that activity is arrested and restarted at one position, there being a single 'decision point' per cycle (Dawkins and Dawkins, in press). For example, the wings of many insects are folded at the end of flight and fresh flight starts from a constant position with an upstroke (Pond 1972). Some insects, however, are able to arrest flight at more than one possible position; dragonflies can start fresh flight with an upstroke or a downstroke (Pond, unpublished). Some libellulids, when on the look-out for prey, perch on a branch with the wings in an extremely depressed position (e.g. *Trithemis* sp.) maintained by tonic contractions in depressor muscles against wing hinge elastic forces. The continuous muscular activity required may serve a thermogenic function and thus allow rapid take-off in pursuit of prey in these dragonflies which, unlike larger gomphids and aeshnids, make little use of thermogenic shivering. The same species may also bask and roost with the wings more or less horizontal. Fresh flight can therefore be initiated from more than one point in the cycle (personal observation).

Bouts of ventilatory activity in many species are usually terminated at the end of inspiration. Inspiration is brought about or assisted by elastic forces and the point of normal arrest therefore corresponds to one of minimal energy expenditure. Ventilation may stop temporarily elsewhere in the cycle, and fresh strokes can start from more than one place. The ability to freeze a rhythmical activity at any moment is most strikingly seen in walking. The oscillators which are believed to pattern the walking output (Pearson and Iles 1970; Delcomyn 1971b) must be capable of being braked or disconnected from the motor output at any point in the cycle, to be restarted or reconnected at that point when walking is resumed. This problem has recently been discussed by Land (1972) and will be mentioned again when we consider the coupling of motor neurones to oscillators.

INTERACTIONS OF OSCILLATORS

Oscillators may interact and affect each other's output when there is some form of coupling or energy flow between them. A useful means of detecting weak interactions between oscillating systems has recently been described by Hughes (1972). Strong interactions may cause oscillators to synchronize to a stable output frequency as may occur in the flight system (Wilson and Wyman 1965; Wilson 1966b). Weaker interactions may be reflected in temporary periods of synchronization or in changes in frequency in one or both oscillators. Frequency division may also occur in which one oscillator cycles at a frequency which is a simple multiple of that of another; for example, locust hindlegs sometimes step at half the frequency of the mid and front legs.

Oscillators may cooperate by acting together through a common output channel. For example, four oscillators may cooperate to produce the courtship sequence of *Gomphocerippus rufus*: one initiates the sequence, a second times the start of each of the three sub-units, a third programmes the slower movements and a fourth the rapid movements (Elsner and Huber 1969; Elsner 1973). In the stridulation of *Gryllus*, two oscillators apparently cooperate: a fast oscillator with a frequency of *c*. 30 Hz controls the pulse output in each chirp and a slow oscillator with a frequency of 2-4 Hz governs the chirp duration and the interchirp interval. The fast oscillator is thought also to control the flight output, while the slow oscillator may also affect walking and ventilation which can occur at the same frequency as the chirps (Fig. 2) (Huber 1960a; Kutsch 1969; Bentley 1969b). When two activities occur at the same frequency, they may both be driven by one oscillator, or their separate oscillators may have become tightly coupled. A further example of oscillators acting through a common channel comes from the locust where the motor neurones supplying the dorsal longitudinal muscles of the abdomen may simultaneously be affected by the flight and the ventilatory oscillators; the units fire at the flight frequency but only during expiration (Hinkle and Camhi 1972). Bentley (1969a) found many units in cricket thoracic ganglia which were capable of responding, but at different times, to the flight and ventilatory oscillators.

In some species oscillators in separate individuals may interact. The remarkable displays of some fireflies (*Pteroptyx* sp.), which congregate in large numbers along Malaysian river banks and accurately synchronize their flashing, provide one example (Buck and Buck 1966; Lloyd 1971). The males and females of some other species of fireflies exchange

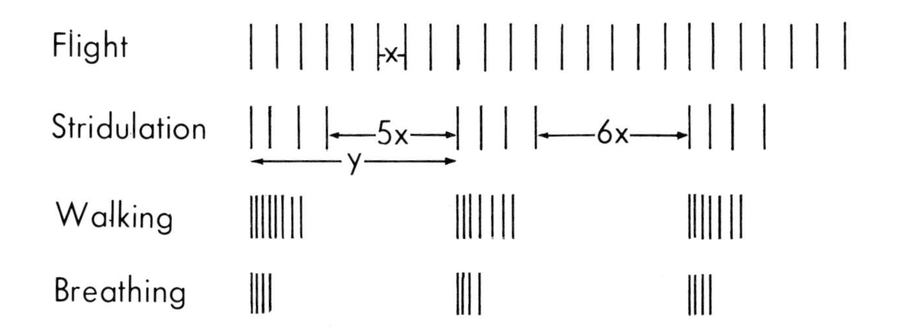

Fig. 2. Diagram illustrating the interrelationship of some rhythms in a cricket. A fast 30 Hz oscillator, x, is responsible for the flight and stridulatory pulse intervals. A slow 3-4 Hz oscillator, y, controls the interchirp interval and the stepping and breathing rhythms. (From Kutsch 1969)

flashes at precise intervals as part of a species recognition procedure. In some bush crickets (*Pholidoptera* sp.), the chirps of two males may alternate or become synchronized (Jones 1966). Many further examples are probably to be found among courtship and mating behaviour.

Displays used in courtship are often important species-isolating mechanisms. They depend, for their success, on accurate identification of the signal by the recipient. Temporally-patterned courtship behaviour employing visual (e.g. firefly flashing or wing waving in Sepsidae), tactile (e.g. Cerambycidae - Michelsen 1967), or auditory (stridulation) communication presumably depends on a pattern-selecting receiver which compares the input with an inherited template. The template may take the form of an oscillator which matches the oscillation of the sender (Ewing 1970; Bailey and Robinson 1971). The observation that a female cricket at a different temperature from the male may not respond to his song suggests that the postulated coding and uncoding oscillators are temperature dependent (Walker 1957). Pulse-interval structure (Zaretsky 1972), or pulse or chirp rate (Walker 1957) may be important for species recognition among cricket species, and in the tettigonid, *Homorocoryphus*, pulse repetition rate (Bailey and Robinson 1971), and in *Drosophila* pulse interval (Bennet-Clark and Ewing 1969) are important. Stout and Huber's (1972) observation, that the maximum response to the male's song recorded in auditory interneurones of the female corresponds to one phase of her ventilatory cycle, suggests some interaction between the ventilatory oscillator and the postulated auditory receiver.

COMMAND SYSTEMS

Unpatterned activity in command interneurones causes the expression of spatially and temporally patterned activity from groups of motor neurones. Such activity may be initiated by excitation or by release from inhibition, and command cells may remain active throughout supplying continuous drive and perhaps controlling the frequency of the output. The imprecise nature of these statements underlines the paucity of information about insect command systems. However, more is known about them in Crustacea. Single units which control abdominal position have been studied in lobsters by Evoy and Kennedy (1967) and by Larimer and Kennedy (1969a,b). The frequency of swimmeret beating in lobsters has been shown to be regulated by five command fibres, some of which control different parts of the frequency range (range fractionation) (Atwood and Wiersma 1967; Davis and Kennedy 1972). Command interneurones which release complex patterned activity receive little direct sensory input, whereas those ordering simpler movements are affected more directly by input. Wiersma (1952) showed that the defence posture of a crab, involving muscles in many parts of the body, was under the control of a single command fibre. The elaborate deimatic response of large mantids when confronted with a predator may perhaps be controlled in the same way, although this aspect has not been studied (Maldonado 1970).

Although some command fibres are believed to arise in the brain, central programmes located in segmental ganglia do not necessarily depend on the brain for their expression. For example, decapitated insects have been shown to be able to fly, walk, swim, jump, groom, copulate, oviposit, moult and breathe; headless crickets can produce calling, courtship and aggressive songs (Kutsch and Otto 1972). Some activities are performed incessantly after brain damage, brain removal or decapitation: for example, heat lesions applied to a small area of the mushroom bodies of crickets release continual calling song (Bentley 1969b), and decapitated locusts (Rowell 1964) or cockroaches (Eaton and Farley 1969) spend more than the normal amount of time on certain forms of grooming. Excitation of a behaviour pattern in some cases may therefore consist of raising a cluster of neurones above a general inhibitory background (Rowell 1969). Bentley (1969a) found many motor neurones in cricket ganglia which received continual barrages of inhibitory post-synaptic potentials (IPSPs), perhaps representing part of a general inhibiting system. The neural machinery which controls flight and stridulation in crickets, and flight in locusts, is perfected some time before the skeletal development is complete; continuous inhibition apparently prevents the inopportune expression of stridulation and flight patterns before the wings are fully grown (Bentley and Hoy 1970; Kutsch 1971; Kutsch, this volume).

Brain stimulation through implanted electrodes can evoke a variety of forms of behaviour (e.g. Otto 1969), but the results cannot readily be interpreted in terms of the activation of specific command fibres and may depend as much on the intensity of current used as on the location of the electrodes. The three forms of cricket song can each be produced by brain stimulation (Fig. 3), and Otto (1971) has suggested

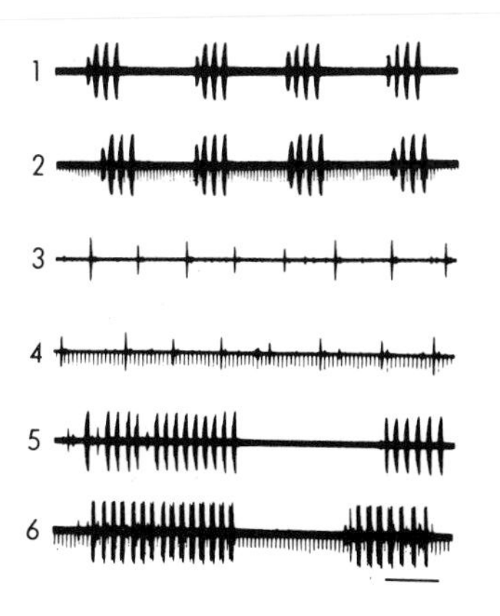

Fig. 3. The three types of song in *Gryllus campestris* are illustrated. In 1, 3 and 5, they are produced naturally. In 2, 4 and 6, they are produced by brain stimulation. (From Otto 1971)

that selection of one song pattern may depend on an intensity switch operated by a group of command fibres which select in order of increasing intensity, first calling, then courtship and finally rivalry song. Stimulation of whole or split connectives may give rise to simpler and more easily analysed activity; for example, particular forms of patterning in leg motor neurones (Elsner 1970) or changes in the ventilatory output (Miller 1966). This approach seems to offer more hope of unravelling the command systems as they run between ganglia.

Arousal is a little understood phenomenon in insects. It may be controlled in part by the totality of sensory input (Hoyle 1970), acting possibly through specific interneurones and affecting many motor systems. In addition, certain highly specific forms of input may promote increased locomotory activity and they may also heighten the general state of arousal perhaps by lowering sensory and motor thresholds. For example, when the female cricket *Scapsipedus* hears the calling song of the male, her first responses are emergence from the burrow and increased locomotory activity. Her subsequent orientation to the male may involve a separate mechanism (Zaretsky 1972). Zaretsky has shown that the songs of different but sympatric species may also produce increased locomotion but not orientation towards the male. If the different species sing at the same time, the resulting increase in locomotion may improve the chances of finding a male of the right species.

Similarly pheromones, which "attract" the opposite sex, may in some cases only stimulate locomotion, other mechanisms being then responsible for the subsequent orientation (Kennedy 1971). A type of command fibre may therefore be envisaged which has a highly selective input but a less specific output affecting the thresholds of the locomotory and perhaps other systems; that is, its activity arouses the insect (cf. Rowell 1971).

In some cases hormones may by-pass a command fibre system and evoke activity directly. One of the best documented examples is the evocation of pre-eclosion behaviour in saturniid pupae which has already been mentioned. Corpus allatum hormone acts directly or indirectly to release mating behaviour in mature adult male *Schistocerca* and some other acridids (see Pener, this volume), and other examples probably await description.

COORDINATING SYSTEMS

Interneurones which relay patterned information from an oscillator to motor neurones are termed coordinating interneurones. The contributions of command and coordinating interneurones have been thoroughly examined in the crayfish swimmeret system (Stein 1971), but in no insect system is there equivalent information. Three ways in which intersegmental coordination can be achieved may be envisaged. Intersegmental reflexes may play a part, for example in the control of walking, but deafferentation of cockroach legs has shown that reflexes are not the sole coordinating mechanism (Pearson and Iles 1970); similarly, intersegmental coordination is not lost in the deafferented flight (Wilson 1965) of breathing systems (Miller 1966). Motor neurone collaterals running between ganglia might coordinate some activities, but although some motor neurone somata lie in the ganglion anterior to that from which their axons emerge (e.g. Lewis, Miller and Mills 1973), there is no evidence that they play a role in intersegmental coordination. In most cases it seems then that interneurones are used to unite the activity of different ganglia into a common pattern.

Strong ventilation in *Schistocerca* is brought about by a stroke which is more or less synchronized in all abdominal segments. However, measurements of the emission of motor bursts to expiratory and inspiratory muscles have shown that anterior segments tend to fire earlier than posterior ones, with the metathoracic ganglion normally in the lead. In many other forms of rhythmic activity, posterior segments fire earlier and activity sweeps anteriorly. This occurs, for example, in cockroach walking, bug swimming, locust flying, dragonfly larval ventilation, crayfish swimmeret beat and *Limulus* gill ventilation; it may also occur during weak locust ventilation. However, in most of these examples, except flight, intersegmental delays are relatively long and activity is metachronal.

Bursts of impulses, phase-locked with the expiratory motor bursts but starting earlier, can be recorded in the abdominal connectives of *Schistocerca* during normal ventilation. They are unchanged after deafferentation, occur in advance of all expiratory motor activity and travel posteriorly passing through abdominal ganglia without delays; individual spikes are identifiable along much of the connective. The bursts, which are abolished by extirpation of the metathoracic ganglion, are believed to occur in two interneurones, one in each connective, which are activated by a burst-forming oscillator in the metathoracic ganglion. They are thought to coordinate the pumping cycle in each abdominal ganglion by exciting expiratory motor neurones and at the same time weakly inhibiting inspiratory neurones. The latter are strongly inhibited directly or indirectly during the expiratory bursts.

Evidence for a similar coordinating system has been obtained from the cricket (Huber 1960b) and cockroach (Farley, Case and Roeder 1967).

MOTOR SYSTEMS

New information about the morphology of motor neurones has recently become available from the use of Procion yellow and cobalt chloride which can be introduced into a neurone from an intracellular electrode immediately after recordings have been taken. Neural structure can therefore be correlated with physiological properties. However, some of the results, discussed elsewhere by Altman and Tyrer in this volume, have so far baffled the physiologist. They show ramifications of some motor neurones which may extend through much of one half of a ganglion and were hardly anticipated from older studies using methylene blue or silver stains. The physiologist is presented with potential areas for synaptic interaction far in excess of the demands of his models (Bentley 1970; Iles 1972; Burrows 1973a,c).

Information about the activity of motor neurones has been gained from the use of electrodes implanted into the muscles of relatively unrestrained insects, and the electromyograms so obtained tell us much about the patterns of impulses each muscle receives, but it does not elucidate the sources of that activity. However, the use of micro-electrodes passed through the ganglion sheath into the somata of neurones, by which attenuated spikes and at least some of the synaptic activity of a cell can be recorded, is beginning to yield more information about events within the CNS (Hoyle 1970; Iles 1972; Hoyle and Burrows 1973; Burrows 1973a,b), and Burrows's work on flight has already been mentioned.

Direct interactions between motor neurones could alone account for the patterning of some central programmes, but the evidence suggests that it only contributes to or reinforces some aspects of output. Electrical or chemical coupling between synergistic flight motor neurones has been reported in some insects (Kendig 1968; Bentley 1969a; Mulloney 1970a), although none was found in the units examined in *Chortoicetes* (Burrows 1973b). Both positive and negative interactions were shown to exist between motor units in the stridulatory system of the cricket (Bentley 1969a,b). Similar interactions may explain some of the features of the neural output to the myogenic flight mechanism of some flies (Wyman 1966; Wyman 1969a,b; Wilson 1966b, 1968a; Mulloney 1970a) which Wilson (1965) has suggested may have evolved from a neurogenic system by the weakening or loss of connections between neurones. The fly flight pattern exhibits three main features: (1) Common frequency trends among several motor neurones probably brought about by common driving from interneurones (Wyman 1970). (2) Phase multistability, in which there are periodic abrupt transitions to new phase relations between syngergists to the same muscle, perhaps explained by strong inhibitory coupling between the units. (3) Phase sliding, in which gradual changes of phase between pairs of units which supply the same muscle occur. Some phase relations are not represented and these may reflect inhibitory zones, caused perhaps by weak inhibitory coupling between the units. Alternatively, some of these aspects may be explained on the basis of coupling via interneurones as in Burrows's model (1973b) (Fig. 1). For example, phase sliding may have evolved from a neurogenic system by the weakening or loss of the action of each delay oscillator on the contralateral neurone so that the pair of homologous motor neurones is partially uncoupled. The presence of weak inhibitory coupling between the units, indicated by the inhibitory zones, remains a problem in both types of explanation and was considered by Wilson (1966b).

It is interesting to note that the provision of a delay, which is an important element in some oscillating systems with antagonistic sets of motor neurones, and is probably undertaken by interneurones in an unexplained way in the neurogenic system, has been transferred to the periphery in the myogenic system where it is provided by a special property of the stretch-activated fibrillar flight muscles (Pringle 1967).

THE COUPLING OF MOTOR NEURONES TO OSCILLATORS

The heading suggests that motor neurones do not themselves contain the oscillators which pattern the output in rhythmic systems, and this has been implied in much of the discussion so far. Evidence for this conclusion in the flight, walking and breathing systems comes firstly from the anti-dromic stimulation of motor axons (not necessarily a reliable technique - Mulloney and Selverston 1972) which may reveal interconnections between motor neurones but fails to perturb the basic pattern. Secondly, it comes from the recording of activity with extra- and intracellular electrodes (Wilson 1964; Kendig 1968; Pearson and Iles 1970; Lewis, Miller and Mills 1973; Burrows 1973b). In the flight system, for example, oscillations of potential without spikes have been observed in some flight motor neurones before or after flight; they may occur at the flight frequency at times when no flight motor activity is detectable and are thought to be explained by the activity of an interneuronal flight oscillator which drives the motor neurones at sub-threshold levels (Bentley 1969a; Burrows 1973b). The flight motor neurones are apparently uncoupled from the oscillator, and Pond (1972) considers that at the start of flight in cockroaches and locusts the initiation of oscillatory activity is a separable phenomenon from the coupling of motor neurones to it. She found that the first few flight cycles were sometimes longer than normal and that some units were not initially coupled whereas others showed abnormal phasing or burst length.

Some stridulatory motor neurones in the cricket may be regularly uncoupled from the oscillator for short periods. This is indicated by the observation that the pause length between chirps corresponds exactly to a multiple of the basic period of 30-40 msec (Fig. 2). Moreover, gaps which appear in the calling song are a whole number of chirps in length, the next chirp coming in at precisely the right moment (Fig. 4) (Bentley 1969b; Kutsch 1969). Both the slow and fast stridulatory oscillators, therefore, free run when uncoupled without a change in frequency. Feedback from motor neurones does not seem to play a part in the maintenance of oscillator frequency, at least for short periods. Similarly, the uncoupled flight oscillator maintained its frequency for many cycles.

Intracellular records from the somata of ventilatory motor neurones in the posterior part of the metathoracic ganglion of *Schistocerca* (Burrows, unpublished), and in the first abdominal ganglion (Mills, unpublished) show that at times some cells are driven by the ventilatory oscillator without spiking. They show undulations of potential at the ventilatory frequency apparently resulting from barrages of synaptic potentials. As ventilation becomes stronger, spikes start to appear on the crests of the depolarisations and the neurones contribute to the motor output.

Fast axons may be recuited into expiratory bursts during hyper-ventilation, and they usually fire later than slow axons. This agrees with Henneman's (1968) size principle for vertebrates which states that

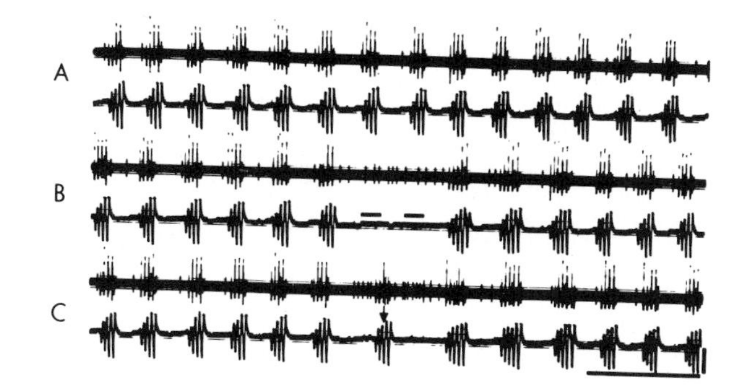

Fig. 4. Intracellular records from a presumed motor nerve in a cricket during the generation of calling song (lower lines), together with extracellular records from motor nerves to stridulatory muscles (upper lines). In B, a gap occurs in which two chirps are missed and the next chirp comes in three intervals later. In C, an out-of-phase chirp occurs (arrow) but the rhythm is resumed as though two chirps had been missed. Horizontal scale, 1 sec. Vertical scale, 10 mV. (From Bentley 1969b)

large motor neurones have higher thresholds than small ones. The same has been found to be true for some arthropod systems (Davis 1971; Hinkle and Camhi 1972). In locusts and other insects, ventilatory recruitment extends to new muscle systems, for example those which bring about longitudinal abdominal pumping in *Schistocerca*, and even fresh segments may be caught up in the rhythm when head and prothoracic pumping appear (Miller 1971a). Motor neurones to mouthparts, legs and wings may also become recuited when ventilation is strong, although their action apparently serves no function in breathing (Hoyle 1964; Bentley 1969a).

With strong ventilation, the locust rhythm becomes more regular and intersegmental delays show less variation (Lewis, Miller and Mills 1973). Similarly, in the gill ventilation of *Limulus* (Fourtner, Drewes and Pax 1971) and in the swimmeret beat of crayfish, intersegmental delays shorten as frequency of beating rises and the output may become more regular. If feedback from motor neurones onto the oscillator plays a part in the stabilisation of the rhythmic output in these slower-acting systems, then the recruitment of additional units may account for the increased regularity which often accompanies a rise in frequency and amplitude.

In rhythmic systems which act through sets of antagonistic motor neurones, one set may be able to remain silent while the other continues to produce regular bursts. For example, the elevators of the locust flight system may continue to fire bursts when the wings are held down, even though the depressors remain silent (Waldron 1967a,b). Likewise, in the cockroach walking system, coxal depressor bursts can occur without elevator bursts (Pearson and Iles 1970), and isolated inspiratory bursts may occur without expiration in the locust (Hustert, personal communication). Courtship song of *Gryllus* is brought about by rhythmic bursts in the opener muscles (subalars and basalars) which produce soft sounds, and there are only occasional bursts in the closers (dorso-ventrals) which give loud clicks (Kutsch 1969). These examples from four systems indicate that alternating action in sets of antagonists is not an essential for the production of the rhythm, and they show a lability of coupling which in the stridulatory system can be used to produce a significantly different behavioural output.

CHANGES IN THE PHASES OF COUPLING OF MOTOR NEURONES TO OSCILLATORS

In several systems, not only is the coupling of motor neurones labile but also the timing or phase of their activity within the cycle can be varied. At the start of some rhythmic activities the phasing of some units seems to be less strictly controlled than it is later, the correct pattern appearing after the first few cycles (e.g. Pond 1972). At the start of cricket stridulation, basalar units may fire synchronously as they do in flight, but they then switch to the correct stridulation pattern in which they fire alternately (Kutsch 1969). Some mantids (Roeder 1937) and stick insects (Graham 1972) occasionally start to walk with the legs of a segment stepping in synchrony, but soon switch to alternate movements. When mantids swim, the legs of a segment are moved synchronously throughout (Miller 1972); at the start meta- and mesothoracic legs sometimes all stroke together before the normal pattern is established in which metathoracic legs stroke 90-180° ahead of mesothoracic legs.

These examples illustrate 'mistakes' which are soon rectified. However, new behaviour patterns may evolve through the preservation of 'mistakes' such as these. For example, changes of coupling may have been the means whereby the rich variety of songs in some katydids arose (Walker and Dew 1972), and they may likewise have allowed the development of two walking gaits in stick insects (Fig. 5) (Graham 1972). 'Incorrect'

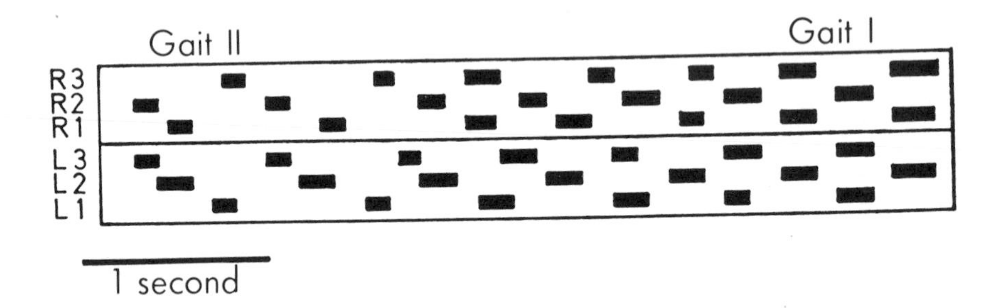

Fig. 5. The stepping patterns in a first instar stick insect. At the start, gait II appears in which diagonal pairs of legs are moved and a metachronal sequence appears in the legs of one side. Gait II then changes to gait I in which three legs are protracted at one time in the alternating tripod pattern which is common in many insects. (From Graham 1972)

phasing of flight motor units at the start of flight may have been preserved in sphingids and some other large insects to provide them with a high-frequency and low-amplitude output suitable for use in warming up before flight (Kammer 1968, 1970). In some sphingids, units which fire antagonistically in flight may act synchronously during warm-up and they slide into the normal phase relationships as warm-up changes into flight; the lability of phase may reappear during flight when turns are being made (Kammer 1971). Separate command systems may be responsible for biassing the system either to warm-up coupling, or to flight coupling (Hanegan 1972). In *Bombus*, which has a myogenic flight system, preflight warm-up is sometimes produced by well-synchronized bursts of spikes in several units, whereas flight patterns consist of unsynchronized but regular firing in the same units (Fig. 6) (Mulloney 1970b; Kammer and Heinrich 1972). In the warm-up of bumblebees, units which would be expected to act as antagonists in a neurogenic flight system fire synchronously and warm-up pattern seems to preserve more of the characteristics of a neurogenic system than does the flight pattern.

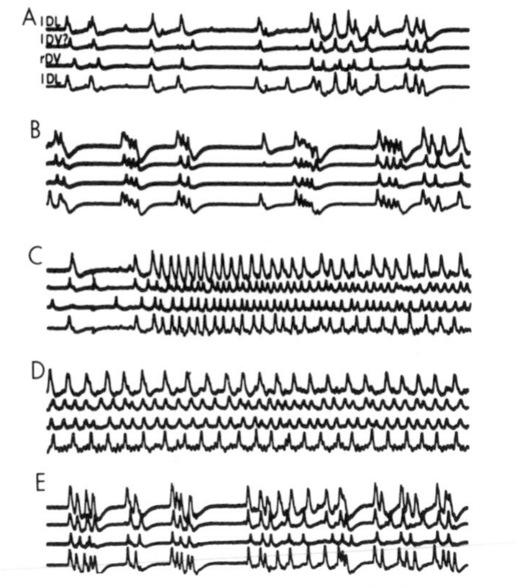

Fig. 6. Electromyograms from the flight muscles of *Bombus sonorus*. A and B, show patterns of pre-flight warming-up activity. C, shows the transition to flight activity. D, flight activity. E, after flight a resumption of shivering activity which warms the thorax. Horizontal scale, 200 msec. (From Kammer and Heinrich 1972)

A further example of a change in coupling with behavioural significance comes from the courtship sequence of *Gomphocerippus*. In sub-unit 1 of the sequence, the head makes regular shaking movements produced by alternate bursts of firing in muscles on either side. At a later stage the movements cease, but recordings show that the motor units are not silent but each fires bursts synchronously with the other at the head-shaking frequency and the head remains stationary (Elsner and Huber 1969; Elsner 1973).

Changes in the phases of spiracle muscle activity during the ventilatory cycle in locusts, cockroaches and mantids can produce reversals of the direction taken by the ventilating air current through the insect. In the cockroach *Blaberus giganteus*, for example, the airstream is normally directed posteriorly and abdominal spiracles open during the expiratory stroke to emit air. Carbon dioxide enhances the flow but perfusion with gas mixtures containing less than about 5% oxygen cause the stream to be reversed and the abdominal spiracles now open with inspiration and admit air. Activity has been examined in the last pair of abdominal spiracles (Nos. 10) where the opener muscles play a major role in controlling the valves (Miller 1973). They are innervated separately by paired axons from the last abdominal ganglion. Inspiratory valve opening is produced by a relatively low-frequency burst of spikes in a fast axon in left and right spiracles. Expiratory opening, however, is caused by a high-frequency burst of spikes in the same axon on either the right or the left side, and only one spiracle of the pair opens. Thus reversal of the ventilating airstream involves both a temporal and a spatial alteration of the coupling of the opener motor neurones to the ventilatory oscillator. Transitions between the two forms of activity may occur, one valve then opening twice per cycle. The asymmetrical motor output during expiratory coupling cannot result from direct inhibitory coupling between the motor neurones because it is not abolished when the motor neurones are separated by saggital section of the ganglion. Two sets of coordinating interneurones running between a metathoracic ventilatory oscillator and the motor neurones have been postulated. One set possesses reciprocal inhibitory coupling and harnesses one of the two spiracles to expiration; the other has no such coupling and causes symmetrical coupling in both spiracles to the inspiratory phase of ventilation.

A quiescent *Blaberus* shows intermittent ventilation, with either the left or right spiracle 10 opening with expiration. Which spiracle shows activity may be determined partly by asymmetrical sensory input. Cercal or leg stimulation has been found to cause the ipsilateral spiracle 10 to open with expiration (i.e. to be dominant). Section of one thoracic connective causes the contralateral spiracle to be continuously dominant, whereas electrical stimulation of a thoracic connective produces ipsilateral dominance. In unstimulated cockroaches one spiracle may remain dominant for long periods, perhaps reflecting a weak innate asymmetry in the CNS. An innate asymmetry in the motor output has been described in the locust flight system (Wilson 1968b; cf. Goodman 1965), in the walking output of *Oncopeltus* (Chapple 1966) and in the stridulatory output of crickets (Kutsch and Huber 1970). In flight and walking, it is believed to be compensated for normally by sensory input. Imperfections in locomotory pattern generators can therefore be tolerated because sensory input acts to trim and adjust the output, and also allows it to compensate for unilateral damage. Wilson (1968b) has shown that in a locust flying in the dark and free to rotate in the longitudinal axis, the removal of one hind wing does not perturb the motor pattern, and the insect rolls in one direction. In the light it attempts to maintain a steady flight visually by cutting down the output to the intact side (Fig. 7).

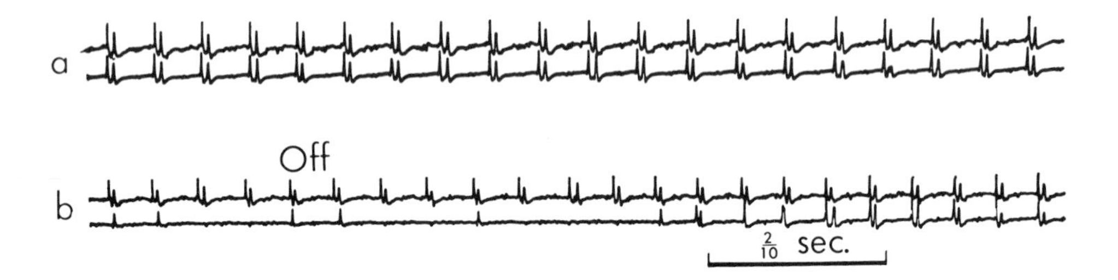

Fig. 7. Electromyograms from the flight muscles (two hindwing subalars) of *Schistocerca*. One hindwing has been removed. a, flight in the dark; the output to both hindwings is symmetrical. b, flight with the light on; the output on the intact side is almost abolished as the locust 'attempts' to keep on an even course. At "off" the light is extinguished and a symmetrical output is soon resumed. (From Wilson 1968b)

In the cockroach spiracle system, however, an innate asymmetry or one derived from sensory input is amplified by the neural connections to produce a strongly asymmetrical output. A similar amplification of asymmetrical input has been described in the gintrap reflex of sphingid pupae by Bate (personal communcation), and is also known in the tettigonid auditory T-interneurones (Suga and Katsuki 1961). Murphey (1971) suggested that reciprocal inhibitory interactions occurred between the sensory paths responding to ripples from prey on the two sides of *Gerris*, producing a response towards the more strongly stimulated side. However, while these examples can be interpreted functionally, the significance of asymmetrical spiracle action in *Blaberus* is unknown, although it may be related to the unilateral expulsion of noxious chemicals at predators from spiracle glands (Roth and Stay 1958).

The coupling of motor neurones to oscillators has, therefore, been shown in these examples to be both variable and labile in phase. As already suggested, new behaviour patterns may have evolved from such alterations. The probable derivation of warm-up activity from flight may have taken place several times in sphingids (Kammer 1970).

Stridulation in crickets has probably arisen out of flight (Huber 1962), and in other insects it may have evolved from different types of rhythmic activity (e.g. Miller 1971c). There are, no doubt, many other examples which await description, in which an activity is lifted from one behavioural context and placed in another as a result of small changes in coupling.

INTERNAL AND EXTERNAL FEEDBACK

Feedback may be derived internally, that is from within the CNS, or externally from receptors at the periphery. Internal feedback has already been mentioned and it forms an important element in Burrows's (1973b) model (Fig. 1). It arises from motor neurone activity and may act on oscillators or other pre-motor cells to adjust or reinforce their output. The role of internal feedback may be tested by passing current across the soma of a motor neurone so as to alter its spike output. Internal feedback probably plays an important part in the execution of many insect central programmes, particularly when precise timing is required as in flight and stridulation (Bentley 1969b). Fresh attention is being paid to the part it may play in vertebrate motor output (Evarts 1971).

External feedback may act on either the motor neurones, the oscillator, or possibly on the command system. Receptor input which is rhythmically patterned by the output may be able to alter the motor discharge phasically within one cycle, as is known for example in the lobster swimmeret system (Davis 1969b,c) and in insect walking (Pearson 1972), or it may affect the next cycle after the one which produces it, as in the synchronization mechanism of flashing fireflies (Buck and Buck 1968). Alternatively, phasic input may have only a tonic effect on the output as for example in the action of the stretch receptors on locust wing beat frequency (Wilson and Gettrup 1963), or in the regulation of forewing twisting by hindwing campaniform sensilla (Gettrup 1966). Tonic unpatterned input can alter the output only in a tonic manner; wind receptors on the locust head for example affect wing beat frequency (Weis-Fogh 1956).

The execution of some central programmes is remarkably unaffected by sensory disturbance or deprivation. For example, the flight, breathing and pre-eclosion programmes can be executed after motor nerves have been cut, and even when ganglia are isolated. The timing and basic patterning of the complex motor output determining courtship in *Gomphocerippus* are hardly affected by amputation, fixation or loading of the hind legs, which normally participate in the behaviour (Loher and Huber 1966; Elsner and Huber 1969). However, visual stimuli can interrupt it at some stages, and gluing the head and prothorax to the pterothorax blocks its expression completely. Similarly, the stridulatory pattern of *Gryllus* is largely unaffected by loading, fixing or removing the tegmina, or even by cutting the motor nerves (Kutsch and Huber 1970). Lacquering the subcostal sensory pads prevents the adoption of the singing position but the motor pattern is unchanged even though no song is produced (Möss 1971).

Some examples where the timing of the oscillation can be altered phasically by input are known. In locust flight, Waldron (1967b) showed that a stroboscope flashing at within 3% of the frequency of natural flight could pace the output, and recently Wendler (1972) has applied forced oscillations to one locust wing at near the flight frequency and finds that they alter the phase of firing of some units to other wings. The cycle in some slower rhythms can more readily be re-set by phasic stimuli. For example the ventilatory cycle can be re-set and paced in dragonfly larvae by electrical stimulation of lateral nerves (Mill 1970), in cockroaches by electrical and mechanical stimuli (Farley and Case 1968), and in mantids by electrical, mechanical and visual stimuli, but only when ventilation is weak (Miller 1971b).

Much attention has recently been paid to the interaction of central and peripheral mechanisms in insect walking, even though a quantitative description of cockroach walking, one of the favourite victims, has only just become available (Hughes 1952; Wilson 1966a; Delcomyn 1971a,b,c). The existence of a central programme for walking in cockroaches was demonstrated by Pearson and Iles (1970) by showing that partially de-afferented cockroaches could produce alternating bursts in leg levator and depressor motor nerves with many of the characteristics of walking in the intact insect, although the high frequencies of cockroach running (25 steps/sec) could not be achieved. Evidence for a central programme in *Oncopeltus* was obtained by Hoy and Wilson (1969) and it may be a feature of most insects. However, such a programme interacts with peripheral input, on which the proper execution of the movements depends, because each step has to be adjusted to an unpredictable terrain. Wendler (1966) showed that the destruction of hairplates in stick insects led to overstepping, and the walking pattern of locusts can largely be disrupted by sensory ablation (Runion and Usherwood 1968). Usherwood and Runion (1970) have shown that the metathoracic tibial extensor muscle of *Schistocerca* is controlled during walking by activity in a slow axon and an inhibitory axon. The output of the slow axon is strongly affected by input from a chordotonal organ and from tarsal receptors; input from the tarsal receptors also affects the inhibitory axon. The two axons fire alternately in walking, the inhibitory axon becoming active just before the flexors fire, thereby hastening the relaxation of the extensor. Similarly, in cockroaches, activity in the three common inhibitory nerves which supply parts of the coxal depressor and other muscles, accelerates the relaxation of the depressor during walking (Iles and Pearson 1971). The phasic input from the trochanteral campaniform sensilla reinforces the motor activity in the slow axon to the coxal depressor, and it also seems to act on the oscillator to slow the cycle, particularly when the cockroach is made to pull a load or walk uphill (Pearson 1972; see also Delcomyn 1971c). In contrast, tonic receptors accelerate the cycle and abbreviate depressor bursts.

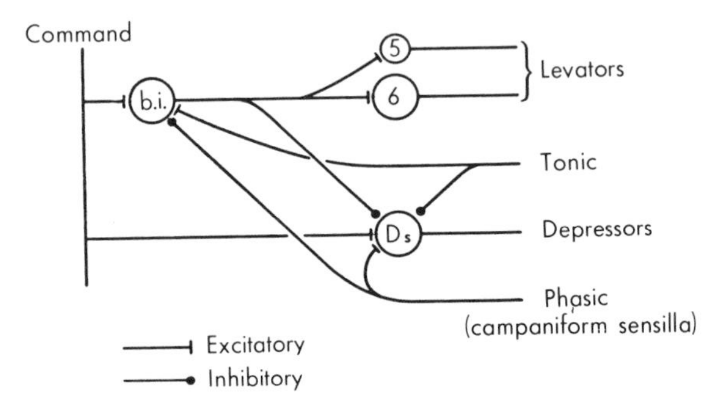

Fig. 8. A model which accounts for some aspects of the discharge patterns to the levator muscle axons 5 and 6, and to the depressor muscle axon D_S during walking in a cockroach. A bursting interneurone (b.i.) simultaneously excites the levator units and inhibits the depressor unit. During the interburst the depressor fires due to post-inhibitory rebound. Trochanteral campaniform sensilla excite the depressor and inhibit the oscillator phasically, while tonic receptors produce the reverse effects. (From Pearson 1972)

Pearson and Iles's (1970) model to explain the basic patterning of cockroach walking consists of a bursting interneurone whose activity is maintained by a command fibre; it simultaneously excites levator muscles and inhibits depressors, and during the interburst the depressors fire as a result of post-inhibitory rebound. Pearson (1972) added some sensory elements to the model and his amended version is shown in Fig. 8. Delcomyn (1971b) postulated a similar model but it depended on a sinusoidal oscillator in place of the bursting interneurone as already mentioned. Similar models with an asymmetrical action on the antagonistic sets of motor neurones have been suggested for cricket stridulation (Bentley 1969b) and for locust ventilation (Lewis, Miller and Mills 1973).

Land (1972) has made an examination of stepping in salticid spiders, particularly during turning responses to moving objects, and has postulated a model which contains no separable pattern-generating oscillator which could programme walking in the absence of input. The model is, therefore, significantly different from that of Pearson and Iles, and of Delcomyn. In Land's model there are four oscillators for each leg, two for forward stepping and two for backward stepping, and they are driven directly by input from proprioceptors. The oscillators are continuously active when the legs are stationary at a level determined by the proprioceptors, and as was pointed out previously the cycle can be stopped and started at any point. The model is shown in Fig. 9 and

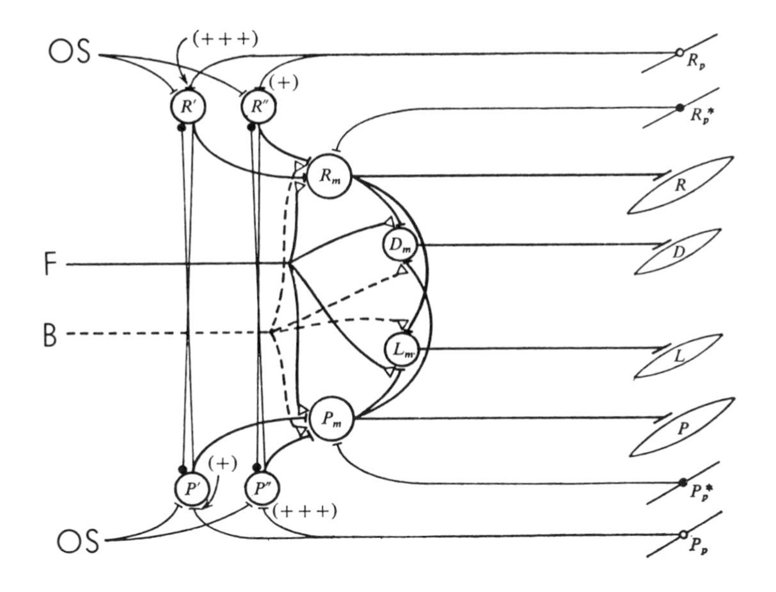

Fig. 9. A model of the neural control of stepping of one leg in a salticid spider. *R*, *D*, *L* and *P* are the retractor, depressor, levator and protractor muscles respectively. R_m, D_m, L_m, and P_m are their motor neurones. R_p, P_p, R_p*, and P_p* are proprioceptors responding to joint position and load. *R'*, *R''*, *P'*, and *P''* are four oscillators which receive input directly from the proprioceptors and from other segments (OS). They act in two pairs which mutually inhibit each other. Command fibres (F, for forwards stepping and B, for backwards stepping) facilitate the synapses (triangles) between oscillators and motor neurones. Other synaptic conventions as in Fig. 8. (From Land 1972)

described further in the legend. Salticid walking is characterised by steps of constant amplitude over a wide range of frequencies and with differing loads; the oscillator must therefore be controlled by proprioceptive input to enable the legs to produce changes in power output corresponding to the changes in frequency. Pearson (1972) showed that feedback helped to answer the same problem in cockroaches.

In a study of the gaits of stick insects, Graham (1972) recognized two distinct patterns which occurred in first instar larvae. One takes the form of alternating tripods with three legs simultaneously protracted and is characteristic of many insects. In the other, which is perpetuated as the normal stepping pattern for the adult, diagonal pairs of legs are moved and four legs remain on the ground at a time. To explain both gaits, Graham provides a model which can be applied equally well to a system driven by a central programme or to one dependent on peripheral input for the basic patterning. The model comprises three independently linked 'relaxation' oscillators on each side, one for each leg, which are connected to each other through delay oscillators. This system can produce the metachronal sequence on each side with considerable unilateral independence.

CONCLUSION

Early in this article I presented a list of the possible components of an oscillatory system - command, oscillator, coordinator, motor and feedback elements - and suggested that they might be represented by separate cell layers in some systems. It may, however, seem that the subsequent review of some features of a number of rhythmic activities has not gone far to justify the division. In many systems, oscillators cannot definitely be separated from coordinating elements or even from motor neurones. Interneurones and motor neurones may be functionally linked so that the concept of the oscillator must include the whole pool of interacting neurones, but in the flight system at least there seems to be a good chance of making a separation. More confidence can be felt in the separate nature of other elements, for example of the command and feedback neurones, even though we do not know much about how they work. The scheme however may still serve a purpose in directing attention to potentially separable elements during the early stages of analysis.

The apparent similarities and suggested common features of several rhythmical behaviour patterns which have been pointed out give more cause for optimism. They suggest that similar organizational principles may underlie various behaviour patterns in arthropods, although activities such as flight and stridulation show many differences from the slower and more variable rhythms such as walking and breathing. Mechanisms of interneural cooperation exhibit some degree of conservatism, although small changes in patterning, phasing, frequency, amplitude and recruitment may allow an activity to evolve new behavioural significance. Therefore a thorough understanding of one rhythmic system, chosen for its amenability to current physiological techniques, may go a long way towards the elucidation of other systems.

ACKNOWLEDGEMENTS

I am grateful to Drs. C. Pond, N. Elsner, R. Hustert and M. Burrows for permission to make use of their unpublished material.

REFERENCES

ATWOOD, H.L., WIERSMA, C.A.G.: Command interneurons in the crayfish central nervous system. J. exp. Biol. 46, 249-261 (1967).

BAILEY, W.J., ROBINSON, D.: Song as a possible isolating mechanism in the genus *Homorocoryphus* (Tettigonioidea: Orthoptera). Anim. Behav. 19, 390-397 (1971).

BENNET-CLARK, H.C., EWING, A.W.: Pulse interval as a critical parameter in the courtship song of *Drosophila melanogaster*. Anim. Behav. 17, 755-759 (1969).

BENTLEY, D.R.: Intracellular activity in cricket neurons during the generation of behaviour patterns. J. Insect Physiol. 15, 677-699 (1969a).

BENTLEY, D.R.: Intracellular activity in cricket neurons during the generation of song patterns. Z. vergl. Physiol. 62, 267-283 (1969b).

BENTLEY, D.R.: A topological map of the locust flight system motor neurons. J. Insect Physiol. 16, 905-918 (1970).

BENTLEY, D.R.: Genetic control of an insect neuronal network. Science, N.Y. 174, 1139-1141 (1971).

BENTLEY, D.R., HOY, R.R.: Postembryonic development of adult motor patterns in crickets: a neural analysis. Science, N.Y. 170, 1409-1411 (1970).

BUCK, J.B., BUCK, E.M.: Biology of synchronous flashing of fireflies. Nature, Lond. 211, 562-564 (1966).

BUCK, J.B., BUCK, E.M.: Mechanism of rhythmic synchronous flashing of fireflies. Science, N.Y. 159, 1319-1327 (1968).

BURROWS, M.: Physiological and morphological properties of the metathoracic common inhibitory neuron of the locust. J. comp. Physiol. 82, 59-78 (1973a).

BURROWS, M.: The role of delayed excitation in the co-ordination of some metathoracic flight motoneurons of a locust. J. comp. Physiol. 83, 135-164 (1973b).

BURROWS, M.: The morphology of an elevator and a depressor motoneuron of the hind wing of a locust. J. comp. Physiol. 83, 165-178 (1973c).

CHAPPLE, W.D.: Motoneuron responses to visual stimuli in *Oncopeltus fasciatus* Dallas. J. exp. Biol. 45, 401-410 (1966).

CHEN, C.F., von BAUMGARTEN, R., TAKEDA, R.: Pacemaker properties of completely isolated neurones in *Aplysia californica*. Nature New Biol. 233, 27-29 (1971).

DAVIS, W.J.: The neural control of swimmeret beating in the lobster. J. exp. Biol. 50, 99-117 (1969a).

DAVIS, W.J.: Reflex organisation in the swimmeret system of the lobster. I. Intrasegmental reflexes. J. exp. Biol. 51, 547-563 (1969b).

DAVIS, W.J.: Reflex organisation in the swimmeret system of the lobster. II. Reflex dynamics. J. exp. Biol. 51, 565-573 (1969c).

DAVIS, W.J.: Functional significance of motoneuron size and soma position in the swimmeret system of the lobster. J. Neurophysiol. 34, 274-288 (1971).

DAVIS, W.J., KENNEDY, D.: Command interneurones controlling swimmeret movements in the lobster. I. Types of effect on motoneurons. II. Interaction of effects on motoneurons. III. Temporal relationships among bursts in different motoneurons. J. Neurophysiol. 35, 1-12, 13-19, 20-29 (1972).

DAWKINS, R., DAWKINS, M.: Decisions and the uncertainty of behaviour. Behaviour (In press).

DELCOMYN, F.: The locomotion of the cockroach *Periplaneta americana*. J. exp. Biol. 54, 443-452 (1971a).

DELCOMYN, F.: The effect of limb amputation on locomotion in the cockroach *Periplaneta americana*. J. exp. Biol. 54, 453-469 (1971b).

DELCOMYN, F.: Computer aided analysis of a locomotor leg reflex in the cockroach *Periplaneta americana*. Z. vergl. Physiol. 74, 427-455 (1971c).

EATON, R.C., FARLEY, R.D.: The neural control of cercal grooming behaviour in the cockroach *Periplaneta americana*. J. Insect Physiol. 15, 1047-1065 (1969).
ELSNER, N.: Command fibres in the central nervous system of the grasshopper *Gastrimargus africanus* (Oedipodinae). Zool. Anz. (Suppl.) 33, 465-471 (1970).
ELSNER, N.: Central nervous control of courtship behaviour in the grasshopper *Gomphocerippus rufus* L. (Orthoptera, Acrididae). Symp. Invert. Neurobiol. Tihany. (1973).
ELSNER, N., HUBER, F.: Die Organisation des Werbegesanges der Heuschrecke *Gomphocerippus rufus* L. in Abhängigkeit von zentralen und peripheren Bedingungen. Z. vergl. Physiol. 65, 389-423 (1969).
EVARTS, E.V.: Central control of movement. Neurosci. Res. Prog. Bull. 9(1), 1-169 (1971).
EVOY, W.H., KENNEDY, D.: The central nervous organisation underlying control of antagonistic muscles in the crayfish. I. Types of command fibres. J. exp. Zool. 165, 223-238 (1967).
EWING, A.W.: The evolution of courtship songs in *Drosophila* (Diptera, Drosophilidae). Rev. Comp. Anim. Behav. 4, 3-8 (1970).
FARLEY, R.D., CASE, J.F.: Sensory modulation of ventilative pacemaker output in the cockroach *Periplaneta americana*. J. Insect Physiol. 14, 591-601 (1968).
FARLEY, R.D., CASE, J.F., ROEDER, K.D.: Pacemaker for tracheal ventilation in the cockroach *Periplaneta americana*. J. Insect Physiol. 13, 1713-1728 (1967).
FOURTNER, C.R., DREWES, C.D., PAX, R.A.: Rhythmic motor outputs coordinating the respiratory movements of the gill plates of *Limulus polyphemus*. Comp. Biochem. Physiol. 38A, 751-762 (1971).
GETTRUP, E.: Sensory regulation of wing twisting in locusts. J. exp. Biol. 44, 1-16 (1966).
GOODMAN, L.J.: The role of certain optomotor reactions in regulating stability in the rolling plane during flight in the desert locust, *Schistocerca gregaria*. J. exp. Biol. 42, 385-407 (1965).
GRAHAM, D.: Behavioural analysis of temporal organisation of walking movements in first instar and adult stick insects, *Carausius morosus*. J. comp. Physiol. 81, 23-52 (1972).
HANEGAN, J.L.: Pattern generators of the moth flight motor. Comp. Biochem. Physiol. 41, 105-113 (1972).
HENNEMAN, E.: Peripheral mechanisms involved in the control of muscle. *In* "Medical Physiology", Vol. II (Ed., V.B. Mountcastle), pp. 1697-1716. St. Louis: The E.V. Mosby Co. (1968).
HINKLE, M., CAMHI, J.M.: Locust motoneurons: bursting activity correlated with axon diameter. Science, N.Y. 175, 553-556 (1972).
HOY, R.R., WILSON, D.M.: Rhythmic motor output in the leg motor neurons of the milkweed bug, *Oncopeltus*. Fedn Proc. Fedn Am. Socs exp. Biol. 28, 588 (1969).
HOYLE, G.: Exploration of neuronal mechanisms underlying behaviour in insects. *In* "Neural Theory and Modeling" (Ed., R.F. Reiss), pp. 346-376. Stanford, Calif.: Stanford Univ. Press. (1964).
HOYLE, G.: Cellular mechanisms underlying behaviour - neuroethology. Adv. Insect Physiol. 7, 349-444 (1970).
HOYLE, G., BURROWS, M.: Neural mechanisms underlying behaviour in the locust *Schistocerca gregaria*. I. Physiology of identified neurons in the metathoracic ganglion. J. Neurobiol. 4, 3-41 (1973).
HUBER, F.: Untersuchungen über die Funktion des Zentralnervensystems und Insbesondere des Gehirnes bei der Fortbewegung und der Lauterzeugung der Grillen. Z. vergl. Physiol. 44, 60-132 (1960a).
HUBER, F.: Experimentelle Untersuchungen zur Nervösen Atmungsregulation der Orthopteran (Saltatoria: Gryllidae). Z. vergl. Physiol. 43, 359-391 (1960b).
HUBER, F.: Central nervous control of sound production in crickets and some speculations on its evolution. Evolution 16, 429-442 (1962).

HUGHES, G.M.: The co-ordination of insect movements. I. The walking movements of insects. J. exp. Biol. 29, 267-284 (1952).
HUGHES, G.M.: The relationship between cardiac and respiratory rhythms in the dogfish *Scyliorhinus canicula* L. J. exp. Biol. 57, 415-434 (1972).
ILES, J.F.: Structure and synaptic activation of the fast coxal depressor motoneurone of the cockroach *Periplaneta americana*. J. exp. Biol. 56, 647-656 (1972).
ILES, J.F., PEARSON, K.G.: Central patterning of motoneuronal activity in the cockroach. J. Physiol. (London) 204, 54-55P (1969).
ILES, J.F., PEARSON, K.G.: Coxal depressor muscles of the cockroach and the role of peripheral inhibition. J. exp. Biol. 55, 151-164 (1971).
JONES, M.D.R.: The acoustic behaviour of the bush cricket *Pholidoptera griseoaptera*. I. Alternation, synchronism and rivalry between males. J. exp. Biol. 45, 15-30 (1966).
KAMMER, A.E.: Motor patterns during flight and warm-up in Lepidoptera. J. exp. Biol. 48, 89-109 (1968).
KAMMER, A.E.: A comparative study of motor patterns during pre-flight warm-up in hawkmoths. Z. vergl. Physiol. 70, 45-56 (1970).
KAMMER, A.E.: The motor output during turning flight in a hawkmoth, *Manduca sexta*. J. Insect Physiol. 17, 1073-1086 (1971).
KAMMER, A.E., HEINRICH, B.: Neural control of bumblebee fibrillar muscles during shivering. J. comp. Physiol. 78, 337-345 (1972).
KENDIG, J.J.: Motor neurone coupling in locust flight. J. exp. Biol. 48, 389-404 (1968).
KENNEDY, J.S.: The relevance of animal behaviour. Inaug. Lect. Sd. Imperial Coll. Lond. (Nov. 1969). (1971).
KUTSCH, W.: Neuromuskuläre Aktivität bei verschiedenen Verhaltensweisen von drei Grillenarten. Z. vergl. Physiol. 63, 335-378 (1969).
KUTSCH, W.: The development of the flight pattern in the desert locust, *Schistocerca gregaria*. Z. vergl. Physiol. 74, 156-168 (1971).
KUTSCH, W., HUBER, F.: Zentrale versus periphere Kontrolle des Gesanges von Grillen (*Gryllus campestris*). Z. vergl. Physiol. 67, 140-159 (1970).
KUTSCH, W., OTTO, D.: Evidence for spontaneous song production independent of head ganglia in *Gryllus campestris*. J. comp. Physiol. 81, 115-119 (1972).
LAND, M.F.: Stepping movements made by jumping spiders during turns mediated by the lateral eyes. J. exp. Biol. 57, 15-40 (1972).
LARIMER, J.L., KENNEDY, D.: Innervation patterns of fast and slow muscle in the uropods of crayfish. J. exp. Biol. 51, 119-133 (1969a).
LARIMER, J.L., KENNEDY, D.: The central nervous control of complex movements in the uropods of crayfish. J. exp. Biol. 51, 135-150 (1969b).
LAUDIEN, G.: Über die Beeinflussung der Häufigkeit deplacierten Putzens bei der Schabe *Blaberus craniifer* (Blattaria) durch Veränderung der Versuchsbedingungen und Läsionen am Körper und am Zentralnervensystem. Z. Tierpsychol. 27, 136-149 (1970).
LEWIS, G.W., MILLER, P.L., MILLS, P.S.: Neuro-muscular mechanisms of abdominal pumping in the locust. J. exp. Biol. 58, (1973).
LLOYD, J.E.: Bioluminescent communication in insects. A. Rev. Ent. 16, 97-122 (1971).
LOHER, W., HUBER, F.: Nervous and endocrine control of sexual behaviour in a grasshopper. Symp. Soc. exp. Biol. 20, 381-400 (1966).
MALDONADO, H.: The deimatic reaction in the praying mantis *Stagmatoptera biocellata*. Z. vergl. Physiol. 68, 60-71 (1970).
MENDELSON, M.: Oscillator neurons in Crustacean ganglia. Science, N.Y. 171, 1170-1173 (1971).
MICHELSON, A.: On the evolution of tactile stimulatory actions in long-horned beetles (Cerambycidae, Coleoptera). Z. Tierpsychol. 23, 257-266 (1967).
MILL, P.J.: Neural patterns associated with ventilatory movements in

dragonfly larvae. J. exp. Biol. 52, 167-175 (1970).
MILLER, P.L.: The regulation of breathing in insects. Adv. Insect Physiol. 3, 279-354 (1966).
MILLER, P.L.: Rhythmic activity in the insect nervous system: thoracic ventilation in non-flying beetles. J. Insect Physiol. 17, 395-405 (1971a).
MILLER, P.L.: Rhythmic activity in the insect nervous system. II. Sensory and electrical stimulation of ventilation in a mantid. J. exp. Biol. 54, 599-607 (1971b).
MILLER, P.L.: A note on stridulation in some cerambycid beetles and its possible relation to ventilation. J. Ent. (A) 46, 63-68 (1971c).
MILLER, P.L.: Swimming in mantids. J. Ent. (A) 46, 91-97 (1972).
MILLER, P.L.: Spatial and temporal changes in the coupling of cockroach spiracles to ventilation. J. exp. Biol. 58, (1973).
MÖSS, D.: Sinnesorgane im Bereich des Flügels der Feldgrille (*Grillus campestris* L.) und ihre Bedeutung für die Kontrolle der Singbewegung und die Einstellung der Flügellage. Z. vergl. Physiol. 73, 53-83 (1971).
MULLONEY, B.: Organisation of flight motoneurons of Diptera. J. Neurophysiol. 33, 86-95 (1970a).
MULLONEY, B.: Impulse patterns in the flight motorneurones of *Bombus californicus* and *Oncopeltus fasciatus*. J. exp. Biol. 52, 59-77 (1970b).
MULLONEY, B., SELVERSTON, A.I.: Antidromic action potentials fail to demonstrate known interactions between neurons. Science, N.Y. 177, 69-72 (1972).
MURPHEY, R.K.: Sensory aspects of the control of orientation to prey by the water strider *Gerris remigis*. Z. vergl. Physiol. 72, 168-185 (1971).
O'SHEA, M.R.: The antennal cleaning reflex in the desert locust, *Schistocerca gregaria* (Forsk.). *In* "Proceedings of the International Study Conference on the Current and Future Problems of Acridology", London, U.K., 6-16 July 1970 (Eds., C.F. Hemming and T.H.C. Taylor), pp. 55-59 (1972).
OTTO, D.: Hirnreizinduzierte komplexe Verhaltensfolgen bei Grillen. Zool. Anz. (Suppl.) 33, 472-477 (1969).
OTTO, D.: Untersuchungen zur zentralnervösen Kontrolle der Lauterzeugung von Grillen. Z. vergl. Physiol. 74, 227-271 (1971).
PAGE, C.H.: Unit responses in the metathoracic ganglion of the flying locust. Comp. Biochem. Physiol. 37, 565-571 (1970).
PAUL, D.H.: Swimming behaviour of the sand crab *Emerita analoga* (Crustacea, Anomura). III. Neuronal organisation of uropod beating. Z. vergl. Physiol. 75, 286-302 (1971).
PEARSON, K.G.: Central programming and reflex control of walking in the cockroach. J. exp. Biol. 56, 173-193 (1972).
PEARSON, K.G., ILES, J.F.: Discharge patterns of coxal levator and depressor motoneurones of the cockroach *Periplaneta americana*. J. exp. Biol. 52, 139-165 (1970).
POND, C.M.: Neuromuscular activity and wing movements at the start of flight of *Periplaneta americana* and *Schistocerca gregaria*. J. comp. Physiol. 78, 192-209 (1972).
PRINGLE, J.W.S.: The contractile mechanism of insect fibrillar muscle. Prog. Biophys. Molec. Biol. 17, 1-60 (1967).
ROBERTS, B.L.: Spontaneous rhythms in the motoneurons of spinal dogfish *Scyliorhinus canicula*. J. mar. biol. Ass. U.K. 49, 33-49 (1969).
ROEDER, K.D.: The control of tonus and locomotor activity in the praying mantis (*Mantis religiosa* L.). J. exp. Zool. 76, 353-374 (1937).
ROEDER, K.D.: Episodes in insect brains. Am. Sci. 58, 378-389 (1970).
ROTH, L.M., STAY, B.: The occurrence of paraquinones in some arthropods, with emphasis on the quinone-secreting tracheal glands of *Diploptera punctata* (Blattaria). J. Insect Physiol. 1, 305-318 (1958).

ROWELL, C.H.F.: Central control of an insect segmental reflex. I. Inhibition by different parts of the central nervous system. J. exp. Biol. 41, 559-572 (1964).
ROWELL, C.H.F.: Central control of an insect segmental reflex. II. Analysis of the inhibitory input from the metathoracic ganglion. J. exp. Biol. 50, 191-201 (1969).
ROWELL, C.H.F.: Antennal cleaning, arousal and visual interneurone responsiveness in a locust. J. exp. Biol. 55, 749-761 (1971).
RUNION, H.I., USHERWOOD, P.N.R.: Tarsal receptors and leg reflexes in the locust and grasshopper. J. exp. Biol. 49, 421-436 (1968).
SEIBT, U.: Beschreibung und Zusammenspiel einzelner Vehaltensweisen von Stielaugenfliegen (Gattung *Diopsis*) unter besonderer Berücksichtigung des Putzverhaltens. Z. Tierpsychol. 31, 225-239 (1972).
STEIN, P.S.G.: Intersegmental coordination of swimmeret motoneuron activity in crayfish. J. Neurophysiol. 34, 310-318 (1971).
STOUT, J.F., HUBER, F.: Responses of central auditory neurons of female crickets (*Gryllus campestris* L.) to the calling song of the male. Z. vergl. Physiol. 76, 302-313 (1972).
SUGA, N., KATSUKI, Y.: Central mechanism of hearing in insects. J. exp. Biol. 38, 545-558 (1961).
SVIDERSKY, V.L.: Central mechanisms controlling the activity of locust flight muscles. J. Insect Physiol. 13, 899-911 (1967).
TRUMAN, J.W.: Physiology of insect ecdysis. I. Eclosion behaviour of saturniid moths and its hormonal release. J. exp. Biol. 54, 805-814 (1971).
TRUMAN, J.W., SOKOLOVE, P.G.: Silk moth eclosion: hormonal triggering of a centrally programmed pattern of behaviour. Science, N.Y. 175, 1491-1493 (1972).
USHERWOOD, P.N.R., RUNION, H.I.: Analysis of the mechanical responses of the metathoracic extensor tibiae muscles of free-walking locusts. J. exp. Biol. 52, 39-58 (1970).
WALDRON, I.: Mechanisms for the production of the motor output pattern in flying locusts. J. exp. Biol. 47, 201-212 (1967a).
WALDRON, I.: Neural mechanism by which controlling inputs influence motor output in the flying locust. J. exp. Biol. 47, 213-228 (1967b).
WALKER, T.J.: Specificity in the response of female tree crickets (Orthoptera: Gryllidae, Oecanthinae) to calling songs of the males. Ann. ent. Soc. Am. 50, 626-636 (1957).
WALKER, T.J., DEW, D.: Wing movements of calling katydids: fiddling finesse. Science, N.Y. 178, 174-176 (1972).
WEIS-FOGH, T.: Biology and physics of locust flight. IV. Notes on sensory mechanisms in locust flight. Phil. Trans. R. Soc. (B) 239, 553-584 (1956).
WENDLER, G.: The coordination of walking movements in arthropods. Symp. Soc. exp. Biol. 20, 229-249 (1966).
WENDLER, G.: Einfluss erzwungener Flügelbewegungen auf das motorische Flugmuster von Heuschreken. Naturwissenschaften 59, 220 (1972).
WIERSMA, C.A.G.: The neuron soma. Neurones of arthropods. Cold Spring Harb. Symp. quant. Biol. 17, 155-163 (1952).
WILSON, D.M.: Relative refractoriness and patterned discharge of locust flight motor neurons. J. exp. Biol. 41, 191-205 (1964).
WILSON, D.M.: The nervous coordination of insect locomotion. *In* "Physiology of the Insect Central Nervous System" (Eds., J.E. Treherne and J.W.L. Beament), pp. 125-139. London / New York: Academic Press (1965).
WILSON, D.M.: Insect walking. A. Rev. Ent. 11, 103-122 (1966a).
WILSON, D.M.: Central nervous mechanisms for the generation of rhythmic behaviour in arthropods. Symp. Soc. exp. Biol. 20, 199-228 (1966b).
WILSON, D.M.: The nervous control of insect flight and related behaviour. Adv. Insect Physiol. 5, 289-338 (1968a).
WILSON, D.M.: Inherent asymmetry and reflex modulation of the locust

flight motor pattern. J. exp. Biol. 48, 631-641 (1968b).
WILSON, D.M., GETTRUP, E.: A stretch reflex controlling wingbeat frequency in grasshoppers. J. exp. Biol. 40, 171-185 (1963).
WILSON, D.M., WYMAN, R.J.: Motor output patterns during random and rhythmic stimulation of locust thoracic ganglia. Biophys. J. 5, 121-143 (1965).
WYMAN, R.J.: Multistable firing patterns among several neurons. J. Neurophysiol. 29, 807-833 (1966).
WYMAN, R.J.: Lateral inhibition in a motor output system. I. Reciprocal inhibition in Dipteran flight motor system. J. Neurophysiol. 32, 297-306 (1969a).
WYMAN, R.J.: Lateral inhibition in a motor output system. II. Diverse forms of patterning. J. Neurophysiol. 32, 307-314 (1969b).
WYMAN, R.J.: Patterns of frequency variation in Dipteran flight motor units. Comp. Biochem. Physiol. 35, 1-16 (1970).
ZARETSKY, M.D.: Specificity of the calling song and short-term changes in the phonotactic response by female crickets, *Scapsipedus marginatus* (Gryllidae). J. comp. Physiol. 79, 153-172 (1972).

AN ANALYSIS OF DIRECTION FINDING IN MALE MOSQUITOES

P. Belton

Pestology Centre, Department of Biological Sciences, Simon Fraser University, Burnaby, B.C., Canada

Males of most of the mosquitoes and midges that mate in swarms (Downes 1969) are attracted over a short range by the sound of the female. The species that show this behaviour have two characters in common: firstly, the antennae of the males are plumose, with whorls of long fibrillae on their distal segments (Fig. 1a) and secondly, there is a marked difference between the sounds made by the sexes in flight, that of the male being nearly an octave higher than that of the female. This is true in some 27 species that I have studied in the field in North America and the West Indies. In all these cases the sounds are produced solely by the displacement of air caused by the movement of the wings in flight.

The sense organs that detect the sound of the female are named after Johnston (1855) who described them, with little evidence, as hearing organs. Johnston's organs are anatomically and functionally similar in both sexes and consist of the enlarged second antennal segments (Fig. 1b) filled with radially-oriented chordotonal sensilla, consisting of two sensory and several associated cells. The flagellum of the antenna is attached to a basal plate recessed deeply into the second segment. The basal plate consists of radially pleated cuticle, invaginated into the second segment as septa and tubular prongs (Fig. 1c). The majority of the sensory units or scolopidia are attached to these prongs. The antennae of the males of the species that have been investigated have about the same sensitivity to airborne sounds as the human ear (Belton 1961). The sensitivity of the female antenna is much less, evidently because the relatively bare flagellum (one-tenth of the surface area of the male according to Clements (1963)) presents a lower mechanical impedance to sound waves.

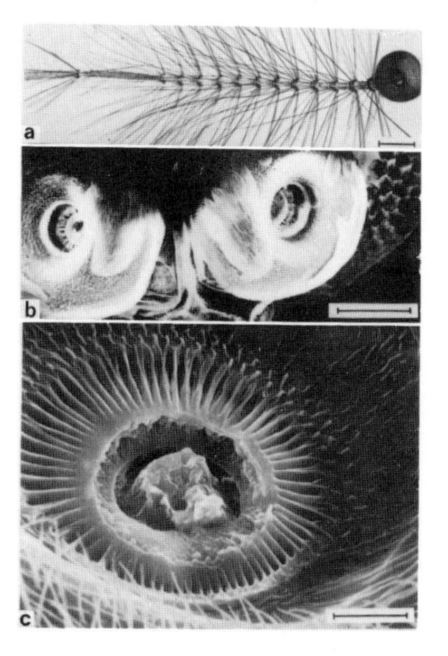

Fig. 1. Micrographs of antennae of male *Aedes aegypti*.

(a) Whole mount of one antenna showing Johnston's organ at the right and the dense "bottle brush" of fibrillae arising from the next 12 segments. Scale 0.1 mm.

(b) Scanning electron micrograph of both Johnston's organs with the flagella plucked out to show the basal plate. Scale 50 μm.

(c) Scanning micrograph of a basal plate with the flagellum removed, at a higher magnification. The 68 radial invaginations of the cuticle that form the prongs inside the segment are shown clearly. Scale 5 μm

In contrast to vertebrate ears, each antenna is potentially highly directional as the flagellum is moved by the displacement of air molecules in line with the source of the sound rather than by pressure changes which, at flight frequencies, convey no directional information.

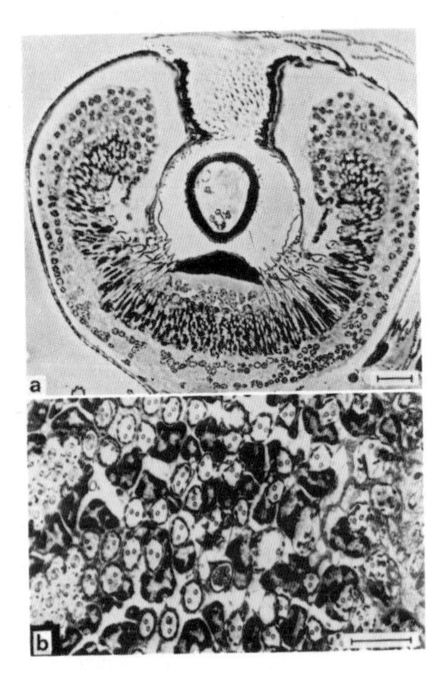

Fig. 2. Micrographs of sections through Johnston's organ of male *Aedes aegypti*.
(a) Light micrograph of a thin section taken obliquely. The flagellum is cut in the centre of the figure as it lies in the recess of the second segment. The section shows the thin pleated cuticle of the basal plate in contrast to the heavy cuticle at the base of the flagellum. Flat prongs radiate from the basal plate, and groups of scolopales can be seen cut in different planes, attached to each one. (Toluidine blue; scale 10 µm).
(b) Electron micrograph showing scolopales cut in transverse section. The left side is close to the point of attachment of two prongs, the right side is close to the periphery of the organ where the nuclei of the sensory cells can be seen in Fig. 2a. There are two ciliary processes surrounded by about ten dense rods in the medial region of each scolopale. Scale 2 µm

The fine structure of Johnston's organ (Fig. 2a) is complex, some 66-70 prongs radiate symmetrically from the basal plate, and some 500 scolopidia are attached to each one (close to Risler's estimate of 30,000 scolopidia, quoted by Schwartzkopff (1962)). This photograph shows the thin pleated cuticle and the flat prongs attached to the base of the flagellum. Groups of sensilla are attached to each prong and these are shown cut traversely in Fig. 2b. The fine structure of the scolopidia has been described by Risler and Schmidt (1967). Cook and I interpret the structure of the scolopidium (Fig. 3) somewhat differently from Risler and Schmidt, and in the material we examined, the apical cap attached to the prong was connected proximally only to the cilia and isolated from the scolopale cell. Also, in our material, the scolopale rods fused into four at the region of the rootlets of the cilia, and the scolopale cell was not surrounded by so distinct a sheath cell as Risler and Schmidt show in their diagram. The precise stimulus that elicits an electrical response in chordotonal sensilla is still a matter for conjecture (Howse and Claridge 1970), although in similar organs in crustaceans responses have been recorded both on stretching and slackening (Whitear 1960).

Although the mechanics of the vibration of the flagellum has received little attention, it seems likely that a point source of sound perpendicular to it will rock the basal plate, stimulating the scolopidia in line with the vibration of the flagellum maximally and those at right angles minimally. Furthermore, as some of the scolopidia on each prong are distal and some proximal with respect to the basal plate, it is likely that a movement of the flagellum away from the source of the sound during this rocking will stimulate one of the groups and movement toward the sound will stimulate the other.

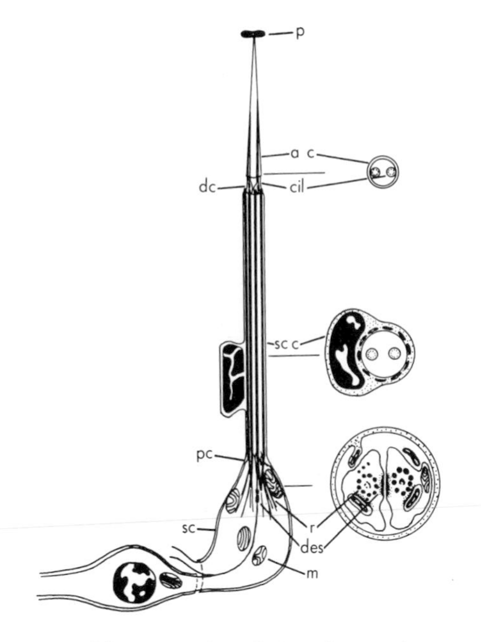

Fig. 3. Diagram of one scolopidium from the Johnston's organ of a male *Aedes aegypti*. Three sections are shown at the level of the horizontal lines. From top to bottom:

p - prong
a c - apical non-cellular cap
cil - cilia
dc - distal "centriole"
sc c - scolopale cell
pc - proximal "centriole"
r - rootlets
des - desmosomes
m - mitochondria
sc - sheath cell

The relationship between the sheath cell and the scolopale cell, and the position of the sensory cell body are not yet clear from our sections

Electrophysiological investigations carried out so far (Tischner 1953, 1954; Tischner and Schief 1955; Keppler 1958a,b; Wishart, van Sickle and Riordan 1962) can be interpreted on the basis of these simple assumptions. However, on the face of it, the views of Tischner, Schief and Keppler appear to be irreconcilable with those of Wishart and his associates. I will enlarge on these differences and present some new evidence to support the statements made in the previous paragraph.

There is no doubt that the Johnston's organ of mosquitoes are sharply tuned to the frequency of the female wingbeat. Wishart showed that this was not based on the mass of the flagellum, by cutting off two-thirds of it. In fact I have found that this procedure lowers the resonant frequency of some antennae rather than raising it as might be expected.

It seems likely that the mechanical properties of the basal plate, and the ball of cells in which its extensions are embedded, determine the resonance of the antenna and, as Keppler (1958b) suggests, the loss of fluid following amputation of the flagellum might in some circumstances lower its resonant frequency.

Frequency doubling has been observed by all previous investigators; that is, the microphonic potential recorded by metal electrodes inserted into Johnston's organ has twice the frequency of the stimulating sound, alternate cycles often being of different amplitude. Tischner at first proposed that this was due to summation of electrical activity at radially opposite parts of the organ, which would indeed explain doubling, whereas Wishart gave the rather vague explanation that the antenna was oscillating. Neither of these explanations fits with Wishart's observation that the phase of the large and small half cycles of the doubled response changes by 180° as the electrode is inserted proximally or distally into Johnston's organ (Wishart, van Sickle and Riordan 1962, fig. 9a,b). It seems more likely that the electrical activity is the sum of the two sets of responses, 180° out of phase, generated by proximal and distal groups of cells, and that the response of the cells nearest the electrode has the greater amplitude.

Tischner (1954) suggested that the relationship of the amplitude of the microphonic at the frequency of stimulus to that of its first overtone*, the doubled response, would give an indication of the direction of the sound. He proposed that the fundamental frequency was produced by a relatively small number of radial scolopidia at the apical part of the organ which would be "compressed" simultaneously by an axial movement of the flagellum away from the head. This would produce one electrical peak for each cycle of sound. Despite a careful reading of Tischner's papers, I have been unable to discover why this would not also apply to the much larger number of radial sensilla attached to the apical part of the prongs.

Keppler (1958a) investigated this possibility in more detail, vibrating the antenna directly with an electromagnetic transducer, and found that the ratio of harmonic to fundamental (the Klirrfaktor, or distortion factor) did in fact increase as the plane of vibration of the flagellum was changed from axial to radial.

When the antennae are stimulated with airborne sounds, however, I have found that the Klirrfaktor shows little change with the direction when a point source of sound is moved in the plane of the flagellum and, perhaps more convincingly, Wishart described an axial "cone of silence" of about 15° where sounds of attractive intensity elicited no detectable electrical activity.

Although it could be argued that the Klirrfaktor is not very meaningful in describing the summation of activities of different functional groups of sensilla, I investigated both resonance phenomena and directionality using a spectrum analyser to measure the amplitudes of fundamental and harmonics of the electrical potential. Both airborne and mechanical stimuli were used to excite the sensilla. Fig. 4 shows the microphonic potential set up by airborne sounds of different frequencies (second trace) immediately below its spectral analysis. Frequency doubling occurred at all the frequencies tested (125-400 Hz) but it is obvious that the Klirrfaktor showed wide changes, from 4:1 at 156 Hz to about 1:2 at 312 Hz. It could be argued that these frequencies are too low, as females at our normal room temperature (20.°C) produce a sound with a fundamental at about 500 Hz; however, allowing for frequency doubling, Wishart, van Sickle and Riordan (1962) found that the male antenna resonated at 350 Hz and Tischner and Schief (1955) recorded a maximal microphonic potential at 320 Hz.

There is little doubt that the insertion of a microelectrode will affect the delicate mechanical resonance of the flagellum, and these phenomena may therefore be shifted down from their natural frequency during such experiments. The mechanical properties of the antennae are likely to be complex because they must be able to reject the vibrations brought about by the wing movements of their owner. It may be more important for the males to reject their own flight frequency than to select the flight sound of the female.

These findings make it likely that the Klirrfaktor hypothesis is oversimplified and that there are at least two phenomena involved in determining the harmonic content of the electrical response namely:

1. Mechanical resonance of the flagellum, determined largely by the properties of the Johnston's organ, and
2. Summation of electrical events occurring in functionally distinct groups of cells by an electrode in a volume conductor.

---

* The term used by Tischner and Keppler, "erster Oberwelle" corresponds to the *second* harmonic of a frequency; Clements (1963) translates this term incorrectly as "first harmonic". The first harmonic usually refers to the fundamental frequency.

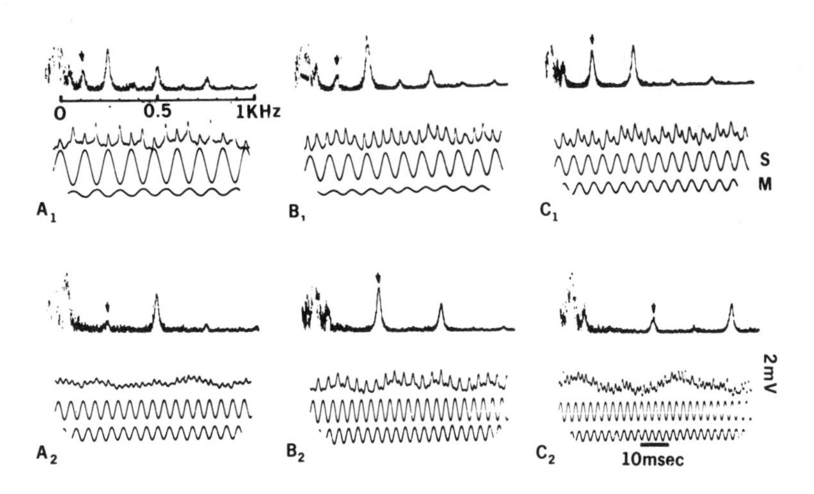

Fig. 4. Receptor potentials recorded from the Johnston's organ of a male *A. aegypti* (second trace). The upper trace is a spectral analysis taken over 10 sec. (Tektronix 1L5 analyser, Dispersion 100 Hz/cm, Resolution 10 Hz, 5 mV rms/cm). The arrow indicates the peak of the fundamental. The bottom trace is the response of a Sony electret condenser microphone close to the preparation (2 mV/, p-p, /cm); the trace above it is the signal fed to the loudspeaker (5 V/, p-p, /cm) A_1 125 Hz, A_2 250 Hz, A_3 500 Hz, B_1 156 Hz, B_2 312 Hz, B_3 624 Hz, C_1 200 Hz, C_2 400 Hz

The electrical activity in response to a brief lateral mechanical deflection of the flagellum demonstrates this summation (Fig. 5).

Fig. 5 covers the same frequencies as Fig. 4, and demonstrates the electrical interaction of components of the response. Let us assume the largest spike-like response in A_1 corresponds to a stretch of the sensilla nearest the electrode, the second upward response corresponds, if my premise is correct, to the stretch of sensilla that are mechanically 180° out of phase, and the third, slightly larger response, corresponds with a second (resonant) stretch of the initial group of cells. As the frequency of stimulating pulses is raised to 400 Hz (C_2), the responses decrease in amplitude owing to electrical or mechanical refractoriness, becoming initially identical in form to the microphonic potential elicited by airborne sounds (cf. Fig. 4, C_1,B_2). At frequencies above 624 Hz (B_3) the second response becomes indistinct as it merges with the larger first response.

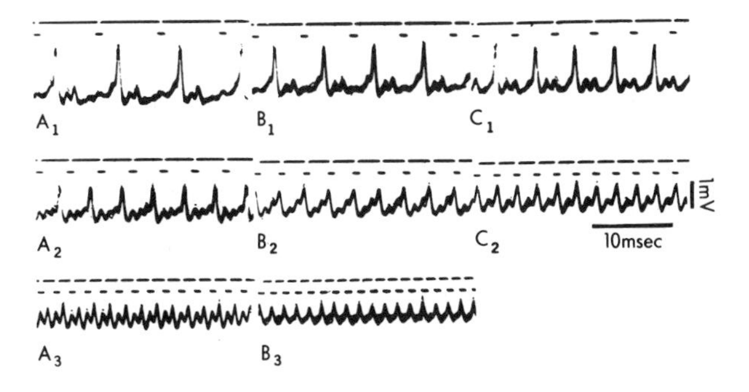

Fig. 5. Receptor potentials elicited by brief (0.5 msec) deflections of the flagellum of the antenna using a piezoelectric transducer. The deflections are repeated at the frequencies used in Fig. 4, i.e. from 125/sec (A_1) to 624/sec (B_3). Note the long latency and complex nature of the response to individual pulses (A_1). The responses become simpler as the repetition rate of the pulse increases. "Frequency doubling" occurs between about 350 and 624 Hz but the fundamental predominates above about 500 Hz. About ten superimposed sweeps

Focal recording with glass-microcapillary electrodes confirms this interpretation, and in fact, the scolopidia behave electrically as conventional mechanoreceptors (Gray 1959).

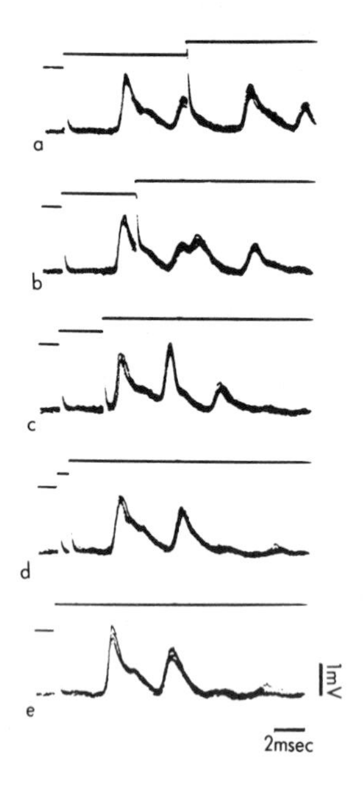

Fig. 6. Receptor potentials elicited by two sequential step deflections of the flagellum using a piezoelectric transducer. The interval between the two steps was reduced from 9.5 msec (a) to zero (e). Recordings were made with a glass capillary electrode and D.C. coupling. There are two responses to each step, indicating mechanical resonance which is largely due to Johnston's organ. Notice the summation, which must be brought about mechanically, when the steps are spaced by 3.4 msec (trace c) corresponding with a resonant frequency of about 300 Hz. About ten superimposed sweeps

In Fig. 6 the flagellum was moved stepwise by successive lateral deflections in the same direction and recordings were made with a glass microelectrode through a D.C. coupled amplifier. After a latency of 3.4 msec a response is evoked that is probably the sum of generator and spike potentials and after a further 3.4 msec a second, smaller response occurs. Assuming the first response represents a stretch of a few scolopales, the second response is probably caused by a smaller (resonant) stretch of the same cells. These responses then correspond with the first and third responses shown in Fig. 5 A_1. The delay between responses corresponds to a frequency of 294 Hz, close to the resonant frequency of the antenna excited by airborne sound. There are significant interactions between the responses shown in Fig. 6. When the interval between deflections decreases from 8 to 4 msec the amplitude of the spike-like response to the second deflection decreases in amplitude. Although this looks like electrical refractoriness of the spike, it is not, because when the interval is reduced further to 3.4 msec (Fig. 6c), the second spike-like response reappears. The amplitude of the second spike at this critical interval appears to be larger than the first, evidently because it has summed with the tail of the generator potential of the first spike. This is unequivocal evidence of mechanical resonance of the antenna. The second response decreases at first because the two deflections are mechanically out of phase, but when the interval between them reaches the resonant frequency (3.4 msec), the two deflections sum. The mechanical resonance is again probably influenced by the damping effect of the recording electrode and some oscillation of the piezoelectric transducer coupled to the stump of the flagellum cannot be ruled out, although it could not be detected with 200x magnification under a compound microscope with stroboscopic illumination.

One more feature of these responses requires comment. There is a small hump on the falling phase of the first 'spike' 1.7 msec after its initiation. This corresponds to double the resonant frequency and thus probably represents the activity of a distant group of cells, responding 180° out of phase with respect to those surrounding the glass electrode.

These findings support Wishart's conclusions that information obtained from the phase of the electrical activity could be used by the mosquito to obtain information on the direction of the source of sound. Perhaps more importantly they illustrate that the Klirrfaktor is a product of coarse recording electrodes and that individual scolopidia (and hence their axons) respond either to a "pull" or a "push" of the flagellum depending on their position, but not to both. This is not to deny that recombination of these responses could not occur in the brain, but it seems unlikely that they would be recombined in the way suggested by Tischner and Keppler.

If Wishart's phase difference hypothesis is to stand up to a similar analysis then one would expect a coarse electrode to reveal amplitude and phase differences according to whether the flagellum of the antenna were "pulled" or "pushed". These differences can in fact be seen when a source of sound is moved radially around a mosquito (Fig. 7); they are more obvious when the flagellum is deflected mechanically. Lateral step deflections of the flagellum in different radial directions can be made using two piezoelectric transducers perpendicular to each other. Recordings of the electrical activity elicited by this technique amply

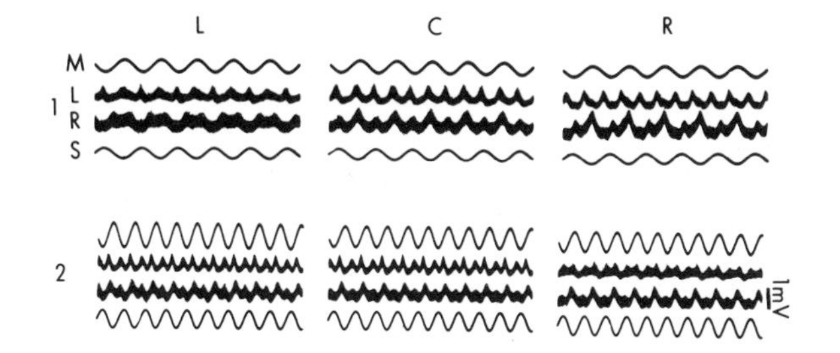

Fig. 7. Phase relationships between the receptor potentials recorded simultaneously from the left and right Johnston's organs. Frequencies were selected to show clear "Frequency doubling": (1) 263 Hz; (2) 460 Hz. The vertical bar represents 1 cm on the oscilloscope trace, and calibration applies to the receptor potentials L and R. Microphone M, 40 uV/cm; voltage to speaker S, 1 V/cm; sweep, 2 msec/cm. Photographs of about 50 superimposed sweeps. The speaker, at constant distance from the mosquito was initially at the centre, C, and was then moved from left, L, to right, R

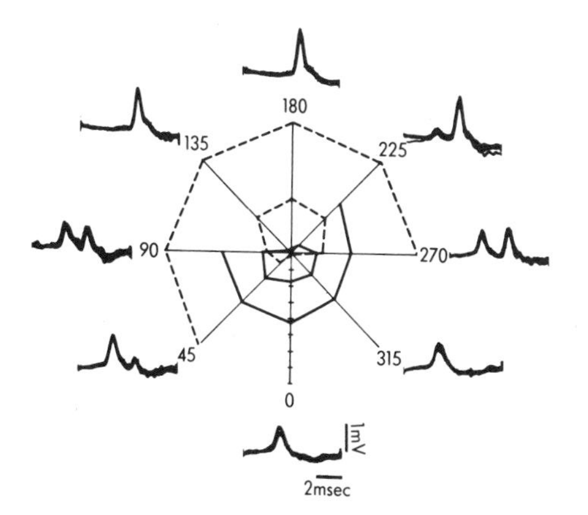

Fig. 8. Responses to a step deflection of the flagellum in directions indicated by the central compass. The relative latencies (outer) and amplitudes (inner) of the first (solid line) and second responses (dashed line) are plotted with polar coordinates. Note the constant latencies (resonance) and the marked change in amplitudes of the responses with the change in radial direction of the stimulus. About ten superimposed sweeps

confirm the hypothesis. An example is shown in Fig. 8. The metal electrode was inserted in the 0° position and the flagellum moved in this plane (0° and 180°), at right angles to it (90° and 270°) and in two intermediate planes. The results of such experiments are so far somewhat variable, since it is difficult to align the transducer, electrode and Johnston's organ, and furthermore, initial offset of the flagellum from its central position has a significant effect on the results. A complete explanation will have to await further experiments.

If we term a deflection towards the electrode a "push" and a deflection away a "pull", then a "push" evokes a response with a short latency and a "pull" a response with a latency 1.7 msec greater. The 1.7 msec delay again indicates a 300 Hz resonance and if my interpretation is correct, the same group of cells is generating both responses. The later response is evidently due to an elastic rebound of the resonant system at 3.4 msec.

A diagram corresponding with push (0°) and pull (180°) of Fig. 8 simplifies this explanation. Only those cells being stretched at a particular time are drawn in (Fig. 9).

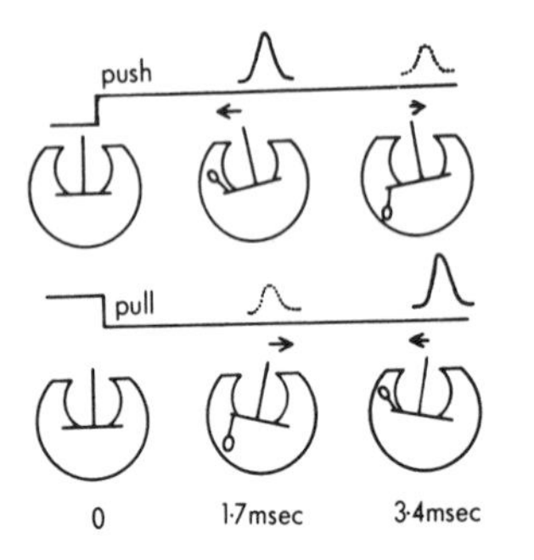

Fig. 9. Diagram corresponding with the push (0°) and pull (180°) responses of Fig. 8. Electrical activity is drawn in above the solid line that indicates the time and direction of a stepwise deflection of the flagellum. Diagrams of Johnston's organ below the lines show the simplest mechanical events that could produce the electrical activity. Cells being stretched by the initial deflection (1.7 msec) and the elastic rebound (3.4 msec) are drawn in

In this particular example the absence of the initial "pull" response could be due to an offset of the flagellum, and the cells may be either stretched or relaxed more than they would be naturally. With sideways stimuli it is likely that the adjacent groups of cells are being stimulated more than those near the recording electrode, and in most experiments these responses are smaller, but equal to each other, in amplitude. There are other possible interpretations of these results, but all of them must invoke resonance of Johnston's organ at about 300 Hz and all of them indicate that the sense cells can detect the phase of a deflection (pull or push).

Finally these observations must be related to the behaviour of a male mosquito. Equipped with two Johnston's organs pointing slightly above its line of flight, and with the antennal flagella at an angle to each other of about 65°, male *Aedes aegypti* can evidently make a rapid assessment of the position of a flying female (Wishart, van Sickle and Riordan 1962). A point source of sound at an angle of greater than 40° (i.e. outside the "cones of silence" of both antennae) would deflect them in phase. Both antennae would be pushed in the same direction at virtually the same time because the wavelength of the sound is much greater than the distance between flagella. A sound directly in front of the antennae would deflect the flagella out of phase, that is, sense cells in identical positions in the two Johnston's organs in the plane of the direction of propagation of the sound would be pulled and pushed by their respective flagella. The mosquito could therefore be almost immediately aware of the position of a source of sound, as the phase of the signals produced by the sensilla would indicate whether the sound was within 30° of the line of flight or greater than 40° from it. The plane of maximal vibration of the flagella would indicate the azimuth of the sound source, and slight divergence of the planes of vibration of the flagella would permit the mosquito to extrapolate these planes and pinpoint the sound in a way analogous to triangulation or radio direction finding. Gillett (personal communication) pointed out that this would not be possible if one of the antennae were removed, and in fact the same situation would arise if one of the flagella were pointed directly at the sound (the sound being in its cone of silence). Faced with this uncertainty, the mosquito would have to alter its direction of flight, and it would be important to discover whether sound presented in the horizontal plane would evoke vertical movement of the mosquito and vice versa as this hypothesis would predict.

Some time ago Roth (1948) showed that mosquitoes can in fact reach a source of sound with only one functional Johnston's organ, and this is not really surprising when its complexity is appreciated. As might be expected, the course of males with only one "ear" is much more erratic than those with two.

Other behavioural observations indicate that triangulation using the plane of vibration of the flagellum is an essential component in locating a sound. It is a common observation (Wishart and Riordan 1959) that in the laboratory males do not respond to a diffuse source of sound, such as a large moving coil loud speaker, and I discovered in field tests (Belton 1967) that males are not attracted from their swarms by a distant source of sound, even though it is at the correct frequency and intensity. These observations almost certainly indicate that the divergence of sound from a point source is being detected. Triangulation is clearly ineffective if the mosquito is present with a broad non-divergent wavefront.

The number of sense cells involved in detecting and locating a female is a flagrant violation of Roeder's principle of neural parsimony (Roeder 1963). There is evidently some redundancy in the directional information provided by these cells, but the sensitivity and directional accuracy of Johnston's organs must explain some of this prodigality. Roeder shows that a moth with a mere four sense cells, can detect and locate a bat, with some chance of avoiding it. Hearing and location must be considerably more important to the male mosquito and this is compelling evidence that, in swarming species, sound is the only modality used in bringing together the sexes. Without it, or perhaps by using it to the insect's disadvantage, we might soon be free from a fascinating but often lethal group of insects.

ACKNOWLEDGEMENTS

I should like to acknowledge the tireless cooperation of Dr. Harry Cook with the transmission electron microscope and thank Dr. Vic Bourne who took the scanning micrographs. Most of this work was supported by operating grants from the National Research Council of Canada. I should like to dedicate this article to the memory of George Wishart who introduced me to the electrophysiology of Johnston's organ.

REFERENCES

BELTON, P.: The physiology of sound reception in insects. Proc. ent. Soc. Ont. 92, 20-26 (1961).
BELTON, P.: Trapping mosquitoes with sound. Proc. Calif. Mosq. Contr. Assoc. 35, 98 (1967).
CLEMENTS, A.N.: "The Physiology of Mosquitoes". New York: Pergamon Press (1963).
DOWNES, J.A.: Swarming and mating flight in Diptera. A. Rev. Ent. 14, 271-293 (1969).
GRAY, J.A.B.: Mechanical into electrical energy in certain mechanoreceptors. Progr. Biophys. biophys. Chem. 9, 286-324 (1959).
HOWSE, P.E., CLARIDGE, M.F.: The fine structure of Johnston's organ of the leaf-hopper, *Oncopsis flavicollis*. J. Insect Physiol. 16, 1665-1675 (1970).
JOHNSTON, C.: Auditory apparatus of the *Culex* mosquito. Q. Jl microsc. Sci. 3, 97-102 (1855).
KEPPLER, E.: Über das Richtungshören von Stechmücken. Z. Naturf. (B) 13, 280-284 (1958a).
KEPPLER, E.: Zum Hören von Stechmücken. Z. Naturf. (B) 13, 285-286 (1958b).
RISLER, H., SCHMIDT, K.: Der Feinbau der Scolopidien im Johnstonschen Organ von *Aedes aegypti* L. Z. Naturf. (B) 22, 759-762 (1967).
ROEDER, K.D.: "Nerve Cells and Insect Behaviour". Cambridge, Mass.: Harvard University Press (1963).
ROTH, L.M.: A study of mosquito behaviour. An experimental laboratory study of the sexual behaviour of *Aedes aegypti* (Linnaeus). Am. Midl. Nat. 40, 265-352 (1948).
SCHWARTZKOPFF, J.: Die akustische Lokalisation bei Tieren. Ergebn. Biol. 25, 136-176 (1962).
TISCHNER, H.: Über den Gehörsinn von Stechmücken. Acustica 3, 335-343 (1953).
TISCHNER, H.: Das Hören der Stechmücken. Attempto (Tübingen) 4, 23-26 (1954).
TISCHNER, H., SCHIEF, A.: Fluggeräusch und Schallwahrnehumng bei *Aedes aegypti* L. (Culicidae). Zool. Anz., Suppl. 18, 453-460 (1955).
WHITEAR, M.: Chordotonal organs in Crustacea. Nature, Lond. 187, 522-523 (1960).
WISHART, G., RIORDAN, D.F.: Flight responses to various sounds by adult males of *Aedes aegypti* (L.) (Diptera: Culicidae). Can. Ent. 91, 181-191 (1959).
WISHART, G., van SICKLE, G.R., RIORDAN, D.F.: Orientation of the males of *Aedes aegypti* (L.) (Diptera: Culicidae) to sound. Can. Ent. 94, 613-626 (1962).

THE DEVELOPMENT OF THE FLIGHT PATTERN IN LOCUSTS

W. Kutsch

Fachbereich Biologie, Universität Konstanz, Konstanz, Germany

One of the major topics of neuroethological studies in insects is the analysis of the motor pattern underlying flight. This has been carried out most successfully with the flight of locusts, a great deal of the progress being the result of the excellent work of The late D.M. Wilson and his co-workers (summary, see Wilson 1968; Hoyle 1970). It is now quite well established that the flight pattern is centrally programmed and that peripheral influences can change the flight motor programme only slightly. It is assumed that the flight pattern is based ultimately on the activity of the central rhythm generator, the so-called "flight oscillator".

All previous studies were carried out with fully mature locusts and give information therefore only about a flight motor programme that is more or less complete or static. Only recently have studies been begun on the dynamics of the central nervous processes associated with the development of distinct behavioural patterns (Bentley 1973). By studying larvae and young adults it should be possible to determine if and how the coordination within the nervous system changes during development. With that approach, neuroethological studies converge with those of developmental biology. Further studies should reveal whether the successive steps leading to a completed motor pattern can be explained in terms of morphological changes, i.e. the formation of distinct neuronal structures such as synapses, or by phenomena such as facilitation or adaptation, detectable only by neurophysiological techniques.

The results presented in this paper represent only a summary of a few experiments which lead into a new field of neuroethology (Kutsch 1971, 1973). The following three questions in particular are to be answered by these experiments: (1) When during the life period of a locust can the first indication of a flight pattern be detected? (2) Is there any change of the motor pattern during the life span? (3) Is the final pattern the result of a pre-programmed endogenous process or do learning processes play a decisive role in the pattern formation?

SOME REMARKS ON THE METHODS

The method on which the studies are largely based is the extracellular recording of the electrical activity of the muscles (Wilson and Weis-Fogh 1962; Kutsch 1969). It is known that the main thoracic musculature which includes the flight muscles is activated neurogenically (Neville 1963), each discharge of a motoneuron being followed by an electrical activation of its innervated muscle part, the muscle unit. This muscle unit contracts only once per motor impulse. If a fast motor unit is involved, the contraction of the muscle fibres is fast and leads to a rapid movement of the appendage. The recording of the appendage's movement or, what is even better, the recording of the muscle potentials provides information on the discharges of the motoneurons. The recording reflects the last step of the process of neuronal integration within the central nervous system (CNS).

In contrast to Wilson and Weis-Fogh (1962), who used a flight balance, I have studied insects only in tethered flight. The locusts are suspended by a copper holder affixed to the pronotum with beeswax and are placed in an air flow of 3.5 m/sec at a temperature of 30±1°C. The method depends on the fact that adult insects which lose tarsal contact with the substrate assume the flight position and fly continuously (tarsal reflex; Fraenkel 1932). Larvae of all stages and adults of the desert locust, *Schistocerca gregaria*, and the migratory locust, *Locusta migratoria*, were tested under identical stimulus situations. The wing-beat frequency was investigated by synchronizing the wing movement with the flashes of a stroboscope (for details, see Kutsch 1971, 1973). The electrical activity was usually recorded from two muscles: subalar (99 and 129; nomenclature after Snodgrass 1929) and first remotor coxae (90 and 119) of the meso- and metathorax. These muscles are antagonistically activated during flight and are innervated by only 2 or 3 fast axons (Kutsch 1970; Kutsch and Usherwood 1970).

DEVELOPMENT OF THE FLIGHT PATTERN

Larvae in the air stream

When first instar larvae of *Schistocerca* are brought into the wind they spread their legs for a short while. The spreading reaction may be repeated a few times in response to changing the stimuli (wind, strobe flashes), but after a few repetitions it is no longer elicited. From the second instar onward larvae assume, upon exposure to stimuli, a typical posture in which the forelegs are folded and drawn to the prothorax, the middle- and hindlegs are extended backwards along the body (Fig. 1).

Fig. 1. Fifth larval instar of *Schistocerca* in "flight" position

This position resembles the posture taken by mature adults during flight and therefore will be called the "flight" position of the larvae. It should be pointed out that the larvae have only immovable wing-pads. The older the larvae are, the sooner they take this position, and the longer they maintain it.

The main thoracic musculature that becomes the flight musculature in adults already exists in *Locusta* and *Schistocerca* in the early larval instars (Wiesend 1957; Bernays 1972). Electrophysiological recording shows that some of these muscles have short periods of activity during the assumption of the "flight" position by the fourth larval instar. The muscle activity is mostly correlated with abrupt changes of the stimuli, such as the shutting off of the air stream, or exposure to intermittent light flashes. In adults these muscles, due to their mode of attachment, have two functions (Wilson 1962): to move both the wings and the legs. When muscle potentials stop in fourth instar larvae the flight position continues to be held. From this it is concluded that the typical leg position during flight posture is due to the activity of the coxal muscles. It is feasible, however, that slow motor units, found in some of the thoracic muscles (Kutsch and Usherwood 1970), may also control leg position during flight.

Fifth instar larvae show complex activity of the "flight" muscles. During longer "flights" we have often found a regular activation of the muscles with a firing frequency of about 20 Hz, but after-discharges are normally missing. The recording of the activity of the subalar and remotor muscles, which are activated antagonistically in adult flight, showed a tendency towards synchrony during "flight" of the fifth larval instar. In instances in which the activation of the muscles is slightly out of phase either can lead the other. This demonstrates that the group of motoneurons of one muscle is loosely coupled with that of the other.

In a few cases a patterned muscle activity suddenly emerges out of the tonic firing in some of the muscles (Fig. 2). Bursts appear that follow at intervals of 100-200 msec. An activation of the thoracic muscles with a frequency of 5-10 Hz is not known from any behavioural pattern but, for reasons described later, it cannot be excluded that this group-generation is the first step in the formation of the flight pattern.

Fig. 2. Muscle activity of fifth larval instar of *Schistocerca* during "flight". Transition from tonic activity to bursting in one muscle (from Kutsch 1971)

Flight on the first day of adult life

If adult *Schistocerca* are suspended in an air stream soon after imaginal ecdysis they assume the flight position, open the wings and flap them. But on the first day the wings are still so soft that they are twisted in the air stream. Because of this, the wing-beat frequency cannot be evaluated by the stroboscopic method and it is necessary to record the muscle potentials. We find (Fig. 3): (1) that there is a phasic activation of the muscles with single or a few discharges per burst; (2) that the two antagonistic muscles of the normal flight alternate regularly on the first day of adult life; (3) that the homologous muscles of the second and third thoracic segments are activated nearly simultaneously; but that the two sets of homologous muscles can be activated either synchronously or with a short interval between the activity of the two sets, in which case either set can lead the other. This indicates that the flight pattern typical of mature adults, in which the hindwing muscles always lead the forewing muscles, is not yet completely fixed.

If the time intervals between successive excitations of a neuromuscular unit are calculated, we see that there are two preferred intervals. The first maximum occurs between 10 and 20 msec and is the interval at which after-discharges occur. The second maximum is found at about 100 msec which represents the interval between successive bursts.

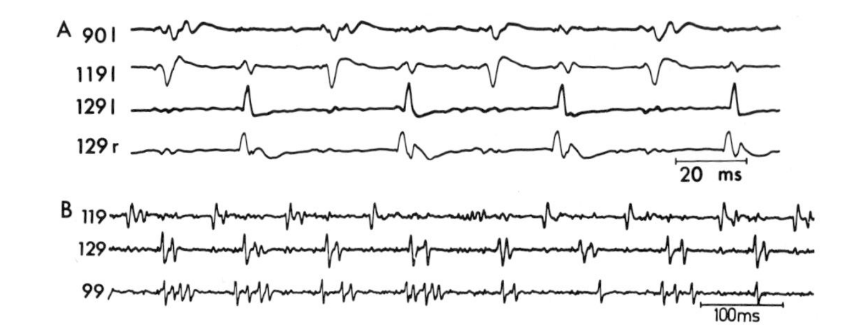

Fig. 3. Activity of subalar (99, 129) and first remotor coxae (90, 119) muscles of meso- and metathorax and left and right side in the fully developed flight motor pattern of a mature adult (A) and the respective pattern on the first day of adult life (B). Note the different time scales (B from Kutsch 1971)

When there is only a single excitation per cycle this long interval equals the whole period of a wing-beat cycle. In fully mature locusts the existence of two maxima has also been demonstrated (Wilson 1964), but these occur at values corresponding to half those in 1-day-old *Schistocerca* adults. Whereas a fully mature *Schistocerca* beats its wings with a frequency of *c.* 20 Hz, the 1-day-old locust flies with a wing-beat frequency of only 10 Hz.

A parameter which is important for the stability of the flight is the phase relationship of the two antagonistic muscles. Their contractions determine how the whole wing-beat cycle is divided into up- and downstrokes. Normally one calculates the phase of the elevator within the depressor cycle (see Fig. 4). By calculating this phase we find

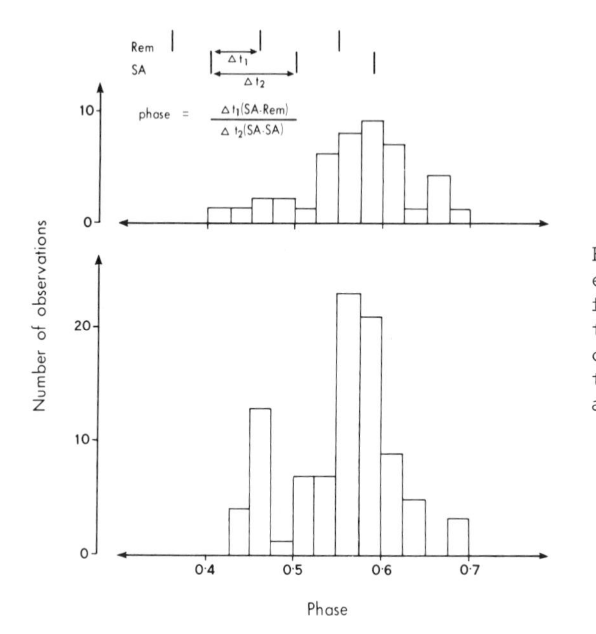

Fig. 4. A histogram of the phase of the elevator within the depressor cycle on the first (upper part of diagram) and the thirtieth day of adult life (lower part) of *Schistocerca*. The inset demonstrates the calculation of the phase (modified after Kutsch 1971)

the proportion of the downstroke within the whole wing-beat cycle. For 1-day-old *Schistocerca* most values are between 0.5 and 0.6, which are similar to those calculated for 30-day-old animals. This means that the downstroke is slightly longer than the upstroke in both the low wing-beat frequency flight of the young adult and the high wing-beat frequency flight of mature animals. In either case the relationship of up- and downstroke is the same. Waldron (1967) claims for mature *Schistocerca* that there is not only a phase maximum but that the phase of the elevator within the depressor cycle remains constant independently of the duration of the whole wing-beat cycle. This is also true for young locusts in which it has been shown that as the interval between successive subalar-excitations increases, the interval between the firing of the subalar and the next remotor-excitation also increases.

It was mentioned that during "flight" of animals in the fifth larval instar, a phasic activation of some muscles sometimes develops out of the normal tonic firing. The opposite is seen during flight of young locusts. Within the normal antagonistic activation of the muscle pair we find transient irregular firings, which show that the motor flight pattern of very young adults is not yet fixed totally.

In summary it may be said that only an approximation to the characteristic flight pattern is seen on the first day of adult life, but that already at this time, the typical features of the mature adult flight pattern exist. The main difference between the flight of young and of old desert locusts is to be found in wing-beat frequency, 10 as against 20 Hz. The wing-beat frequency of the 1-day-old locust resembles the low frequency bursting of some muscles during the "flight" of the fifth instar larvae. Therefore, we interpret this bursting as the first indication of a flight pattern.

Development of the flight pattern during the first weeks of adult life

If we follow the wing-beat frequency by making measurements daily during the first weeks of adult life, we find a marked increase in frequency during the first few days. The daily increase then gets smaller and after about three weeks the wing-beat frequency remains at a nearly constant level (Fig. 5). For *Schistocerca* this constant level is at 20 Hz, for *Locusta* males at 24 Hz, and for *Locusta* females at 21 Hz. The daily measured wing-beat frequencies fit the following exponential function quite well:

$$f_t = F - (F-f_o)\, e^{-kt}$$

f_t = wing-beat frequency on day t, F = final frequency, f_o = initial frequency, k = rate-constant, t = day of adult life. By a curve-fitting programme (after Chandler 1965) the three unknown constants (f_o, F, k) of this function can be computed for each test group. We calculate for *Locusta* - males: f_o = 9.8, F = 23.8, k = 0.12; females: f_o = 9.5, F = 21.5, k = 0.14.

The theoretical starting frequency (f_o), that is the wing-beat frequency at the moment of the last ecdysis, is the same for males and females, the rate-constants of the two sexes are quite similar, but the final frequencies are significantly different. The analysis of the results demonstrates that at least in males a measurable frequency increase is still occurring after 30 days. Theoretically after the sixth week of adult life the difference between the mean of the wing-beat frequency and the calculated final value should be smaller than 0.1 Hz.

Continuous daily measurements of the wing-beat frequencies of individuals up to the time of their death revealed several trends. The period during which wing-beat frequency increases is followed by one in which there are only small variations about the final value. Then in many cases there follows a period during which there are larger day to day variations (up to 4 Hz) in frequency. In addition a slight decrease of the mean wing-beat frequency can be observed. It is, however, quite surprising how regular the flight of many locusts is even on the day of their death.

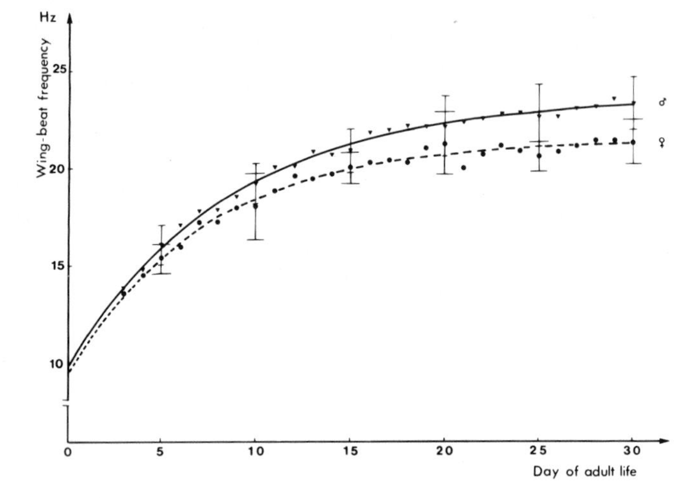

Fig. 5. Increase of the wing-beat frequency with age, *Locusta*. Means and S.D. of continuous daily measurements of ♂♂ and ♀♀ (from Kutsch 1973)

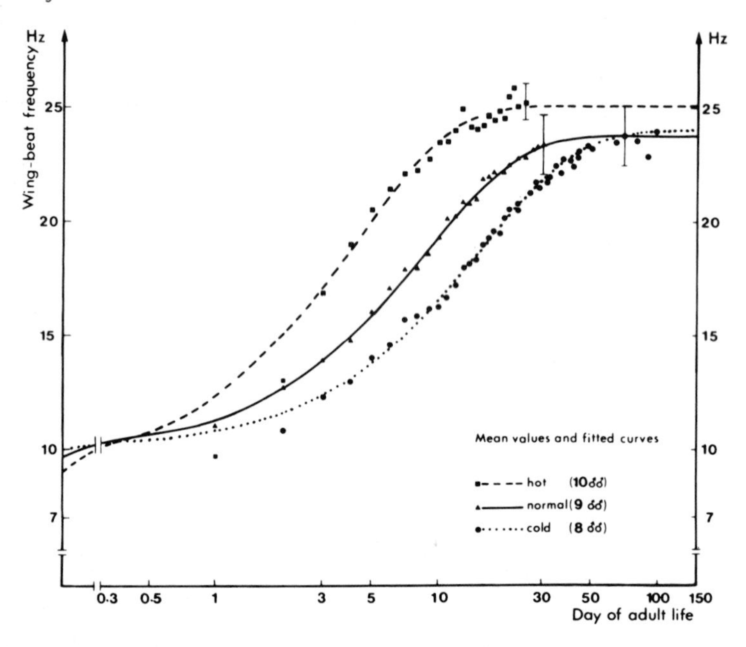

Fig. 6. Increase of the wing-beat frequency with age in *Locusta* reared under different culture conditions (from Kutsch 1973)

Influence of the culture temperature on the wing-beat frequency increase

Starting with the fifth larval instar, males of *Locusta* were reared under three different culture conditions: (1) cold (mean temperature t = 23°C), (2) normal (31°C), (3) hot (37°C). The extremes of relative humidity were 55% (cold night) and 30% (normal and hot day). In all three test groups a clear increase of the wing-beat frequency during the first weeks of adult life can be seen (Fig. 6). The initial frequency (f_o) is found to be in the range of 9-10 Hz for all three groups, and the final frequency (F) in the range of 24-25 Hz, but a clear difference between the rate-constants of the three groups was demonstrated. From these and other experiments it appears that the initial and the final frequencies are pre-programmed, and that only the time period required to reach the final frequency can be influenced, e.g. by diverse culture conditions.

COMMENTS AND PROSPECTS

As mentioned in the introduction, it is certain that many of the problems connected with the development of behavioural patterns will be solved by a careful analysis of morphological changes within the CNS. For *Schistocerca* we know (Sbrenna 1971) that the number of neurons within the thoracic ganglia remains constant throughout the post-embryonic period. The considerable increase of the ganglion size during the development is not only due to the increase of the number of glial cells but also due to the immense increase of arborization of the individual neurons. This was shown for the Australian locust, *Chortoicetes terminifera* by Altman and Tyrer (this volume), who used the newly-developed cobalt chloride method (Pitman, Tweedle and Cohen 1972) for an improved staining of neuronal branches. As the extent of arborization increases so does the possibility of new neuronal connections. If, in fact, the number of interactions between central neurons increases, one could speculate that the number and/or complexity of motor patterns increases during development. The aim of workers in this field of neurobiology will be to correlate the formation of new synapses with a new or a changed motor pattern. An example of such a correlation would be the discovery of fine structural changes within the CNS which parallel the transition from the "flight" of the fifth larval instar to the adult flight pattern (subalar and remotor muscles: tonic and synergistic activation → phasic and antagonistic pattern).

The results on the development of the flight pattern in locusts and crickets (Bentley and Hoy 1970; Weber, unpublished) reveal that there are remarkable differences between the two types of insects. In crickets (*Teleogryllus commodus*) the successive steps up to the adult motor pattern are as follows: in the seventh larval instar some muscles discharge a few times when insects assume the flight position; in later larval stages more motoneurons are recruited and the trains become progressively longer. The whole flight pattern is complete in the last larval instar. In *Schistocerca* a slow transition such as this could not be demonstrated. It has yet to be elucidated whether these differences between crickets and locusts are qualitative or only quantitative. It should be mentioned here that cricket postembryonic development with its 10-12 moults takes much longer than locust development with only 5 larval instars. Perhaps the comparable development of the flight pattern in locusts is compressed to a few days around imaginal ecdysis.

In the introduction, the question was raised as to whether learning processes play a role during flight pattern generation. Weber (1972) found that in cricket larvae the variance of the interval between muscle spikes decreases with successive "flight" tests and he explains this result by "experience of the motor system" resulting in a stabilization of the flight pattern rhythm. With *Schistocerca* I tried to determine whether flight tests influence the increase of the wing-beat frequency. The animals were divided into two groups. The locusts of the first group were made to fly 5-10 min every day, while in the second test group the animals were made to fly only once in their life at various times after the last ecdysis. The measured wing-beat frequencies for the two groups overlap completely; therefore, one has to assume that at least the rate of increase of the wing-beat frequency is not affected by flight experience.

The main difference in flight pattern of young and mature adult locusts was found to be in the output frequency of the flight oscillator. As early as 1961, Wilson postulated that central oscillators are involved in flight, but up to now the flight oscillator or oscillators are still hypothetical. Svidersky (1967) recorded the activity of some spontaneously firing cells within the CNS of *Locusta*, but he did not claim to have found the "flight oscillator". During flight of *Schistocerca*, Page (1970) recorded the activity of neurons within the metathoracic ganglia. Some fired with the frequency of the wing-beat, but he classified them as motoneurons. At present we can only speculate about the physiology and the morphological structure of this rhythm generator. It appears that its morphological basis will be found within the thoracic ganglia.

How can the discharge frequency of this oscillator be increased? The increase of the wing-beat frequency with age seems to be a general feature in insect flight. It was found to occur in locusts (Kutsch 1971, 1973); Chadwick (1953) mentions some unpublished observations of Weis-Fogh on *Schistocerca*, in cockroaches (Farnworth 1972), in *Drosophila* (Chadwick 1953), in *Phormia* (Levenbrook and Williams 1956). I wish to mention here two possible explanatory systems which it is hoped are applicable also to the insects (Diptera) with myogenic flight muscle:

(1) We know (Novak 1966) that the protein metabolism increases after imaginal ecdysis due to hormonal influence of the growing corpora allata. It is conceivable that during the early adult period some metabolic processes directly raise the activity of the flight oscillator.

(2) It is assumed that the oscillator is driven by different inputs, such as those from spontaneously active neurons in the CNS (Svidersky 1967) or from sensory organs. It has been demonstrated (Wilson and Gettrup 1963) that the wing-beat frequency is reduced to half its normal value after destruction of all four stretch receptors, one of which is situated in each wing hinge. These receptors are single-cell organs which send impulses to the CNS in response to stretching during the upstroke of the wing. In contrast to the central neurons an important part of the sensory system develops during the post-embryonic period (Wigglesworth 1953). Svidersky (1969) detected changes in the activity of some wind-sensitive receptors on the frons of *Locusta* even during early adult life. Recordings from the appropriate nerve ($N1D_2$; Campbell 1961) demonstrated that a sensory organ, presumably the abovementioned stretch receptor, is excitable at least in the fifth larval instar (Kutsch 1971). Also a large axon was found in this nerve in the fifth instar larvae which corresponds to that which is thought to be the stretch receptor axon in adults (Gettrup 1962). The morphological and physiological existence of the stretch receptor system does not necessarily mean that its connections with the central flight system are complete in 1-day-old adults and similarly it is possible that the connections of the flight system with other sensory organs and central neurons may also be incomplete. It could well be that the increase of the wing-beat frequency might be explained by a successive completion of all required inputs to the flight oscillator.

ACKNOWLEDGEMENTS

These studies were supported in part by a Deutsche Forschungsgemeinschaft grant to Prof. Dr. W. Rathmayer. I would like to thank Dr. L.

Murdock for his numerous suggestions throughout the work and his help in preparing the English text.

REFERENCES

BENTLEY, D.R.: Postembryonic development of insect motor systems. *In* "Developmental neurobiology of arthropods" (Ed., D. Young), pp. 147-177. London: Cambridge University Press (1973).

BENTLEY, D.R., HOY, R.R.: Postembryonic development of adult motor patterns in crickets: a neural analysis. Science, N.Y. 170, 1409-1411 (1970).

BERNAYS, E.A.: The muscles of newly hatched *Schistocerca gregaria* and their possible functions in hatching, digging and ecdysial movements (Insecta: Acrididae). J. Zool. (London) 166, 141-158 (1972).

CAMPBELL, J.I.: The anatomy of the nervous system of the mesothorax of *Locusta migratoria migratorioides* R.& F. Proc. Zool. Soc., Lond. 137, 403-432 (1961).

CHADWICK, L.E.: The motion of the wings. *In* "Insect Physiology" (Ed., K.D. Roeder), pp. 577-614. New York: John Wiley & Sons (1953).

CHANDLER, J.P.: Quantum Chemistry Exchange Program. Bloomington, Indiana (1965).

FARNWORTH, E.G.: Effects of ambient temperature, humidity, and age on wing-beat frequency of *Periplaneta* species. J. Insect Physiol. 18, 827-839 (1972).

FRAENKEL, G.: Untersuchungen über die Koordination von Reflexen und automatisch-nervösen Rhythmen bei Insekten. Z. vergl. Physiol. 16, 371-393 (1932).

GETTRUP, E.: Thoracic proprioceptors in the flight system of locusts. Nature, Lond. 193, 498-499 (1962).

HOYLE, G.: Cellular mechanisms underlying behavior-Neuroethology. *In* "Advances in Insect Physiology" (Eds., J.W.L. Beament, J.E. Treherne, and V.B. Wigglesworth), 7, 349-444. London / New York: Academic Press (1970).

KUTSCH, W.: Neuromuskuläre Aktivität bei verschiedenen Verhaltensweisen von drei Grillenarten. Z. vergl. Physiol. 63, 335-378 (1969).

KUTSCH, W.: Nervöse Kontrolle der Aktivität eines thorakalen Muskels der Heuschrecke *Schistocerca gregaria* Forskål. Zool. Anz. (Suppl.) 33, 489-493 (1970).

KUTSCH, W.: The development of the flight pattern in the desert locust, *Schistocerca gregaria*. Z. vergl. Physiol. 74, 156-168 (1971).

KUTSCH, W.: The influence of age and culture-temperature on the wing-beat frequency of the migratory locust, *Locusta migratoria*. J. Insect Physiol. 19, 763-772 (1973).

KUTSCH, W., USHERWOOD, P.N.R.: Studies of the innervation and electrical activity of flight muscles in the locust, *Schistocerca gregaria*. J. exp. Biol. 52, 299-312 (1970).

LEVENBROOK, L., WILLIAMS, C.M.: Mitochondria in the flight muscles of insects. III. Mitochondrial cytochrome c in relation to the ageing and wing-beat frequency of flies. J. gen. Physiol. 39, 497-512 (1956).

NEVILLE, A.C.: Motor unit distribution of the dorsal longitudinal flight muscles in locusts. J. exp. Biol. 40, 123-136 (1963).

NOVAK, V.J.A.: "Insect Hormones", 3rd Ed. London: Butler and Tanner (1966).

PAGE, C.H.: Unit responses in the metathoracic ganglion of the flying locust. Comp. Biochem. Physiol. 37, 565-571 (1970).

PITMAN, R.M., TWEEDLE, C.D., COHEN, M.J.: Branching of central neurons: intracellular cobalt injection for light and electron microscope. Science, N.Y. 176, 412-414 (1972).

SBRENNA, G.: Postembryonic growth of the ventral nerve cord in

Schistocerca gregaria Forsk. (Orthoptera: Acrididae). Boll. Zool. 38, 49-74 (1971).
SNODGRASS, R.E.: The thoracic mechanism of a grasshopper, and its antecedents. Smithson. misc. Collns 82, 1-111 (1929).
SVIDERSKY, V.L.: Central mechanisms controlling the activity of locust flight muscles. J. Insect Physiol. 13, 899-911 (1967).
SVIDERSKY, V.L.: Receptors of the forehead of the locust, *Locusta migratoria* in ontogenesis. J. Evol. Biochem. Physiol. 5, 482-490 (1969). [In Russian]
WALDRON, I.: Mechanisms for the production of the motor output pattern in flying locusts. J. exp. Biol. 47, 201-212 (1967).
WEBER, Th.: Stabilisierung des Flugrhythmus durch "Erfahrung" bei der Feldgrille. Naturwissenschaften 59, 366 (1972).
WIESEND, P.: Die postembryonale Entwicklung der Thoraxmuskulatur bei einigen Feldheuschrecken mit besonderer Berücksichtigung der Flugmuskeln. Z. Morph. Ökol. Tiere 46, 529-570 (1957).
WIGGLESWORTH, V.B.: The origin of sensory neurones in an insect, *Rhodnius prolixus* (Hemiptera). Q. Jl microsc. Sci. 94, 93-112 (1953).
WILSON, D.M.: The central nervous control of flight in a locust. J. exp. Biol. 38, 471-490 (1961).
WILSON, D.M.: Bifunctional muscles in the thorax of grasshoppers. J. exp. Biol. 39, 669-677 (1962).
WILSON, D.M.: Relative refractoriness and patterned discharge of locust flight motor neurons. J. exp. Biol. 41, 191-205 (1964).
WILSON, D.M.: The nervous control of insect flight and related behavior. *In* "Advances in Insect Physiology" (Eds., J.W.L. Beament, J.E. Treherne, and V.B. Wigglesworth), 5, 289-338. London / New York: Academic Press (1968).
WILSON, D.M., GETTRUP, E.: A stretch reflex controlling wing-beat frequency in grasshoppers. J. exp. Biol. 40, 171-185 (1963).
WILSON, D.M., WEIS-FOGH, T.: Patterned activity of coordinated motor units, studied in flying locusts. J. exp. Biol. 39, 643-667 (1962).

INSECT FLIGHT AS A SYSTEM FOR THE STUDY OF THE DEVELOPMENT OF NEURONAL CONNECTIONS

J.S. Altman and N.M. Tyrer

Department of Neurobiology, Research School of Biological Sciences, Australian National University, Canberra, A.C.T., Australia

Behavioural acts result from patterns of activity in the nervous system and this implies a certain ordering of connections between neurones. The developmental process by which correct contacts between neurones are established must require considerable accuracy and, by investigating the way in which these connections are formed during development, it may be possible to determine the rules which regulate the growth of nerve terminals in the neuropile. In order to do this, it is necessary first to make an exact description of a set of connections and their development. The appearance of a new behaviour pattern implies that new neuronal circuitry has become functional, and this should provide a basis for making this description. This review examines the development of the neural control of locust flight to determine whether it is a suitable system for such an investigation.

A second purpose of studying the development of a behaviour pattern is to use the development of coordination as a tool for examining the integrating processes involved in the adult pattern. This aspect has been the subject of two recent reviews, by Bentley (1973) and Kutsch (this volume). The two aims, however, differ more in conceptual attitude than in experimental technique and so both depend to some extent on the same results.

Orthopteran flight has a number of features which make it attractive for both these aims. Development is direct through a number of larval stages to a final moult producing a winged adult. Because the neural circuits controlling flight are being added to an already functioning nervous system, the developmental changes will be limited and should therefore be easier to study. The neural output in adult locust flight is well known (Wilson 1964, 1968) as a rhythmical pattern of alternate firing of the motor neurones to antagonistic flight muscles (Fig. 1). The pattern is generated within the thoracic ganglia (Wilson 1961) and the final motor pathway to the flight muscles involves a total of about eighty motor neurones, twenty in each half segment (Bentley 1970). Several studies have been made of the development of the characteristic motor output of adult flight in Orthoptera (in crickets: Bentley and Hoy 1970; Weber 1972; in locusts: Kutsch 1971 and this volume; Altman, in prep. a,b; - see also review by Bentley 1973). Although both the time course and sequence of events in locusts differ from those in crickets, the pattern in both appears gradually and perhaps depends on the maturation of interconnections between motor neurones.

Recent technical advances have made it feasible to push the analysis into the central nervous system. Intracellular recording from insect cell bodies (Hoyle and Burrows 1973) and intracellular injection of dyes and metal ions (Stretton and Kravitz 1968; Pitman, Tweedle and Cohen 1972) provide a means of describing identified neurones both physiologically and morphologically and should enable us to compare a particular neurone in individuals of different ages. It should therefore be possible to examine changes in the flight neurones as they acquire their adult function. Here progress to date is reviewed and future prospects assessed.

DEVELOPMENT OF THE PATTERNED MOTOR OUTPUT

In both crickets and locusts, the adult flies when contact with the tarsi is removed and a stream of air passes over the wind-sensitive sensilla on the head. A characteristic posture is adopted with the legs tucked up and the abdomen flexed dorsally. Although larvae do not move their wings and cannot fly, elements of this adult behaviour can be elicited with the same stimuli, the posture becoming more adult and being maintained for longer periods as the animal gets older. In locusts, which have 5 larval instars, the response is present from the first instar (Kutsch 1971; Altman, in prep.a), whereas in crickets, where there are 9 or 10 larval instars, Bentley and Hoy (1970) observed it only from the seventh instar.

During these 'flight' responses in the larvae, activity which foreshadows the adult flight motor pattern can be recorded in the flight muscles. In the cricket, *Teleogryllus commodus*, the flight muscles begin to function one at a time during the last four larval instars. The muscles start to fire in their correct phase relationships and the complete adult pattern is established by the last larval stage even though the wings cannot move (Bentley and Hoy 1970). In locusts, investigations in two species, *Schistocerca gregaria* (Kutsch 1971) and *Chortoicetes terminifera* (Altman, in prep.a), confirm that the adult pattern is fully developed only after the final moult. Records from the last two larval instars show that there is rhythmic activity in both elevator and depressor muscles at about flight frequency. The antagonists, however, are not coordinated into the alternating pattern of the adult but fire approximately in synchrony (Fig. 1). In *Chortoicetes*, in a detailed study of animals of exactly known ages, Altman (in prep.a) has shown that the phase relationships of the adult pattern are established gradually during the final two days of the last larval instar (A - 1) and the first three days of adult life (Fig. 1b-e). There is, however, considerable variation in the time of onset and completion of this development between individuals, so the process probably takes rather less than 5 days. Coupling between the free running motor neurones must develop gradually about the time of the final moult, and there is also some indication that coupling between elevators and depressors is weak at first and improves in the first days of adult life (Altman, in prep.b).

This sequence is different from that in crickets, but this may only be the result of different time scales. The underlying developmental processes are probably very similar since, in both, the motor neurones become entrained into the oscillating network, but in the cricket this happens sequentially over several instars whereas in the locust coupling between all the neurones is compressed into one short period. This must depend entirely on an intrinsically determined developmental programme, as the pattern is fully established in crickets before the wings become functional and in the locust *Chortoicetes* it has been shown that practice is not necessary for the establishment of the correctly phased alternation of antagonists, as the patterned motor output is still produced even if the wings are fixed immovably at the time of fledging (Altman, in prep.b).

An economical hypothesis is that motor neurones become entrained as a result of the maturation of interconnecting links between them. This could be the result of growth of new nerve processes, establishment of new synapses or a change in thresholds in elements of existing circuits. If we are to discover the mechanism of this development, we must compare nervous elements in the larvae and late adult, both structurally and functionally. The following investigations have all been made

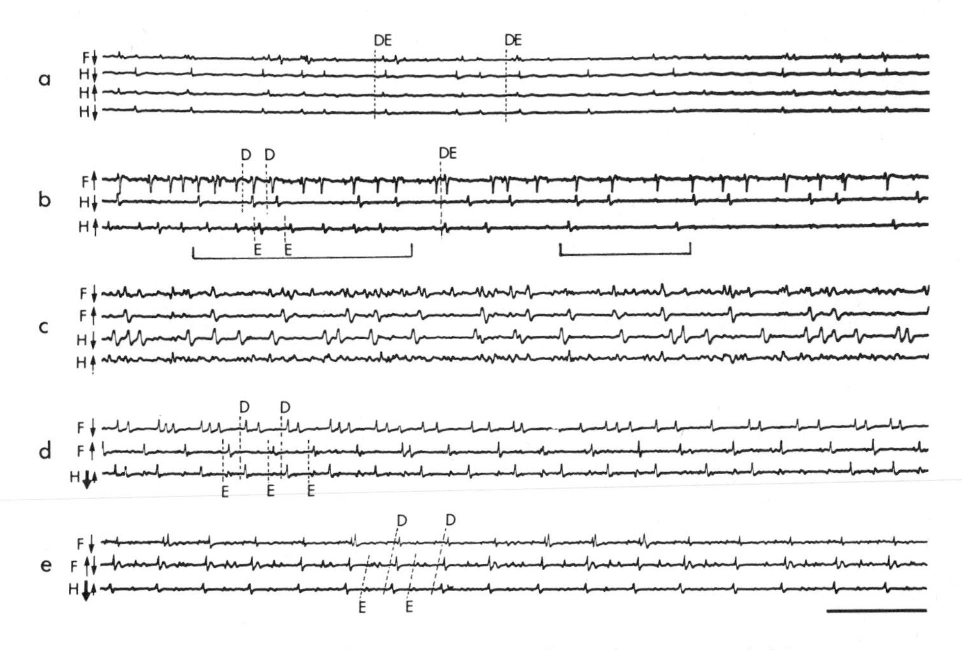

Fig. 1. Myograms from the flight muscles of *Chortoicetes terminifera* to show the progressive development of the patterned motor output about the time of the final moult. 5 separate preparations.

(a) late (A - 2) instar: There is firing in all the muscles at about flight frequency, but all units are approximately synchronous.
(b) late (A - 1) instar: Firing is still usually synchronous, but for short periods (marked with bracket) antagonistic units fire alternately.
(c) 5 hr after final moult (A + 5 hr): The motor pattern is irregular but some alternation of antagonists is apparent, especially in the hindwing. The phase shift between fore- and hindwings is not consistent.
(d) A + 12 hr: The alternating motor pattern is present but with no phase shift between fore- and hindwings.
(e) A + 26 hr: Both alternation and phase shift are well established.

Time mark: 100 msec
D - depressor muscle = ↓
E - elevator muscle = ↑
F - forewing
H - hindwing

in *Chortoicetes*, although some comparisons with the cricket *Teleogryllus oceanicus* are available (Bentley 1973). The interneurones involved in flight, where the main developments might be expected to occur, are generally not directly accessible to investigation, so we are necessarily limited to observing changes in the motor and sensory neurones. From their properties, however, it should be possible to infer some of the characteristics of the interneurones.

MORPHOLOGY OF ADULT FLIGHT NEURONES

The morphology of the main sensory and motor neurones involved in the control of the wings in *Chortoicetes* has now been described from cobalt preparations, by filling identified flight neurones with cobalt chloride from their peripheral ends (Tyrer and Altman, in press) and by intracellular injection (Burrows 1973a). Bentley (1973) also gives a preliminary account of flight neurones filled with cobalt in the cricket *Teleogryllus oceanicus*. The cobalt is precipitated as a sulphide which gives a picture of the morphology of single neurones, which appear black against a colourless background (Fig. 2). A good preparation is very similar to a Golgi impregnated profile but it has the advantage that it can be examined in a whole mount of the ganglion. Cobalt preparations are not always consistent, however, and care has to be taken with their interpretation (Tyrer and Altman, in press).

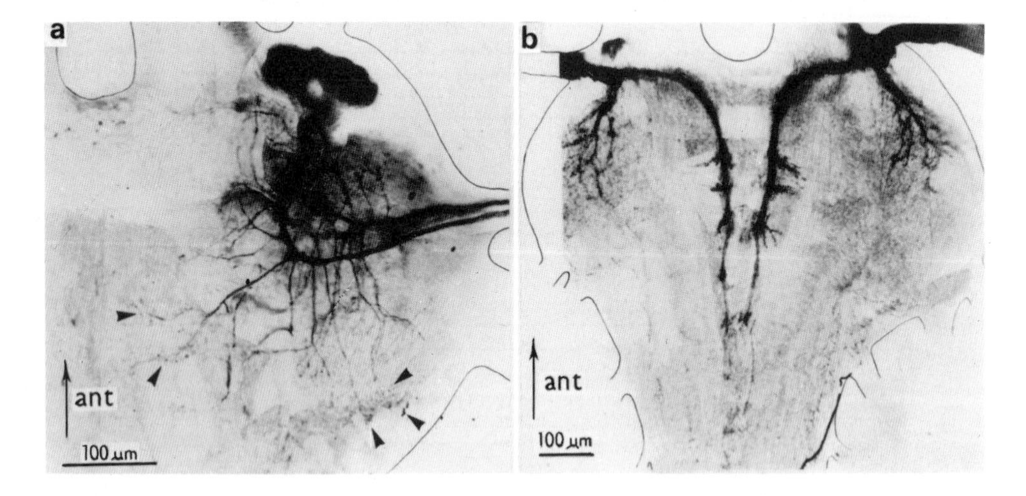

Fig. 2. Montages of neurones in whole mounts, photographed by transmitted light.

(a) Mesothoracic ganglion: the two motor neurones of the tergosternal muscles (83 and 84), photographed from the dorsal surface. Despite the montaging technique, problems with the depth of focus make it impossible to demonstrate all the branches in one picture. The finest dendrites in the adult terminate in small nodules some of which can be seen here (arrowed). These structures are not so obvious in the juveniles.

(b) Metathoracic ganglion: the sensory projections from both metathoracic 1C nerves, which include the axons from the campaniform sensilla on the hindwings and the hair sensilla of the tegulae. Two main components of the 1C projection are visible here, both near the dorsal surface of the ganglion. The midline bundle is fairly typical of the projections from the sense organs concerned with flight (cf. Fig. 8b) with its compact main bundle giving rise to discrete zones of branches which lie more dorsally. The particular origins of the two projections have not yet been identified

Motor neurones

Many muscles in the thorax of the locust are used both in flight and in walking (Wilson 1962). Not only should the muscles with a dual role have more complex inputs but their development will be more difficult to analyse because they are functional in walking from the time of hatching. In this article, therefore, we will restrict detailed description to the motor neurones of two muscle groups involved solely in flight. These are muscles 83 and 84 (Snodgrass 1929), the mesothoracic tergosternal muscles, the chief elevators of the forewing, and the muscle 112, the metathoracic dorsal longitudinal muscle, the indirect depressor of the hindwing.

The cell bodies of the motor neurones of these muscles, demonstrated by cobalt filling, are almost identical in position to those shown for

Schistocerca gregaria, using intracellular injection of the dye Procion yellow (Bentley 1970). The tergosternal muscles are innervated by two motor neurones located on the ipsilateral side of the mesothoracic ganglion, whereas there are five neurones to the dorsal longitudinal muscle, four on the ipsilateral side of the mesothoracic ganglion and one on the contralateral side of the metathoracic ganglion (Fig. 3) (Neville 1963; Guthrie 1964; Bentley 1970; Tyrer and Altman, in press). The distribution of the cell bodies of the motor neurones to the dorsal longitudinal muscles is unusual, since all other flight motor neurones are both on the ipsilateral side of the ganglion and in the same segment as the muscles they innervate. This suggests that the dorsal longitudinal muscles may have a special function in the integration of the flight pattern (see also Bentley 1973).

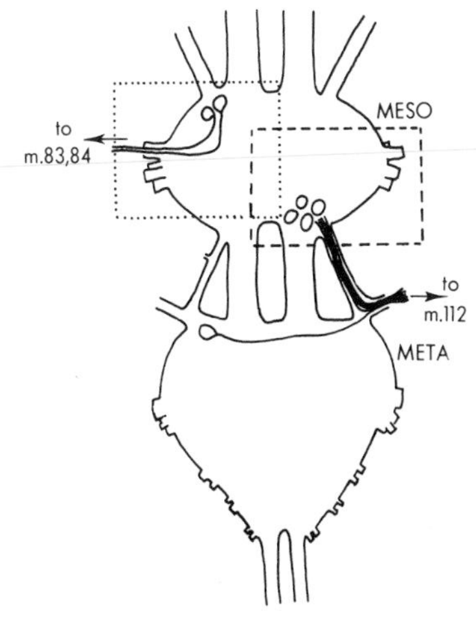

Fig. 3. The location of the cell bodies of the motor neurones described in Figs. 4, 5, 6 and 9. The mesothoracic tergosternal muscles, 83 and 84, are supplied by two motor neurones in the anterior part of the mesothoracic ganglion and the area enclosed in the dotted lines is that covered by Figs. 2a, 6 and 9. The metathoracic dorsal longitudinal muscle, 112, has five motor neurones, one on the contralateral side of the metathoracic ganglion. The area surrounded by dashed lines is represented in Fig. 4 and 5

Cobalt profiles of all the flight motor neurones reveal a very complex morphology (Figs. 4, 6; Burrows 1973a; Tyrer and Altman, in press) with large numbers of branches ramifying extensively through the dorsal part of the neuropile ipsilateral to the muscles they innervate. This is unexpected, as the Procion yellow fillings of flight neurones (Bentley 1970; Tyrer - unpublished) show a comparatively simple structure. Evidently Procion demonstrates only the cell body, neurite, axon and some of the largest branches while leaving much of the dendritic tree unfilled (see Fig. 4). The terms we use for the parts of the neurone are shown in Fig. 4. Features revealed by cobalt, which are not apparent in the Procion filled preparation, are the multiple subdivision of the major branches and their extensive ramification throughout the dorsal neuropile in the posterior two-thirds of the ganglion.

Variation in motor neurones

Comparison of identified neurones in different individuals shows considerable variation, both in cell body position and branching pattern. A range of the variation in the relative positions of cell bodies, and in the geometry of the primary branches of the mesothoracic motor neurones of muscle 112, is demonstrated in Fig. 5. In some preparations the cells are close together, in others they are widely separated. Similarly the neurites and first order branches of the four neurones may be so close together that they cannot be separated visually in whole mounts, or they may lie quite apart.

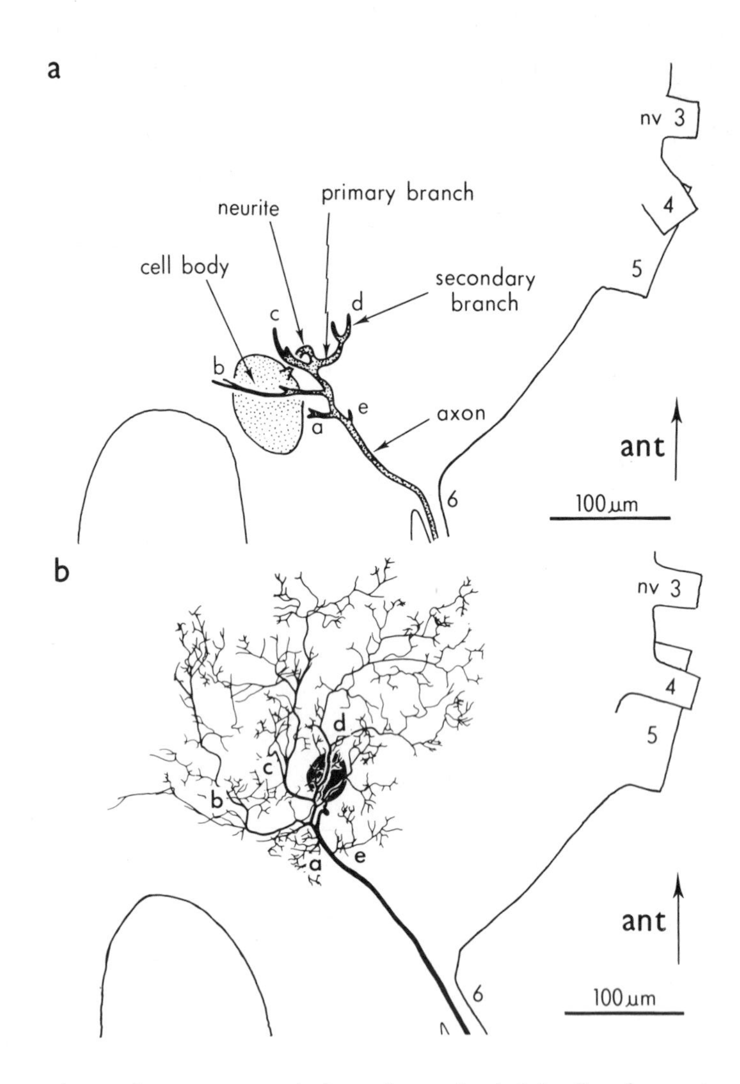

Fig. 4. Comparison of neurone morphology demonstrated by Procion and cobalt filling. The terms used in this paper for the parts of the neurone are also defined in these figures.

(a) One of the motor neurones of muscle 112 in the mesothoracic ganglion injected with Procion yellow. The drawing is a dorsal view reconstructed from serial sections.

(b) A motor neurone in the same group filled antidromically with cobalt and viewed in whole mount from the dorsal side.

Both methods show the major features of the neurone, the cell body, neurite and main branches, and the axon leaving in the peripheral nerve. The primary branches a-e are comparable in the two specimens. In the Procion filled neurone, however, the branches are quite simple, most appear to end abruptly and none are less than 3 µm in diameter. The cobalt filled cell, in contrast, shows several subdivisions of these branches, to form a large dendritic tree in which twiglets of less than 0.25 µm can be resolved.

These neurones have their cell bodies on the ventral surface of the ganglion and the neurite rises almost vertically through the neuropile to the dorsal surface of the ganglion. The branches of the dendritic tree all lie high in the dorsal neuropile. Apart from two small groups which run to the midline, all the dendritic branches are in the posterior two-thirds of the ipsilateral side of the ganglion

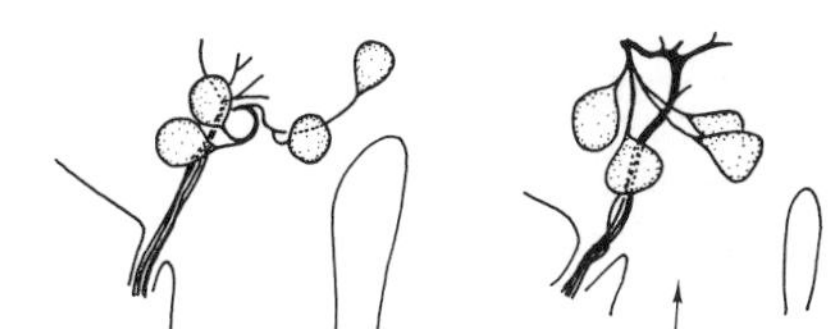

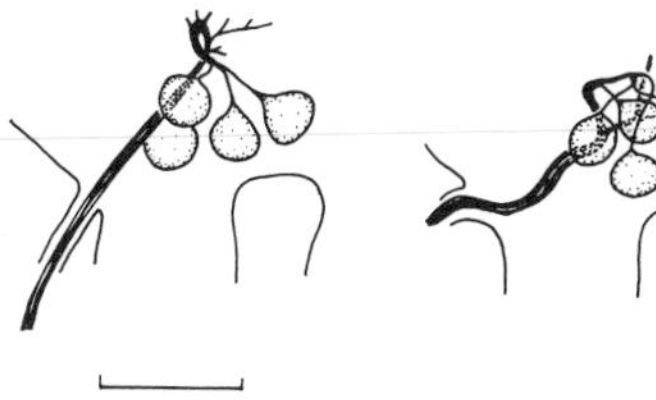

Fig. 5. Variation in cell body position is demonstrated in this comparison of the mesothoracic motor neurones of muscle 112 in six preparations. The drawings are from the ventral surface of the ganglion in whole mounts, made using camera lucida. As well as cell body position, there is also variation in the geometry of the neurites and primary branches

The variation of cell body positions probably occurs during the expansion which the thoracic ganglia undergo during postembryonic growth. There is considerable evidence that the full complement of neurones is present at the time of hatching since neuroblast divisions cease before the end of embryonic development (Hong 1968; Sbrenna 1971 for *S. gregaria*; also Gymer and Edwards 1967; Panov 1966 for the house cricket, *Acheta domesticus*). The ganglia continue to grow until well into adult life but the change in volume of the cell cortex is far smaller than that of the neuropile. In the mesothoracic ganglion of *Schistocerca*, for example, the cortex volume increases to 12-13 times its size at hatching while the neuropile increases by about 24 times (Sbrenna 1971). As the cortex surrounds the neuropile, this will result in a stretching which will separate and displace the cell bodies. This effect probably differs between individuals, with the result that, although the region of the ganglion in which a particular cell lies is consistent, its exact location is not absolutely predictable. Similar variation has been shown for the cockroach metathoracic ganglion (Crossman, Kerkut, Pitman and Walker 1971; Cohen and Jacklet 1967) and for the lobster abdominal ganglia (Otsuka, Kravitz and Potter 1967) and is probably of fairly general occurrence. Its physiological significance is probably trivial. In *Chortoicetes*, the main branches of the motor neurones are established by the A - 3 (see below and Fig. 9) whereas the major phase of expansion of the ganglion appears, from toluidine blue stained wholemounts (Altman and Bell, in prep.), to be in the A - 2 and A - 1 instars, as it is in *Schistocerca*, where quantitative studies have been made (Sbrenna 1971). However, variation in position may have important practical consequences if one is aiming to impale a particular small cell body with a microelectrode.

A second kind of variation, which is of greater significance, is in the branching pattern of the motor neurones within the neuropile. This is a feature of all the flight neurones examined in detail. Each neurone can be characterized by the general shape of the path of its axon through the ganglion and of its dendritic tree. For example Fig. 6 shows the morphology of one of the mesothoracic tergosternal neurones in four adult

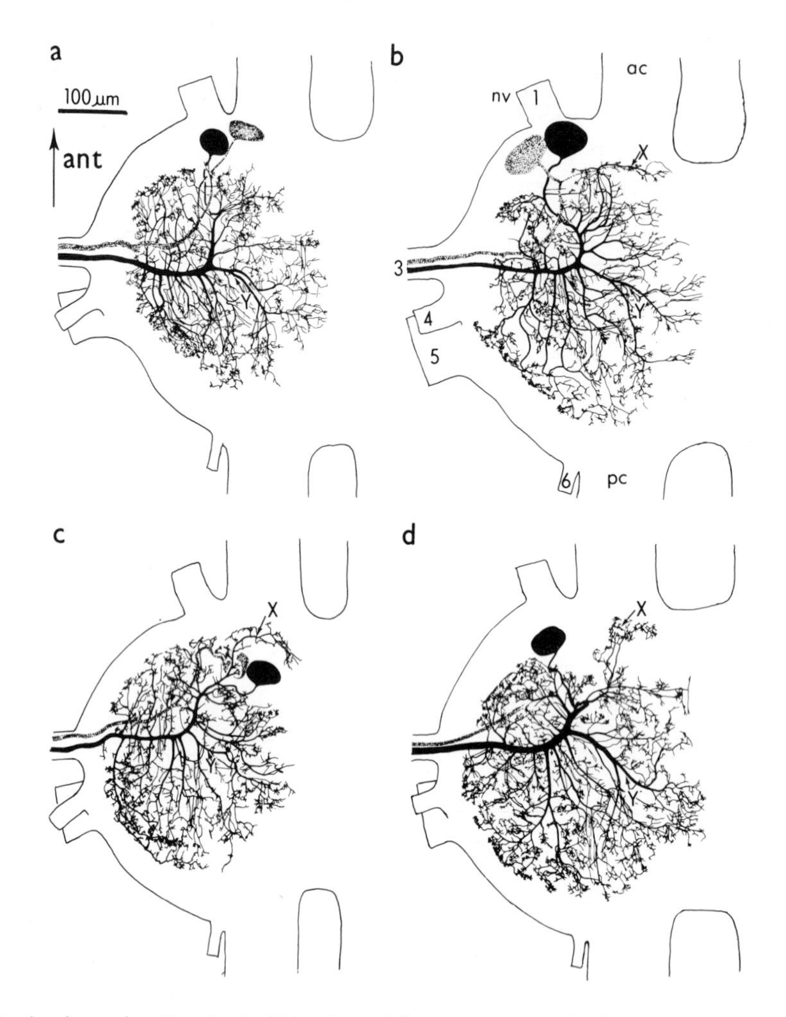

Fig. 6. Variations in the dendritic branching patterns of the neurone to one of the mesothoracic tergosternal muscles (m.83 and 84) in four adults. The preparations have been drawn from whole mounts using camera lucida. Only the cell with the more medial axon path through the ganglion has been drawn in detail; the cell body and axon of the second cell are shown stippled.

The primary branches can be identified approximately from one individual to another, and features such as the branches labelled X and Y can be recognised fairly consistently. (The absence of X in (a) could possibly be due to a failure to fill with cobalt.) In general, however, there is marked variation in the numbers and positions of the branches. This variation becomes even more marked in the higher order branches. In functional terms it might be important only that the neurone is represented in particular areas of the neuropile and the means by which this is achieved may be of secondary importance.

The cell body is ventro-lateral, close to the root of nerve 1 and the neurite, which is relatively short, runs obliquely up to the dorsal neuropile, where most of the dendritic branches lie. Fine twigs occur throughout the area covered by the dendritic tree. The apparent concentration at the junction between cortex and neuropile results from viewing the branches in depth as they follow the curvature of the neuropile.

ac - anterior connective
pc - posterior connective
1 - 6 - nerve roots entering ganglion

individuals. This neurone can repeatedly be recognised by its more medial, more right-angled and more dorsal axon path, although we have not been able to ascertain which of the pair of muscles 83 and 84 it innervates. The large forked branch, marked Y in Fig. 6, is a consistent feature of this neurone, and the branching, X, close to the cell body is seen in most preparations. There is, however, considerable variation in the points at which even main branches leave the axon, and still more so in their subsequent divisions. In comparison with other accounts of motor neurone geometry (Stretton and Kravitz 1968; Selverston and Kennedy 1969; Nicholls and Purves 1970; Pitman, Tweedle and Cohen 1972) this variability seems to be unusual. These authors emphasize the constancy of cell shape, although several note 'some heterogeneity' in the finer branches. Where comparative diagrams of the same cell in different preparations are available, e.g. cockroach leg motor neurones (Iles 1972) and locust common inhibitor neurone (Burrows 1973b), variation is obvious, suggesting that it is not restricted to the flight neurones and may be of fairly common occurrence. It may originate from the different genetic backgrounds of the animals studied. As the population we used was inbred but not genetically homogeneous this could not be tested.

The range of shapes of identified motor neurones implies that their growth may not follow a predetermined pattern, or blue-print, but rather be guided only by a few simple rules. Such variation leads one to speculate on the organization of the motor neuropile. If there are orderly connections between the motor neurones and interneurones these cannot depend on a rigid morphology of the motor neurone. We know so little, however, about the structure of the neuropile in insect thoracic ganglia that we cannot say what degree of variation is important in practical terms. Functionally, it may only be necessary that particular areas of the neuropile contain representatives of the neurone, regardless of the exact geometry of the branches. We do not even know which parts of the motor neurone are zones of synaptic contact, or whether all branches have functional synapses. Nor do we know how accurate the connections in a motor neuropile must be in order to generate a motor output correctly. In general, motor neuropiles have a far less ordered appearance than sensory or association neuropiles (Young 1965). Some sensory neuropiles, particularly insect optic lobes, are highly ordered and have recently been shown to be extremely accurately wired (Braitenberg 1970; Horridge and Meinertzhagen 1970) but it is not safe to extrapolate from this to other neuropiles, particularly to those such as the motor areas of the thoracic ganglia where there is no obviously ordered structure.

Sensory neurones

Associated with each wing there are four main groups of sense organs which are involved in the fine control of the motor output concerned with maintaining stable flight (Gettrup 1966; Wilson and Gettrup 1963). All these receptors send their axons into the ganglion of the same segment through the ipsilateral nerve 1 (Fig. 7). This nerve has two sensory branches: 1C which includes axons from the campaniform sensilla of the wings and from the hair sensilla of the wings and tegula, and $1D_2$ which contains those from the stretch receptor and from the scolopidial organ (nerve and branch classification following Campbell 1961; see Fig. 7). Because of this anatomical division of the innervation, the central projections of these two groups can be examined separately using the cobalt technique. This has been successfully accomplished in the metathoracic ganglion (Fig. 8) although we have not yet achieved fillings from the individual sense organs. The two projections in the mesothoracic ganglion have proved more difficult to separate, as they are closely

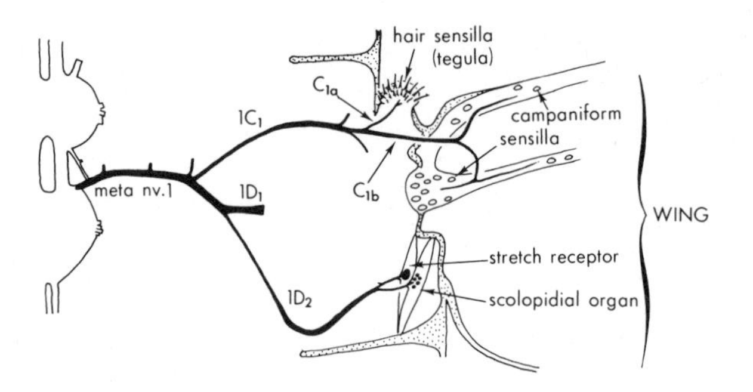

Fig. 7. The sensory branches of metathoracic nerve 1 from the sense organs associated with flight. Branch 1C includes axons from the campaniform sensilla and hairs of the wings and hairs of the tegula. The axons of the stretch receptor and scolopidial organ run in branch $1D_2$. Branch $1D_1$ contains motor neurones innervating the dorsal longitudinal muscles. The branches of mesothoracic nerve 1 are very similar. (After Campbell 1961; Gettrup 1962, 1966)

associated and more complex. Tentative descriptions are given in Fig. 8. Further details can be found in Tyrer and Altman (in press), and Bentley (1973) gives an outline description of these projections in the cricket.

There are several characteristics of these projections which have important implications. Firstly, they are all restricted to the ipsilateral side of the ganglion, with the exception perhaps of a few fine fibres of the ventral plexus of the metathoracic $1D_2$ projection. This is consistent with independent control of each wing and with the ipsilateral location of the flight motor neurones. Secondly, the $1D_2$ projections in general occupy a more ventral region of the neuropile than that containing the dendritic branches of the motor neurones. The main part of the 1C projection lies nearer the midline of the ganglion than most of the motor neurone branches, although there are some places where dendrites of the motor neurones appear to intermingle (Bentley 1973; Burrows 1973a; Tyrer and Altman, in press). It is not yet clear how close is the association between them. The third point which has an experimental rather than a functional significance, is that the sensory projections are on the whole both fairly compact and appear to be much less variable, from animal to animal, than the motor neurones. The branching of the fine terminals may not be the same from one animal to the next but because they occupy very small regions of the neuropile (Figs. 2, 7), it should be much more feasible to identify the same component of the projection in different individuals. This means that high resolution studies to find synaptic zones and to follow the changes in them during development should be possible.

CHANGES IN NEURONE MORPHOLOGY DURING DEVELOPMENT

Both in the cricket, *Teleogryllus oceanicus* (Bentley 1973), and in *Chortoicetes*, major components of the flight circuitry can be filled with cobalt before the adult flight pattern can be detected. In *Chortoicetes*, it is more difficult to get good fillings in earlier instars. Motor neurone cell bodies have been filled only from the beginning of the A - 3 instar and sensory projections first appear later than this. In the cricket, Bentley (1973) has found a neurone which appears to be present only in the larval stages, and he speculates that this might be associated with the suppression of the adult behaviour in the larvae. So far, no cell has been found in larval locusts which has not also been seen in adults.

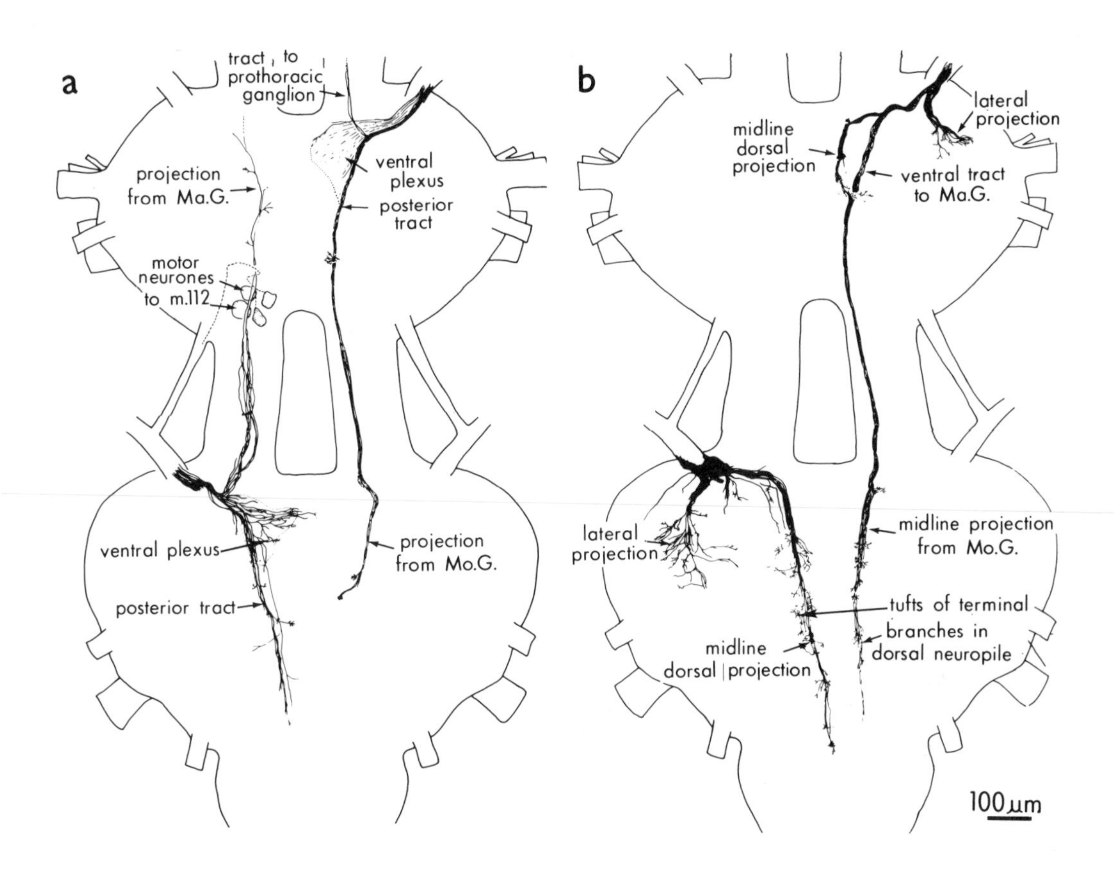

Fig. 8. Semi-diagramatic drawings of the central projections of the sensory nerves from the sense organs involved in flight (see Fig. 7).

(a) Branch $1D_2$: left, metathoracic; right, mesothoracic.
(b) Branch 1C : left, metathoracic (see also Fig. 2b); right, mesothoracic.

The $1D_2$ projections lie ventrally and have three main components: a plexus of fine fibres deep in the ventral neuropile, a tract of fibres running posteriorly through the ganglion and a tract running through the anterior connective. Mesothoracic $1D_2$ also has a tract in the posterior connective and a projection in the ventral neuropile of the metathoracic ganglion. The 1C projections each have two main components, both in the dorsal neuropile. One is lateral, close to the nerve root and is loosely branched, similar to motor neurone dendrites. The other is a compact tract running posteriorly close to the midline of the ganglion, giving off tufts of fine terminals into the neuropile dorsal to the tract. From mesothoracic 1C, a tract runs ventrally with the $1D_2$ projection, through the posterior connective to the metathoracic ganglion where it has a similar midline projection, close to that of metathoracic 1C. There is no anteriorly directed tract from metathoracic 1C.

In general the nerve 1C projections stain very much more darkly with cobalt than the $1D_2$, which suggests they are composed of coarser fibres. The ventral plexuses of the $1D_2$ projections, especially in the mesothoracic ganglion, are composed of numerous very fine fibres and terminal branches and have only been represented diagramatically. As more fibres are seen in the connectives than can be distinguished in the next ganglion, the projections between the ganglia are probably stronger than suggested here.

Abbreviations: Mo.G - mesothoracic ganglion
Ma.G - metathoracic ganglion

Once the form of an identified neurone has been determined in the adult, the development of its gross morphology can be examined using the cobalt technique. If particular consistent changes are observable, it should be possible to investigate the fine structural features of changes during development.

Motor neurones

The earliest stage at which cobalt filling of individual motor neurones has been successful in *Chortoicetes* is the second half of the A - 3 instar (Fig. 9). The developmental sequence described here has been seen in several motor neurones but we have concentrated on one of the two motor neurones of the mesothoracic tergosternal muscles (muscle 83/84). By this stage, this neurone is already extensively branched, with most of the primary branches seen in the adult filled, and third-order and even higher branches present. The numerous fine ramifications appear less complex and some branches, such as those in the anterior lateral quadrant, are not visible. No firm conclusions can be drawn from these differences as the structures may be present but have failed to fill with cobalt or, in the case of the fine ramifications, the processes may be too slender to be resolved with certainty in whole mounts. Again, the range of variation described in the adult makes differences in the larva difficult to interpret.

In the later stages, there is an increase in the length of the branches and probably in their diameter as well. The branching apparently becomes more complex. Although this may only be the result of better cobalt filling as fibre diameter increases, it is probably due to continuing growth. Reduced silver staining of the neuropile also reveals progressively greater detail in older animals. This supports the idea that there is a continuing differentiation of fine processes since developing nerve fibres are usually much more difficult to impregnate with silver. By the A - 1 instar both cobalt and reduced silver preparations display the same order of complexity as in the adult, although in cobalt wholemounts the terminal twigs of the dendrites (Fig. 2a) appear to be less dense.

Sensory projections

A preliminary investigation of the development of the central projections from the hindwing sense organs has been made by filling metathoracic nerve 1 from early in the A - 3 instar to immediately before the final moult. The peripheral branches 1C and $1D_2$ are both present in all larval stages but sensory projections in the ganglion have not been seen before the second half of the A - 3 instar. Part of the $1D_2$ projection can first be detected at this stage but the 1C projection is not seen before the beginning of the A - 1 instar. Five satisfactory fillings of each of the stages described have been obtained. In the younger stages it is very difficult to resolve the details of the projections since the individual fibres are very fine and fill only faintly, but the general extent of the structures is quite clear. Details of the complexity of the terminal zones can only be determined from sectioned material. This analysis is still in progress.

Towards the end of the A - 3 instar the sensory ventral root of nerve 1 and some of the fine ventral plexus of the $1D_2$ projection can be discerned (Fig. 10a). In the A - 2 instar the $1D_2$ projection becomes clearly visible, although the ventral plexus and the fibres in the anterior connective are apparent earlier than the fibres running posteriorly through the ganglion (Fig. 10b,c). By early in the A - 1 instar,

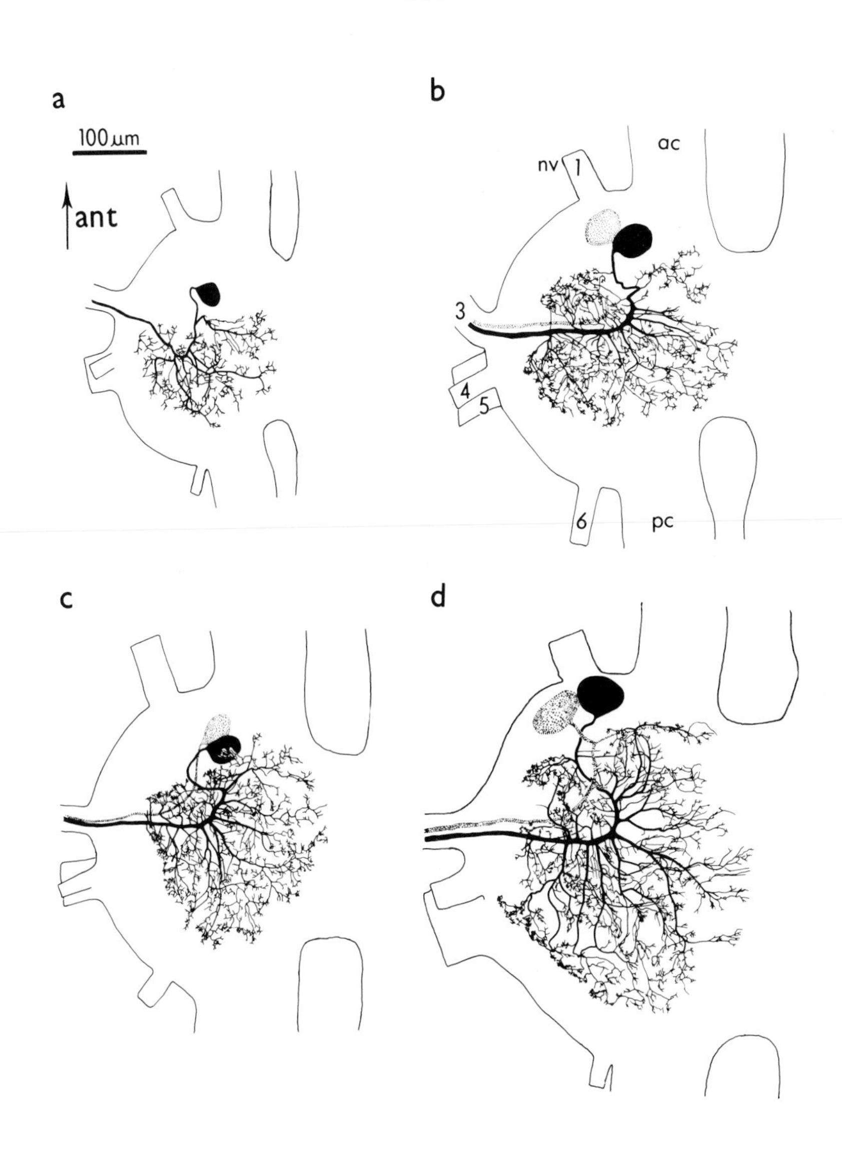

Fig. 9. Camera lucida drawings from whole mounts of the same tergosternal motor neurone as in Fig. 6, in different stages of development. The second neurone, where present, is shown stippled.

(a) late A - 3 instar: The neurone already has its basic adult characteristics, with branches extending over most of the ipsilateral side of the ganglion. The branches in the anterior lateral quadrant and many of the higher order branches may not yet have grown, or have not filled, or are too fine to resolve.
(b) A - 2 instar
(c) A - 1 instar
(d) adult

From A - 2 onwards the neurone is very similar to that in the adult, and there is only an increase in the complexity of the fine terminals as the animal develops. In the adult the terminal twiglets appear to be more dense than in the A - 1 instar.

The size of the ganglion increases markedly from A - 3 to adult but the region covered by the dendritic branches remains fairly constant. All the established branches must therefore continue to grow throughout this period. The cell body volume in the adult is also nearly double that of the A - 3 instar

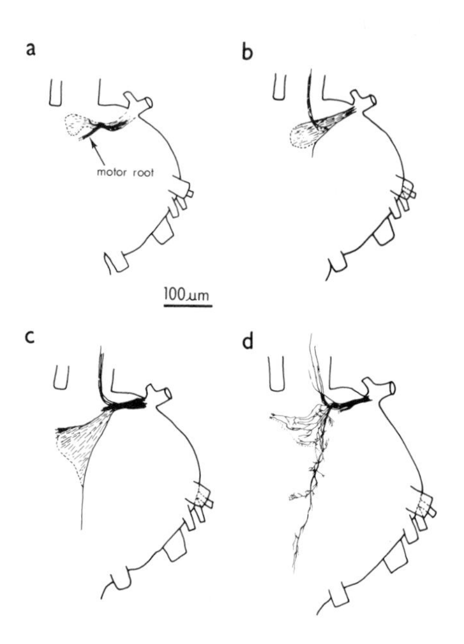

Fig. 10. Diagrams of the development of the central projection of metathoracic sensory nerve $1D_2$ made using camera lucida. The details of the projections are very difficult to resolve in the younger animals and so only the general form of the projection is shown.

(a) late A - 3 instar: The root of nerve $1D_2$ fills and the beginnings of the ventral plexus are visible.
(b) early A - 2 instar: The tract in the anterior connective fills and the ventral plexus is larger.
(c) late A - 2 instar: The ventral plexus is now extensive and some fibres can be traced to the midline of the ganglion. The tract running posteriorly through the neuropile is seen faintly.
(d) early A - 1 instar: The projection resembles that of the adult. The only further developments are in the complexity of the branching of the fine terminals

this projection is very similar to that in the adult (Fig. 10d) although the terminal branching may continue to increase in complexity. In contrast, the medial tract of the 1C projection cannot be seen before the beginning of the A - 1 instar when it first fills faintly (Fig. 11b). Later it appears darker (Fig. 11c) but even at the end of the instar the terminal zones are not as bushy as in the adult (Fig. 11d). In the second half of the instar the other branches of the 1C projection first become visible.

Again, the failure of a projection to fill does not necessarily mean that it is absent. It is interesting that the fine plexus of the $1D_2$ projection is one of the earliest structures seen, as this indicates that small fibre diameter may not be the reason that other parts do not fill at this stage. This sequence also correlates with the functional development of the sense organs. The stretch receptor and scolopidial organs in the thorax, whose axons run in nerve $1D_2$, are present in at least the last two larval instars, and responses to distortion of the wing pads can be recorded in the nerve in these stages (Kutsch 1971 and personal communication). In adult crickets Möss (1971) found that the equivalent stretch receptors are active during respiration and walking, so possibly these receptors are already functional in larval locusts. On the other hand, nerve 1C contains axons from the sense organs of the wings, which develop only in the last two larval instars when the wing buds first appear as separate flaps. They cannot become properly functional until the wings are spread after the final moult.

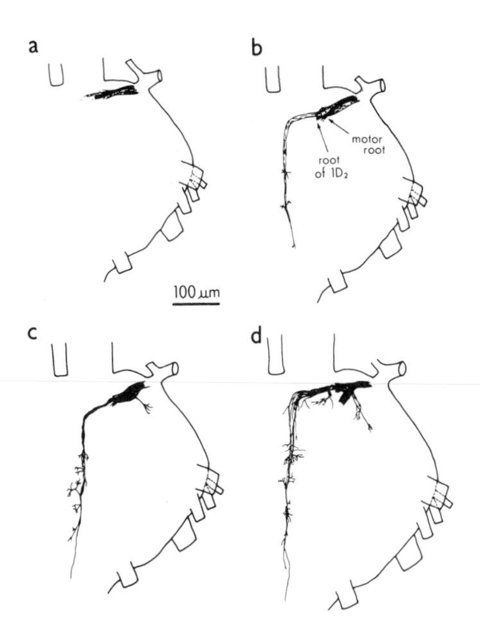

Fig. 11. Development of the central projections of metathoracic sensory nerve 1C.
(a) late A - 2 instar: The motor and $1D_2$ projections fill (cf. Fig. 10c) but the root of nerve 1C is only faintly visible.
(b) early A - 1 instar (1/6 days): The midline tract with a few small terminal tufts is discernible.
(c) mid A - 1 instar (4/6 days): Individual fibres of the midline tract can be followed and the terminal tufts are more extensive. The lateral component is visible for the first time.
(d) late A - 1 instar (6/6 days): The projection is similar to the adult but the terminal branching of both the midline tract and the lateral component are still not complete

PHYSIOLOGICAL STUDIES

Using intracellular recording techniques, it should be possible to investigate the synaptic inputs of the motor neurones and the interactions between them, both when the flight motor pattern is fully established and as it develops. So far this aim has proved difficult to realise.

For accurate studies of development, it is essential that the units investigated can be reliably identified from one preparation to another. Consequently, recordings must be made in the cell body which is the most consistently accessible part of the insect neurone. As most of the flight motor neurones have their somata on the ventral surface of the thoracic ganglia, the most convenient approach for recording is through a window cut in the ventral side of the thorax. The animal is mounted upside down and an air stream is directed over its head (Kendig 1968; Page 1970; Burrows 1973c).

With a few exceptions (Crossman *et al.* 1971), the soma membranes of insect neurones appear to be electrically inexcitable (Hoyle 1970), and this is true of the flight neurones in *Chortoicetes*. The action potential invades the cell body electrotonically and this results in a depolarization of between 2 and 10 mV when the neurone fires (Fig. 12). In the experimental situation, *Chortoicetes* often responds only poorly to flight stimuli and firing in the flight neurones is frequently irregular (Fig. 12a,b) although spiking at a frequency similar to that of normal flight can be recorded.

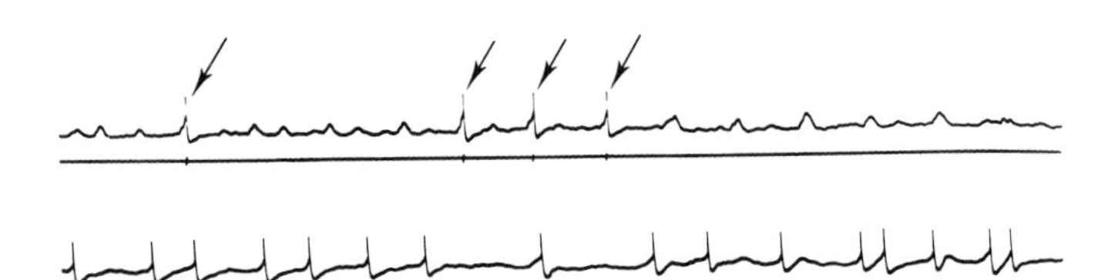

Fig. 12. Intracellular records from one of the 112 motor neurones in adult *Chortoicetes terminifera*, in response to blowing on the head. Top trace: microelectrode record. Bottom trace: myogram from muscle 112 (a and b only).

(a) Excitatory post-synaptic potentials summing to produce rather irregular depolarizations of 1 and 2 mV, but at a frequency similar to normal flight. No response to these depolarizations is seen in the muscle. When the neurone is excited above threshold an attenuated spike of about 4 mV (arrowed) is recorded in the cell body which precedes a spike in the muscle.
(b) A very irregular train of spikes in response to blowing on the head.
(c) Regular firing at flight frequency, probably as a result of damage caused by the microelectrode.

Scale: horizontal - 100 msec; vertical - 4 mV

As might be expected from the size of the dendritic tree, there is an abundant synaptic input to the flight motor neurones. Much of the input is difficult to identify or manipulate and appears to be from interneurones. Bentley (1969) shows that individual excitatory post-synaptic potentials can usually be seen only when excitation is low. Moreover, because of the extensiveness of the dendritic tree, many of the synaptic inputs may be a considerable distance from the recording electrode, so that the events detected in the cell body may only partially reflect the subtleties of integrative activity in the distant branches. There could be significant developmental changes which might go undetected when recording from the soma. It does not seem likely, therefore, that we will be able to identify, reliably, new synaptic inputs as the system develops, merely by recording in the cell bodies.

The most useful information produced, so far, on the nature of the circuitry involved in the generation of the flight motor pattern is Burrows' demonstration of a delayed excitation coordinating the activity of the motor neurones of the left and right metathoracic elevator muscles, the tergosternals (m.113) (Burrows 1973c). When the neurone on one side spikes, there follows excitation of both itself and its contralateral partner with a delay similar to that of the frequency of normal flight. The spiking of the tergosternal motor neurone can also excite the neurone of an ipsilateral antagonistic muscle, the basalar (m.127), but there is no reciprocal excitation of the tergosternal neurone from the basalar. He postulates that the excitation could be mediated with the correct time delays by circuits of interneurones.

It seems unlikely at present, however, that these techniques can be used for studying the development of interneurone circuits of this type. Firstly, the technical problems associated with obtaining stable intracellular records from two cells simultaneously increase considerably in younger, smaller animals. Secondly, even in adults the coupling could only be demonstrated in 20% of the animals tested and seemed to depend on the animals being in a 'flight mood'. Failure to detect similar interactions in younger animals would not necessarily indicate that a circuit was not functional. Next we intend to try to test the development of inputs from the wing sense organs to the flight motor neurones by using an isolated thorax preparation in which the flight response may be elicited by artificial stimulation, as used for example by Wilson and Wyman (1965).

DISCUSSION

The introduction of Procion yellow as a tool for correlating the physiological responses of an identified neurone with its morphology was a breakthrough which provided a method for analysing the tangle of the neuropile and for defining neuronal pathways responsible for particular behaviour patterns (Stretton and Kravitz 1968; Selverston and Kennedy 1969). It has been seen as the key to determining the actual neuronal circuit diagram concerned with a specific piece of behaviour. The improvement in the technique resulting from the use of cobalt (Pitman, Tweedle and Cohen 1972), however, raises serious doubts as to whether this aim is realizable and has several important implications for studies of the development of connections in the flight system.

The most striking feature revealed in our study is the complex morphology of the flight motor neurones. This is particularly surprising when the cobalt pictures are compared with Procion pictures of the same neurones (Fig. 4). Such complexity makes the prospect of producing a complete anatomical description of a neuronal circuit for the flight system much more remote. Even if we could identify all the neurones making functional contact with only one of the motor neurones, it would probably be impossible to describe their connections comprehensibly.

There has been a tendency to think of flight as a relatively simple motor pattern (e.g. Hoyle 1970) and so the Procion picture of the flight motor neurones was quite acceptable. Recent work with locusts on the sensory control of flight stability (Camhi 1970a,b) and regulation of flight speed (Gewecke 1972, and this volume) as well as steering (Camhi 1970a) and coordination with other body functions such as respiration (Hinkle and Camhi 1972), all indicate that stable flight in Orthoptera requires a complicated and subtle integration of many systems. When interneurone systems for the generation of the patterned motor output (Burrows 1973c) are also taken into account, the extent of the dendritic tree of the flight motor neurones which we have described is not unreasonable. It also suggests that much of the integration involved happens in the motor neurones, not at interneurone level.

Studies of the development of synaptic connections with the flight neurones are hampered not so much by the complexity of the branching pattern as by its variability. An accurate study requires that one particular set of inputs can be identified reliably in any individual. Unless a region of the dendritic tree can unambiguously be associated with a special function in the adult, it will not be possible to follow the development of a known input. The variability of motor neurone branching in the adult makes it unlikely that these criteria will be met, although mapping the inputs to the motor neurones may reveal some simple ordering of inputs which would lessen this objection. Furthermore, any such investigation will ultimately require an ultrastructural study to identify synaptic connections between the inputs and the motor neurones, but both the variability and the extensiveness of the dendritic branches make this technically almost impossible. Moreover, as the dendritic tree seems to be well developed before the adult flight pattern is established, new connections may well be widely distributed over the neurone rather than restricted to newly formed branches. The chance of identifying the same set of synapses at different stages of development is therefore fairly slender. Finally, the gradual appearance of the adult flight pattern (Bentley and Hoy 1970; Altman, in prep.a; Weber 1972) suggests increasingly efficient coupling between motor neurones, perhaps through an increase in the number of effective synapses. But it may not be possible to distinguish functional from non-functional synapses on structural criteria alone (Marotte and Mark 1970).

The sensory input from the wings appears to be more promising for a developmental study than the unknown inputs to the motor neurones. Projections like the dorsal midline bundle from metathoracic nerve 1C (Figs. 2b, 8) are fairly compact and each projection probably only contains endings from one type of sense organ. This should make it easier both to locate a group of synapses of known origin and to follow their development. Consequently our efforts are being concentrated on the sensory rather than the motor aspects of the flight system.

The study of the developing flight system, however, still has much to offer as a means of dissecting the mechanism by which the adult flight motor pattern is generated. There is still valuable information to be gained using the myogram technique (Altman, in prep.b) and considerably more qualitative data can be gathered by careful use of cobalt. But the most important results will now probably come from the development of a preparation for intracellular recording in which the flight pattern can routinely be elicited. Then the different developmental stages of the flight pattern, especially in the cricket where the pattern is built up gradually, can be used to analyse the mechanisms underlying particular sets of interactions.

REFERENCES

ALTMAN, J.S.: Changes in the flight motor pattern during the development of the Australian plague locust, *Chortoicetes terminifera*. I. Development of adult behaviour. (In prep. a)

ALTMAN, J.S.: Changes in the flight motor pattern during the development of the Australian plague locust, *Chortoicetes terminifera*. II. Establishment of the alternating motor pattern. (In prep. b)

ALTMAN, J.S., BELL, E.M.: A rapid technique for the demonstration of nerve cell bodies in invertebrate central nervous systems. (In prep.)

BENTLEY, D.R.: Intracellular activity in cricket neurons during the generation of behaviour patterns. J. Insect Physiol. 15, 677-699 (1969).

BENTLEY, D.R.: A topological map of the locust flight system motor neurons. J. Insect Physiol. 16, 905-918 (1970).

BENTLEY, D.R.: Postembryonic development of insect motor systems. *In* "Developmental Neurobiology of Arthropods" (Ed. D. Young), pp. 147-177. London: Cambridge University Press. (1973).

BENTLEY, D.R., HOY, R.R.: Postembryonic development of adult motor patterns in crickets: a neural analysis. Science, N.Y. 170, 1409-1411 (1970).

BRAITENBERG, V.: Ordnung und Orientierung der Elemente im Sehsystem der Fliege. Kybernetik 7, 235-242 (1970).

BURROWS, M.: The morphology of an elevator and a depressor motoneuron of the hind wing of a locust. J. comp. Physiol. 83, 165-178 (1973a).

BURROWS, M.: Physiological and morphological properties of the metathoracic common inhibitory neuron of the locust. J. comp. Physiol. 82, 59-78 (1973b).

BURROWS, M.: The role of delayed excitation in the co-ordination of some metathoracic flight motoneurons of a locust. J. comp. Physiol. 83, 135-164 (1973c)

CAMHI, J.M.: Yaw-correcting postural changes in locusts. J. exp. Biol. 52, 519-531 (1970a).

CAMHI, J.M.: Sensory control of abdomen posture in flying locusts. J. exp. Biol. 52, 533-537 (1970b).

CAMPBELL, J.I.: The anatomy of the nervous system of the mesothorax of *Locusta migratoria migratorioides* R.& F. Proc. zool. Soc. Lond. 137, 403-432 (1961).

COHEN, M.J., JACKLET, J.W.: The functional organization of motor neurons in an insect ganglion. Phil. Trans. R. Soc. (B) 252, 561-572 (1967).

CROSSMAN, A.R., KERKUT, G.A., PITMAN, R.M., WALKER, R.J.: Electrically excitable nerve cell bodies in the central ganglia of two insect species *Periplaneta americana* and *Schistocerca gregaria*. Investigation of cell geometry and morphology by intracellular dye injection. Comp. Biochem. Physiol. 40, 579-594 (1971).

GETTRUP, E.: Thoracic proprioceptors in the flight system of locusts. Nature, Lond. 193, 498-499 (1962).

GETTRUP, E.: Sensory regulation of wing twisting in locusts. J. exp. Biol. 44, 1-16 (1966).

GEWECKE, M.: Die Regelung der Fluggeschwindigkeit bei Heuschrecken und ihre Bedeutung für die Wanderflüge. Verh. dt. zool. Ges. 65, 247-250 (1972).

GUTHRIE, D.M.: Observations on the nervous system of the flight apparatus in the locust *Schistocerca gregaria*. Q. Jl microsc. Sci. 105, 183-201 (1964).

GYMER, A., EDWARDS, J.S. The development of the insect nervous system. I. An analysis of postembryonic growth in the terminal ganglion of *Acheta domesticus*. J. Morph. 123, 191-197 (1967).

HINKLE, M., CAMHI, J.M.: Locust motoneurons: bursting activity correlated with axon diameter. Science, N.Y. 175, 553-556 (1972).

HONG, C.: Descriptive and experimental studies on the embryonic development of *Schistocerca gregaria* (Forskål). Ph.D. Thesis, University of London (1968).

HORRIDGE, G.A., MEINERTZHAGEN, I.A.: The accuracy of the patterns of connexions of the first- and second-order neurons of the visual system of *Calliphora*. Proc. R. Soc. (B) 175, 69-82 (1970).

HOYLE, G.: Cellular mechanisms underlying behavior - neuroethology. Adv. Insect Physiol. 7, 349-444 (1970).

HOYLE, G., BURROWS, M.: Neural mechanisms underlying behavior in the locust *Schistocerca gregaria*. I. Physiology of identified neurons in the metathoracic ganglion. J. Neurobiol. 4, 3-41 (1973).

ILES, J.F.: Structure and synaptic activation of the fast coxal depressor motoneurone of the cockroach, *Periplaneta americana*. J. exp. Biol. 56, 647-656 (1972).

KENDIG, J.J.: Motor neurone coupling in locust flight. J. exp. Biol. 48, 389-404 (1968).

KUTSCH, W.: The development of the flight pattern in the desert locust, *Schistocerca gregaria*. Z. vergl. Physiol. 74, 156-168 (1971).

MAROTTE, L.R., MARK, R.F.: The mechanism of selective reinnervation of fish eye muscle. II. Evidence from electronmicroscopy of nerve endings. Brain Res. 19, 53-62 (1970).

MÖSS, D.: Sinnesorgane im Bereich des Flügels der Feldgrille (*Gryllus campestris* L.) und ihre Bedeutung für die Kontrolle der Singbewegung und die Einstellung der Flügellage. Z. vergl. Physiol. 73, 53-83 (1971).

NEVILLE, A.C.: Motor unit distribution of the dorsal longitudinal flight muscles in locusts. J. exp. Biol. 40, 123-136 (1963).

NICHOLLS, J.G., PURVES, D.: Monosynaptic chemical and electrical connexions between sensory and motor cells in the central nervous system of the leech. J. Physiol., Lond. 209, 647-667 (1970).

OTSUKA, M., KRAVITZ, E.A., POTTER, D.D.: Physiological and chemical architecture of a lobster ganglion with particular reference to gamma-aminobutyrate and glutamate. J. Neurophysiol. 30, 725-752 (1967).

PAGE, C.H.: Unit responses in the metathoracic ganglion of the flying locust. Comp. Biochem. Physiol. 37, 565-571 (1970).

PANOV, A.A.: Correlations in the ontogenetic development of the central nervous system in the house cricket *Gryllus domesticus* L. and the mole cricket *Gryllotalpa gryllotalpa* L. (Orthoptera, Grylloidea). Ent. Rev., Wash. 45, 179-185 (1966).

PITMAN, R.M., TWEEDLE, C.D., COHEN, M.J.: Branching of central neurons: intracellular cobalt injection for light and electron miscroscopy. Science, N.Y. 176, 412-414 (1972).

SBRENNA, G.: Postembryonic growth of the ventral nerve cord in *Schistocerca gregaria* Forsk. (Orthoptera: Acrididae). Boll. Zool. 38, 49-74 (1971).

SELVERSTON, A.I., KENNEDY, D.: Structure and function of identified nerve cells in the crayfish. Endeavour 28, 107-113 (1969).

SNODGRASS, R.E.: The thoracic mechanism of a grasshopper and its antecedents. Smithson. Misc. Collns 82, 1-111 (1929).

STRETTON, A.O.W., KRAVITZ, E.A.: Neuronal geometry: determination with a technique of intracellular dye injection. Science, N.Y. 162, 132-134 (1968).

TYRER, N.M., ALTMAN, J.S.: Motor and sensory flight neurones in a locust demonstrated using cobalt chloride. Brain Res. (In press).

WEBER, T.: Stabilisierung des Flugrhythmus durch "Erfahrung" bei der Feldgrille. Naturwissenshaften 59, 366 (1972).

WILSON, D.M.: The central nervous control of flight in a locust. J. exp. Biol. 38, 471-490 (1961).

WILSON, D.M.: Bifunctional muscles in the thorax of grasshoppers. J. exp. Biol. 39, 669-677 (1962).

WILSON, D.M.: The origin of the flight-motor command in grasshoppers. *In* "Neural Theory and Modeling" (Ed. R. Reiss) pp. 331-345. Stanford: Stanford University Press (1964).

WILSON, D.M.: The nervous control of insect flight and related behavior. Adv. Insect Physiol. 5, 289-338 (1968).

WILSON, D.M., GETTRUP, E.: A stretch reflex controlling wing-beat frequency in grasshoppers. J. exp. Biol. 40, 171-185 (1963).

WILSON, D.M., WYMAN, R.J.: Motor output patterns during random and rhythmic stimulation of locust thoracic ganglia. Biophys. J. 5, 121-143 (1965).

YOUNG, J.Z.: Centres for touch discrimination in *Octopus*. Phil. Trans. R. Soc. (B) 249, 45-67 (1965).

Note added in proof:

More recent experiments have revealed that the sensory projections from the wings are more extensive than reported here. In the metathoracic ganglion the posteriorly directed bundles extend right through thr ganglion but do not enter the abdominal connectives. The mesothoracic bundles have more zones of tufting than shown here. The ventral bundles of nerves 1C and $1D_2$ send tufts into the dorsal neuropile at specific points. Endings of both the nerve 1C and nerve $1D_2$ projections in each ganglion come into close physical proximity with certain branches of the motor neurones. A full description is being published (Tyrer and Altman, in press - see References).

DESIGN AND FUNCTION IN THE INSECT BRAIN

P.E. Howse

Department of Biology, The University, Southampton, England

Aristotle (in *De Anima*) supposed the mind of insects was located between the head and abdomen, and was sometimes multiple, which explained why an insect continued to live when cut in two. The behavioural autonomy of the headless insect is illustrated in recent experimental studies. *Gryllus campestris* can be made to sing calling and courtship songs on stimulation of the thoracic connectives (Otto 1971); a male mantid can complete a sequence of mating behaviour with its head eaten away (Roeder 1963) and a *Hyalophora cecropia* pupa, without a brain, can undergo a sequence of activities enabling it to escape from the cocoon (Truman 1971).

The brain, then, may perhaps be thought of as switchgear which operates the appropriate nervous machinery in the ventral ganglia, and we can think of insects as little machines in the Cartesian sense. Indeed, it is rarely necessary, when describing insect behaviour, to use intervening motivational models and concepts that are legends to almost every picture of vertebrate behaviour. Dethier (1966) once went to great lengths to argue that physiological quantities add up to a drive state, but the tendency to use what Kennedy (1967) has called psychological concepts for processes that are better explained in physiological terms dies hard.

Where insect behaviour reaches its greatest complexity, as in the nest-building of social insects or in the foraging and provisioning behaviour of aculeate Hymenoptera, this complexity can be resolved into a sequential linking of discrete activities. Sometimes it is clear that activities are linked by sensory stimulation of a very specific kind, and without this the behaviour may just stop. When provisioning its nest, the hunting wasp *Liris nigra* normally stings the cricket in four ganglia, of which the suboesophageal ganglion is the last to be stung. Steiner (1962), found that a decapitated cricket caused the wasp immense frustration, which sometimes lasted for more than an hour while the animal appeared to search for the site of the missing ganglion.

Another possibility is that behaviour may recycle if it is disrupted, and there are many observations of this in the literature on hunting wasps. For example Peckham and Peckham (1898) described a spider hunter, *Pompilus quinquenotatus*, which, after capturing and paralysing a spider, hung it on a plant while it reopened its nest hole. The Peckhams replaced the spider with a similar, unparalysed but immobile one. The wasp would not accept this, but flew off and returned with another spider, which it hung up again, and then proceeded to dig a new hole next to the existing one: "Foolish little wasp ... truly, if you are endowed with energy beyond your fellows you are but meagrely furnished with reason." Fabre (1880) wrote on a sphegid wasp that takes Ephippigers for prey. This species positions its prey close to the burrow, enters the burrow and then reappears to drag the cricket into the burrow by its antennae. Fabre moved the cricket away from the hole while the wasp was in its burrow. The wasp reappeared, left the burrow and returned the cricket to the edge once more and descended again, alone. Fabre re-elicited this response about forty times before he lost patience.

The fact that the behaviour sequence, or parts of it, recycles, suggests that internal factors are mainly responsible for the forging of links between the successive activities in question.

Any concept which envisages insect behaviour as a triggering of successive stored subroutines is quite inadequate if it fails to take into account the after-effects of one activity upon another. This phenomenon is clearly seen in the 'dances' of honeybees and in the settling movements of hemileucine moths (Bastock and Blest 1958) which are both performances governed by a measure of preceding flight activity. Tendencies of aphids to fly or to land, or to feed are all interrelated, the duration of one activity affecting the duration of others or the probability of them occurring (Kennedy 1965). Pflumm (1969) showed that there was a reciprocal relationship between tendencies of foraging honeybees to drink or depart from a food source, and that the frequency of grooming could be predicted from measurements of these tendencies. Relationships of a similar kind were found between grooming and tendencies to feed at a food source already used, or at an unexploited one.

It is therefore reasonable to suppose that the following functions accrue to the insect brain: (a) triggering of motor centres, (b) linking of activities into sequences, and (c) transference of effects resulting from the performance of one activity to the controlling mechanism of another. In addition, it must be remembered that some insects are capable of learning complex foraging routes, an ability which may lay claim to a relatively large portion of the brain.

NEUROANATOMY OF THE INSECT BRAIN

Connections of the mushroom body

The paired mushroom-shaped bodies, formed of compact and ordered neuropile, stand out conspicuously in sections of the insect brain (Fig. 1). They were first described by Dujardin (1850) who considered them to be the equivalent of the cerebral cortex because they appeared so large in honeybees. Later, von Alten (1910) found that they were larger in social than in non-social Hymenoptera, and that the calyces were also more cup-shaped in the former. Suggestions that they were substrates for visual integration could not be accepted because well-developed mushroom bodies were found in blind ants (Forel 1928).

A mushroom body typically consists of one or two calyces supported on a short stalk (pedunculus) with two roots. The root formed by a continuation of the stalk was called the β lobe by Vowles (1955), and the lateral outgrowth, which is unusually long in termites and cockroaches was named the α lobe. The calyces are piled high with neurone cell bodies. These cells, whose processes form nearly all the mushroom body tissue, were first described in detail by Kenyon (1896) and I shall refer to them as 'Kenyon cells'. A typical Kenyon cell has a branch, which enters the calyx wall and arborises there, and a long collateral. The collaterals from all the Kenyon cells pour out of the bottom of the calyx to form the stalk. They continue as the β lobe giving (in the honeybee) symmetrically arranged branches to form the α lobe.

Connections between the lobes and other brain areas are difficult to trace and often more apparent than real, as comparison of selectively and non-selectively stained sections show, but Vowles (1955) and Pearson (1971) believe that the β lobe connects with the suboesophageal ganglion. Vowles has also claimed that the α lobe connects with sensory centres and has proposed that there is a feed-back system, whereby the sensory effects of motor output are led into the α lobe and regulate the output

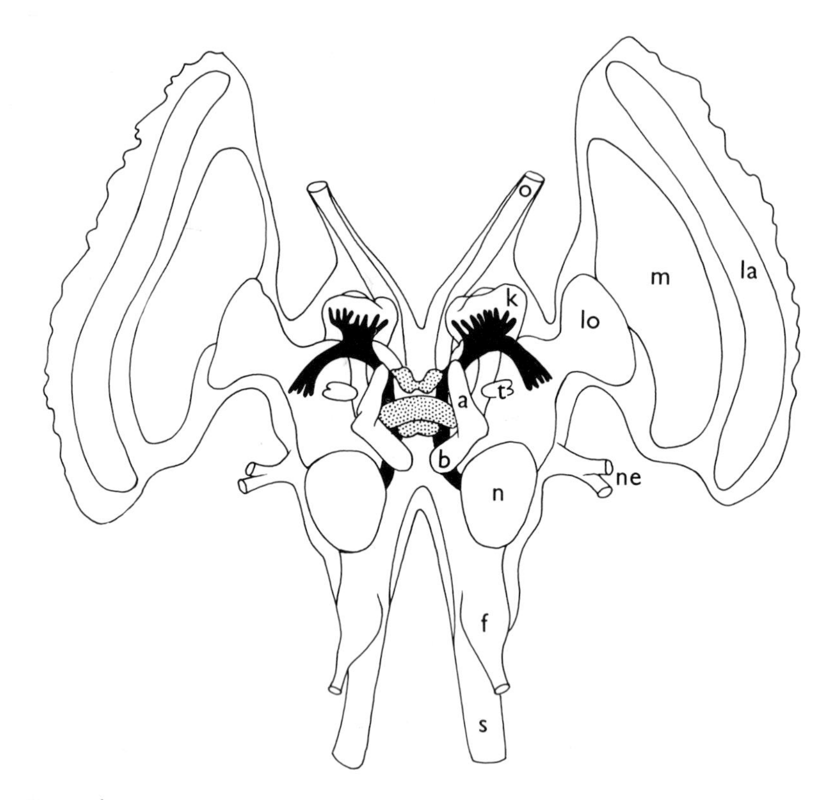

Fig. 1. The main features of the brain of the locust, *Schistocerca gregaria* seen in frontal section.
k - calyx of mushroom body, a - α lobe, b - β lobe, la - lamina of optic lobe, m - medulla, lo - lobula, n - antennal lobe, ne - antennal nerve, t - optic tubercle, s - circumoesophageal commissure, f - frontal nerve, o - ocellar nerve. The antennal complex is lightly stippled, and the olfactorio-globularis tracts running from the antennal lobes to the calyces and ending in the lateral neuropile are shown in black. (Redrawn from Williams 1972.)

of the β lobe. Huber (1960) has neurophysiological evidence that the mushroom bodies are a selection apparatus in crickets. Electrical stimulation of the stalk and lobes produced, according to the site and intensity of the stimulation, aggressive, calling or courtship songs. Stimulation in the region of the central body, on the other hand, produced a new, atypical, song pattern. Huber originally concluded that the choice of action was made by the mushroom bodies which then passed their decision to the central body where the instructions for the song were formulated. Otto (1971) subsequently discovered that a headless cricket could be made to sing normal calling or courtship songs on stimulation of the neck connective nerves and that atypical songs resulted from high intensity, or non-physiological stimulation of the mid-brain. Another role therefore has to be suggested for the central body.

Size variations

A study of comparative anatomy shows that the central body is unlikely to be a storehouse of motor patterns. The volume of this structure was compared in a series of insects chosen to form a spectrum from those with a small behavioural repertoire, e.g. butterflies and sawflies, to those, such as social Hymenoptera and termites, with a rich repertoire. It was found that the volume varies in direct relation with the head width of the insects concerned ($P < .01$, $n = 17$) and not in relation to the apparent complexity of their behaviour. If the volume of the mushroom bodies relative to that of the central body is compared for such a range of insects, it is found to be highest in the social and subsocial Hymenoptera and in termites (Fig. 2). There is, however,

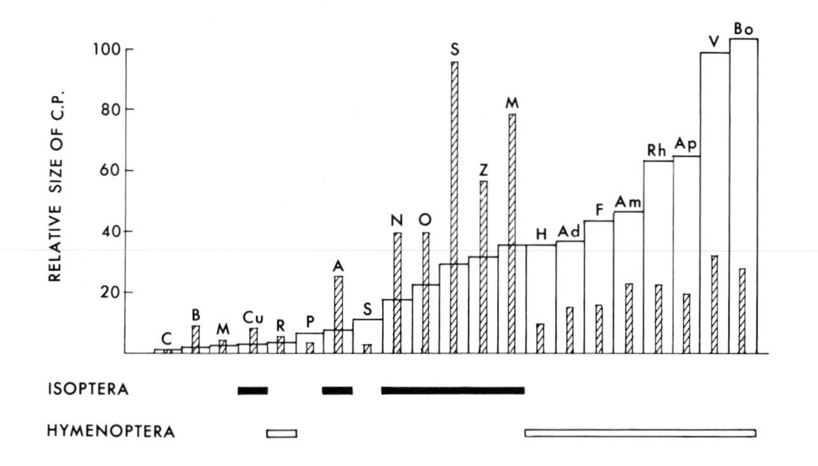

Fig. 2. Relative volumes of the corpora pedunculata of a series of insects differing in their behavioural repertoire. The relative size is obtained by dividing the volume of the mushroom bodies by the volume of the central body in each species and multiplying this by a factor to allow for the relative size of globuli cell bodies in the brain. The volume of the calyces is shown in open blocks and that of the stalk and lobes in hatched blocks.
Species: C - *Chrysops*, B - *Blaberus*, M - *Mantis*, Cu - *Cubitermes*, R - *Rhogogaster*, P - *Pieris*, A - *Apicotermes*, S - *Schistocerca*, N - *Neotermes*, O - *Odontotermes*, S - *Schedorhinotermes*, Z - *Zootermopsis*, M - *Macrotermes*, H - *Hemipepsis*, Ad - *Andrena*, F - *Formica*, Am - *Ammophila*, Rh - *Rhyssa*, Ap - *Apis*, V - *Vespa*, Bo - *Bombus*

a notable difference between the two orders: in the Hymenoptera it is the calyces that are large, while in the termites it is the stalk and lobes that account for by far the greater mass of tissue. Hence, it is likely that one function of the calyces is tied to vision. Insects which have good vision but which show little evidence of a well developed visual memory (e.g. Odonata, syrphids, mantids) have very small mushroom bodies, which indicates that a large proportion of the calyx tissue in Hymenoptera is concerned with visual memory operative in foraging. The lobes, on the other hand, which are universally large in social insects, may be involved in the organization of behaviour sequences. The fact that blind insects can have well-developed mushroom bodies can be explained partly by this dual function of the mushroom bodies and partly by differences in the histology of the calyces and their presumed function (see below). There is a curious paradox that applies to termites, for as behavioural complexity increases - this can be measured in terms of refinement of nest architecture - the size of the brain and the number of neurones in it tends to decrease.

Neuronal architecture of the mushroom bodies

The calyx wall of the honeybee is not an homogenous structure; it has three well-defined regions and even these may be further subdivided. The description that follows differs in some respects from those given by Vowles (1955) for bees and ants, Goll (1967) for ants, and Strausfeld (1970) for honeybees. It is taken from material stained with selective and non-selective silver methods and with the osmic acid / ethyl gallate method of Wigglesworth (1957).

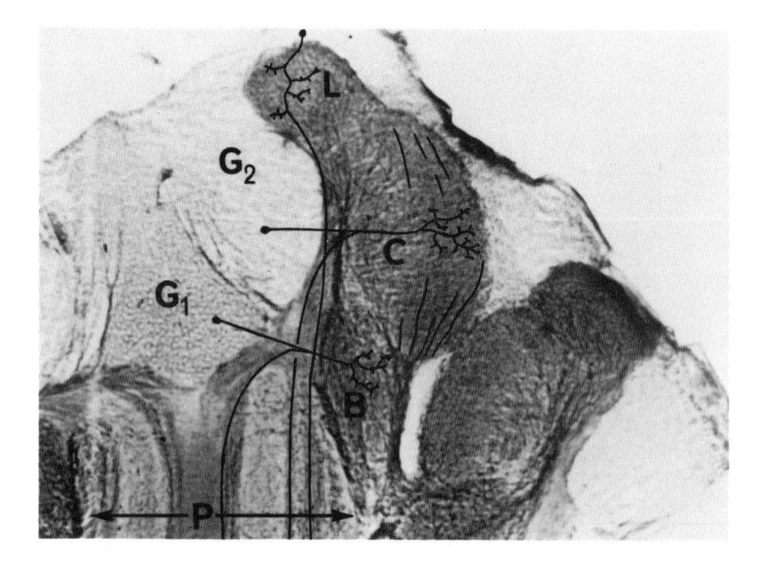

Fig. 3. Photomicrograph of part of the calyx of *Apis mellifera*.
L - lip, C - collar, B - basal ring, P - pedunculus, G_1 - inner group of globuli cells, G_2 - outer group of globuli cells.
The courses of typical neurones entering the calyx wall are shown and the orientation of the fibres of presumed optic origin is also indicated

The basal ring of the calyx (Fig. 3) is distinguished by many chromophilic glomeruli. It receives Kenyon cell fibres from the central cell group of calyx neurones and fibres from the olfactorio-globularis tract which runs from the antennal lobe. Some of the fibres of this tract pass directly to a well-defined lip (*Randwulst*), but many appear to enter the basal ring before doing so. The middle portion of the calyx wall, or collar, receives, particularly in its distal part, some fibres from the continuation of the olfactorio-globularis tract. Its main input, though, comes from an optic lobe tract that enters the calyx between the basal ring and the collar and breaks up into fibres that run parallel with the wall. All parts of the calyx are pervaded by highly arborescent Kenyon cell endings. In the collar, these run roughly at right angles to the fibres with presumed optic function which appear as a form of grid around them.

Fibres from the calyx wall pass into the stalk in an orderly fashion (Fig. 4a) that finds expression in a series of light and dark bands appearing as concentric circles in transverse section (Fig. 4b). The outer band derives from Kenyon cells of the lip, the middle band from the Kenyon cells of the collar, and the inner band from those of the basal ring. The calyces are double each side, and the two stalks fuse a little below the calyx bases. This is achieved by the two cylinder

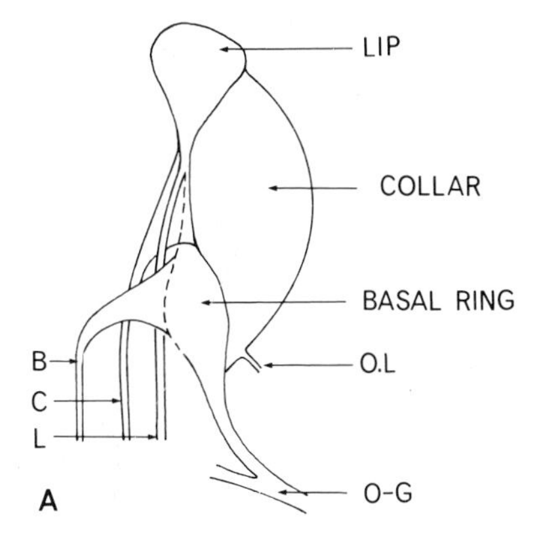

Fig. 4.

A. Diagram to show the structure of the calyx wall of a bee, and the origin of the darkly staining fibre tracts.
OL - presumed optic tract
O-G - olfactorio-globularis tract
B - tract from the basal ring
C - tract from the collar
L - tract from the lip.

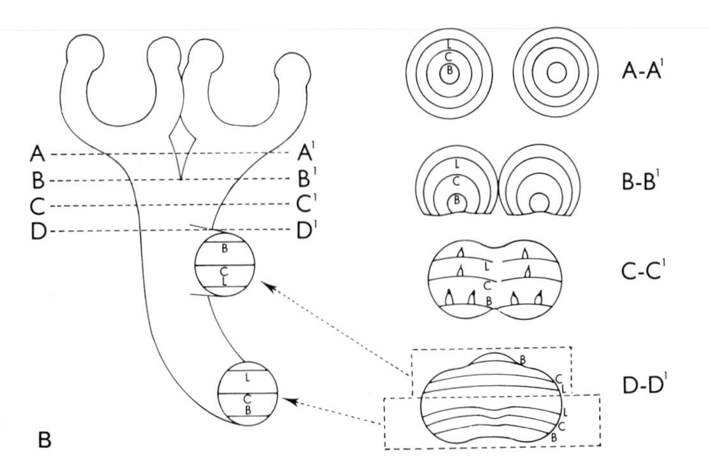

B. Diagram to show the origin of the bands in the α and β lobes from a series of concentric cylinders of tissue in the stalk

systems opening, as if by a longitudinal split, then joining with one another and straightening so that each pair of initially cylindrical bands becomes a single, more or less straight, confluent band. At the level where the cylinders open, finger-like processes of the stalk tissue force outwards to form the α lobe.

The α and β lobes are near identical copies in the honeybee; not only do the Kenyon cells preserve their spatial relationships in each lobe (as in *Formica rufa*) (Goll 1967) but the banding pattern that evolves from them is preserved in each lobe.

Insects without a highly developed visual memory have a different calyx structure. In the wood wasp, *Xeris spectrum*, and in termites, cockroaches and Orthoptera the whole of the calyx wall is glomerular and resembles only the basal ring of the bee brain. This glomerular region is innervated by the olfactorio-globularis tract and its function must lie primarily in the integration of antennal input. Most social and subsocial Hymenoptera have a collar and lip added on to the basal ring. The lip is strikingly large in ants and the collar is strikingly large in bees and wasps (Fig. 5). This, in consideration with the existence of visual fibres in the collar, suggests that the latter is a substrate for the storage of visual memory. The lip, which receives fibres from the basal ring and continuation of the olfactorio-globularis tract should then be concerned with olfactory memory, which we expect to

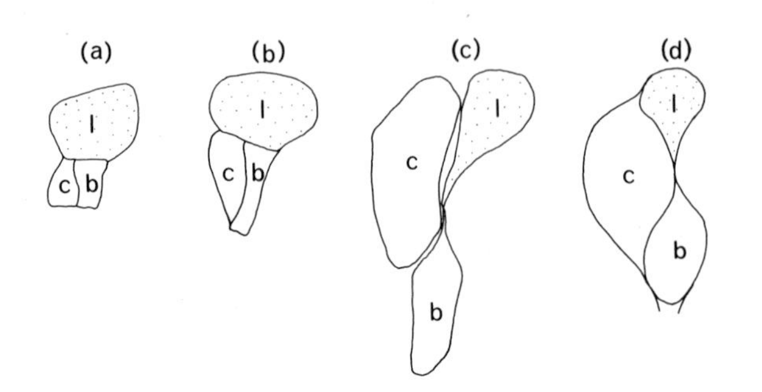

Fig. 5. Sections of the calyx wall of (a) the wood ant (*Formica rufa*), (b) a leaf-cutting ant (*Atta cephalotes*), (c) a potter wasp (*Synagris spiniventris*), and (d) a honeybee (*Apis mellifera*) to show the relative development of the lip (l), collar (c) and basal ring (b)

be of greater importance than visual memory in the foraging of ants.

Any band in the α or β lobe can now be construed as a projection of calyx wall of one level, extending over two calyces, each band having a specific function. There will, however, be some overlap between functions. For example, different banding patterns in the collar can be correlated with the number and distribution of endings which diverge from the tract going to the lip and entering the collar. This may allow an association of olfactory and visual stimuli, represented by a sub-division of a main band in the α and β lobes.

There are about three main darkly-staining bands in the lobe system of bees. In adult cockroaches and termites there are between ten and twelve. In Diptera the bands are in the form of three concentric circles and in *Schistocerca gregaria* the α lobe is largely composed of five nerve bundles, almost circular in cross section (Williams 1972). These bands appear to represent areas dense in synaptic endings and selectively stained sections show that the sometimes very arborescent endings of extrinsic cells terminate in them. Frontali (1968) has demonstrated the existence of catecholamines in the α and β lobes of *Periplaneta americana*, which show as bright transverse bands under the fluorescent microscope. He interprets these bands as zones of extrinsic fibre endings running at right angles to the main axis of the lobes. This interpretation is almost certainly wrong: clear transverse bands can be seen in sections of the cockroach brain especially when the α and β lobes are sectioned in the same plane. They derive from the many longitudinal bands and are artifacts of the plane of section. The existence of the fluorescent bands indicate that the α and β lobes share common transmitter substances which are not found in the calyx and which are probably located in the darkly-staining bands. In the honeybee, extrinsic fibres enter each lobe near its base, in the axillary region, and often have twig-like endings in more than one band. The innervation of the two lobes is, to some degree, symmetrical.

Odonata, Diptera and Hemiptera often have large and highly differ-entiated optic and antennal lobes, but very small mushroom bodies. The calyces, therefore, cannot form a blank screen on which a detailed picture of the outside world can be projected. On the contrary, it is likely that they are organized to respond to specific patterns of stimulation which will 'unlock' certain behaviour patterns. In social Hymenoptera, especially, it is possible that such essentially innate patterns, which are responsible for 'preferred' shapes, scents, etc., can be modified as a result of experience, that is to say the insect can learn to respond to particular features of its environment.

It is possible that memory traces are established in the arborescent Kenyon cell endings. As Vowles (1955) has noted, neurones extrinsic to the mushroom bodies can tap the excitation of bands of Kenyon cell branches in the lobes. Let us suppose that a particular extrinsic cell with endings in the β lobe is a command fibre to a lower motor centre (Fig. 6). It could be made to operate only when its branches at specific

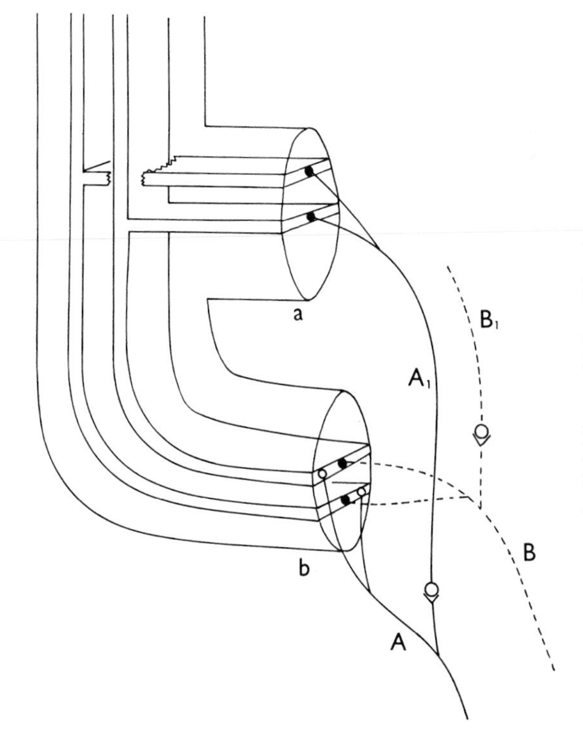

Fig. 6. Diagram of supposed connections between the α lobe (a) and the β lobe (b). Excitation carried down the β lobe by Kenyon cell neurones overcomes the threshold of command neurone A. A collateral from A excites parts of the Kenyon cell bands of the α lobe. This excitation is carried through the symmetrically placed branches of the Kenyon cells to the corresponding positions of the β lobe. Here it lowers the threshold for command neurone B, which will require a different sensory input

points on different bands are excited. This might mean, for example, that only one particular combination of olfactory and visual inputs could excite this command fibre. Sequences of behaviour are now accounted for by supposing that each command fibre has a collateral branch that excites a further interneurone with endings in the α lobe. These endings will lower the threshold for excitation at different points on the α lobe. Since each Kenyon cell has a symmetrically placed branch in each lobe, the thresholds will be reduced at corresponding points on the β lobe, which will have a priming effect of a second command fibre, B. In this way, a sequence of actions, A-B-C etc., can be largely preselected in the central nervous system. If there is no change in the appropriate sensory stimulation, a transition to an activity B becomes increasingly likely the longer A is activated. Both the sensory inflow associated with an action and the duration of the performance could thus affect in a qualitative and quantitative fashion the performance of the next action. Vowles (1955) has given evidence that the β lobe has motor connections and the α lobe sensory connections, although this is disputed by Pearson (1971). Whichever it is (and it may be that both lobes have mixed connections) makes no difference to this theory. If the density of staining in the lobes represents the concentrations of synaptic terminals, then a darkly staining band could have electrical continuity (i.e., electrotonic spread of impulses) from one side to the other. This would reduce the need for point to point connections of extrinsic neurones in the scheme outlined above, and would also mean that the activity of a whole band could be modulated by relatively few endings, thereby affecting the thresholds of many behaviour patterns simultaneously.

The orientation of termites depends very much upon chemical trails; not upon vision or, as far as is known, upon learning. On the above theory it is not difficult to see why they should have small calyces and long lobes. A system of long lobes with many longitudinal bands offers great possibilities for 'wiring in' long behaviour sequences where little flexibility of behaviour is required. There is evidence that in termite species in which the behaviour is apparently relatively complex, the links in sequences of behaviour are more rigid (Howse 1970). The apparent simplicity of the brains of 'higher' termites can thus be explained by postulating that, because of redundancies created in 'better' wiring of a long lobe system, many interneurones have been lost.

Neuronal architecture of the central complex

The central body is the largest part of the central complex, which also includes the protocerebral bridge, the ellipsoid body and the ventral tubercles (Fig. 7). There have been many claims, usually based on material stained with general tissue stains, that direct connections are present between the central body and other zones of 'differentiated'

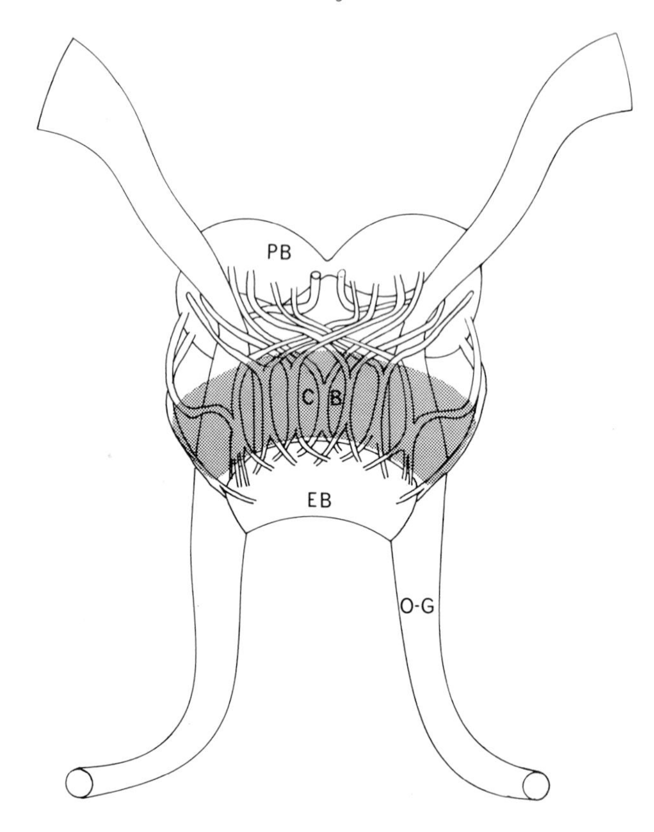

Fig. 7. Diagram to show the relationships between the protocerebral bridge (PB), the central body (CB, stippled), the ellipsoid body (EB), and the olfactorio-globularis tract (OG), in the locust. (After Williams 1972.)

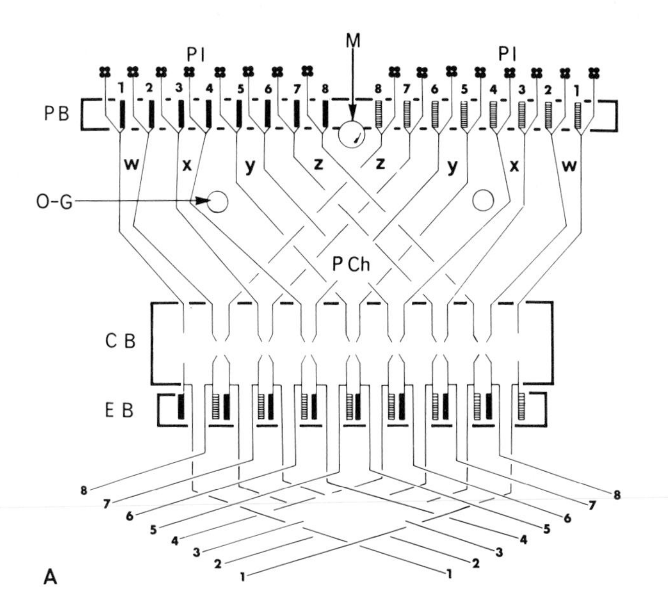

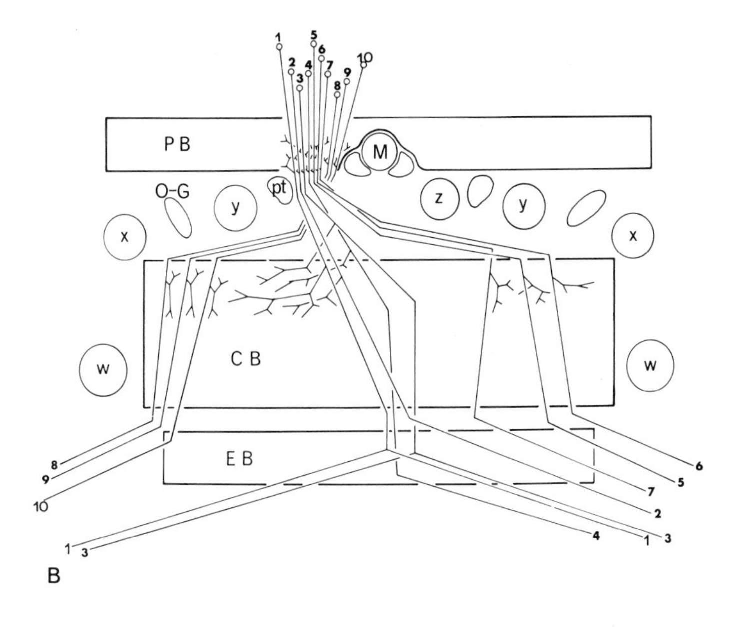

Fig. 8.

A. Diagram to show the distribution of a set of 64 neurones which contribute to the central complex. The cell bodies of these neurones lie in the pars intercerebralis (PI) and the fibres run in bundles of four. Each bundle is represented by a single line. Arborisation occurs in both the protocerebral bridge (PB) and the ellipsoid body (EB) and is represented by the solid and hatched blocks. Between the protocerebral bridge and the central body (CB) the bundles associate in four pairs on each side forming components of the w, x, y and z bundles.
M - median ocellar nerve, PCh - posterior chiasma of the central body, and
O-G - olfactorio-globularis tract.

B. Diagram illustrating the course of 10 pairs of neurones on the right hand side which contribute to the z bundle of the central complex.
pt - protocerebral tract. (After Williams 1972.)

neuropile, such as the α and β lobes. A detailed study of the locust mid-brain carried out in this laboratory has, however, failed to show that any such direct connections exist. This, and the uniformity of structure of the central body throughout the Insecta, argues against Huber's earlier hypothesis that executive decisions made in the mushroom bodies are passed directly to the central body which formulates further instructions.

The striking features of the central complex (see Figs. 7 and 8) are the multiplicity and harmonious order of the connections between its component parts, the existence of several pairs of large fan-like fibres that extend throughout the central body in one plane, and the apparent separation of the whole complex from direct brain pathways. Williams (1972) found that a conspicuous feature of the central complex in locusts is a system of 64 neurones which run in groups of four between the central body, protocerebral bridge and ellipsoid body, arborising in the two latter structures (Fig. 8A). Along part of their route, between the protocerebral bridge and central body, the eight bundles associate to form part of four paired bundles (w, x, y and z). Fig. 8B shows the distribution of ten other neurones forming another subsystem that also contributes to the z bundle and the central body. It is likely that similar subsystems also contribute to the w, x and y bundles, but this is still under investigation.

We have already suggested that the relative constancy of size of the central complex points to a generalised function such as arousal. The many connections between the central body and the other parts of the central complex provide a host of potential reverberatory circuits, which could have the function of maintaining a steady state of nervous excitation. Because of the lack of direct connections with the mushroom bodies or the optic or antennal lobes, this steady state would not be changed by transient events in these structures, but only by relatively persistent or particularly violent changes in their activity. The olfactorio-globularis tract runs very close to the central body and the median ocellar tract runs very close to the protocerebral bridge. It is tempting to suppose that there are connections between the ocelli and the bridge, but in a detailed study, Williams (1972) failed to find any direct connections between the two in the locust.

There have been a number of claims that the central body connects with the ventral nerve cord. We have not been able to confirm this with certainty, but the possibility remains open that the central complex exercises a direct regulatory influence on the excitability of the lower motor centres. An alternative site of possible regulatory input is the stalk, and it is interesting to note that extrinsic neurones with large fields have been described from the stalk of ants (Goll 1967), and they also exist in locusts (Williams 1972).

BRAIN LESION EXPERIMENTS

Table 1 summarises the results of lesion experiments (unpublished) carried out on the brains of locusts (*Schistocerca gregaria*) and honey-bees (*Apis mellifera*). Calyx lesions were made to investigate the role of the mushroom bodies and the split brain experiments were done primarily to study the role of the central complex. The supraoesophageal ganglion was exposed by cutting away a flap of cuticle between the compound eyes. Calyces were removed with fine forceps or a cut was made in the mid-line in the saggital plane, with a glass knife, to separate the ganglion into two. In honeybees, the brain was either split or a series of cuts were made in the calycal region. The flap of cuticle was then replaced. Lesions were also made by drilling into the

Table 1. Summary of the results of brain lesion experiments on *Schistocerca gregaria* and *Apis mellifera*.

LOCUSTS	Split Brain	Calyx Lesions
Atypical leg movements	1/11 (9%)	15/19 (79%)
Atypical flight reflex	0/11 (0%)	14/16 (88%)
HONEYBEES	**Split Brain**	**Mushroom Body Lesions**
Unable to fly	35/35 (100%)	16/27 (59%)
Unable to walk	13/35 (37%)	3/27 (11%)
Disinhibition of competitive reflexes	15/35 (43%)	25/27 (93%)
Continuous sting movements	4/35 (12%)	0/27 (0%)

head capsule with a fine entomological pin. These lesions were small and localised compared with the cuts but less successful in producing recognisable behavioural disturbances.

The mushroom bodies

In both species, calyx lesions led to disturbances in behaviour that can be interpreted as the result of disinhibition of one or more competing reflexes. For example, calyx-damaged locusts fluttered their wings when air was blown on the head, even when all their tarsi were in contact with the substrate. These flight movements resulted in the wings hitting against the legs, as the insects rarely attempted to move from a sitting position. Several locusts, however, lifted their forelegs into the normal flight posture and also lifted their hindlegs until they were almost horizontal. Disturbances in stance and locomotion were common. Some locusts held one hindleg almost vertically, as if in a defensive kicking movement, but the leg moved through a certain angle in common with the other legs when the insect was walking. Others held one or both hindlegs flexed with the tarsi off the ground. Another common feature was hesitation in stepping movements, taking the form of a slow tremor, especially of the forelimbs. In contrast, only one of the split-brain locusts showed clearly atypical leg postures; here one middle leg was held permanently off the ground.

A similar picture emerged from the studies on honeybees, but there was less disparity between split-brain and calyx-damaged insects. The likely explanation for this is that it is almost impossible to split the honeybee brain without damaging the inner calyces and the β lobes, while such an operation is quite easy in locusts.

More than twice as many bees with mushroom body lesions showed clear evidence of disinhibition of competitive reflexes compared with split-brain bees. The same insect could show, at one time, protrusion of the sting, extension of the proboscis, walking and antennal or abdominal preening. Permanent changes in posture or walking patterns were also common among calyx damaged bees. These changes included lifting of the head with the proboscis extended as in trophallaxis, lowering of the hind legs as in flight and landing, and flexion of the abdomen under the thorax.

Experiments on other insects give results that support the hypothesis that the mushroom bodies are operative in selection of motor patterns and in the formation of behaviour sequences.

Van der Kloot and Williams (1953) found that lesions in the region of the mushroom bodies prevented *Cecropia* larvae from carrying out the full sequence of reactions involved in cocoon spinning. Huber (1960) found that complete removal of the mushroom bodies eliminated acoustic behaviour in crickets and grasshoppers, but mechanical damage to part of the calyx or stalk released continuous singing in crickets. Otto (1971) has been able to show that electrical stimulation of the mushroom bodies of free-moving crickets at a single locus could produce complex sequences of behaviour in which foraging, digging and singing occurred. These behaviour patterns followed one another but singing also occurred during locomotion and digging. In one insect, three different song types could be initiated by varying the intensity of stimulation at one point. Sensory information was found to play a role in sequence formation: for example, hunger and the amount of food offered determined the threshold of responses subsequent to foraging. Otto (personal communication) also observed disinhibition of normally incompatible reflexes in *Gryllus campestris* on stimulation of the mushroom bodies, especially in the calycal regions. He obtained (in different animals) singing and preening, singing and locomotion, singing and feeding, and courtship singing with aggressive antennal lashing.

The central complex

Anatomical evidence suggests a regulatory function for the central complex and this is supported by the results of lesion and stimulation experiments. These are, however, never very satisfactory, for it is impossible with current techniques to reach the central complex without passing through nerve tracts whose function and importance are unknown. The results of central lesions are conflicting in the sense that some experimenters report excitatory effects upon behaviour, others inhibitory effects. In honeybees and locusts (see above and Table 1) a spectrum of responsive states was obtained by mid-brain lesions, with extremes of hyperactivity and inertness that were very striking. Four bees began to sting as soon as they recovered from the operation and stung themselves to death. By contrast, some other bees became very unresponsive, even to very strong stimulation. None of the split-brain bees was able to fly, whereas about half those with calyx lesions could fly. Similarly, many fewer split-brain bees could be induced to walk compared with calyx-damaged bees (Table 1).

Drescher (1960) found that damage to the central complex of *Periplaneta americana* by a medial cut resulted in lowered thresholds of responsiveness to sensory stimuli, rapid running around, continuous feeding and autophagy. Roberts (1966) also observed hyperactivity and disturbances to circadian rhythmicity of locomotion in *Periplaneta americana* with split brains, although he attributed the changes to damage to the pars intercerebralis. Drescher (1960) also found that catalepsy could not be induced in split-brain cockroaches, and Steiniger (1933) found the same for stick insects in which the central complex had been cut out. Lesions to the central body of crickets and grasshoppers led to a reduction in locomotion, after an initial increase, and resulted in the elimination of acoustic behaviour (Huber 1960). *Mantis religiosa* with a split brain showed a reduction in locomotory activity but a heightened responsiveness to visual stimuli coupled with an inability to discriminate between food and non-food items (Roeder 1937).

Some of the disparity in these findings may be due to authors drawing general conclusions from the behaviour of only a few animals. This being so, a likely hypothesis is that the central complex is involved in setting the level of responsiveness of the insect and that interference with the machinery of the complex can leave the insect at one of a wide range of possible settings.

SUMMARY

A study of the brains of insects differing widely in the richness of their behavioural repertoires shows that the mushroom body volume is related to behavioural complexity, but that of the central body is not.

The calyces of aculeate Hymenoptera are especially well developed and divided into lip, collar and basal ring regions. Only the latter is found in other insects and it receives extensive input from the antennal lobes. The collar is relatively much larger in bees and wasps than in ants, but the lip is the major part of the ant calyx. Therefore the lip may be concerned with olfactory memory and the collar with visual memory. The stalk and lobes are especially well-developed in termites and are therefore thought to be concerned in the organisation of behaviour sequences.

In the hymenopteran brain, the spatial order of the calyx elements is preserved and translated into a series of bands in the α and β lobes. A simple model is proposed to explain how the structure can generate transitions between behavioural activities, and sequences of behaviour.

Lesion and stimulation experiments on the mushroom bodies support the hypothesis that these are concerned with selection and sequential organisation of behaviour.

The central complex is a highly ordered system comprising many repetitively branching neurones, and does not have direct connections with other neuropile areas. The results of lesion experiments are consistent with the hypothesis that the central complex is involved in setting levels of behavioural responsiveness.

ACKNOWLEDGEMENTS

The unpublished work described here was carried out with the support of a grant from the Science Research Council of Great Britain, to whom I am grateful. It is a pleasure also to thank T.N. Madgwick and J.L.D. Williams for their help and I am grateful for permission of the latter to quote from his unpublished thesis. A Travel Grant from the Royal Society made it possible for me to present this paper at the 14 International Congress of Entomology.

REFERENCES

von ALTEN, H.: Zur phylogenie des Hymenopterengehirns. Jena. Z. Naturw. 46, 511-590 (1910).

BASTOCK, M., BLEST, A.D.: An analysis of behaviour sequences in *Automeris aurantiaca* Weym (Lepidoptera). Behaviour 12, 243-284 (1958).

DETHIER, V.G.: Insects and the concept of motivation. Neb. Symp. Motiv. 14, 105-136 (1966).

DRESCHER, W.: Regenerationsversuche am Gehirne von *Periplaneta americana* unter Berücksichtigung von Verhaltensänderung und Neurosekretion. Z. Morph. Ökol. Tiere 48, 576-649 (1960).

DUJARDIN, F.: Mémoire sur le système nerveux des Insectes. Annls Sci. nat. (Zool.) 14, 195-206 (1850).
FABRE, J.H.: "Souvenirs Entomologiques", Series 1. Paris. (1880).
FOREL, A.: "The Social World of the Ants Compared with that of Man". London: Putnams (1928).
FRONTALI, N.: Histochemical localization of catecholamines in the brain of normal and drug-treated cockroaches. J. Insect Physiol. 14, 881-886 (1968).
GOLL, W.: Strukturuntersuchungen am Gehirn von *Formica*. Z. Morph. Ökol. Tiere 59, 143-210 (1967).
HOWSE, P.E.: "Termites: a study in social behaviour". London: Hutchinson (1970).
HUBER, F.: Untersuchungen über die Funktion des Zentralnervensystems und insbesondere des Gehirnes bei der Fortbewegung und der Lauterzeugung der Grillen. Z. vergl. Physiol. 44, 60-132 (1960).
KENNEDY, J.S.: Co-ordination of successive activities in an aphid. Reciprocal effects of settling on flight. J. exp. Biol. 43, 489-509 (1965).
KENNEDY, J.S.: Behaviour as physiology. *In* "Insects and Physiology" (Eds., J.W.L. Beament and J.E. Treherne). London: Oliver & Boyd (1967).
KENYON, F.C.: The brain of the bee. A preliminary contribution to the morphology of the nervous system of the Arthropoda. J. comp. Neurol. 6, 133-210 (1896).
OTTO, D.: Untersuchungen zur zentralnervösen Kontrolle der Lauterzeugung von Grillen. Z. vergl. Physiol. 74, 227-271 (1971).
PEARSON, L.: The corpora pedunculata of *Sphinx ligustri* L. and other Lepidoptera: an anatomical study. Phil. Trans. R. Soc. (B) 259, 477-516 (1971).
PECKHAM, G.W., PECKHAM, E.G.: "On the Instincts and Habits of Solitary Wasps". Wisconsin Geol. & Nat. History Survey. Madison. (1898).
PFLUMM, W.: Beziehungen zwischen Putzverhalten und Sammelbereitschaft bei der Honigbiene. Z. vergl. Physiol. 64, 1-36 (1969).
ROBERTS, S.K.deF.: Circadian activity rhythms in cockroaches. J. Cell. Physiol. 67, 473-486 (1966).
ROEDER, K.D.: The control of tonus and locomotor activity in the praying mantis (*Mantis religiosa* L.). J. exp. Zool. 76, 353-374 (1937).
ROEDER, K.D.: "Nerve Cells and Insect Behaviour". Cambridge, Mass.: Harvard University Press (1963).
STEINER, A.: Etude du comportement prédateur d'un Hyménoptère Sphégien, *Liris nigra* V.d.L. (= *Notogonia pompiliformis* Pz.). Annls Sci. nat. (Zool.) Ser. 12, 4, 1-126 (1962).
STEINIGER, F.: Die Erscheinung der Katalepsie bei Stabheuschrecke und Wasserlaufen. Z. Morph. Ökol. Tiere 26, 591-708 (1933).
STRAUSFELD, N.J.: Variants and invariants of cell arrangements in the nervous systems of insects. (A review of neuronal arrangements in the visual system and corpora pedunculata.) Verh. dt. zool. Ges. 1970, 97-108 (1970).
TRUMAN, J.W.: The physiology of insect ecdysis. 1. The eclosion behaviour of saturniid moths and its hormonal release. J. exp. Biol. 54, 804-814 (1971).
Van der KLOOT, W.G., WILLIAMS, C.M.: Cocoon construction by the Cecropia silkworm. Behaviour 5, 141-174 (1953).
VOWLES, D.M.: The structure and connexions of the corpora pedunculata in bees and ants. Q. Jl Microsc. Sci. 96, 239-255 (1955).
WIGGLESWORTH, V.B.: The use of osmium in the fixation and staining of tissues. Proc. R. Soc. (B) 147, 185-199 (1957).
WILLIAMS, J.L.D.: Some observations on the neuronal organisation of the supra-oesophageal ganglion in *Schistocerca gregaria* Forskål with particular reference to the central complex. Unpublished Thesis, University of Wales, Cardiff. (1972).

LEARNING AND MEMORY IN THE HONEYBEE

R. Menzel, J. Erber and T. Masuhr

Zoologisches Institut der Technische Hochschule, Darmstadt, W. Germany

The understanding of the physiology of learning is dominated by two basically different hypotheses. The deterministic view, following Hebb's (1949) concept of the memory engram, presupposes a memory groove which is built during memory formation by the adaptive change of a relatively small number of reacting sites or switch points. These so-called 'switch-point theories' or 'place theories' assume that memory involves a discrete set of cells reserved for the special function of information storage (Young 1964; Eccles 1964; Ungar 1970). The non-deterministic or statistical theory is based on Lashley's (1950) findings which suggest that all, or nearly all, stored information is distributed throughout the whole association cortex rather than by distinct association paths or centres. The individual neuronal switch points may then be involved in the storage of many different memory traces (John 1967, 1972). The two views are similar in that they take the adaptivity of single synapses between neurones as the basic modifiable component of the nervous system (Eccles and McIntyre 1953; Eccles 1964; Ungar 1970; John 1972). They differ, however, in their conception of the gross structure of the memory system. The crucial problem, then, is to locate the stored information. The spatio-temporal pattern of activity during memory formation produces a localised change in the excitability of specific neurones. It should be possible to find such neurones using the same techniques as have been employed for the location of units in the sensory integration centres.

The failure to find such specific neurones has led proponents of non-deterministic theories to conclude that the informational significance of an event is represented by the average behaviour of a responsive neuronal ensemble, rather than by the exclusive behaviour of any specifiable neurone in the ensemble. However, the failure to find the neurones predicted by the deterministic theories cannot be regarded as conclusive evidence against these theories. The vertebrate brain consists of 10^{10} neurones each with as many as 10^3-10^4 synapses. This complexity is so great that it may be impossible to locate with existing recording methods a single specific neurone, the parameters of whose responses are initially unknown.

A more fruitful approach could be to use simple nervous systems or parts of nervous systems containing relatively few neurones. Some, e.g. *Aplysia* have already been used successfully in the study of synaptic adaptivity (Kandel and Tauc 1965; Kandel and Spenser 1968; Kandel *et al.* 1970). Others describe adaptive changes on the level of single neurones (e.g. leg movement preparation in cockroaches and locusts; Horridge 1962, 1968; Hoyle 1965; Eisenstein 1972). These simple systems show habituation, sensitation, hetero- and homo-synaptic facilitation and some kind of associative adaptivity. But it is still open to question whether these are examples of 'true' learning. Again and again attempts to define learning have pointed up the difficulties in distinguishing between sensory or motor adaptation, sensitation, habituation and associative learning (see Thorpe 1963). Difficulties arise even where adaptive changes of neuronal elements are studied, but where the details of the behavioural changes caused are not known. We find that, by reducing the complexity of the nervous system to allow a more successful approach to the study of storage mechanisms, the possibilities for both learning and memory themselves may be lost. Simple systems as mentioned above, may be not only quantitatively different from a complex structure, but may have different and new rules of integration and storage. Therefore we must demonstrate learning and memory, without any doubt, by behavioural studies *before* detailed analyses are undertaken. On the other hand it is necessary to search for an animal with less complex behaviour than most vertebrates, more dominance of innate behaviour patterns and less variety in learning.

We know that several insect species show learning in their behavioural repertoire. For example, bees, wasps, ants, etc. learn orientation marks rapidly and remember them for a long time (Markl and Lindauer 1965). Most of the behaviour patterns in insects are genetically fixed and learning plays only an insignificant role compared with the variety of innate behaviour patterns. This is so even in the most highly evolved insects, the Hymenoptera. We now understand the function of the insect brain as a serial arrangement of sensory integration centres and parallel controlling and coordinating boxes for special action patterns (Huber 1965, 1967; Maynard 1967; Markl and Lindauer 1965; Hoyle 1970). It is possible that integration, control and coordination follow functionally fixed neuronal pathways which lack plasticity in the sense of long lasting changes as a result of experience. Learning may, therefore, be restricted to neuronal circuits linking the controlling and coordinating boxes. The plasticity of these linking circuits is small in nearly all insect species, but is more developed in social insects such as bees, wasps and ants.

The whole insect brain contains 10^5-10^6 neurones (Witthöft 1967) many of which (about 80%) are distributed in peripheral lobes or ganglia (Hoyle 1970). Consequently, a relatively small number of neurones must be involved in storage mechanisms. We can, therefore, postulate the existence of a very economical learning system. In recent years the deterministic approach has been used very successfully for the explanation of neuronal mechanisms underlying such fixed behaviour patterns in insects, as flying, walking, breathing and singing (Huber 1965, 1967; Hoyle 1970).

In this paper we will present a selection of our experiments concerning the learning process and memory in the honeybee. We chose this insect because it is one whose sensory capabilities and learning behaviour have been extensively studied. Also, we have chosen its learning of spectral colours as our specific example as there is already detailed information on colour vision in bees (von Frisch 1914; Daumer 1956; Autrum and von Zwehl 1964; Menzel 1967; von Helversen 1972) and the physical parameters of spectral light can be controlled easily. Our first step was to perform experiments designed to define quantitatively the parameters of visual learning within the bee's normal behaviour. In these the test animals were unrestricted and were free to fly between hive and experimental apparatus depending only on their motivation to collect food. The next set of experiments included learning tests in which the animals were fixed to allow direct access to the brain. The results of these tests could be compared with those obtained with unrestrained bees to determine the influence of the experimental method.

The greatest barrier to the analysis of the learning process in the insect brain is our lack of knowledge of its functional structure. The ingenious experiments of Huber and his co-workers (Huber 1965; 1967) on the function of parts of the brain controlling different song patterns in crickets are practically the only work being done in this area (compare Bullock and Horridge 1965; Hoyle 1970). Much of the related work is reviewed in this volume by Howse who also gives a model of the interactions between the mushroom bodies and the sensory and motor centres in the brain and will not be repeated here.

Indirect evidence that the mushroom bodies may be association centres comes from the observations (1) that their volume increases with the complexity of the behaviour (von Alten 1910; Holmgren 1916; Pietschker 1911; Hanström 1928; Pandazis 1930; Gossen 1949), (2) that the number of cells in the mushroom bodies is much higher in the worker bee than in the drone bee (40% and 24% respectively of all brain cells) (Witthöft 1967), and (3) that the density of synapses in mushroom bodies is high especially in species with complex behaviour (Vowles 1955; Goll 1967; Schürmann 1970, 1972). Our recent experiments give evidence that long-term memory is located within the mushroom bodies. During the establishment of long-term memory the sensory ganglia also are involved. This means that the location of memory 'traces' changes with time after the initial association, and that information on the time dependent processes is needed to understand the succession of storage places.

THE BEE TRAINING METHOD AND LEARNING CURVES

In our standardised procedure, we train freely flying bees in almost natural conditions. Their high flower specificity and persistent searching behaviour enable us to study their learning of artificial signals. The experimental arrangement and procedure is described elsewhere (Menzel 1967). Briefly, bees are fed on a ground glass disc containing sugar solution 30-50 m away from the hive. The disc lies in the middle of a large round grey table and can be illuminated from beneath with spectral colours. Two other ground glass discs are placed on either side of the one containing food. These two can be illuminated with two alternative colours. During training, the bee has access only to the middle disc on which it finds a sugar solution as reward; in the test situation only the two alternative discs are illuminated with the test wavelengths and no reward is given. All experiments are carried out with individual marked bees. The test bee is a new-comer which has been directed to the experimental area by the dances performed by a group of bees which collect continuously from the disc. The motivation of such new-comers to search and learn varies little between animals and is almost constant in the same animal because the animal comes to the food source only if its motivation is sufficiently high. The group of dancing bees is removed from the test area during the experiment. During each reward, the test bee is allowed to suck its fill of 2M sugar solution and then return to the hive. Each reward consists of a series of actions during the foraging cycle, namely leaving the hive, landing on the disc, sucking, flying back to the hive and regurgitating.

Pretraining which familiarizes the animal with the experimental arrangement and the reward situation consists of three rewards on an unlit disc. After pretraining, the spontaneous choice preference to the two alternative colours is tested. The spontaneous choice level depends on the wavelengths of the alternative colours and their radiation density. By varying the radiation density of both spectral colours, one can manipulate the spontaneous choice level so that each colour is chosen about equally. The radiation densities for pairs of wavelengths to elicit about a 50% choice (45-55%) is very much the same in different animals (Menzel 1967). All *tests* last 4 min during which the test bee makes up to 40 "choices"; a choice being defined as a direct approach to within 2 cm of one of the coloured discs.

After the spontaneous choice test, the bee is rewarded on one of the colours (1st reward) which is later to be displayed on one of the outer two discs and tested again after returning from the hive. Bees are tested after each of several rewards on the colour to be learned (λ_+) and the results are expressed quantitatively by a learning curve for each test bee. Fig. 1 shows some individual learning curves and the

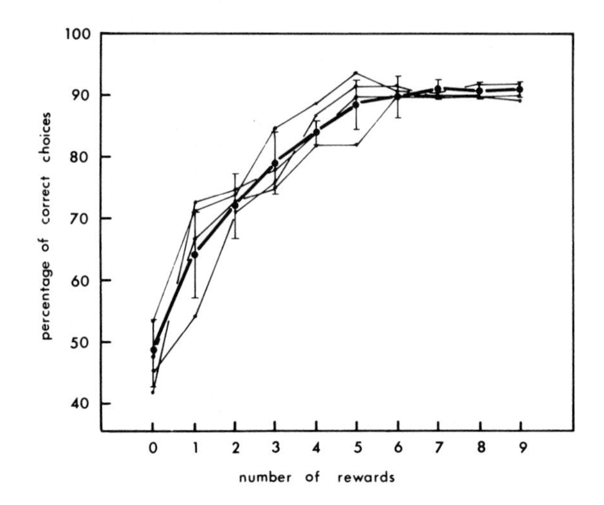

Fig. 1. Learning curves of 4 individual bees (thin lines) and the average learning curve of 71 bees (number of decisions n = 21,300; thick lines with standard deviations). The colour rewarded is λ_+ = 532 nm; the alternative colour in the test is λ_- = 413 nm. The first point on the abscissa (0) is the result of the spontaneous choice test, after a pretraining of 3 rewards on an unilluminated ground glass disc. The bees need 4 to 5 rewards with sugar solution on the green (532 nm) ground glass before they make a high percentage (>90%) of correct choices

average curve for λ_+ = 532 nm (alternative colour in test, λ_- = 413 nm). Within 4 rewards λ_+ is learned and chosen with high probability. Such a learning curve can be established in about two hours as the bee makes regular flights every 3-10 min.

Learning curves are found to be unaffected by the following parameters: temperature, time of day, season, social factors (e.g. food requirements of the hive), age of the test bee, genetic differences in flight activity, quantity and quality of reward (Menzel 1967; Pflumm 1969a,b; Rau 1970; Menzel and Erber 1972; Menzel, Freudel and Rühl 1973). On the other hand, learning curves are affected by other parameters: nature of learning signal, number and duration of rewards, number of interruptions during reward, genetic factors influencing learning disposition, and temporal relationship between stimulus and reward (Opfinger 1931; Lindauer 1970; Menzel 1967, 1968, 1969; Menzel and Erber 1972; Menzel, Freudel and Rühl 1973; Erber 1972).

It should be emphasized that concentration and availability of sugar solution over a wide range have no effect on the learning performance during the first few rewards (Menzel and Erber 1972). Those factors not influencing the learning process will not be discussed further. We shall concentrate on the influence of the nature of the learning signal, the duration and sequence of reward and the temporal relationship between stimulus and reward. An understanding of the dependence of learning on these parameters allows the development of certain testable hypotheses.

SPECTRAL COLOUR LEARNING

Using the method described in the section above, we trained many individual bees to spectral colours, choosing complementary colours as λ_+ and λ_- (Daumer 1956). The average learning curves are given in Fig. 2. Bees need be rewarded only once on a violet illuminated disc (413,

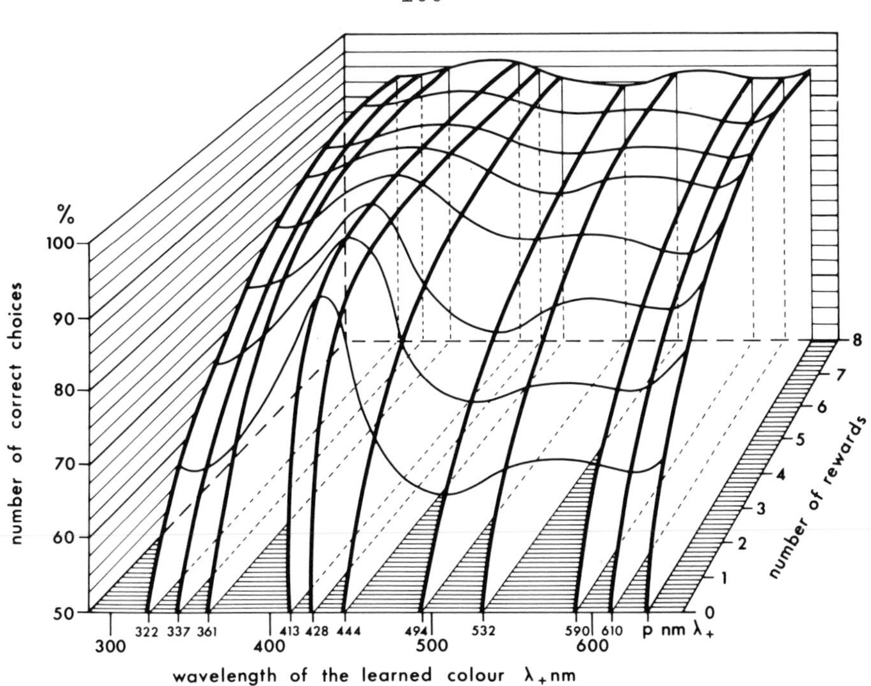

Fig. 2. Average learning curves of honey bees for spectral colours in the range 300-600 nm. In the test situation the alternative colour λ_- is always the complementary colour of λ_+. The intensities of both colours are adjusted so that the animals spontaneously choose both colours with equal frequency. The small differences in the average spontaneous preference for different colours (<±5%) are ignored here. "p" (x-ordinate) shows purple for bees. This is a mixture of about 12% UV and 88% yellow (see text). Total number of choices is n = 54,712. The bees learn violet (413, 428 nm) most quickly and bluish green (494 nm) most slowly. The small second maximum for the learning rate in green is not significant, but the maximum in violet is significantly different from all other wavelengths tested

428 nm) to make them choose this colour with high probability (85%). Bluish green (494 nm) is learned most slowly; after the first reward it is chosen only in 59% of the decisions. Blue (444 nm) is learned significantly faster than yellow and orange. The learning curves of the other colours tested rise equally steeply, all differing significantly from violet, bluish green and blue. After more than 6 rewards all learning curves reach a plateau level between 89% (λ_+ = 494 nm) and 93% (λ_+ = 413, 428 nm).

It is important to know if these differences in the learning curves for different wavelengths of λ_+ depend on the experimental procedure used or if these differences can be understood purely on the basis of receptor properties. First we showed that the *pretraining* does not change the learning rate. Then we determined whether the alternative colour λ_- influences the learning curves. Tests were made with 5 different wavelengths of λ_+ each compared with 10 different wavelengths of λ_-. The learning curves for each λ_+ are almost identical, as long as the alternative colours are distinguishable to the bee. Thus λ_- does not influence the characteristic learning curve for λ_+ and the complementary colour, therefore is representative of all other colours appearing unlike λ_+.

The dependence of the learning curve on the intensity of λ_+ was also investigated to determine whether differences in the subjective brightness of the spectral colour influences learning rate. To test this, we obtained learning curves for 5 different λ_+ as the radiation density on the glass disc was decreased. The learning rate is independent of the radiation density of λ_+ over a very wide range. The learning curves flatten out only when the intensity approaches perceptual threshold. Near perceptual threshold and near discrimination threshold between λ_+ and λ_- the characteristic differences in the learning curves for different wavelengths can still be seen. The learning plateau is much lower, but this plateau is reached for violet after one or two rewards, and for other wavelengths after several rewards. These experiments show that the differences in the learning rate for different λ_+ is independent of the experimental procedure. Furthermore, these results indicate a very small if any dependency on receptor parameters.

A more detailed comparison of the learning system with the colour vision system is possible, since we know the properties of the colour receptors from single cell recordings and the result of the colour integration system from behavioural determinations of the spectral threshold and spectral discrimination function (Fig. 3). The three different types of colour receptors have their sensitivity maxima at

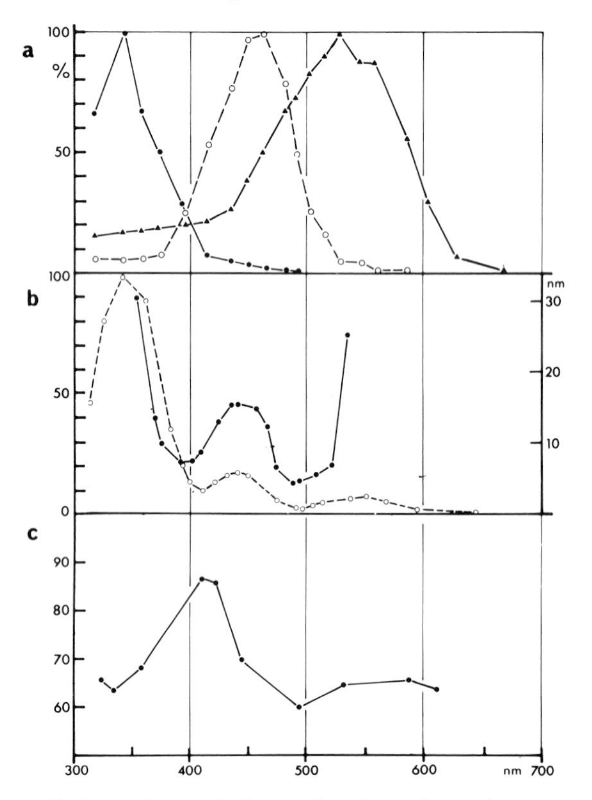

Fig. 3. Comparison of the colour vision and colour learning system in bees.
(a) The spectral sensitivity of single photoreceptors in bees measured by intracellular recordings (Autrum and von Zwehl 1964). There are 3 main types of receptors: UV (-●-), blue (-O-) and green (-▲-) receptors.
(b) The dashed line shows the spectral threshold function (left ordinate) and the continuous line shows the spectral discrimination function. $\Delta\lambda$ (right ordinate) is the minimum wavelength difference between two spectral colours which can be distinguished by the bee with a reliability of 70% correct responses (von Helversen 1972). Spectral sensitivity is highest in UV and lowest in bluish green, whereas spectral discrimination is best in the violet region (around 400 nm $\Delta\lambda$ = 8 nm) and bluish green (around 440 nm $\Delta\lambda$ = 5 nm). Both functions prove that the bee has a trichromatic colour vision system based on the colour receptors given in a.
(c) The percentage of correct choices after one reward on a colour given on the abscissa. As in Fig. 2 the alternative colour in the test is the complementary colour. Learning rate is highest in violet (413, 428 nm) and lowest in bluish green

340 nm (UV), 420-460 nm (blue) and around 530 nm (green) (Autrum and von Zwehl 1964; Menzel, unpublished). As each ommatidium contains two or all three colours receptor types the bee has a fine grain colour vision system (Snyder, Menzel and Laughlin, in press). The spectral threshold and spectral discrimination function (Fig. 3) reflect the trichromatic colour vision system. Both functions were measured with freely flying bees tested with about the same experimental arrangement as that used in our experiments (von Helversen 1972; Fig. 3). The colour appearing brightest to the bee is UV. Discrimination is highest in bluish green and violet. The colour learned most quickly however is violet; bluish green is the colour learned most slowly; UV and green are learned with about the same rate.

The comparison of the bee's colour vision with the colour learning shows that learning does not depend on a simple receptor mechanism or on one or the other of the colour integration mechanisms only. Violet - the colour learned best - is the region where colour discrimination is high *and* subjective brightness is still very high. In bluish green discrimination is highest but subjective brightness is poorest, and the learning rate is slowest for bluish green. This comparison shows that information for colour learning is not extracted from simple sensory mechanisms but requires complex integration processes which take colour discrimination and subjective brightness into account.

The biological background for these findings seems to be that the bee's visual world is divided into several distinct parts, which would be an economical way of managing all the visually controlled behaviour patterns. UV is the colour of the sky for the bee. Orientation to the polarisation pattern of the sky is mediated by a specialised UV-receptor (von Frisch 1967; Snyder, Menzel and Laughlin, in press; Menzel, unpublished). UV is very useful for navigation, because polarisation of skylight is more pronounced in UV and less affected by atmospheric distortions (von Frisch 1967). Bees use mainly the green receptors for the regulation of flight (Kaiser 1972; Kaiser and Liske 1972; Menzel 1973). The green receptors are very useful for this purpose because they have a very broad spectral sensitivity function and they are a receptor type found most frequently in the eye (Menzel and Snyder, in press). Bees may use mainly the long wavelength region from blue to yellow for orientation to landmarks because in nature the arrangement of trees, grassland areas, sand and rocks guide the bees on their foraging flights. The high colour discrimination found in bluish green will be of great importance here.

On the other hand, flowers that bees visit show a broad variety of different colours, but very few reflect UV or red only. The largest number of insect-visited flowers are violet, blue or "bee-purple" (Daumer 1958, see von Frisch 1967 for further details). "Bee-purple" is a mixture of UV and yellow (e.g. 12% UV and 88% yellow) which is distinguished by the bee from UV and yellow and appears to be a colour quality of its own to the bee (Daumer 1956). Pure UV would be an unreliable colour signal for flowers since their appearance would change considerably depending whether they are in sun or in deep shade.

Colouration of the blossom and the colour vision system of flower visiting insects is the result of co-evolution (Sprengel 1793, cited after von Frisch 1967). In addition our results show that not only the colour vision system but also the learning system has phylogenetically adapted to the colouration of the nectar bearing blossoms. This "phylogenetic pre-learning" enables bees to learn more quickly those colours which are more likely to be food signals.

In summary, then, it appears that the information for the colour learning signal is not extracted from receptor inputs by a simple mechanism but by complex integration processes. These processes are understood as "phylogenetic pre-learning", which increases learning rate for food signals of higher probability.

LONG-TERM MEMORY

Bees learn colours quickly and they retain them in their memory for a long time. Lindauer in 1963 observed that trained bees came to the same training place after a winter period of 173 days. If the hive is displaced laterally to the line of flight by only 2-3 m, bees return to the old place within the next 12 days. If they do not fly out at the new place and therefore do not learn new orientation marks, they can remember the old location for up to 30 days (Viullaume 1959; von Frisch 1967, p. 467f). Therefore we would expect to find a long lasting memory.

We have studied the retention of spectral colours after differing numbers of initial rewards. If test bees are rewarded three or more times on any spectral colour they do not forget it for at least a fortnight, provided they are prevented from flying out and learning new cues (Menzel 1968). If they are rewarded only once for 30 sec on a certain colour they will forget it within 2-6 days (Fig. 4). The colours which are learned more quickly are retained longer than those learned slowly, e.g. after one reward on λ_+ = 444 nm, bees revert to spontaneous preference after 6 days, whilst choices for λ_+ = 590 nm, a colour learned more slowly, reach spontaneous choice level after 2 days.

The retention curves in Fig. 4 demonstrate 3 different phases of retention. In the first two hours after the reward the percentage of correct choices increases. Then the retention remains constant for from several hours up to 1 day, and finally there is an exponential memory decay. The initial rise of retention is also found in learning experiments with vertebrates (e.g. McGaugh 1966). Since this period coincides with the time span of greatest pharmacological susceptibility of stored information (to puromycin, cycloheximide; e.g. Agranoff 1967; Miller 1967), it is thought to be a consolidation period and the initial improvement is interpreted as the transition from short-term to long-term memory. We studied this question in the experiments discussed in the next section.

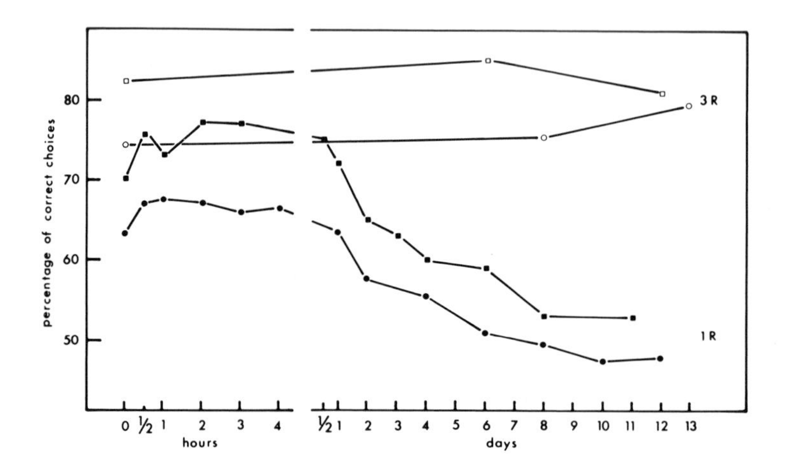

Fig. 4. Long term retention curves for λ_+ = 444 nm (-■- and -□-) and λ_+ = 590 nm (-●- and -O-). The two lower curves indicated by 1R give the percentage of correct choices after one reward on the colour to be learned. The abscissa indicates the interval between the last reward and testing. The two upper curves (3R) show that after 3 rewards both colours are well remembered during the following two weeks

TIME-DEPENDENCE OF THE STORAGE PROCESS

A basic problem in memory formation is that of the transfer of information between processes occupying different time courses. The responses of the receptors and sensory interneurones last milliseconds or seconds, but the establishment of a long-lasting memory trace may take from many minutes up to some hours. A maintaining system is therefore necessary as a bridge between these time dependent processes, to retain the transient activity with its high information density, until the transformation into the slow running processes is completed. In addition, it is very often necessary for the animal to recall the learned cues when the formation of a long lasting memory trace is not yet completed.

There is much experimental evidence that the storage process is a multiphasic process in both vertebrates and invertebrates (e.g. Chen, Aranda and Luco 1970; McGaugh 1966, 1969; Messenger 1971; Riege and Cherkin 1971; Young 1970; Sanders and Barlow 1971). It is possible to distinguish between different kinds of memory processes by differences in their time course and sensitivity to disruptive procedures. A frequently used classification is the differentiation between a "sensory memory", a short-term and a long-term memory.

We have examined the time dependence of the learning process and the effect of different disruptive procedures on the performance of freely flying bees. First we showed that bees learn a colour signal only in a

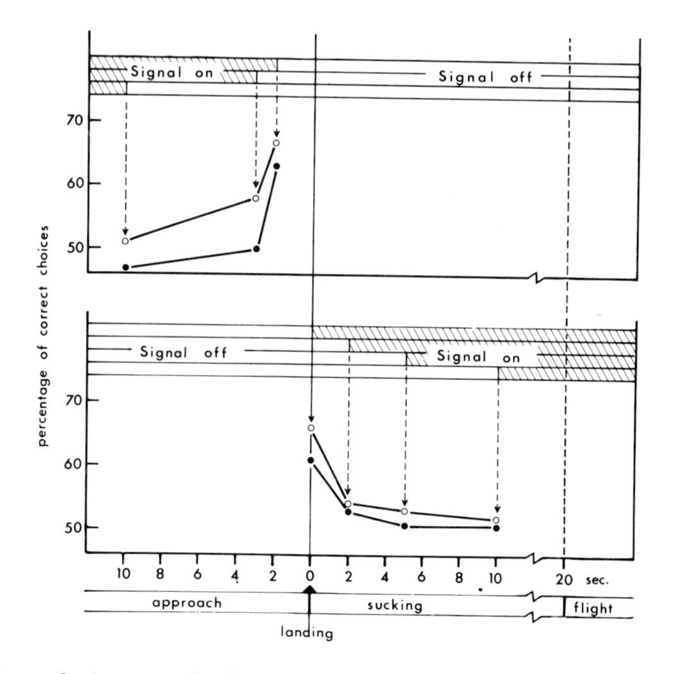

Fig. 5. Dependence of the association process on the time relation between presentation of the colour signal and reward. In the first experiment (upper graph) the bee approaches the food source when the colour signal is switched on (-O- $\lambda+$ = 444 nm; -●- $\lambda+$ = 590 nm). The colour is switched off a short time (10, 3, 2 sec) before the bee lands and starts to suck. The bee is then rewarded for 20 sec on an unilluminated disc. The curves show the percentage of correct choices during the following test. The colour signal is associated with food only if the bee sees the signal 2 sec before landing. In the second experiment (lower graph) approaching bees see no colour. The signal is switched on the moment the bee lands or some seconds later. Here too, the reward lasts for 20 sec. The curves for both colours (444 nm, 590 nm) show that association with reward takes place only if the colour is seen in the first second of sucking

short period during their approach to the colour signal. In this experiment (Fig. 5, left side) the bee approached the illuminated ground glass disc, then the colour signal was switched off a short time (10, 3 and 2 sec) before the bee landed and started to suck. We found that,if the colour signal was seen 2 sec before the reward started, it was learned with the same success as if the bee could see the signal during approach, sucking period and flying off. If the colour signal was switched off more than 2 sec before landing it was not learned. This experiment showed that bees first store the colour signal for 2 sec in a "sensory memory" where the signal is associated with the reward (Menzel 1968). If bees approach an unilluminated ground glass disc and the colour signal is switched on at the same time as or after landing, it can be shown that the association between signal and reward occurs within a very short time before and during landing, and extension of the proboscis. If the colour is switched on after landing the bee does not learn it even though it may suck for 26-28 sec on the illuminated disc and can see the colour while flying away. These results quantify an earlier observation by Opfinger (1931) which showed that, for a signal to be learned, it must be seen by the bee during its approach.

This test situation is excellent for the study of the time dependence of memory storage. Since freely flying bees need only one reward to learn the training colour with a high percentage of success, the instant of association can be defined with an accuracy of 1-2 sec and the quantity of perceived information can be calculated. The results described above and elsewhere (Menzel and Erber 1972) have shown that the quantity of sugar solution given in the first reward has no influence on the storage process. On the other hand the duration of the reward does influence the time course of the retention. This was demonstrated as follows.

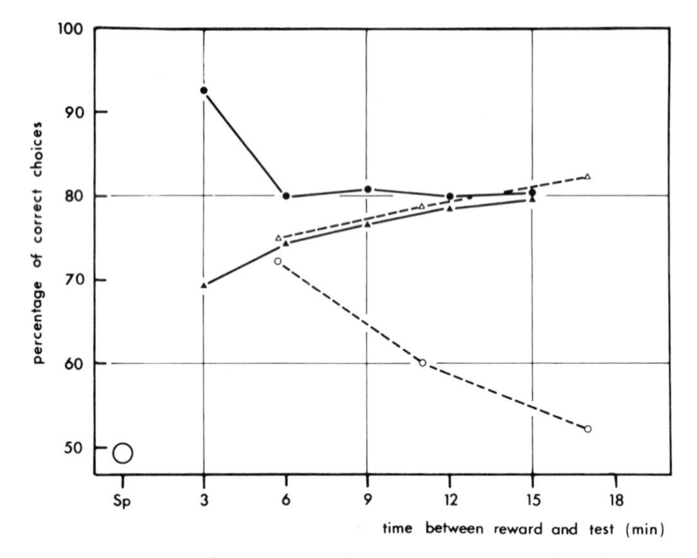

Fig. 6. Dependence of retention on the duration of reward. In one experimental series the animals are tested only once for 2 min (continuous lines). In another series, animals are repeatedly tested (extinction group, dotted line). Sp indicates the spontaneous choice level. If the bees are only rewarded for 2 sec (-●- and -O-) the percentage of correct choices is first very high but decreases. In the extinction group, percentage correct choices reaches the spontaneous choice level after about 15 min. However, if the bees suck sugar solution for 15 sec (-▲- and -Δ-) they continuously improve their choice in the following minutes. In contrast to the retention after a 2 sec reward, the time course is the same for both experimental groups after a 15 sec reward. This experiment leads to the prediction that the storage process may consist of two mechanisms which can be distinguished by altering the duration of reward and the influence of experimental extinction

Bees were allowed to suck sugar solution for different periods ranging from 2 to 30 sec (Fig. 6) and were tested at different intervals after the reward. In one group, each bee was tested only once for 2 min (Fig. 6, solid lines); in another group each bee was tested repeatedly, each test lasting 2 min (extinction procedure, Fig. 6, dotted lines). After a short reward (2 sec) the test bees chose correctly with high probability during the first 3-4 min after being rewarded. Later the proportion of correct responses decreased. The same decrease was much more pronounced in the extinction experiment where the percentage of correct responses reached the spontaneous choice level after 12-15 min. By contrast, after a reward lasting 15 sec the percentage of correct responses was first low and then increased with time. This time course holds true for both experimental groups (single test and extinction group). During the time after the reward, no new information on the rewarding situation is received. The increase in correct responses, therefore, reflects a consolidation process. We have described the consolidation process for different programmes of reward (Erber 1972) and have found that interruptions of sucking lead to an increase in the amplification factor of the time-dependent consolidation function. Such experiments show that at least 2 different storage mechanisms can be distinguished as a function of reward duration. These mechanisms differ in their time course and in their dependence on experimental extinction: a short reward leads to a phasic retention curve while a reward of 10 sec or longer results in a gradually increasing retention.

Another series of experiments using external disruptive methods tested whether the two mechanisms are based on two physiologically different memory storage processes. It is already well known that, in vertebrates, a variety of external disruptive factors interfere with newly stored information and cause complete or gradual amnesia (e.g. McGaugh *et al.* 1972; Dawson and McGaugh 1969; Herz 1969; Maldonado 1968, 1969; Paolino and Levy 1971; Alpern and McGaugh 1968; Alpern and Kimble 1967). In our experiments the test animals were exposed to 4 different disruptive procedures at variable time intervals after the reward:
(1) Electroconvulsive shocks (ECS) of 30V, 25 Hz were applied between a wire on the frons of the head and needle forceps in the neck. The animals immediately reacted to the shock with a tetanus but all recovered within a few minutes and displayed normal behaviour.
(2) Test animals were narcotised with 100% carbon dioxide, or
(3) 100% nitrogen for 60 sec. The narcosis occurred after 7-10 sec.
(4) Other animals were cooled to 1°C for 3 min; after 1-2 min the bees showed no motor activity.
All these four treatments had no effect on behaviour: after about 20 min all bees showed normal flight behaviour and motivation for searching and collecting.

The strength of the ECS treatment was not critical for its effect. As a measure of ECS strength we used the integral of current with respect to time (Fig. 7). Bees were trained to $\lambda_+ = 444$ nm and shocked immediately after the reward with different ECS strengths. Fig. 7 shows that the stored information is completely destroyed if the ECS is stronger than 300 μA·sec although the bees were able to withstand a much stronger shock (1400 μA·sec). Thus in the succeeding tests we always maintained an ECS strength between 300 and 1000 μA·sec.

The effect of the interference of these 4 disruptive treatments on the storage of information is given in Fig. 8. We varied the time between reward and treatment and tested the animals after they came back from the hive, that is about 20-30 min after the reward. Control experiments are shown in the histogram. The result represented by bar C proves that ECS does not operate as a punishment stimulus (negative stimulus). This was shown by shocking bees shortly after landing on the

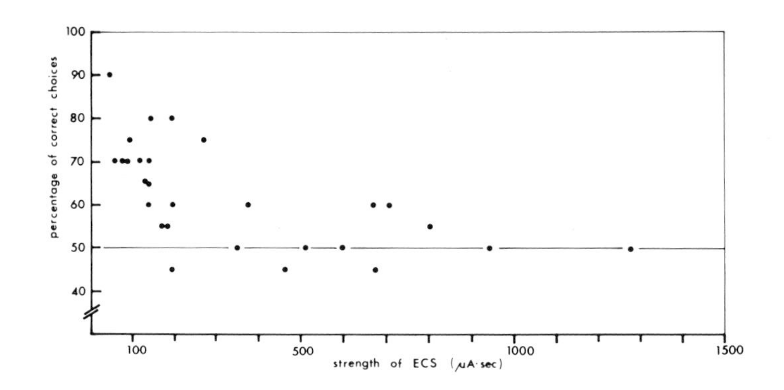

Fig. 7. Influence of the strength of electric shock (ECS) on the disruption of stored information. The test animals are rewarded once on a blue (λ_+ = 444 nm) ground glass disc. They are treated with ECS, varying in strength, 15 sec after the 20 sec long reward. The integral of current with respect to time is used as a measure for the ECS strength. The ECS is a voltage with positive going pulses of a frequency of 25 Hz. The positive electrode is the wire on the head surface; the reference electrode is a pair of needle forceps which hold the animal by the neck. The figure shows that the stored information is destroyed if the ECS is stronger than 300 μA·sec. Even with much higher current flow (1400 μA·sec) all animals recover within minutes and come back to search for sugar solution. Each point represents one animal

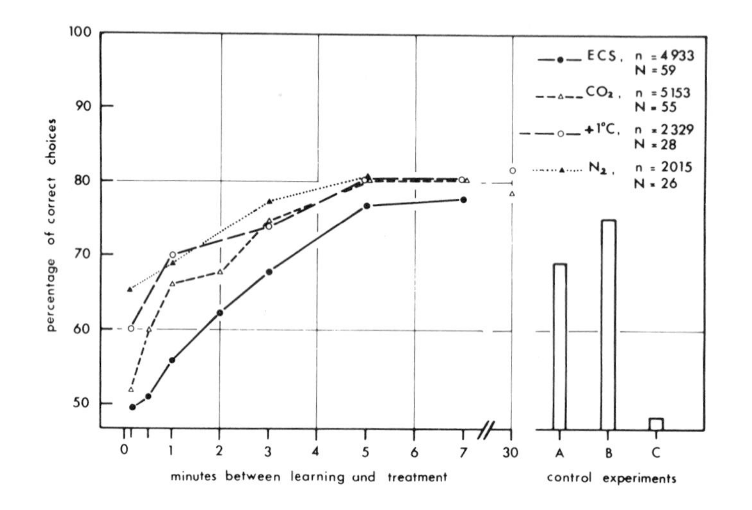

Fig. 8. Time course of the disruptive effect of 4 different treatments on newly stored information. The test animals are rewarded once on λ_+ = 444 nm with one reward lasting 20 sec and then treated with ECS, CO_2, N_2 or cooling at different times after the reward (abscissa). The curves show the percentage of correct choices (ordinate) in the next test. ECS has the strongest effect, N_2 the weakest. None of these treatments influences the long-term storage of the learned signal if given more than 5 min after the reward. Control experiments are shown on the right side. A shows the percentage of correct choices of untreated animals but with the same manipulations as in the ECS series. B show the percentage of correct choices of animals after restraining for 30 min after the reward. C demonstrates that the ECS treatment does not operate as a punishment stimulus (negative stimulus); the bees are shocked while sitting on a coloured disc but without reward. In the following tests the two alternative colours are chosen with the same percentage as in spontaneous choice experiments, i.e., the "shocked" colour is not chosen less frequently. N gives the number of bees tested, n the number of decisions

coloured disc. They showed no change in their spontaneous preference for the alternative colours in the next test. All 4 treatments destroyed the stored information if they were applied shortly after learning. Their effect decreases with increasing delay between learning and treatment showing no effects if they were applied more than 7 min after the reward. ECS treatment produces the greatest disruption, probably because it acts immediately, whereas the cooling and narcosis take more than 1 min to narcotise the animal. As the first minute is very important it is to be expected that the effect of cooling or narcosis will be less than that of ECS. The different effects of the two gases CO_2 and N_2, shown in Fig. 8 is not yet understood because the animals seem to be narcotised within the same time.

In comparing our results with others (mainly on vertebrates) it must be recognized that the memory storage process is being examined largely by behavioural techniques rather than by direct investigation of relevant neural events, and therefore that inference about the underlying process must be made only with considerable caution. From our present knowledge of the physiological basis of memory formation, it is most likely that the pattern of excitation in the participating neurones causes a short-lasting change in the synapses. The synthesis of specific proteins is initiated concurrently or as a result of such activity, and this causes a long lasting change in the effectiveness of the synapses (for instance Kandel *et al.* 1970; Barondes 1970; Hyden and Lange 1970). The synthesis of a m RNA containing 100 trinucleotides takes 7 sec (Manor, Goodman and Stent 1969; Griffith 1970). Certainly the synthesis of longer RNA, transcription of a protein, its transport to the synapses and its incorporation into the membrane would take many minutes, possibly some hours. Inhibitors of protein synthesis block the transfer of memory to a long lasting storage mechanism from the first 30 min up to some hours.

Our experiments show, for the first time in an insect, that there is a short-term retention, which differs from long-term retention in its sensitivity to various disruptive factors. Also we have demonstrated that recall from short-term memory is more accurate than from long-term memory and that an experimental extinction has a stronger effect on short-term memory. We can assume from these results that the short-term memory is dependent on orderly neural activity. It has not been determined whether this activity requires a dynamic mechanism involving circulation of impulses through closed interneurone pathways or only continuing plastic changes in the functional effectiveness of synapses. One mechanism may be heterosynaptic plastic facilitation, which lasts up to 40 min in *Aplysia* (Kandel and Tauc 1965). It has been found in vertebrates that artificial changes in neural excitability, such as those caused by ECS, cooling and concussion, interfere with stored information during this period (McGaugh 1966; Jarvik 1968). However, there is the following evidence that spike activity is more important for short-term memory: (1) Tauc and Epstein (1967) showed that hetero-synaptic facilitation in *Aplysia* is not affected by cooling or by substituting Li^+ for Na^+; (2) it is known from vertebrates that cooling, polarising current and spreading depression changes strongly the neuronal activity (Lippold 1970) but that the synaptic properties are apparently unchanged (Gartside 1968; Bureš and Burešová 1970); (3) spike activity could not be recorded in the bee brain during CO_2-narcosis or cooling to 1°C (Leutscher-Hazelhoff and Kuiper 1966).

In insects, the transfer from short-term to a long-term storage mechanism depends on continuing protein synthesis as in vertebrates. This was shown by Brown and Noble (1967, 1968) who demonstrated that the application of cycloheximide greatly reduced the learning performance in the cockroach leg preparation. However, it is very unlikely that the coupling between electrical excitation and protein synthesis is the same as in vertebrates because of the different nature of the soma. In contrast to vertebrates, where the somata are intimately involved in electric transmission, the soma of insect neurones do not receive synapses and may not participate in neuronal excitation.

We hope that the analysis of learning performance under natural conditions combined with the use of specific inhibitory substances and treatments, will provide further information on transfer mechanisms between short-term and long-term memory.

THE SEARCH FOR THE LOCATION OF THE MEMORY TRACE

It has been thought for a long time that the mushroom bodies in the protocerebrum are responsible for the "intelligence" of social insects (von Alten 1910; Hanström 1928; Pietschker 1911; Howse, this volume). This suggestion is supported by the structural regularity of the globuli-fibres (Vowles 1955; Goll 1967) and the high density of the synapses (Schürmann 1970, 1972). Other indirect evidence comes from the observation that specific turning habits taught to the larva of *Tenebrio* are retained through metamorphosis (Alloway 1972). He and other authors conclude that the memory must be stored in neuronal tissue which persists through metamorphosis, such as the mushroom bodies (Jawlowski 1936). More indirect evidence comes from experiments on learning and memory after lesions were made in the brain. In training experiments with ants, Vowles (1964a,b, 1967) cut the connections between the optic or antennal lobes and the α-lobes of the mushroom bodies, ipsi- or contralaterally. He postulated that an engram is built up by reverberating circuits between the appropriate sensory ganglia (lobes) and the α-lobes. As lesions in the mushroom bodies affected neither olfactory nor visual learning performance, Vowles concluded that memory storage is distributed throughout a relatively large neuronal network including the sensory ganglia. This is supported by the finding that, especially in visual learning tasks, the contralateral sensory ganglion is involved in memory formation. Since Vowles studied only complex behaviours in his trained ants, it is possible that secondary fibres which are not involved in learning and memory were affected, e.g. fibres which choose the right pattern of behaviour (command fibres).

We used a simple conditioned response to examine the problem of transfer of stored information in the visual and olfactory systems of bees. Our method allows a direct and reversible blocking of activity in localised areas in the brain. Bees were fixed by the head in a small tube in such a way that the antennae and mouthparts were free to move. After a paired presentation of a colour or odour signal with application of sugar solution to the antenna, the bee will later extend its proboscis when exposed to the colour or odour alone (Kuwabara 1957). In 70%-80% of bees, one pairing of an odour signal and reward is sufficient to elicit the proboscis response. For colours, however, a long training with many reinforcements is necessary.

We used blue as a signal during colour training. Only one eye was illuminated (the learning eye) and on the first day learning curves were plotted for this eye. On the second day either the learning or the test eye was trained further. No difference was found in the learning performance whichever eye was used on the second day (Fig. 9). There was also no difference if both eyes were trained simultaneously. (For more details see Masuhr and Menzel 1972.) We concluded that during learning with one eye, information was stored equally in both sides of the brain. We still have not determined whether the contralateral side is involved in the actual learning process or whether ipsilaterally stored information is later transferred to it.

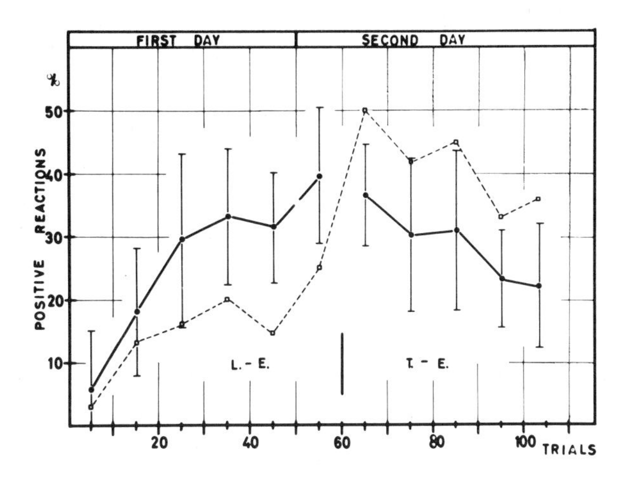

Fig. 9. Test of side specific information storage in the optic system. In contrast to the experimental series given in Fig. 10, here the bees are not freely flying but are fixed to a holder. The proboscis reflex is conditioned to broad band blue light (Schott filter BG 25). The left part of the diagram ("first day") gives the average increase of positive responses for one eye (the learning eye, L.-E. -●-) with the number of trials. On the second day (right part of the diagram) the other eye (test eye, T.-E.) was trained. The response curve for the test eye starts at the level which was reached with the learning eye the day before. This proves that the information stored during training of the "learning eye" is available to the other eye. The following decrease in performance with further trials does not depend on the special transfer situation, but is a result of overtraining which also occurs in a test situation where both eyes are trained (-O-). The vertical bars give the standard deviations. Each point is the average percentage of positive responses of 10 trials. Number of bees for each curve is 18

Odour learning was found to be quite different. The conditioned response was only elicited when the trained antenna was stimulated (Fig. 10). The same result was obtained from odour training experiments with freely flying bees when one antenna was trained while the other was covered with a removable barrier cream. This result is astonishing as there are inter-hemispheric tracts between antennal lobes, between mushroom bodies and from antennal lobe to mushroom bodies (Vowles 1955, 1967; Pareto 1972; Howse, this volume). A detailed analysis of the experiments (Fig. 10) shows that some transfer must have taken place as further training of the "test antenna" to the same odour produced a lower percentage of correct responses than the first training of the trained antenna. We interpret this as meaning that the spatial parameter in odour learning is so important that the engram consists of "odour on the right side, no odour on the left side", when only the right antenna is trained. If the right antenna is cut off or the odour applied only to the left antenna, the test situation is so different from the engram that the conditioned response is not elicited. This means that the training of the test antenna with the same odour given to the first trained antenna is a reversal learning which produces a lower percentage of correct choices for the same number of rewards (compare Menzel 1969). This confirms the important role of spatial specificity of odour information for the bee. Forel (1910) called this the "topochemical" character of the olfactory sense and Martin (1964) has examined it in detail.

A more direct approach in our search for the location of memory is training the proboscis reflex of fixed bees and locally cooling small areas of the brain. The cooling reversibly interrupts neuronal activity. This enables us to examine the time course of information integration and storage in different parts of the brain. The method used in these

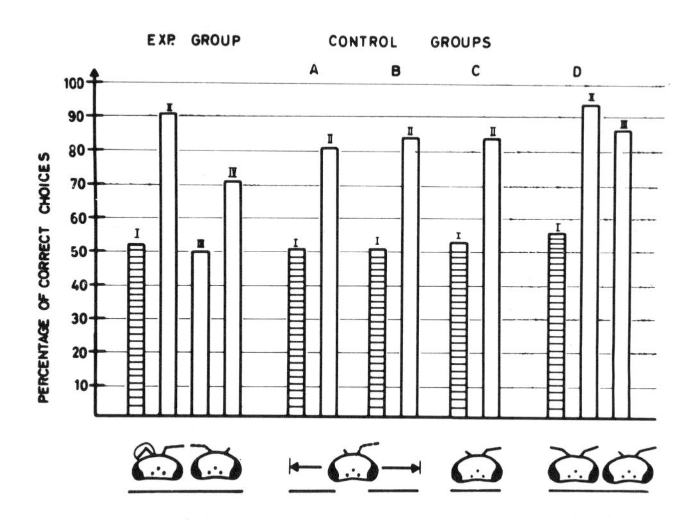

Fig. 10. Test of side-specific information storage in the olfactory system. Bees in the experimental group are first tested for their spontaneous preference (I - striped bar). Then one antenna is covered with a barrier cream and the animal trained to an odour with 6 rewards. Bar II shows that then the odour is chosen with a high percentage using the "learning antenna". If the "learning antenna" is cut off and the untrained antenna exposed, the choices are similar to the spontaneous choice preference (bar III). This proves that the stored information can not be recalled by stimulation of the untrained antenna. The following experiment shows that this result is not explained by supposing only ipsilateral information storage (bar IV): Training of the "test antenna" with 6 rewards to the same odour results in a significantly lower percentage of correct responses than the initial training (experimental group bar II) and the control group A (II). Control groups: in all cases the striped bar gives the spontaneous choice preference. A and B give the results of training (6 rewards) of one antenna, the other having been cut off previous to learning. The remaining antenna was first covered for short time (A) or for several hours (B) with the cream and then cleaned again. The result is the same in both cases which proves that the barrier cream does not affect the olfactory receptors. C as A and B except that the antenna tested was not treated with cream before testing. D, bees are trained (6 rewards) with both antennae and then tested (bar II). After this test one antenna is cut off and the animals tested once more (III)

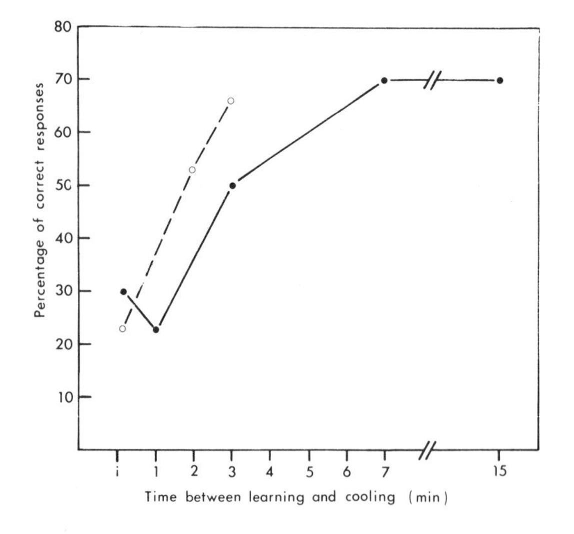

Fig. 11. Time course of the effect on the information storage caused by local cooling in two different parts of the brain. After removing the frontal head capsule and exposing the brain the bee was conditioned by one trial to an odour stimulating both antennae. Between 1 to 15 min later (abscissa) two thin, cooled wires were placed on the two antennal lobes (-O-) or the two α-lobes of the mushroom bodies (-●-), i: cooling immediately (5 sec) after the trial. Each point in both curves is the average of 30 bees. A comparison with Fig. 8 shows that the time course for the α-lobe cooling is exactly the same as the time course of the ECS-treatment. The curve for the cooling of the antennal lobes is significantly steeper

experiments is very similar to that already described. The proboscis reflex can be conditioned to an odour by one trial even when the head and thorax are fixed to a holder, and the anterior head capsule between the complex eyes is removed and the brain exposed (Masuhr, unpublished). For conditioning to an odour, one antenna or both were stimulated with this odour. Small parts of the brain were cooled to 0°C for 1 min with 0.5 mm Ø wire either immediately or at a set time interval after training. Fig. 11 shows the effect of cooling the antennal lobe and the region of the α-lobe of the mushroom bodies. In each case right and left lobes were cooled simultaneously. The effect of cooling the α-lobes shows a similar time course to that after the cooling of the freely flying bee (see Fig. 8). Information storage is blocked by this reversible cooling only in the first minutes after training. After more than about 5 min, the cooling did not affect the storage process. It is of special interest to note that storage is also blocked by cooling the antennal lobes alone, but the effect occurs only within 2-3 min after training. This indicates that the antennal lobes are not just sensory centres, but must play a part in information storage. Otherwise the time course of the effect would be limited to the time course of the excitation caused by stimulation of the antenna by an odour.

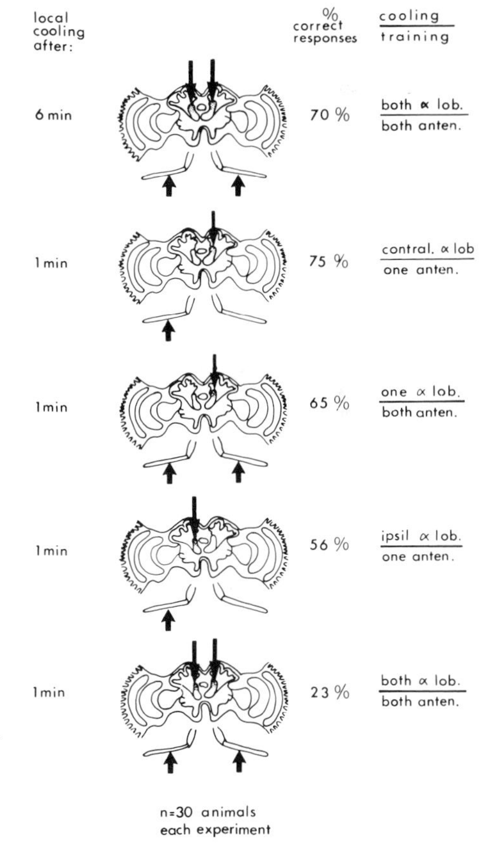

Fig. 12. Test of information storage in the ipsi- and contralateral α-lobes of the mushroom bodies. An odour stimulus is given to one or both antennae (arrows pointing to antennae); cooling needles are set on the ipsi- or the contralateral α-lobes or both (arrows pointing to α-lobes, the dotted area within the mushroom bodies). The left column gives the time interval between trial and cooling; the right column describes cooling procedure above the line and training situation below the line. The percentage of correct responses is the result of the test after one trial learning. Each experiment was carried out with 30 test bees.

The uppermost experiment is a control experiment, which shows that the effect studied by local cooling of α-lobes interfered only with the short-term memory: the response level of 70% is similar to control experiments on animals without any cooling and without opening the head capsule. This experiment also proves that the experimental procedure itself does not inhibit memory formation. In all other experiments the α-lobes were cooled one minute after the trial. Cooling of the contralateral α-lobe does not reduce responsiveness (75% responses). Some decrease is found if both antennae are stimulated but only one α-lobe is cooled. A significant reduction of responsiveness can be seen during cooling of the ipsilateral α-lobe. Cooling of both lobes when both antennae are stimulated completely blocks memory formation (23% last experiment, spontaneous response level is 25-30%). The significantly higher responsiveness in the second last experiment compared with the last one shows that memory formation takes place not only in the ipsilateral α-lobe but also in the contralateral (see text)

The involvement of the contralateral side in learning an odour signal is shown in experiments in which either one or both α-lobes are cooled (Fig. 12). When the contralateral α-lobe is cooled within 1 min after training of only one antenna, the bees respond with the same level of probability as untreated controls. If the ipsilateral α-lobe is cooled the response is clearly reduced, but at 56% is still higher than if both α-lobes are cooled after training both antennae (23%). Although all these differences are highly significant, these data are not yet sufficient to produce a functional diagram of the interactions between antennal lobe and mushroom bodies during conditioning. The conclusions that can be drawn from these experiments are:
(1) Neither the contralateral antennal lobe nor α-lobe are necessary for learning an odour signal. The major functional areas lie ipsilaterally.
(2) If the ipsilateral α-lobe is blocked during training of one antenna some association still occurs, most probably in the contralateral mushroom body.
(3) During short-term memory interactions between the ipsilateral antennal lobe, calyx and α-lobe are the most important, but the final storage occurs in the mushroom body. The interaction with the antennal lobe could maintain a specific pattern of excitation sufficient to produce a long lasting memory in the mushroom bodies.

CONCLUSION

The information so far collected on the learning mechanism in the honeybee does not enable us to support either one or the other of the hypotheses discussed in the introduction. The differences in the colour integration system and the colour learning system should allow us to distinguish between elements in the visual integration system and in the learning system by probing with microelectrodes in the bee's brain. The knowledge of the time span of sensory memory should give further help during these studies.

The most important supposition, however, for a successful approach to the crucial question of memory location is the exact knowledge of the time course of different storage mechanisms, which may indicate that different storage mechanisms occupy different parts of the brain. Short-term memory disruption with ECS of a colour learning task in freely flying bees gives the same time course as bilateral cooling of the α-lobes in the mushroom bodies during an olfactory learning task in fixed bees. This supports strongly the idea of a similar mechanism for short-term to long-term memory transcription in the 2 modalities. Our hypothesis is that information stored in the short-term memory is coded in the spatio-temporal pattern of neuronal activity, which activates pathways connecting sensory integration centres with the mushroom bodies, during several minutes following the reward. This idea, is in principle, consistent with Vowles's (1955; 1967) hypothesis of memory formation in insects but with the important exception that we restrict circulating activity between sensory lobes and mushroom bodies *only* to short-term memory. We have as yet no clear idea which of the intrinsic and extrinsic pathways of the mushroom bodies are involved in short-term memory of olfactory and visual tasks. However, the localised and reversible blocking of small brain areas should be an appropriate method for more detailed studies. The more precise knowledge of involved brain structures which carry the spatially and temporally ordered activity pattern during short-term memory should be accessible to a deterministic analysis. As sensory excitation reaches and passes the sensory integration centres in a spatio-temporally ordered pattern, it seems a logical strategy to follow the information processing with the hope of finding or excluding the mysterious transition from deterministic to non-deterministic brain mechanisms.

ACKNOWLEDGEMENTS

We gratefully acknowledge the support and helpful criticism of our colleagues Prof. H. Markl, Drs. J. Altman, I. Meinertzhagen and J. Kien. We especially thank Miss J. Kien and Miss M. Blakers for their help with the English text. This research was supported by the Deutsche Forschungsgemeinschaft throughout the last 5 years. In the last 2 years this research was part of a research project "Zentralnervöse Verarbeitung von Sinnesdaten" supported by the Deutsche Forschungsgemeinschaft under "Forschergruppe Markl und andere".

REFERENCES

AGRANOFF, B.W.: Agents that block memory. *In* "The Neurosciences. A Study Program" (Eds., G.C. Quarton, T. Melnechnk and F.O. Schmitt). pp. 756-764. New York: The Rockefeller University Press. (1967).

ALLOWAY, T.M.: Retention of learning through metamorphosis in the grain beetle (*Tenebrio molitor*). Am. Zoologist 12, 471-477 (1972).

ALPERN, H.P., KIMBLE, D.P.: Retrograde amnesic effects of diethylether and bis.(*trifluorethyl*) ether. J. comp. physiol. Psychol. 63, 168-171 (1967).

ALPERN, H.P., McGAUGH, J.L.: Retrograde amnesia as a function of duration of electroshock stimulation. J. comp. physiol. Psychol. 65, 265-269 (1968).

von ALTEN, H.: Zur phylogenie des Hymenopterengehirns. Jena. Z. Naturw. 46, 511-590 (1910).

AUTRUM, H., von ZWEHL, V.: Die Spektrale Empfindlichkeit einzelner Sehzellen des Bienenauges. Z. vergl. Physiol. 48, 357-384 (1964).

BARONDES, S.A.: Multiple steps in the biology of memory. *In* "The Neurosciences: Second study program" (Ed., F.O. Schmitt). New York: Rockefeller University Press. pp. 272-278 (1970).

BROWN, B.M., NOBLE, E.P.: Cycloheximide and learning in the isolated cockroach ganglion. Brain Res. 6, 363-366 (1967).

BROWN, B.M., NOBLE, E.P.: Cycloheximide, amino acid incorporation and learning in the isolated cockroach ganglion. Biochem. Pharmac. 17, 2371-2374 (1968).

BULLOCK, T.H., HORRIDGE, G.A.: "Structure and Function in the Nervous System of Invertebrates", Vol. 2, 1719 pp. San Francisco: Freeman (1965).

BUREŠ, J., BUREŠOVÁ, O.: Plasticity in single neurones and neural populations. *In* "Short-term Changes in Neural Activity and Behaviour" (Eds., G. Horn and R.A. Hinde), pp. 363-403. London: Cambridge University Press (1970).

CHEN, W.Y., ARANDA, L.C., LUCO, J.V.: Learning and long- and short-term memory in cockroahces. Anim. Behav. 18, 725-732 (1970).

DAUMER, K.: Reizmetrische Untersuchung des Farbensehens der Bienen. Z. vergl. Physiol. 38, 413-478 (1956).

DAUMER, K.: Blumenfarben, wie sie die Bienen sehen. Z. vergl. Physiol. 41, 49-110 (1958).

DAWSON, R.G., McGAUGH, J.L.: Electroconvulsive shock effects on a reactivated memory trace: Further examination. Science, N.Y. 166, 525-527 (1969).

ECCLES, J.C.: "The Physiology of Synapses", 316 pp. Berlin-Göttingen-Heidelberg: Springer (1964).

ECCLES, J.C., McINTYRE, A.K.: The effects of disuse and of activity on mammalian spinal reflexes. J. Physiol. 121, 492-516 (1953).

EISENSTEIN, E.M.: Learning and memory in isolated insect ganglia. Adv. Insect Physiol. 9, 111-181 (1972).

ERBER, J.: The time-dependent storing of optical information in the honeybee. *In* "Information Processing in the Visual System of Arthropods" (Ed., R Wehner), pp. 309-314. Berlin, Heidelberg, New

York: Springer-Verlag (1972).
FOREL, A.: "Das Sinnesleben der Insekten". München: E. Reinhardt (1910).
von FRISCH, K.: Der Farbensinn und Formensinn der Bienen. Zool. Jb. (Abt. Allgem. Zool. Physiol.) 35, 1-182 (1914).
von FRISCH, K.: "The Dance Language and Orientation of Bees", 566 pp. Cambridge, Mass.: The Belknap Press of Harvard Univ. Press (1967).
GARTSIDE, J.B.: Mechanisms of sustained increases of firing rate of neurones in the rat cerebral cortex after polarization: reverberating circuits or modification of synaptic conductance. Nature, Lond. 220, 382-383 (1968).
GOLL, W.: Strukturuntersuchungen am Gehirn von Formica. Z. Morph. Ökol. Tiere. 59, 143-210 (1967).
GOSSEN, M.: Untersuchungen an Gehirnen verschieden grosser, jeweils verwandter Coleopteren- und Hymenopterenarten. Zool. Jb. (Abt. Allgem. Zool. Physiol.) 62, 1-64 (1949).
GRIFFITH, J.S.: The transition from short to long-term memory. *In* "Short-term Changes in Neural Activity and Behaviour" (Eds., G. Horn and R.A. Hinde). London: Cambridge University Press (1970).
HANSTRÖM, B.: Vergleichende Anatomie des Nervensystems der Wirbellosen Tiere. Berlin: Springer (1928).
HEBB, D.O.: "The Organisation of Behaviour". New York: J. Wiley (1949).
von HELVERSEN, O.: Zur spektralen Unterschiedsempfindlichkeit der Honigbiene. J. comp. Physiol. 80, 439-472 (1972).
HERZ, M.J.: Interference with one-trial appetitive and aversive learning by ether and ECS. J. Neurobiol. 1, 111-122 (1969).
HOLMGREN, N.: Zur vergleichenden Anatomie des Giehirns von Polychaeten, Onychophoren, Xiphosuren, Arachniden, Crustanceen, Myriapoden und Insekten. K. svenska Vetensk-Akad. Handl. 56, 1-303 (1916).
HORRIDGE, G.A.: Learning of leg position by the ventral nerve cord in headless insects. Proc. R. Soc. 157, 33-52 (1962).
HORRIDGE, G.A.: "Interneurones", 436 pp. London: W.H. Freeman (1968).
HOYLE, G.: Neurophysiological studies on "learning" in headless insects. *In* "The Physiology of the Insect Central Nervous System" (Eds., J.E. Treherne and J.W.L. Beament), pp. 203-232. London / New York: Academic Press (1965).
HOYLE, G.: Cellular mechanisms underlying behaviour - Neuroethology. Adv. Insect Physiol. 7, 349-444 (1970).
HUBER, F.: Brain controlled behaviour in Orthopterans. *In* "The Physiology of the Insect Central Nervous System" (Eds., J.E. Treherne and J.W.L. Beament), pp. 233-246. London / New York: Academic Press (1965).
HUBER, F.: Central control of movements and behaviour in invertebrates. *In* "Invertebrate Central Nervous Systems" (Ed., C.A.G. Wiersma), pp. 333-354. Chicago: Chicago University Press (1967).
HYDÉN, K., LANGE, P.W.: Protein changes in nerve cells related to learning and conditioning. *In* "The Neurosciences. Second study program" (Ed., F.O. Schmitt), pp. 278-289. New York: Rockefeller University Press (1970).
JARVIK, M.E.: Consolidation of memory. *In* "Psychopharmacology: A Review of Progress" (Eds., D.E. Efron *et al.*), pp. 885-889. Washington: U.S. Government Printing Office, PHS Publ. 1839. (1968).
JAWLOWSKI, H.: Über der Gehirnbau der Käfer. Z. Morph. Ökol. Tiere 32, 67-91 (1936).
JOHN, E.R.: "Mechanisms of Memory". New York: Academic Press (1967).
JOHN, E.R.: Switchboard versus statistical theories of learning and memory. Science, N.Y. 177, 850-864 (1972).
KAISER, W.: A preliminary report on the analysis of the optomotor system of the honeybee: single unit recording during stimulation with spectral light. *In* "Information Processing in the Visual Systems of Arthropods" (Ed., R. Wehner), pp. 167-170. Berlin, Heidelberg, New York: Springer (1972).

KAISER, W., LISKE, E.: A preliminary report on the analysis of the optomotor system of the honeybee: behavioural studies with spectral lights. *In* "Information Processing in the Visual Systems of Arthropods" (Ed., R. Wehner), pp. 163-166. Berlin, Heidelberg, New York: Springer (1972).

KANDEL, E.R., TAUC, L.: Heterosynaptic facilitation in neurones of the abdominal ganglion of *Aplysia depilans*. J. Physiol. (Lond.) 181, 1-27 (1965).

KANDEL, E.R., CASTELLUCCI, V., PINSKER, H., KUPFERMANN, I.: The role of synaptic plasticity in the short-term modification of behaviour. *In* "Short-term Changes in Neural Activity and Behaviour" (Eds., G. Horn and R.A. Hinde), pp. 281-322. London: Cambridge University Press (1970).

KANDEL, E.R., SPENSER, W.A.: Cellular neurophysiological approaches in the study of learning. Physiol. Rev. 48, 65-134 (1968).

KUWABARA, M.: Bildung des bedingten Reflexes von Pavlovs Typus bei der Honigbiene (*Apis mellifica*). J. Fac. Sci. Hokkaido Univ. (VI: Zool.) 13, 458-467 (1957).

LASHLEY, K.S.: In search of the engram. Symp. Soc. exp. Biol. 4, 454-482 (1950).

LEUTSCHER-HAZELHOFF, J.T., KUIPER, J.W.: Clock-spikes in the *Calliphora* optic lobe and a hypothesis of the organisation of the compound eye. *In* "The Functional Organisation of the compound eye" (Ed., C.G. Bernhard), pp. 483-492. Oxford: Symp. Publ. Press, Oxford. (1966).

LINDAUER, M.: Allgemeine Sinnesphysiologie, Orientierung im Raum. Fortschr. Zool. 16, 58-140 (1963).

LINDAUER, M.: Lernen und Gedächtnis, Versuche an der Honigbiene. Naturwissenschaften 57, 463-467 (1970).

LIPPOLD, O.C.J.: Long-lasting changes in the activity of cortical neurones. *In* "Short-term Changes in Neural Activity and Behaviour" (Eds., G. Horn and R.A. Hinde), pp. 405-432. London: Cambridge University Press (1970).

MALDONADO, H.: Effects of electroconvulsive shock on memory in *Octopus vulgaris* Lamarck. Z. vergl. Physiol. 59, 25-37 (1968).

MALDONADO, H.: Further investigations on the effect of electroconvulsive shock (ECS) on memory in *Octopus vulgaris*. Z. vergl. Physiol. 63, 113-118 (1969).

MANOR, H., GOODMAN, D., STENT, G.S.: RNA chain growth rates in *Escherichia coli*. J. molec. Biol. 39, 1-29 (1969).

MARKL, H., LINDAUER, M.: Physiology of insect behaviour. *In* "The Physiology of Insecta" (Ed., M. Rockstein), Vol. II, pp. 1-122. New York: Academic Press (1965).

MARTIN, H.: Zur Nahorientierung der Biene im Duftfeld Zugleich ein Nachweis für die Osmotropotaxis bei Insekten. Z. vergl. Physiol. 48, 481-533 (1964).

MASUHR, T., MENZEL, R.: Learning experiments on the use of side-specific information in the olfactory and visual system in the honey bee (*Apis mellifica*). *In* "Information Processing in the Visual Systems of Arthropods" (Ed., R. Wehner), pp. 315-322. Berlin, Heidelberg, New York: Springer-Verlag (1972).

MAYNARD, D.M.: Organisation of central ganglia. *In* "Invertebrate Nervous Systems" (Ed., C.A.G. Wiersma), pp. 231-258. Chicago: Chicago University Press (1967).

McGAUGH, J.L.: Time-dependent processes in memory storage. Science, N.Y. 153, 1351-1358 (1966).

McGAUGH, J.L.: Facilitation of memory storage processes. *In* "The Future of the Brain Sciences" (Ed., S. Bogoch), pp. 355-370. New York: Plenum Press (1969).

McGAUGH, J.L., ZORNETZER, S.F., GOLD, P.E., LANDFIELD, P.W.: Modification of memory systems: some neurobiological aspects. Q. Rev. Biophys. 5, 163-186 (1972).

MENZEL, R.: Untersuchungen zum Erlernen von Spektralfarben durch die Honigbiene (*Apis mellifica*). Z. vergl. Physiol. 56, 22-62 (1967).
MENZEL, R.: Das Gadächtnis der Honigbiene fur Spektralfarben. I. Kurzzeitiges und Langzeitiges Behalten. Z. vergl. Physiol. 60, 82-102 (1968).
MENZEL, R.: Das Gadächtnis der Honigbiene fur Spektralfarben. II. Umlernen und Mehrfachlernen. Z. vergl. Physiol. 63, 290-309 (1969).
MENZEL, R.: Spectral response of moving detecting and "sustaining" fibres in the optic lobe of the bee. J. comp. Physiol. 82, 135-150 (1973).
MENZEL, R., ERBER, J.: The influence of the quantity of reward on the learning performance in honeybees. Behaviour 41, 27-42 (1972).
MENZEL, R., FREUDEL, H., RÜHL, U.: Rassenspezifische Lernunterschiede bei der Honigbiene. Apidologie 4, 1-24 (1973).
MENZEL, R., SNYDER, A.: The colour vision system in the honey bee. Single cell recording in retinula cells. J. comp. Physiol. (In press).
MESSENGER, J.B.: Two stage recovery of a response in Sepia. Nature, Lond. 232, 202-203 (1971).
MILLER, A.: Vergleich der Vergessenskurven für Reproduzieren und Wiederkennen von sinnlosem Material. Z. exp. angew. Psychol. 7, 29-38 (1967).
OPFINGER, E.: Über die Orientierung der Biene an der Futterstelle. Z. vergl. Physiol. 15, 431-487 (1931).
PANDAZIS, G.: Über die relative Ausbildung der Gehirnzentren bei biologisch verschiedenen Ameisenarten. Z. Morph. Ökol. Tiere 18, 114-169 (1930).
PAOLINO, R.M., LEVY, H.M.: Amnesia produced by spreading depression and ECS: Evidence for time-dependent memory trace localization. Science, N.Y. 172, 746-749 (1971).
PARETO, A.: Die zentrale Verteilung der Fühlerafferenz bei Arbeiterinnen der Honigbiene (*Apis mellifica*). Z. Zellforsch. mikrosk. Anat. 131, 109-140 (1972).
PFLUMM, W.: Beziehungen zwischen Putzverhalten und Sammelbereitschaft bei der Honigbiene. Z. vergl. Physiol. 64, 1-36 (1969a).
PFLUMM, W.: Stimmungsänderungen der Biene während des Aufenthalts an der Futterquelle. Z. vergl. Physiol. 65, 299-323 (1969b).
PIETSCHKER, H.: Das Gehirn der Ameise. Jena. Z. Naturw. 47, 43-114 (1911).
RAU, G.: Zur Steuerung der Honigmagenfüllung sammelnder Bienen an einer künstlichen Futterquelle. Z. vergl. Physiol. 66, 1-21 (1970).
RIEGE, W.H., CHERKIN, A.: One trial learning and biphasic time course of performance in the goldfish. Science, N.Y. 172, 966-968 (1971).
SANDERS, G.D., BARLOW, J.J.: Variation in retention performance during long term memory formation. Nature, Lond. 232, 203-204 (1971).
SCHÜRMANN, F.W.: Über die Struktur der Pilzkörper des Insektengehirns. I. Synapsen im Pedunculus. Z. Zellforsch. mikrosk. Anat. 103, 365-381 (1970).
SCHÜRMANN, F.W.: Über die Struktur der Pilzkörper des Insektengehirns. II. Synaptische Schaltungen im a-Lobus des Heimchens *Acheta domesticus* L. Z. Zellforsch. mikrosk. Anat. 127, 240-257 (1972).
SNYDER, A., MENZEL, R., LAUGHLIN, S.: Structure and function of the fused Rhabdom. J. comp. Physiol. (In press)
TAUC, L., EPSTEIN, R.: Heterosynaptic facilitation as a distinct mechanism in *Aplysia*. Nature, Lond. 214, 724-725 (1967).
THORPE, W.H.: "Learing and Instinct in Animals", 2nd Edition. London: Methuen (1963).
UNGAR, G.: Role of proteins and peptides in learning and memory. *In* "Molecular Mechanisms in Memory and Learning" (Ed., G. Ungar), pp. 149-175. New York: Plenum Press (1970).
VOWLES, D.M.: The structure and connexions of the corpora pedunculata in bees and ants. Q. Jl microsc. Sci. 96, 239-255 (1955).

VOWLES, D.M.: Olfactory learning and brain lesions in the wood ant (*Formica rufa*). J. comp. physiol. Psychol. 58, 105-111 (1964a).
VOWLES, D.M.: Models and the insect brain. *In* "Neural Theory and Modeling" (Ed., R.F. Reiss). Stanford Calif.: Stanford University Press (1964b).
VOWLES, D.M.: Interocular transfer, brain lesions, and maze learning in the wood ant, *Formica rufa*. *In* "Chemistry of Learning" (Eds., W.C. Corning and S.C. Ratner), pp. 425-447. N.York: Plenum Press (1967).
VUILLAUME, M.: La rètention mémorique chez *Apis mellifica*. Annls Inst. natn. Rech. agron., Paris 2, 159-170 (1959).
WITTHÖFT, W.: Absolute Anzahl und Verteilung der Zellen im Hirn der Honigbiene. Z. Morph. Ökol. Tiere. 61, 160-184 (1967).
YOUNG, J.Z.: "A Model of the Brain". London: Oxford University Press. (1964).
YOUNG, J.Z.: Short and long memories in *Octopus* and the influence of the vertical lobe system. J. exp. Biol. 52, 385-393 (1970).

PERIODICITY IN THE ACTIVITY AND LEARNING PERFORMANCE OF THE HONEYBEE

R. Koltermann

Zoologisches Institut der Universität, Frankfurt, West Germany

All processes in living organisms are rhythmical. Well known examples are the photoperiodicity of plants and the circadian, lunar and annual periodicities in animals. In man, the disturbance caused by jet flight from east to west and *vice versa* is now well documented and recently it has been shown by Aschoff, v. Saint Paul and Wever (1971) that flies (*Phormia terrae novae*) which experienced light-shifts simulating eastwards or westwards movement died earlier than individuals which experienced no light-shift.

It has long been known that bees direct their activity towards feeding sources at fixed times of the day and that this activity is controlled by an endogenous clock. This was first reported by Forel (1910). Von Frisch and especially Beling (1929) laid the foundation for a large amount of research work into the characteristics of the clock. It was found by Beling (1929) that bees would learn to come to food sources at three times during the day. Wahl (1932) later succeeded in training them to come as many as five times a day. Also, Wahl (1932) trained bees to collect food in the morning at one place and in the afternoon at another and demonstrated that they could remember the hour at which the concentration of sugar solution offered was higher than at other times of day. Kleber (1935) and Körner (1939) showed that this sense of time was of advantage to the bees since it saved them from wasting energy searching for nectar or pollen at times when it was not produced by flowers. In nature, foraging times are positively correlated with the hours at which the flowers produce these commodities.

Research on this general subject up to 1967 has been summarised by von Frisch in his book "The Dance Language and Orientation of Bees" and is therefore readily accessible to English speaking scientists. I came to realise during the 14 International Congress of Entomology in Canberra (Australia) in 1972 that, because of the language barrier, much of the German work in this field since 1967 is relatively unknown to English speakers. I have decided therefore to review in this paper recent work which has been published in German together with some which is yet to be published.

INTERNAL AND EXTERNAL FACTORS

The earlier work having established that bees possess a time sense, the next step was to investigate the factors and mechanisms involved. Renner (1957) transported time trained bees across the ocean (from Paris to New York) and showed that, at least on the first day after their arrival, their time sense was independent of longitude, thereby demonstrating that they have an endogenous circadian periodicity. In further investigations Renner (1959) showed that light had an influence on the circadian rhythm. Beier (1968), Beier, Medugorac and Lindauer (1968), and Beier and Lindauer (1970) made a detailed investigation of the effects of light in laboratory experiments in which they shifted the phase of light and darkness. In these, bees were trained for about a week to find food at a fixed hour of the day and were then tested to see whether the training had been successful. After two more days of training the light-dark regime was shifted by 3.5 hr. It was found that it took the bees three days before they fully adjusted to the new photoperiodic regime. Aschoff and Wever (1963, 1966) demonstrated that it took the chaffinch (*Fringilla coelebs*) five days to re-synchronise after a shift of 6 hr. In neither animal therefore did the endogenous clock immediately adjust to the shift of photoperiod and it appears that bees adjust rather more rapidly than birds.

Using other time-shifting experiments Beier was able to show that the biological clock of the bee has two components, one of which remained at the original hour of training and one which was near the simulated new time. It was concluded that there were two mechanisms of time measuring, one influenced by external factors and the other linked more closely to endogenous factors. Synchronization was shown to be achieved more rapidly when the photoperiodic regime was advanced than when retarded.

Beier (1968) also carried out experiments to determine the range of conditions to which bees would become entrained when the total of the light and dark periods was greater or less than the normal 24 hr. He found that the range for entrainment was between 20 and 26 hr. He showed that no training occurred when the period was 19 hr and that little success was obtained at 20 and 26 hr. It was shown further that the period of the endogenous circadian oscillation was 23.4 hr (see also Bennet and Renner 1963). Under natural conditions the 24 hr regime effects a continuous synchronization.

CO_2 NARCOSIS AND THE SOCIAL TIME INDICATOR

Medugorac (1967) and Medugorac and Lindauer (1967) narcotised bees with CO_2 in order to characterise their endogenous clock. They trained bees to forage at a fixed hour of the day for a period of 5 days and after each training period the bees were narcotised with CO_2 at a concentration of greater than 20% for longer than 2 hr. On the day of test the bees displayed two visiting maxima at the training table which on this occasion was left without food. The first peak corresponded to the time of training and the second was delayed by a time which approximately equalled the period of anaesthesia. It was found also that the period of postponement of the second peak was influenced not only by the duration of anaesthesia but also by the concentration of CO_2. Again we have evidence for the existence of one time-measuring mechanism which is relatively stable and a second which can be influenced by external factors. The two peaks were obtained not only in the air-conditioned laboratory but also outdoors even though under these latter conditions the bees would have had the opportunity to use the sun for making time adjustments. In laboratory trials the insects were not influenced by artificial changes of light and darkness or by constant light. Furthermore, similar results were obtained irrespective of whether the bees were held at 22°, 26° or 33°C during the period of anaesthesia. Thus the timing of the second peak was influenced therefore only by CO_2 anaesthesia and not by a variety of other factors.

It was clear that the timing of the second maximum of visits after CO_2 narcosis was strictly correlated with the duration of anaesthesia, but the possibility existed that the timing of the first peak was due to the collecting activities of the non-narcotised bees in the hive. To investigate this possibility an experiment was carried out in which time trained bees which had been anaesthetised with CO_2 were put into a hive containing bees which had been trained to forage at a different time. The narcotised bees when tested the following day showed three peak times for visiting the training table which as before now lacked food. The first corresponded to the time to which they had been trained, the second was delayed by the period of the anaesthesia and the third corresponded to the collecting time of the hive into which they had been introduced. The last peak was due to a 'social time indicator', the community of the second hive having imprinted its time of collecting activities on the guest bees. A similar result was obtained when the time trained group of bees of the second hive was removed before the introduction of the guest bees. This finding indicates that the time of collecting was imprinted on the whole colony and that the non-trained bees communicated the time to the guest bees. If, however, the second hive had not been time trained at all, the guest bees displayed no third visiting maximum.

LUNAR PERIODICITY

Oehmke (1971) investigated another periodic activity in bees. Over a period of 4 months he recorded the number of exits and entrances made by a colony of bees (*Apis mellifera intermissa*) in the field in Morocco. Their activity, as measured by the number of exits and entrances, changed with the phase of the moon so that maxima occurred at the semi-lunar periods. The same rhythm was observed in a second colony of Moroccan bees which was observed at the same time. A colony of *Apis mellifera carnica*, also in Morocco, displayed a lunar rhythm but in this maximum activity occurred at new moon and minimum activity at full moon.

A colony of *A. mellifera carnica* kept in the laboratory at Frankfurt (Germany) also showed lunar phases of activity even when light, temperature and humidity were kept constant and food was continuously available.

A colony of *A. mellifera intermissa* which had shown semi-lunar maxima of flight activity in the field in Morocco changed to a lunar periodicity in which the maximum period of activity was a full moon when kept under constant conditions in the laboratory at Frankfurt. This change provided the first hint that lunar periodicity might depend on the geographical longitude and latitude and the associated changes in gravity and inclination of the earth's magnetic field. The results of another experiment confirmed this supposition. During the spring and summer of 1969 two hives of *Apis mellifera carnica* showed lunar periodicity of activity but the periods of maximal and minimal activity of the two colonies were not synchronized to the same lunar phase. One which was oriented in a north-south direction in an air conditioned room had its maximum period of flight activity at full moon whereas the other which was oriented in an east-west direction had its maximum at new moon. Further information came from another experiment. In this, bees (*A. mellifera carnica*) which had started their winter rest period and were put into constant laboratory conditions, changed their lunar flight activity rhythm into a semi-lunar one. During this period changes occurred in the intensity of the earth's magnetic field.

Despite these excellent experiments, it is clear that the analysis of the factors involved in the lunar and semi-lunar patterns of flight activity is still in its very early stages and that much further work is required. Lindauer and Martin (1968) showed that the earth's magnetic field influences the orientation of bees and it now seems that changes in the earth's magnetism may determine whether they display lunar or semi-lunar periodicity in their flight activity.

TIME LINKED LEARNING AND REMEMBERING OF FOOD SIGNALS

The work discussed above has demonstrated convincingly that bees show a circadian rhythm in remembering their training to a food (sugar solution) source. Our own research since 1967 has consisted of investigations to determine whether bees store information in their long term memory in a time linked process, not only about the location and quality of food, but also about signals, such as scent and colour, associated with the food source, and whether they remember these food signals strictly in a 24 hr cycle (Koltermann 1969, 1971).

Olfactory signals

The general method for experiments to investigate the time linked learning and remembering of olfactory signals was as follows. A group of individually marked (von Frisch 1967) bees (*Apis mellifera carnica*) are fed 10 times over a period of 50 min with 2M sucrose which is not marked with scent. The bees are then presented three times over a total period of 15 min with dishes of sucrose solution which have surrounding them filter paper bearing geraniol. After the training period the bees are again presented with sucrose solution without scent until the next training period and so on.

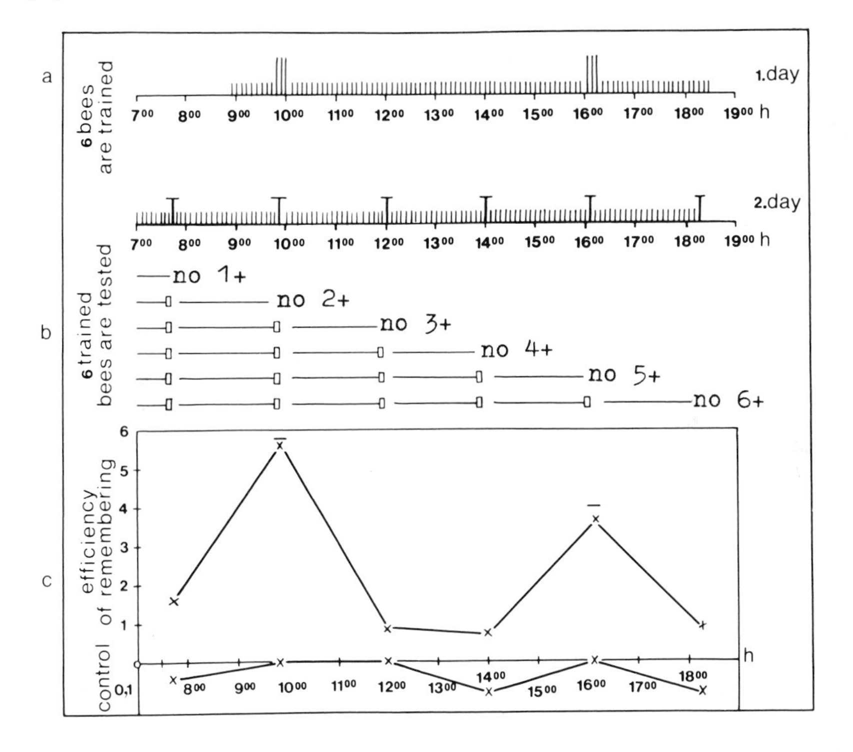

Fig. 1. (a) The procedure on day 1 for training bees to geraniol at two times of the day (viz., between 0945 and 1000 hr and between 1600 and 1615 hr). Each vertical line on the time axis represents one full feeding reward. The taller lines indicate the occasions on which the scent was present and the shorter ones those on which there was no scent.

(b) The procedure on day 2 for testing the trained bees. Bees number 1 to 6 are tested singly, the rest of the group being held in a small box (▯) during the test. The bees have to choose between the learned scent geraniol, and an unknown scent, fennel. Each bee is tested once and then discarded. Again the shorter vertical lines represent full feeding rewards without scent. T indicates the time of test.

(c) The efficiency of remembering, the data being from two 24 hr cycles

On the following day the bees are fed sugar solution with no scent present except at the time of testing. In the test situation each bee is tested separately and is required to make a choice between the learned scent, geraniol, and an unfamiliar one, fennel, the dishes being empty of sugar solution. Fig. 1 (a and b), which refers specifically to the situation discussed below in which bees were given two training periods, illustrates the general method. Because sugar solution was present at all times of day the method excludes the possibility that the bees were being trained to the food itself and ensures that the training was to the associated scent. Furthermore a social communication of the type described by Medugorac and Lindauer (1967) is impossible, each trained bee has to rely on its individual memory of the food signal. Furthermore, each bee was tested separately.

In the first of these experiments bees were trained to geraniol at only one time of day and were thereafter given sugar solution without scent. Fig. 2 shows that the efficiency of remembering, as measured by the number of times the bees alighted on each of the two scent-marked sources, follows a 24 hr periodicity. The dotted line was obtained from experiments with bees which, instead of being allowed to feed on 2.0 M sucrose, were kept in the hive between training and testing. The result for these bees was similar except that the efficiency of remembering was somewhat higher because in this instance there was no extinguishing effect of feeding them sugar solution without scent. Each point in Fig. 2 is the mean of about 20 different individually tested bees. The positive ordinate shows the mean number of bees which chose geraniol within the testing period of 5 min and the negative ordinate indicates the number of them which chose fennel.

The question may now be asked whether bees are able to associate a scent with a fixed time more than once a day. An experiment was performed in which bees were trained to geraniol at two times separated by 6 hr during which they were fed sugar solution which was not marked with scent. Next day between tests also they were fed sugar solution without scent. Fig. 1c shows that on the day after this the bees remembered the two times at which they had been trained to geraniol, the differences being highly significant.

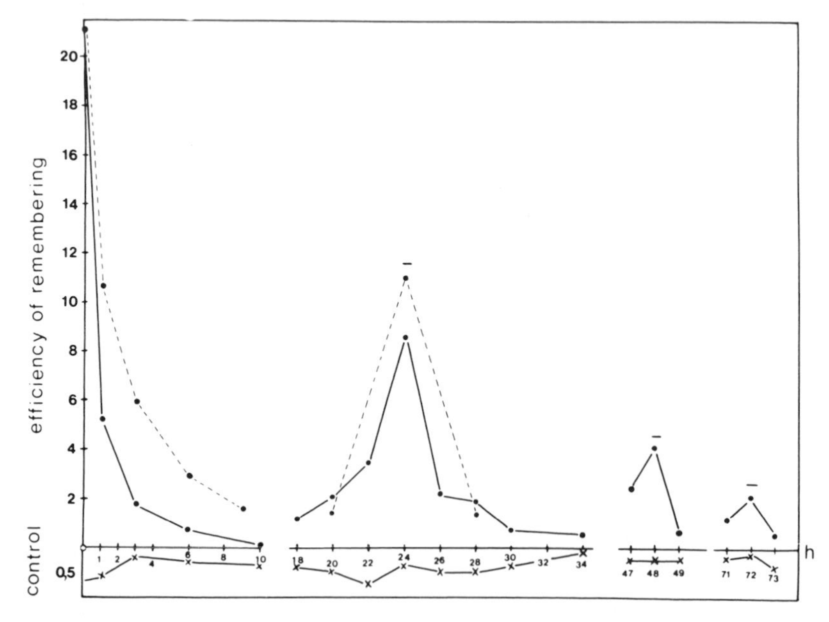

Fig. 2. The efficiency of remembering over a period of 4 days following that on which the bees had been trained to geraniol at one time of day. The training consisted of 3 full feeding rewards over 15 min

This experiment was followed by one in which bees were trained to geraniol at a third time intermediate between the two used in the previous experiment. Here there were three memory peaks coming exactly 24 hr after each of the three training times. In another experiment there were five training times separated by 2 hr and again the bees remembered the geraniol odour exactly 24 hr after each of the five training times. In this experiment the interval between tests was 1 hr instead of the 2 hr used in the previous experiments in which the bees were trained to fewer times. The finding that there were five significant peaks in the efficiency of remembering means that the bees were able to distinguish two times separated by only one hour. These peaks were more pronounced than in previous experiments, a fact which may be explained by the higher number of training acts (5x3) during training to the geraniol scented source.

Wahl (1932) also succeeded in training bees to a food source at five times during the day but our experiments differed from his in three ways. First, Wahl's time training was carried out for 6-8 days whereas in our experiments training was completed in only one day. Secondly, Wahl fed his bees only at the training times, the period of feeding in each case being 1.5-2 hr. In our experiments sugar solution was given to the bees throughout the day and the input time for each food signal was only 15 min. Thirdly, Wahl found that bees were unable to distinguish between training times that were separated by less than 2 hr. In our experiments the training times were separated by two hours but we found that, when the bees were tested next day, they were able to distinguish time 1 hr before or after that corresponding to the time of training.

Finally, we determined whether the same scent could be associated 9 times with a fixed time of day and whether the food signal was still remembered next day in an exactly 24 hr cycle. Before attempting this however we had to examine another problem which arises from the fact that bees can be trained to more than 5 times in a day only if the interval between the times is shortened to less than 2 hr. With bees trained to 9 times the intertrial interval is only 45 min. Before proceeding with the main experiment therefore we had to determine whether bees were able to distinguish two 24 hr points which were separated from one another by a 45 min interval. It was found that not only were they able to separate the two points but that the performance at these times differed significantly from neighbouring points. Thus the bees were able to separate two points which were separated by only about 20 min. Von Frisch (1967) cites experiments of Haidl which showed that bees which could not see the sun and which were constantly dancing in the hive were able to measure to an exactness of about 5 min. Similar observations were made later by Lindauer (1954, 1957) and others.

The results of our preliminary experiments made it reasonable to attempt to determine whether bees could be trained to geraniol at 9 different times in a day. Since there were three feeding acts at each training period and therefore 27 acts of learning, it seemed possible that the bees would next day remember the training regime as one with equal intensity throughout the day with no minima between the 24 hr point. The results obtained with 285 singly tested bees which made a total of 1023 decisions showed that the bees were able to learn and distinguish 9 different times and to remember them in a 24 hr cycle (Fig. 3).

In the experiments described above the intervals between the training times were equal and it may be argued therefore that the equality of the intervals may have influenced the learning performance of the bees. To determine whether this was so we trained bees to geraniol four

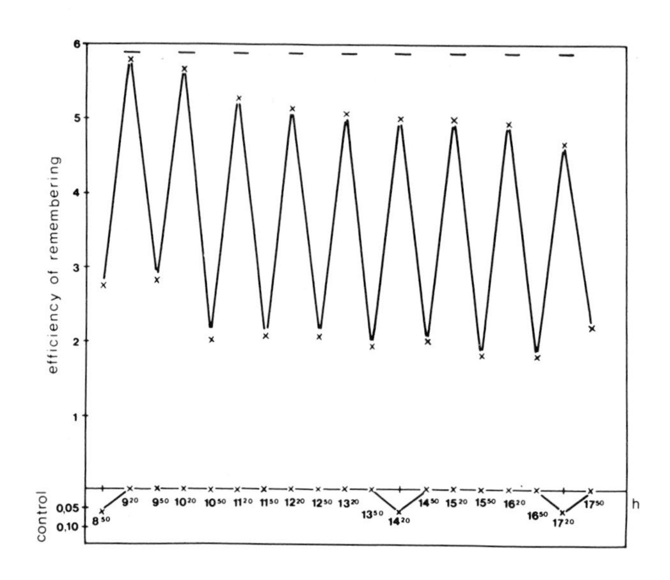

Fig. 3. Efficiency of remembering on the day following that on which the bees had been trained to geraniol at 9 different times of day. Each training consisted of 3 learning acts

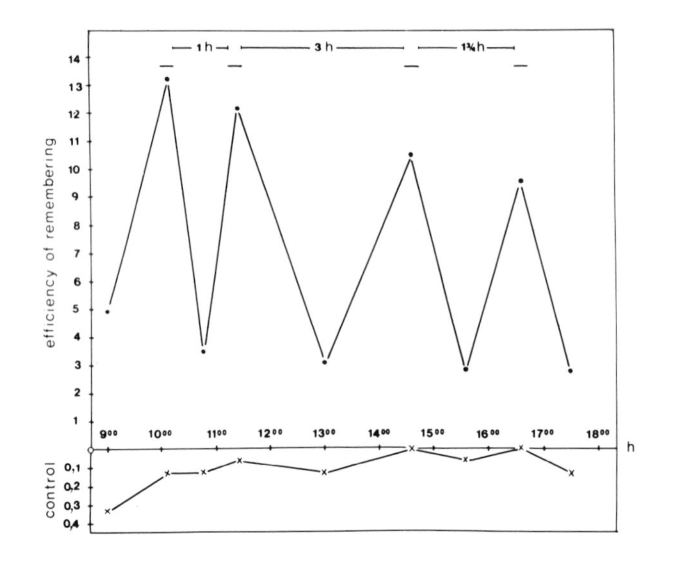

Fig. 4. Efficiency of remembering on the day following that on which bees had been trained to geraniol at 4 different times separated by irregular intervals. Each training consisted of 3 learning acts

times in a day with different intervals between the times. There was one of 1 hr between the first and second training times, 3 hr between the second and third and 1¾ hr between the third and fourth. Under these conditions the bees remembered the scent just as precisely in a 24 hr cycle as in the previous experiment with equal intervals. Equality of intervals is not therefore necessary or even beneficial to the learning performance (Fig. 4).

The next problem was to determine whether there is a precise 24 hr memory when bees are trained to two different scents. In our experiment the interval between training times was 6 hr and the methods were similar to those described previously. In the morning the bees were trained to the odour of geraniol with food through 3 acts of feeding; 6 hr later they were similarly trained to the odour of fennel. Between and after these times they received sugar solution the location of which was not marked with any scent. The two learned scents were presented simultaneously next day at regular intervals. It was found that the bees chose each of the two food signals more often 24 hr after being trained to it but that they chose geraniol for a longer period and at a higher rate than expected. There seemed to be two possible explanations for this; the first is that the results might be influenced by the particular hour of day at which the bees were trained, and the second is that geraniol is inherently more attractive to bees than fennel. The first possibility was eliminated by the result of an experiment in which bees were trained to fennel in the morning and geraniol in the afternoon. It must be concluded therefore that geraniol is more attractive. A paper dealing with the relative attractivenesses of a range of scents will be published soon.

The above experiment was repeated using geraniol and thyme oil which had previously been found to be equally attractive (Koltermann 1969). The results (Fig. 5) show that the bees' 24 hr memory was quite precise,

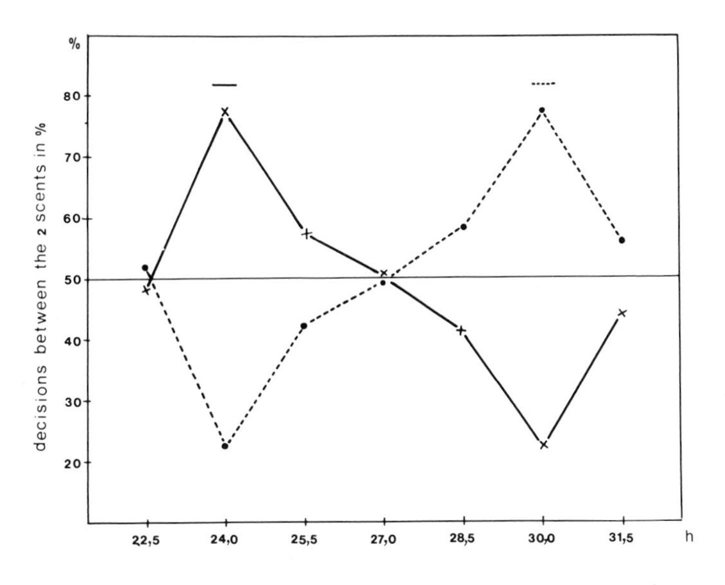

Fig. 5. Efficiency of remembering on the day following that on which bees had been trained to two different scents (geraniol and thyme oil) in the morning and in the afternoon respectively. Each training consisted of 5 learning acts

the peak for the two scents being equally high at the appropriate times. Midway between the times of training the bees chose the two scents equally. We can conclude from this experiment that the choice made by bees between the two scents depends on (i) a 24 hr memory cycle, and (ii) the inherent attractiveness of the scent. The first of these is the more important factor, the second may have the effect of modifying the effects of the memory.

Visual signals

Our previously described experiments have shown that bees associate olfactory signals with the hour of day and remember them in a 24 hr cycle. Our next step was to examine the possibility that bees may also associate visual signals with the hour of day. In order to do this we applied the same training methods as those we used with scents. Bees received 6 feeding acts between 1000 and 1030 hr with sugar solution presented on a piece of green cardboard 14 x 14 cm. Thereafter until 1530 hr food was supplied on a white piece of cardboard. From 1530 to 1600 hr they were similarly trained to violet cardboard. Again after this training period food was supplied on white cardboard. Next day at regular intervals the two colours were presented simultaneously. The bees chose green in 64.7% tests between 1000 and 1030 hr and violet in 67% of those between 1530 and 1600 hr. Thus bees can associate visual signals with the hour of day and can remember them in a 24 hr cycle.

CONCLUSION

The experiments reviewed in this paper have shown that the long term memory of bees is strictly time linked for a number of different kinds of food signal. They have been shown to learn, in a time linked process different feeding times (Beling 1929, Wahl 1932), different qualities of food (Wahl 1933), different locations of food (Wahl 1932, Finke 1958), and olfactory and visual signals. There has been demonstrated therefore a fundamental characteristic of the learning of information about food sources by bees. The time linked learning process has no doubt evolved as an adaptation to the regular nectar and pollen production of flowers. The mechanisms demonstrated enable bees to exploit nectar and pollen sources with a minimum of wasted energy and hence with maximum efficiency, since they fly to sources only when the material is being produced.

The physiological basis for this type of behaviour is as yet unknown. One approach, however, to the elucidation of this may be provided by the concept of state dependent learning (John 1967), in which bees would associate a particular food signal with a certain inherent physiological state. Further work is certainly required and it is probable that the next step may be the special task of biochemists.

REFERENCES

ASCHOFF, J., v. SAINT PAUL, U., WEVER, R.: Die Lebensdauer von Fliegen unter dem Einfluß von Zeitverschiebungen. Naturwissenschaften 58, 574 (1971).

ASCHOFF, J., WEVER, R.: Resynchronisation der Tagesperiodik von Vögeln nach Phasensprung des Zeitgebers. Z. vergl. Physiol. 46, 321-335 (1963).

ASCHOFF, J., WEVER, R.: Circadian period and phase-angle difference in Chaffinches (*Fringilla coelebs* L.). Comp. Biochem. Physiol. 18, 397-404 (1966).

BEIER, W.: Beeinflussung der inneren Uhr der Bienen durch Phasenverschiebung des Licht-Dunkel- Zeitgebers. Z. Bienenforsch. 9, 356-378 (1968).

BEIER, W., MEDUGORAC, I., LINDAUER, M.: Synchronisation et dissociation de "L'Horloge interne" des abeilles par des facteurs externes. Annls Epiphyt. 19, 133-144 (1968).

BEIER, W., LINDAUER, M.: Der Sonnenstand als Zeitgeber für die Biene. Apidologie 1, 5-28 (1970).

BELING, I.: Über das Zeitgedächtnis der Bienen. Z. vergl. Physiol. 9, 259-338 (1929).

BENNET, M.F., RENNER, M.: The collecting performance of honey bees under laboratory conditions. Biol. Bull. mar. biol. Lab., Woods Hole 125, 416-430 (1963).
FINKE, I.: Zeitgedächtnis und Sonnenorientierung der Bienen. Lehramtsarbeit, Naturw. Fak. Univ. München (1958).
FOREL, A.: "Das Sinnesleben der Insekten". Munchen: E. Reinhardt (1910).
von FRISCH, K.: "The Dance Language and Orientation of Bees". Cambridge, Mass.: Belknap Press of Harvard Univ. Press (1967).
JOHN, R.E.: "Mechanisms of Memory". New York / London: Academic Press (1967).
KLEBER, E.: Hat das Zeitgedächtnis der Bienen biologische Bedeutung? Z. vergl. Physiol. 22, 221-262 (1935).
KOLTERMANN, R.: Lern- und Vergessensprozesse bei der Honigbiene - aufgezeigt anhand von Duftdressuren. Z. vergl. Physiol. 63, 310-334 (1969).
KOLTERMANN, R.: 24-Std-Periodik in der Langzeiterinnerung an Duft- und Farbsignale bei der Honigbiene. Z. vergl. Physiol. 75, 49-68 (1971).
KÖRNER, I.: Zeitgedächtnis und Alarmierung bei den Bienen. Z. vergl. Physiol. 27, 445-459 (1939).
LINDAUER, M.: Dauertänze im Bienenstock und ihre Beziehung zur Sonnenbahn. Naturwissenschaften 41, 506-507 (1954).
LINDAUER, M.: Sonnenorientierung der Bienen unter der Äquatorsonne und zur Nachtzeit. Naturwissenschaften 44, 1-6 (1957).
LINDAUER, M., MARTIN, H.: Die Schwereorientierung der Bienen unter dem Einfluß des Erdmagnetfeldes. Z. vergl. Physiol. 60, 219-243 (1968).
MEDUGORAC, I.: Orientierung der Bienen in Raum und Zeit nach Dauernarkose. Z. Bienenforsch. 9, 105-119 (1967).
MEDUGORAC, I., LINDAUER, M.: Das Zeitgedächtnis der Bienen unter dem Einfluß von Narkose und sozialen Zeitgebern. Z. vergl. Physiol. 55, 450-474 (1967).
OEHMKE, M.: Lunarperiodische und tagesrhythmische Flugaktivität der Bienen. Dissertation d. Naturw.Fak. Univ. Frankfurt/M. (1971).
RENNER, M.: Neue Versuche über den Zeitsinn der Honigbiene. Z. vergl. Physiol. 40, 85-118 (1957).
RENNER, M.: Über ein weiteres Versetzungsexperiment zur Analyse des Zeitsinnes und der Sonnenorientierung der Honigbiene. Z. vergl. Physiol. 42, 449-483 (1959).
WAHL, O.: Neue Untersuchungen über das Zeitgedächtnis der Bienen. Z. vergl. Physiol. 16, 529-589 (1932).
WAHL, O.: Beitrag zur Frage der biologischen Bedeutung des Zeitgedächtnisses der Bienen. Z. vergl. Physiol. 18, 709-717 (1933).

MASSED TRAINING AND LATENT HABITUATION OF THE DEIMATIC RESPONSE IN THE MANTID, *STAGMATOPTERA BIOCELLATA*

J.C. Barrós-Pita

Centro de Biofisica y Bioquimica, Instituto Venezolano de Investigaciones Cientificas, Caracas, Venezuela

The study of habituation is of great general interest, for it seems to represent the most elementary form of behavioural plasticity (Grossman 1967). An adequate theory of this phenomenon may therefore provide a basis for increased understanding of other more complex forms of learning (Grossman 1967; Groves and Thomson 1970; Harris 1943; Thomson and Spencer 1966).

Extensive research on habituation in intact organisms and simplified nervous system preparations has been done mainly on vertebrates. It is clear, however, that invertebrates offer a valuable alternative as they are "naturally simplified" preparations. Their behaviour is simpler, very stereotyped and of limited repertoire; hence any behavioural change is more easily detectable.

Mantids, when faced with insectivorous birds, display a frightening reaction, the Deimatic Response (DR) (Maldonado 1970). The work carried out by Balderrama and Maldonado (1971) on the habituation of this response is a good example of the use of an invertebrate for the elucidation of basic problems in behaviour. This work demonstrated that upon repeated presentation of a specific stimulus (a live bird) to mantids, the response elicited decreases with each successive presentation. This decrement of the DR is long-lasting and has most of the parametric characteristics of habituation in mammals (Balderrama and Maldonado 1971; Thomson and Spencer 1966). This indicates that habituation in mantids parallels many behavioural modifications which are characteristic of higher animals. The understanding of the process involved in the habituation of the DR in mantids might therefore provide insights into more complex mechanisms of habituation and memory in vertebrates. The study of habituation of the DR in mantids is also very important because this habituation, as well as being an interesting phenomenon in its own right, poses the very interesting problem of habituation to the specific signal of a predator. This is also reported to occur in other animals (Hinde 1954; Martin and Melvin 1964; Melvin and Cloar 1969; Melzack 1961), but it seems to be in overt contradiction with the survival principle (Hinde 1954; Thorpe 1950). Balderrama and Maldonado (1971) argue that the constancy of the environment in which the specific stimulus was repeatedly presented, accounts for such habituation. This explanation seems to be an over-simplification because it is far from clear as to how the constancy of the environment can convert, for the experimental animal, such a specific signal as a predator into an indifferent one. The reason for this kind of habituation must be related somehow to the general characteristics of the stimulus presentation. In this regard it is worth noting that Maldonado (1970) reported a DR that lasted 6 hr without decrement (i.e. until the observation was discontinued) when the stimulus was presented continuously. In this case the stimulation time (6 hr) was 2 hr longer than the total stimulation time (about 4 hr) reported by Balderrama and Maldonado (1971), found to be necessary for the production of habituation when the stimulus was presented repeatedly with intervals between each successive presentation. This suggests that the DR might be displayed for at least as long as the stimulus is shown and that, under these conditions, no habituation ensues. It is also known that, in higher animals, temporarily spaced training produces better learning than massed training (Hinde 1954; Woodworth and Schlosberg 1964). Therefore to determine some of the causal factors that take part in the process of habituation, the effects of massed as compared to spaced training in the habituation of the DR of mantids to the presentation of live birds are investigated.

REARING AND PRE-TREATMENT OF MANTIDS

The subjects were experimentally naive adult female mantids *Stagmatoptera biocellata* that had reached the imago stage 20 days before. All mantids had been reared in individual cages in a laboratory vivarium at a constant temperature of 29°C during day time and 24°C during night time. The illumination time was 12 hr per day to reproduce the day-night cycle. After the last moult and before the experimental period, the mantids were given six feeding sessions constituted and distributed as follows: 15 sarcophaga flies (*Parasarcophaga argyrostoma*) the first time, 10 flies the second, 5 flies the third, 10 flies the fourth, fifth and sixth times. From the first to the fourth, the feeding sessions were given every fourth day; from the fourth to the sixth, every third day. This feeding programme is very important because it permits the running of the experimental period without feeding the mantids (which could introduce interfering signals) and without causing them any ill effects. The day after the last feeding session the mantids were mounted in the mantid holder (Fig. 1) and brought to the experimental apparatus (Fig. 2).

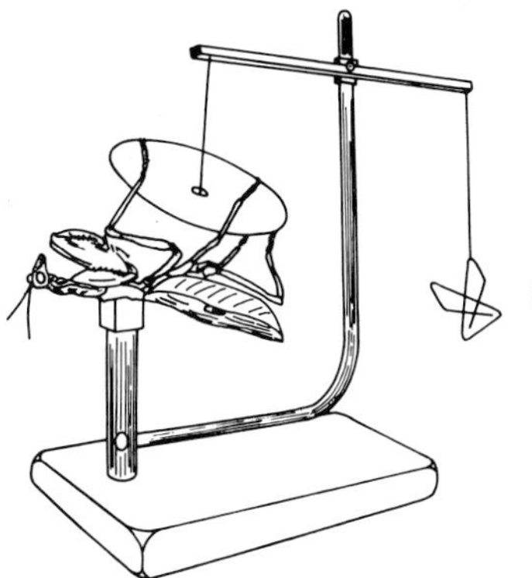

Fig. 1. Mantid in rest position mounted on mantid holder

APPARATUS

A schematic drawing of the experimental set-up is shown in Fig. 2. This apparatus, which could contain at one time 12 mantids and 12 birds, is a wooden structure 1.53 m wide, 1.85 m high, 0.40 m deep with 12 mantid compartments 30 x 20 x 10 cm each arranged in two rows, one above the other. Each mounted mantid was placed in one compartment facing a bird cage, from which it was separated by a sliding screen. Birds enclosed in individual cages with the front end of transparent lucite, were put on two shelves, exactly facing the sliding screen on the opposite side of the mantid compartments (Fig. 2). The mantid compartments were individually illuminated and heated with small glow lamps; the temperature and light cycle in the mantid compartments being the same as those of the vivarium. The rear of each mantid compartment had a small window (not shown in Fig. 2) covered with red glass through which the mantid's reaction could be observed.

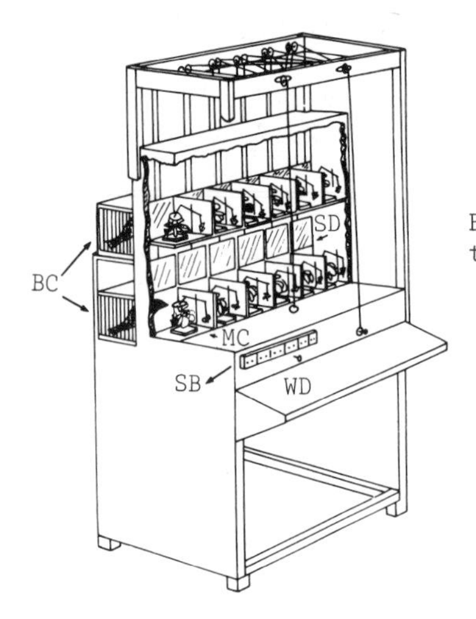

Fig. 2. Experimental apparatus with rear part removed to show inside.
MC: mantid compartments
SD: sliding screen
BC: bird cages
SB: switch board
WD: writing desk

PROCEDURE AND EXPERIMENTAL DESIGN

A trial began when the lights that illuminated the row of compartments containing the mantids to be tested were turned off and the sliding screens were lifted, by means of a pulley system. This allowed the mantids to see the birds (male shiny cowbirds - *Molothrus bonairensis*). During a trial, 6 mantids were watched at the same time and their responses recorded. The responses were registered as 1 or 0 according to whether or not the mantids raised their wings and tegmina (Fig. 3). When a trial was over, the conditions were returned to their previous state. The bird compartments were continuously illuminated by fluorescent tubes throughout the experimental sessions.

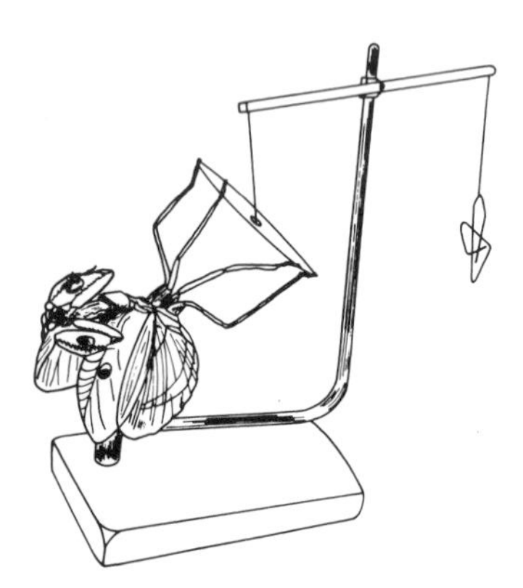

Fig. 3. Mantid displaying the deimatic response (DR) as it is stimulated with the visual presentation of a live bird. The raising of wings and tegmina is the sole criterion for the DR

In Table 1 a summary of the experimental groups and design is shown. The experimental subjects (58) were randomly divided into groups K and D each containing 29 individuals. The mantids forming group K were mounted on day 1 (Fig. 1) and were put in the mantid compartment of the experimental apparatus (Fig. 2) and left undisturbed for the following two days (days 2 and 3). Group K was stimulated with massed training in a daily session-trial 40 min long (which is equivalent to a daily session of 20 trials, 2 min long each, without intertrial intervals) on days 4 and 5. The intersession interval corresponded to the over-

Table 1. A. Description of Experimental Groups.
B. Experimental Design.

A.

Groups	No. of Mantids Accepted per Group	No. of Mantids Rejected per Group	Total No. of Mantids per Group	Acceptance Level per Group
K≡Δ	23 ♀	6 ♀	29 ♀	79%
D	23 ♀	6 ♀	29 ♀	79%

B.

Groups \ Days	1	2	3	4	5	6	7
K 23 Mantids ♀	Mounted	-	-	Massed Training	Massed Training		
Δ is formed by the same Mantids of Group K						Spaced Training	Spaced Training
D 23 Mantids ♀	Mounted	-	-	-	-	Spaced Training	Spaced Training

Massed Training - 1 single trial 40 min long = 20 massed trials in which a live bird shiny cowbird - *Molothrus bonariensis* ♂) is presented for 2 min per trial.

Spaced Training - 20 spaced trials in which a live bird (shiny cowbird - *Molothrus bonariensis* ♂) is presented for 2 min per trial, with intertrial intervals of 12 min.

night rest period. In order to determine the mnemonic condition of group K after the two days of massed training, it was stimulated with spaced training on days 6 and 7 in a daily session of 20 spaced trials 2 min long each, with intertrial intervals of 12 min. For the purpose of this spaced training, group K was re-designated group Δ. The mantids forming group D were handled as those of group K, but were left undisturbed in the mantid compartments for 4 days after mounting (days 2 to 5). On days 6 and 7 group D was stimulated with spaced training in a daily session of 20 spaced trials, 2 min long each with intertrial intervals of 12 min. Cross comparisons of the habituation curves of groups K, Δ and D were made to evaluate the effect of massed training on habituation of the DR as compared to that of spaced training.

RESULTS

In each group, 6 out of 29 animals failed to display the DR upon the presentation of the live birds at any time during the experimental sessions. The data relating to these animals were therefore discarded and only data provided by the other 23 per group were used for computation. The data corresponding to the responses of the 23 mantids in each group were pooled and computed as percentage of DRs per trial. These scores were then plotted as percentage of the DR (on the ordinate axis) against the number of trials (on the abscissa axis). Fig. 4A shows the results for group K stimulated with massed training for two days (data K_1 for the first day, and K_2 for the second) and those of group D stimulated also for two days but with spaced training (data D_1 for the first day, and D_2 for the second). The scores for group K show that, on both days, all mantids sustained the DR throughout the presentation of the stimulus, the steady response level being 100% throughout.

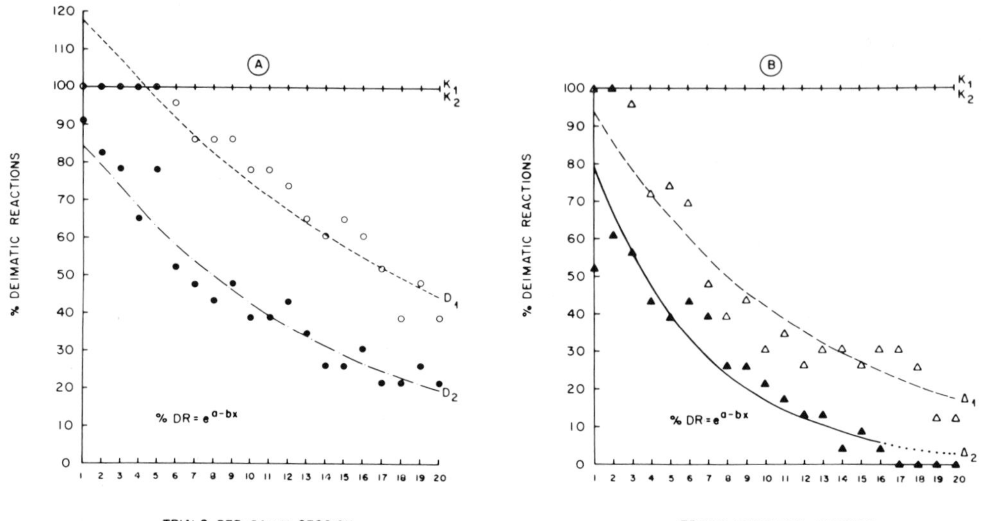

Fig. 4. A. Comparison of the results of group K stimulated with massed training for two days (data K_1 and K_2) with those of group D also stimulated for two days (data D_1 and D_2) but with spaced training. All regression curves are of the form % DR = e^{a-bx}.

Performance of group K during massed training:

K_1 first day) - crosses: actual scores; straight solid line: "regression curve"
K_2 second day)

Performance of group D during spaced training:

D_1 first day - open circles: actual scores; dashed line: regression curve
D_2 second day - closed circles: actual scores; dash-dotted line: regression curve

B. Comparison of the results of group K (massed trained, same as in A) with those of the same group but when it was stimulated on the following 2 days with spaced training. The group K subjected to spaced training was at this time called group Δ. All regression curves are of the form % DR = e^{a-bx}.

Performance of group Δ during spaced training:

Δ_1 first day - open triangles: actual scores; dashed line: regression curve
Δ_2 second day - closed triangles: actual scores; solid line: regression curve; dotted line: extrapolation region of the regression curve as the data from 17 to 20 could not be computed because the actual scores equalled zero

The data for group D, on the other hand, show a clear-cut decrement of response which indicated that there was both within session habituation and a build-up of habituation across two days. The response decrement shown by group D on each of the two days follows a negative exponential law and the data were regressed to the best fitting curve which is of the form % DR = e^{a-bx}. Fig. 4B shows the comparison between the original results of group K and those for the same group, now called group Δ, when stimulated with spaced training during 2 days immediately after the two days of massed training (data Δ_1 for the first day, and Δ_2 for the second). The scores of group Δ(Δ_1 and Δ_2) show a response decrement which follows a negative exponential law indicating within session habituation and build-up of habituation across the two days, as was seen in group D. The habituation of group Δ proceeded at a much faster rate than that of group D, and was maximal on the second day for the last four trials (17 to 20) when the score response was zero. These data were also regressed to the best fitting curve which is also of the form % DR = e^{a-bx}. To obtain the regression curve of Δ_2, only the scores of trials 1 to 16 could be computed because the scores of 17 to 20 were zero. The dotted line, which is a continuation of the solid line, was obtained therefore by extrapolation. The regressed curves as

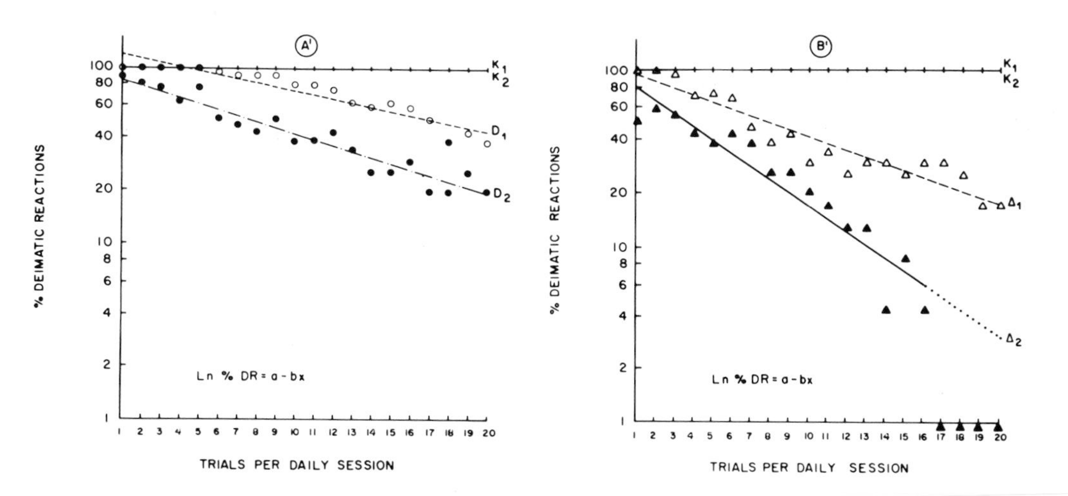

Fig. 5. The actual scores and regression lines of groups $K(K_1$ and $K_2)$, $D(D_1$ and $D_2)$ and $\Delta(\Delta_1$ and $\Delta_2)$ are plotted here in semilog graphs (ordinate axis ln scale) to obtain straight lines of the form ln % DR = a-bx in which the regression coefficient is the slope b. In A' the actual scores and straightened lines for comparison of groups $K(K_1$ and $K_2)$ and $D(D_1$ and $D_2)$ are shown. In B' the actual scores and straightened lines for comparison of groups $K(K_1$ and $K_2)$ and $\Delta(\Delta_1$ and $\Delta_2)$ are shown

well as the actual scores for groups $K(K_1$ and $K_2)$ and $D(D_1$ and $D_2)$ are again plotted in Fig. 5A' but here the percentage of DRs are represented in the ordinate axis on a ln scale. The regression curves, now straight lines, are of the form ln % DR = a-bx, in which the slope b is the regression coefficient. In Fig. 5B' the regressed curves and actual scores for groups $K(K_1$ and $K_2)$ and $\Delta(\Delta_1$ and $\Delta_2)$ are plotted in the same manner as for those of Fig. 5A'. It can be seen from Fig. 5A' and B' that the position and slopes of the straight lines for K_1 and K_2 (group K) and D_1 and D_2 (group D) and for Δ_1 and Δ_2 (group Δ) are different. In order to know the significances of the differences between every pair of straight lines, the F-test for differences between pairs of regression

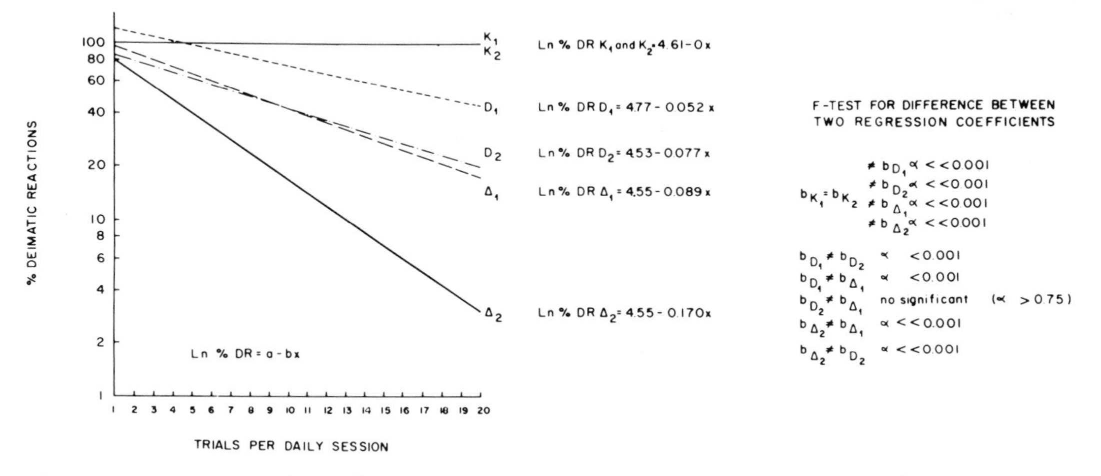

Fig. 6. All the straight lines representing the performance of groups $K(K_1$ and $K_2)$, $D(D_1$ and $D_2)$ and $\Delta(\Delta_1$ and $\Delta_2)$ are plotted in the same semilog graph allowing overall comparison of the performance of all groups. F tests for differences between pairs of regression coefficients (b) were used to determine the significances of the differences between every pair of straight lines. The coefficient pairs compared and the results obtained are shown at the right side of the graph

coefficients (b) was made. For convenience all lines are plotted on the same graph (Fig. 6). The results obtained indicate that the group receiving massed training (group K) exhibited no intrasession or intersession habituation (slope of curves K_1 and K_2 : $bK_1 = bK_2 = 0$) whereas the group receiving spaced training (group D) exhibited intrasession and intersession habituation (slope of curves D_1 and D_2 : bD_1 and $bD_2 < 0$, and also $bD_1 > bD_2$, $\alpha < .001$), despite the fact that both groups K and D received the same number of training trials. On the other hand, when group K became group Δ and thus received spaced training, it also exhibited intrasession and intersession habituation (slope of curves Δ_1 and Δ_2 : $b\Delta_1$ and $b\Delta_2 < 0$; and also $b\Delta_1 > b\Delta_2$, $\alpha << .001$). Furthermore group Δ exhibited greater habituation for the first and second days of spaced training sessions than group D for the same training sessions. Habituation of group Δ the first day of spaced training (Δ_1) is significantly greater than habituation of group D for the first day of spaced training (D_1) ($b\Delta_1 < bD_1$, $\alpha < .001$) and statistically equal to habituation of group D the second day of spaced training (D_2) ($b\Delta_1 = bD_2$, $\alpha > .75$). Habituation of group Δ the second day of spaced training (Δ_2) is significantly greater than habituation of group D for the second day of spaced training (D_2) ($b\Delta_2 < bD_2$, $\alpha << .001$).

DISCUSSION

The massed trained group did not exhibit habituation even though the group which received the same number of trials in spaced training did. This lack of habituation of the massed trained group might be due to the fact that the DR displayed by mantids is "effective" in restraining the birds from attacking, which is what happens in actual long confrontations between birds and mantids (Maldonado 1970). No habituation of a defensive response to a predator should ensue if the response is effective in averting or bringing about the disappearance of the threatening stimulus (Fabricius 1966). By contrast, when mantids are subjected to spaced training, the threatening stimulus appears and disappears according to a strict temporal programme which has no relationship to the defensive response elicited and which inflicts no harm. It is highly improbable that this would ever happen in the naturally repetitive occurrence of a "true" danger signal. It is likely therefore that stimuli presented in this strictly temporal manner would become "meaningless" and that this would result in a waning of the defensive behaviour, i.e. it would habituate.

On the other hand, when the massed trained group (group K) received spaced training (becoming then group Δ), it exhibited on the first day (Δ_1) as much habituation as did group D (spaced trained) on the second training day (D_2). On the second day of spaced training (Δ_2) the habituation of group Δ was total for the last 4 training trials (17-20). This level of habituation surpasses by far that exhibited by the spaced trained group D for the second training day (D_2). Group D would probably have required a third day of spaced training to match the performance of group Δ on the second day of its spaced training (Δ_2). This can be inferred from the trend of the data, and is also in accord with the results reported by Balderrama and Maldonado (1971). The habituation performance exhibited by group K when it became group Δ, suggests that retention of the stimulus (the sighting of a moving bird) could build up in mantids throughout massed training even though there was no sign of habituation. It seems therefore justified to call this phenomenon Latent Habituation.

Habituation as well as Latent Habituation might be the expression of the "meaning" of the memory content (Grossman 1967) in mantids. The retention of a "true" threatening bird builds up in mantids during massed training. Under those conditions, of course, no habituation should be shown, which is what actually happens. However, when these mantids are subsequently subjected to spaced training, the stimulus becomes a "false" threatening bird and the response habituates. This habituation proceeds at a faster rate than in naive mantids, probably because the change of "meaning" of the memory content (from a "true" to a "false" threatening bird) requires less training time than the build-up of such a memory.

In higher animals, spaced training produces better learning than massed training because intertrial intervals seem to be periods during which memory is better fixed because the experimental subjects are free from interfering stimuli (Alloway 1972; Woodworth and Schlosberg 1964). Nevertheless, the main effect of the spaced training (in a strict temporal programme) of mantids with a signal of danger is seemingly the impairment of the "naturality" of the stimulus. The habituation of the response elicited by this type of stimulus is a consequence of its "unnaturality". However, the possibility that the mnemonic process in mantids subjected to massed training is impaired cannot be ruled out (Alloway 1972) but if any interference exists, it seems that the blocking action is not very great.

CONCLUSIONS

The results presented in this paper indicate that habituation of the DR in mantids requires spaced training while the mnemonic process does not. Further research is needed to confirm these findings and also to verify the theoretical interpretations proposed here to explain the phenomena discussed. The DR of mantids is a promising behavioural response for the elucidation of the problem of habituation to a very specific signal of "danger" and of habituation in general.

Because research planned for the future requires better experimental methods, a special cinematographic apparatus (The Kinopsitheriotron) has been designed and built for the visual stimulation of mantids with films of silhouettes of shiny cowbirds. This method will replace the cumbersome use of live birds and will ensure stimulus constancy with the same releasing power for all the subjects. The description, research uses and possibilities of this equipment are published elsewhere.

SUMMARY

The sight of moving birds releases in mantids a defensive behaviour: the Deimatic Response (DR). In order to study the dynamic characteristics of the habituation of the DR, 2 mantid groups, D and K, each consisting of 23 animals, were visually stimulated with the sight of live birds.

Groups D and K received the same kind and duration of stimulation. Group D was stimulated with spaced training in a daily session of 20 trials, 2 min long each with 12 min of intertrial interval, for two days. Group K was stimulated with massed training in a daily session-trial 40 min long (which is equivalent to a daily session of 20 trials, 2 min long each without intertrial intervals), for two days. Group D rapidly habituated, whereas the group K did not show any sign of habituation.

In order to study the mnemonic condition of group K after massed training, this group was called group Δ and subjected to spaced training on the day following the second day of massed training, with the same stimulation programme as that of group D. Habituation in group Δ occurred more rapidly and attained a higher degree than in group D. This is interpreted as a memory build-up of the stimulus during massed training which itself produced no habituation; hence the term Latent Habituation. Habituation and Latent Habituation are interpreted as a result of the different "meaning" of the memory content in mantids.

REFERENCES

ALLOWAY, T.M.: Learning and memory in insects. A. Rev. Ent. 17, 43-56 (1972).

BALDERRAMA, N., MALDONADO, H.: Habituation of the deimatic response in the mantid (*Stagmatoptera biocellata*). J. comp. physiol. Psychol. 75, 98-106 (1971).

FABRICIUS, E.: "La Conducta de los Animales". pp. 97-98. Buenos Aires: Editorial Universitaria de Buenos Aires (1966).

GROSSMAN, S.P.: "A Textbook of Physiological Psychology". pp. 641-642. New York: John Wiley & Sons (1967).

GROVES, M.P., THOMSON, R.F.: Habituation. A dual process theory. Psychol. Rev. 77, 419-446 (1970).

HARRIS, J.D.: Habituation response decrement in the intact organism. Psychol. Bull. 40, 385-422 (1943).

HINDE, R.A.: Factors governing the changes in strength of a partially inborn response as shown by the mobbing behavior of the Chaffinch (*Fringilla coelebs*): II. The waning of the response. Proc. R. Soc. (B) 142, 331-358 (1954).

MALDONADO, H.: The deimatic reaction in the praying mantid *Stagmatoptera biocellata*. Z. vergl. Physiol. 68, 60-71 (1970).

MARTIN, R.C., MELVIN, K.B.: Fear response of the bob-white quail (*Colinus virginianus*) to a live and model red-tailed hawk (*Buteo jamaicensis*). Psychol. Forsch. 27, 323-336 (1964).

MELVIN, K.B., CLOAR, F.T.: Habituation of responses of quail (*Colinus virginianus*) to hawk (*Buteo swaisoni*): Measurement through an "innate suppression" technique. Anim. Behav. 17, 468-473 (1969).

MELZACK, R.: On the survival of mallard ducks after "habituation" to the hawk shaped figure. Behaviour 17, 1-16 (1961).

THOMSON, R.F., SPENCER, W.A.: Habituation: A model phenomenon for the study of neuronal substrates of behavior. Psychol. Rev. 73, 16-43 (1966).

THORPE, W.H.: The concept of learning and their relation to those of instinct. *In* "Physiological Mechanisms in Animal Behavior". Symp. Soc. exp. Biol. 4, 387-408 (1950).

WOODWORTH, R.S., SCHLOSBERG, H.: "Experimental Psychology". pp. 786-794. New York: Holt Rinehart & Winston (1964).

CHEMICAL INFLUENCE ON FEEDING BEHAVIOR OF *LEPTINOTARSA* BEETLES

T.H. Hsiao

Department of Zoology, Utah State University, Logan, Utah, U.S.A.

Among the various interactions that occur between insect and plant, the phenomenon of host plant specificity is particularly intricate and important. Many plant-feeding insects depend on very restricted groups of taxonomically related plant species for their existence. The monophagous and oligophagous feeding habits of these insects reflect the unusual modes of adaptation that have evolved in phytophagous insects.

An understanding of the mechanisms of host selection and specificity in plant-feeding insects is important in explaining the ecological and evolutionary relationships among particular insects and plants. Such understanding is also essential in developing practical ways to manipulate the two groups of organisms to our advantage: such as breeding crops resistant to insect attack and achieving biological control of noxious weeds.

Several recent reviews (Beck 1965; Fraenkel 1969; Dethier 1966, 1970; Schoonhoven 1968) and symposia (de Wilde and Schoonhoven 1969; Van Emden 1972; Rodriguez 1972) have covered a broad range of topics related to insect and plant relationships. Much of the current literature stresses the important roles of plant chemicals in regulating host selection by phytophagous insects. Behavioral analysis along with electrophysiological studies of selected groups of insects such as the silkworm, *Bombyx mori* L. (Ishikawa, Hirao and Arai 1969), the tobacco hornworm, *Manduca sexta* (Johan.) and the cabbage worm, *Pieris brassicae* L. (Schoonhoven 1969; Ma 1972), have provided evidence of the high degree of chemosensory specificity of many phytophagous insects at the receptor level. The diversity of the chemicals in any green plant that influences insect feeding behavior has given rise to various opinions as to the nature of chemical influence in host selection (see above reviews).

Fraenkel (1959, 1969) stressed the essential role of secondary plant substances in host selection by phytophagous insects and further asserted that the secondary plant substances in plants exist solely to attract or repel animals or insects. Most of the current experimental evidence supporting this thesis was derived from investigations of single insect species. Investigations with a group of related insect species that feed on botanically related hosts were expected to further clarify the ecological and evolutionary significance of various plant chemicals.

The genus *Leptinotarsa* (Chrysomelidae, Coleoptera) is indigenous to America, extending southward from the foothills of the Rocky mountains to South America. Of the 44 species that have been described (Blackwelder 1946), many feed on Solanaceae, especially *Solanum*, which is also widely distributed in America. The specificity of the genus *Leptinotarsa* to Solanaceae suggests close ecological and evolutionary relationships.

The Colorado potato beetle, *Leptinotarsa decemlineata* (Say), which is the most destructive pest of the genus, has been the subject of several investigations in our laboratory (Hsiao and Fraenkel 1968a,b,c,d; Hsiao 1969, 1972). Our findings about the mechanism of host selection and specificity of *L. decemlineata*, and the techniques developed in our laboratory provided the basis for a comparative study of the mechanisms of host selection and feeding behavior of a number of species of the genus. Our comparative study included eight *Leptinotarsa* species distributed in the southwestern United States. All species were Solanaceae feeders. The highlights of our results are presented here and illustrate certain relationships between plant chemicals and the feeding behavior of these species. The details of the study will be reported in subsequent papers.

HOST PLANT SPECIFICITY

The food spectra of *Leptinotarsa* species were investigated initially to determine the nature of the insect-plant relationships. Eight *Leptinotarsa* species were collected throughout the southwestern United States during the summers of 1969 and 1970. Natural hosts of each species were recorded and identified. Laboratory cultures of these species were established to facilitate various phases of research. The rearing was held at 26.5°C with a photoperiod of 18 hr light and 6 hr darkness. Under these conditions, the incidence of adult diapause was minimal. Several of the species have been reared for many consecutive generations on their natural or favorite host plants in the laboratory.

Table 1 summarizes the results of laboratory rearing of 8 species on their natural or favorite hosts. All species have similar life histories and developmental requirements. The period from egg hatching to adult emergence averaged 20 to 23 days. The size of each species as indicated by pupal weight is also characteristic of these insects. The *Leptinotarsa* beetle species collected from the field were not equally amenable to laboratory rearing. Only the six most adaptable species were investigated at this stage.

Some 20 species of Solanaceae that were found in the natural habitats of the *Leptinotarsa* beetles or were available locally were included in our evaluation of food preferences. Some plants were field collected and others were grown in a field and in a greenhouse. The acceptability of plants was first tested with 4th instar larvae by a method described by Hsiao and Fraenkel (1968d). Plants accepted by these larvae were investigated to determine their suitability for support of growth and reproduction. Newly hatched first instar larvae were reared

Table 1. Growth and development of *Leptinotarsa* species on their food plants.

Species	Food plant	No. insects	Development duration (day) ± S.E.	Mean pupal weight (mg) ± S.E.	% adult emergence
L. decemlineata	*S. tuberosum**	53	20.7 ± 0.19	140.3 ± 2.8	78.0
L. defecta	*S. elaeagnifolium**	26	21.1 ± 0.18	95.1 ± 2.4	92.3
L. haldemani	*Physalis wrightii**	46	19.6 ± 0.14	156.2 ± 2.8	80.4
L. juncta	*S. dulcamara**	55	23.2 ± 0.26	170.5 ± 5.5	52.7
L. libatrix	*S. dulcamara*	38	20.2 ± 0.18	105.8 ± 2.9	81.6
L. rubiginosa	*S. dulcamara*	61	20.3 ± 0.14	162.2 ± 2.5	80.3
L. texana	*S. elaeagnifolium**	47	21.9 ± 0.15	119.3 ± 3.8	93.6
L. tumamoca	*Physalis wrightii**	6	20.3 ± 0.48	97.4 ± 2.2	66.7

* natural host for the species.

Table 2. Food spectra of *Leptinotarsa* species among solanaceous plants.

Plant species	*decemlineata*	*texana*	*juncta*	*defecta*	*haldemani*	*rubiginosa*
Solanum carolinense	++	++	+++		-	-
S. deflexum	-	-	-	-	+++	+++
S. douglasii	-	-	-	-	++	-
S. dulcamara	+++	++	+++	++	+++	+++
S. elaeagnifolium	++	+++	-	+++	-	-
S. heterodoxum	+	-	-	-	-	-
S. jamesii	-	-	-	-	-	-
S. lumholtzianum	++	+	-		-	-
S. melongena	+++	++	++	+	-	-
S. nigrum	-	-	-	-	++	-
S. rostratum	+++	++	+		+	-
S. tuberosum	+++	+	-	-	+++	+++
S. villosum	+	-	-	-	++	-
Asclepias speciosa *	+	-	-	-	+	+
Datura stramonium	-	-	-	-	-	-
Hyoscyamus niger	++	-	-	-	++	++
Lycium halimifolium	-	-	-	-	+	-
Lycopersicon esculentum (early bell variety)	+	-	-	-	+++	+++
Petunia hybrida	-	-	-	-	-	-
Physalis hederaefolia	-	-	-	-	-	-
P. pubescens	-	-	-	-	+++	+++
P. subglabrata	-	-	-	-	-	-
P. wrightii	-	-	-	-	+++	+++

+++ optimal feeding and growth
++ moderate feeding and growth
+ some feeding and slow growth
- not acceptable and no growth
* Family Asclepiadaceae

on these plants throughout the entire developmental stages. The rate of development, per cent mortality, and pupal weight were our criteria for suitability. The details of this work will be published elsewhere, but the information that is essential for the present discussion is summarized in Table 2. Of the six *Leptinotarsa* species that we investigated in detail, *L. decemlineata* was the most polyphagous species of the group; it was able to feed on 9 *Solanum* species and 3 other plant species. The species *L. texana* Schaeffer, *L. juncta* Guérin, and *L. defecta* Stål were highly host specific, being restricted to several species within the genus *Solanum*. *L. haldemani* Rogers and *L. rubiginosa* Rogers fed on several *Solanum* species as well as species of the genera *Physalis*, *Hyoscyamus* and *Lycopersicon*. The common nightshade, *S. dulcamara*, was the only plant species tested that was acceptable to all *Leptinotarsa* species studied, which suggests that it exerts no deterrent or toxic effects on the test insects.

Outside the family of Solanaceae, acceptable food plants for these *Leptinotarsa* beetles are rare. A detailed study of larval food acceptance of the Colorado potato beetle, *L. decemlineata* (Hsiao and Fraenkel 1968d) showed that of 87 species belonging to 38 plant families, only four plant species, *Asclepias syriaca*, *A. tuberosa*, *Capsella bursa-pastoris*, and *Lactuca sativa* var. *romana*, supported growth and development. Jermy (1961, 1966) reported a high degree of food preference in his tests with the adults of this species. Our survey (Table 2) also indicates that *Leptinotarsa* beetles are highly host specific, showing a definite pattern of food preferences that is unique to each species. This raises the obvious question of what constitutes the basis of feeding specificity in each species. Our previous investigations of the host plant selection of the Colorado potato beetle have established that two classes of plant chemicals, primary nutrients and secondary plant substances, exert major influences on the feeding behavior. These chemicals affect the sequence of behavioral responses of the beetles associated with the selection and acceptance of, and continuous feeding on, host plants.

We have selected several compounds in each group of these plant chemicals to determine the nature of specificity and chemosensory differences among different *Leptinotarsa* species. Findings obtained from these studies are discussed in the following sections.

THE ROLES OF PRIMARY NUTRITIVE SUBSTANCES

Chemosensory responses of the Colorado potato beetle are highly specific. In the absence of a phagostimulant, potato beetle larvae will not feed on a substrate such as filter paper, elder pith, or agar medium. An agar-cellulose medium incorporating various individual chemicals was used to achieve an accurate bioassay of the phagostimulative properties of a given compound. Hsiao and Fraenkel (1968a) tested the phagostimulative effects of some 26 sugars, 27 amino acids, 9 vitamins, 2 sterols, 7 lipids and 12 inorganic salts on the potato beetle larvae. Only one sugar (sucrose), several amino acids (L-alanine, γ-aminobutyric acid, L-serine, etc.), and three phospholipids (lecithin, phosphatidyl inositol, phosphatidyl L-serine), elicited marked feeding responses.

We have now tested these phagostimulants on four other *Leptinotarsa* species to compare the similarity of chemosensory responses. A total of nine nutrient compounds was tested individually in an agar-cellulose medium. An artificial diet that had proved adequate to support feeding and growth of the Colorado potato beetle (Hsiao and Fraenkel 1968c) was also tested. Each compound was tested at 0.01 and 0.1 molar concentrations with 15 to 20 fourth instar larvae in the manner described by Hsiao and Fraenkel (1968a). The results were scored on the basis of number of larvae responding to the tested substances and the average number of fecal pellets deposited by each larva. Chemicals that elicited intensive feeding responses from all the tested larvae at the lower molar concentration were ranked as highly stimulative. Those that only stimulated some feeding at the high concentration were considered slightly stimulative. Chemicals that elicited no feeding response were listed as non-stimulative.

Among the four sugars investigated (Table 3) only sucrose elicited a response from all five *Leptinotarsa* species. Sucrose is an effective phagostimulant for many insects. *L. decemlineata* was highly stimulated by sucrose. *L. juncta* and *L. rubiginosa* were moderately stimulated. *L. texana* and *L. haldemani* were only slightly stimulated by sucrose. The other three sugars, glucose, fructose, and melezitose, elicited slight or moderate feeding responses from *L. decemlineata*, but were non-stimulative to the other insect species. All the insect species responded to the three amino acids, L-alanine, γ-aminobutyric acid, and L-serine. These amino acids were potent phagostimulants for *L. decemlineata*, but elicited only slight to moderate feeding responses from the other *Leptinotarsa* species. Lecithin was a moderate phagostimulant for *L. decemlineata*, slightly stimulative to *L. haldemani*, and non-stimulative to other species.

Table 3. Chemosensory specificity of *Leptinotarsa* species to various compounds tested at 0.1 and 0.01 molar concentrations in an agar-cellulose medium.

Compounds	*decemlineata*	*texana*	*juncta*	*haldemani*	*rubiginosa*
Sucrose	+++	+	++	+	++
Glucose	+	-	-	-	-
Fructose	+	-	-	-	-
Melezitose	++	-	-	-	-
L-alanine	+++	++	++	+	++
γ-aminobutyric acid	+++	++	++	-	++
L-serine	++	+	-	-	+
Lecithin, vegetable	++	-	-	+	-
Chlorogenic acid	++	-	-	-	-
Artificial diet *	+++	+++	+++	+++	+++
Control medium	-	-	-	-	-

+++ highly stimulative; ++ moderately stimulative; + slightly stimulative
- non-stimulative
* artificial diet contained sucrose, glucose, casein, casein hydrolysate, salt mixture, ascorbic acid, cholesterol, B-vitamins, choline chloride.

An artificial diet which had previously proved suitable for rearing the fourth instar larvae of *L. decemlineata* (Hsiao and Fraenkel 1968c) was also highly phagostimulative to the four *Leptinotarsa* species recently tested. Of the nutrient chemicals present in this diet, sucrose was the only phagostimulant common to all species. The reason for the high degree of acceptance of the diet can be explained by synergistic effects of sucrose with other diet components. The importance of additive and synergistic effects of primary nutrients in inducing optimal feeding response in *L. decemlineata* has been pointed out by Hsiao and Fraenkel (1968a) and recently further investigated by Hsiao (1972). The additive effect was observed when two phagostimulants were combined. A synergistic effect was noted when a phagostimulant was combined with an inorganic salt, e.g., potassium salts.

Primary nutrient compounds are sufficiently important at the feeding level that their mere presence was sufficient to induce continuous feeding by *L. decemlineata*. The Colorado potato beetle has been reared successfully, but was smaller than normal in size and required a long period for development, on an artificial diet with only primary nutritive substances (Wardojo 1969a,b). In this diet, the amount of sucrose was high and its removal made the diet much less palatable to the larvae.

THE ROLES OF SECONDARY PLANT CHEMICALS

Our investigation of the chemical basis of host specificity of different *Leptinotarsa* species utilized many of the parameters established in the earlier study of the Colorado potato beetle. Hsiao (1969) described at least five types of chemical stimuli that regulated the feeding behavior of this species. These are olfactory attractants, sign stimulants, feeding stimulants, feeding co-factors, and inhibitors including repellents, deterrents, and toxicants. The primary nutrient chemicals are feeding stimulants and co-factors, although they are by no means the sole chemical stimuli, since secondary plant chemicals such as chlorogenic acid also stimulate feeding. Because green leaves contain such diverse chemicals, the role of any particular group of chemicals in relation to host selection of phytophagous insects must be determined by evaluating insect behavioral responses. The agar-cellulose medium mentioned in the preceding section was used to determine the phagostimulative properties of leaf powders and extracts of a selected group of solanaceous plants. An artificial diet (Hsiao 1972) was used to evaluate the deterrent or toxic effects of Solanaceae alkaloids on the *Leptinotarsa* species. This diet was moderately acceptable to all *Leptinotarsa* species tested. Fourth instar larvae of each species were allowed to feed individually on the diet and weight gains were recorded at 24- and 48-hour intervals, with the food changed every 12 hours. The differences in larval weight gains provide an accurate measurement of insect response, as demonstrated by the study of *L. decemlineata* by Hsiao and Fraenkel (1968c).

To identify the nature of the chemical stimulation exerted by various plants, leaf powder was prepared by stepwise solvent extractions, first with ethyl ether and then 80% ethanol. The leaf powder, ether extract, and alcohol extract, were separately incorporated at 5% concentrations, into the agar-cellulose medium as described by Hsiao and Fraenkel (1968c). A total of 6 *Solanum* species and two other plants were examined in this manner with three *Leptinotarsa* species (Table 4). The acceptability of fresh plants generally coincided with responses to leaf powders in the agar-medium. Plants such as *S. heterodoxum*, *Hyoscyamus niger*, and *Asclepias speciosa* were not acceptable as fresh leaves by *L. texana* and were non-stimulative in leaf powder form. Although *L. rubiginosa* did not feed on *S. rostratum* and *S. heterodoxum* as fresh

Table 4. Chemical basis of food plant specificity of *Leptinotarsa* species for some solanaceous and non-solanaceous plants.

Plant species	*decemlineata*				*texana*				*rubiginosa*			
	L	LP	EE	AE	L	LP	EE	AE	L	LP	EE	AE
Solanum tuberosum	+++	+++	+	+++	+	++	-	+	+++	++	-	++
Solanum elaeagnifolium	++	+++	+	++	+++	+++	-	+++	-	+	-	++
Solanum dulcamara	+++	+++	+	+++	++	++	-	+	+++	+++	-	+++
Solanum rostratum	+++	+++	+	+++	++	+++	-	++	-	++	-	++
Solanum heterodoxum	+	++	-	+++	-	-	-	+	-	++	-	+++
Hyoscyamus niger	++	++	+	+++	-	+	-	-	++	++	+	+++
Asclepias speciosa	+	++	-	+++	-	+	-	-	+	++	-	++

L - fresh leaf; LP - leaf powder; EE - ether extract; AE - 80% ethanol extract.
+++ highly stimulative; ++ moderately stimulative; + slightly stimulative;
\- non-stimulative.

leaves, they readily accepted medium containing leaf powders of these plants. Hsiao and Fraenkel (1968c) suggested that when leaf powders but not fresh plant leaves are acceptable, the plant may contain volatile repellents that are lost in producing the leaf powders. It is also possible that the amounts of leaf powder incorporated into the agar medium were not sufficient to produce a high deterrent effect. When responses to fresh leaves and leaf powders are identical, chemicals present in the leaf must be responsible for the observed effects on feeding.

Examination of the feeding responses to leaf powder extracts showed that the ether extracts of the plants are, as a whole, non-stimulative to feeding. Only the Colorado potato beetle showed slightly positive responses to this fraction, probably due to the phospholipids, which are feeding stimulants for this species. The ethanol extract of the tested plants was, however, either highly stimulatory or inhibitory to feeding. For plant species acceptable as fresh leaves, the ethanol extract of the leaf powder was invariably highly stimulative. For plant species not acceptable as fresh leaves, the alcohol extract was found to contain deterrent chemicals. Water soluble compounds are clearly responsible for the deterrent effect of the alcohol extract. Because both phagostimulative and deterrent compounds were present in the alcohol soluble fraction, it is difficult to identify common stimulants in different solanaceous plants. Behavioral evidence from the early studies of *L. decemlineata* has, however, strongly suggested that certain phagostimulants and oviposition attractants present in most of the *Solanum* species investigated are secondary plant chemicals. Detailed chemical isolation would be needed to identify the specific substances.

The role of chlorogenic acid as a phagostimulant for the Colorado potato beetle is especially interesting (Table 3). This phenolic acid was originally isolated from the potato leaf (Hsiao and Fraenkel 1968b) and is known to be widely distributed among plants. The chlorogenic acid had no stimulative effects on three other *Leptinotarsa* species tested, indicating a unique chemosensory specificity to *L. decemlineata*. Chemosensory receptors specifically responding to chlorogenic acid were recently identified by Schoonhoven and his co-worker (personal communication) in *L. decemlineata*. Our findings suggest that other *Leptinotarsa* species have not evolved sensory mechanisms for detection of this compound.

The role of deterrent chemicals in determining the host preferences of *Leptinotarsa* species is somewhat easier to demonstrate. Solanaceous plants contain a diverse class of alkaloids that is characteristic of the family (Schreiber 1968). The Colorado potato beetle has been investigated in some detail as to the role of various alkaloids in relation to its feeding and survival (Table 5). Schreiber (1958) and Buhr, Toball and Schreiber (1958) incorporated the alkaloids on the surface of potato leaves. Hsiao and Fraenkel (1968c) incorporated the alkaloids into an artificial diet and expressed their results by larval weight gain which provides a good quantitative measurement. With the exception of solanine, chaconine, and solasonine, steroid glycoalkaloids such as demissine, leptines, and tomatine are highly inhibitory to feeding. Nicotine is toxic to this species. The tropane alkaloids from several genera of Solanaceae such as atropine and scopolamine are also feeding deterrents for the Colorado potato beetle.

Solanaceae alkaloids from commercial sources were evaluated in the artificial diet to determine the exact concentrations that deter feeding. Eight compounds have been tested so far with 4 *Leptinotarsa* species. An interesting pattern of effects emerged from this comparative study (Table 6). Tomatine, demissine and atropine strongly deter *L. decemlineata*, but

Table 5. Secondary plant substances from Solanaceae serve as feeding deterrents or toxicants to the Colorado potato beetle.

Chemical	Plant origin	Degree of inhibition
Solanine	*Solanum tuberosum*	-
Chaconine	*S. tuberosum, S. chacoense*	-
Demissine	*S. demissum, S. jamesii*	+++
Leptine	*S. chacoense*	+++
Soladucine	*S. dulcamara*	++
Solacauline	*S. acuale, S. caulescens*	++
Solamargine	*S. aviculare, S. sodomeum*	-
Solanigrine	*S. nigrum*	++
Solasonine	*S. sodomeum, S. carolinense*	-
Tomatine	*Lycopersicon esculentum*	+++
Capsaicin	*S. capsicum*	++
Nicotine	*Nicotiana tabacum, N. rustica*	toxic
Nicandrenone	*Nicandra physalodes*	+++
Atropine	*Atropa belladonna*	+
Scopolamine	*Datura* spp.	++

+++ strong; ++ moderate; + slight; - no deterrent effects.

Data from Schreiber 1958; Buhr, Toball and Schreiber 1958; Hsiao and Fraenkel 1968c; Schreiber 1968.

Table 6. Effects of Solanaceae alkaloids on feeding and growth of *Leptinotarsa* species. Each chemical was incorporated into the basic diet at 0.01 and 0.1% levels.

Chemical	*decemlineata*	*texana*	*haldemani*	*juncta*
Tomatine	+	-	+++	-
Solanine	+++	-	+++	-
Demissine	+	-	-	-
Solasodine	+++	+++	+++	+++
L-Hyoscyamine	++	+	-	-
Hyoscine	++	-	+	
Tropine	+++	+++	+	
Atropine	+	+		
Basic diet alone	+++	+++	+++	+++

+++ normal feeding and growth, no deterrent effect
++ moderate feeding and growth, some deterrent effect
+ slight feeding and growth, strong deterrent effect
- no feeding or growth, complete deterrent effect

5 other alkaloids, viz. solanine, solasodine, L-hyoscyamine, hyoscine and tropine have no deterrent effect. These responses coincide with the plant species that are acceptable to this insect species. Only solasodine and tropine did not deter feeding of *L. texana* and *L. juncta*. *L. haldemani* can feed on tomato and can also tolerate considerable amounts of tomatine in its diet. It was deterred by demissine, L-hyoscyamine, hyoscine and tropine. The data obtained from this preliminary survey indicate that the alkaloids present in Solanaceae coincide with the relative acceptance of plant species by *Leptinotarsa* beetles.

Although additional study would be needed to determine the amounts of specific alkaloids present in each plant species, the available evidence suggests that many species might contain several alkaloids at different concentrations to provide added protection against insects, including *Leptinotarsa* beetles. Only insect species that have evolved mechanisms for reducing the deterrent and toxic effects of the Solanaceae alkaloids could utilize such plants.

One interesting observation is the general acceptance of *Solanum dulcamara* by the *Leptinotarsa* beetles. This plant is indigenous to Europe and was introduced to North America. The lack of deterrency of *S. dulcamara* corresponds to the absence of long term interactions with the *Leptinotarsa* beetles. The present findings tend to support the view that continual interactions between insects and plants are important to the evolution of secondary plant substances that attract or repel insects.

CONCLUSION

Considerable research effort has been directed toward isolating and identifying from plants, various chemical attractants and repellents that regulate behavioral responses of phytophagous insects. These studies have helped to explain the mechanisms of host selection and feeding behavior for individual phytophagous species. However, they have not provided sufficient insight into the ecological and evolutionary relationships that exist between insects and plants. These intricate relationships can be elaborated only by a comparative study centered on a group of related insect and plant species. The specificity of *Leptinotarsa* beetles for Solanaceae provides an ideal system for such experimental investigations. Our findings with these species have revealed several interesting and unique biological relationships between these groups of organisms.

The oligophagous feeding habits exhibited by the *Leptinotarsa* beetles toward Solanaceae under natural conditions and in laboratory studies have confirmed an intimate biological relationship. Comparison of the food preferences of six *Leptinotarsa* species indicated that their feeding habits could be divided into three groups. *L. decemlineata* has the most extensive food range of the three groups. This species feeds on 9 out of 13 *Solanum* species tested and in addition consumes several species outside the genus *Solanum*. Because of this adaptability in feeding habits, *L. decemlineata* is the only species in the genus that has reached pest status. The species *L. defecta*, *L. texana*, and *L. juncta* are more restricted, being limited to the genus *Solanum*. The first two species of this group feed exclusively on *S. elaeagnifolium*, which is widely distributed in southwestern United States and in Mexico. *L. juncta* feeds mainly on *S. carolinense* and *S. dulcamara*, which are found in the eastern and central United States. Both *L. haldemani* and *L. rubiginosa* feed on plants of the genera *Solanum*, *Physalis* and *Lycopersicon*. These two *Leptinotarsa* species feed readily on potato leaves in the laboratory, but have not been observed to feed on potato leaves in the field. This would indicate that other ecological factors are also important in determining their host range.

The differences in feeding preferences among the *Leptinotarsa* beetles correlate with their chemosensory specificity. For example, a larger group of plant chemicals serves as feeding stimulants for the more general feeders, e.g., *L. decemlineata*, as compared to those affecting the more restricted feeders, e.g., *L. texana* and *L. juncta*. Obviously the additive and synergistic interactions of feeding stimulants and co-factors are of special importance to the restricted feeders in inducing an adequate level of feeding response. This was clearly demonstrated in our work when the artificial diet was more effective in eliciting feeding response than were individual feeding stimulants. The restricted feeders were also more dependent upon the presence of host specific chemicals to induce an optimal feeding response.

All *Leptinotarsa* species were highly sensitive to the repellent and deterrent chemicals present in many plants. Evidently, secondary plant chemicals that occur in most plant families are effective as repellents or deterrents to attacks by *Leptinotarsa* beetles. Even among Solanaceae, many secondary plant chemicals inhibit insect feeding and growth. Tropane alkaloids, nicotine, capsaicin, and nicandrenone are but a few of the alkaloids from several genera of Solanaceae that are toxic or inhibitory to insects. Within the genus *Solanum*, the steroidal glyco-alkaloids are decisive in regulating feeding behavior and host selection. Comparison of the responses of the different *Leptinotarsa* species tested showed that the restricted feeders tended to be more sensitive to the deterrent effects; whereas the general feeders could tolerate a wide variety of these compounds. The available data indicates that the qualitative and quantitative aspects of the steroidal glycoalkaloids present would influence the acceptability of the plant. In many cases, the presence or absence of a single alkaloid would determine whether a plant was susceptible or immune to attack by *Leptinotarsa* beetles. These data support the thesis that secondary plant chemicals play a decisive role in regulating host selection by and feeding behavior of oligophagous plant-feeding insects.

SUMMARY

The genus *Leptinotarsa* (Chrysomelidae) is indigenous to America and many species feed exclusively on Solanaceae, especially *Solanum*. Food specificity has been investigated in eight species to elucidate the modes of chemical influences on feeding behavior. At least four types of plant chemicals are responsible for initiation and regulation of feeding responses. These are sign stimulants (host specific chemicals in Solanaceae), feeding stimulants (sugars, amino acids, phospholipids), feeding co-factors (potassium and other inorganic salts) and deterrents (alkaloids in Solanaceae). These chemicals act singly or jointly to stimulate or inhibit feeding in a manner peculiar to each insect species. The nature of oligophagy in *Leptinotarsa* species is determined by their relative sensitivity to and tolerance of the various types of plant chemicals. Species that respond to the largest number of feeding stimulants and co-factors and at the same time exhibit the greatest tolerance to deterrents are correspondingly more diverse in their host range.

REFERENCES

BECK, S.D.: Resistance of plants to insects. A. Rev. Ent. 10, 207-232 (1965).

BLACKWELDER, R.E.: Checklist of the coleopterous insects of Mexico,

Central America, The West Indies, and South America. Bull. U.S. natn. Mus. 185, 673 (1946).

BUHR, H., TOBALL, R., SCHREIBER, K.: Die Wirkung von einigen pflanzlichen Sonderstoffen, insbesondere von Alkaloiden, auf die Entwicklung der Larven des Kartoffelkäfers (*Leptinotarsa decemlineata* Say). Entomologia exp. appl. 1, 209-224 (1958).

DETHIER, V.G.: Feeding behavior. *In* "Insect Behaviour" (Ed., P.T. Haskell). Symp. R. ent. Soc. Lond. 3, 46-58 (1966).

DETHIER, V.G.: Chemical interactions between plants and insects. *In* "Chemical Ecology" (Ed., E. Sondheimer and J.B. Simeone), pp. 83-102. New York and London: Academic Press (1970).

FRAENKEL, G.: The raison d'être of secondary plant substances. Science, N.Y. 129, 1466-1470 (1959).

FRAENKEL, G.: Evaluation of our thoughts on secondary plant substances. Entomologia exp. appl. 12, 473-486 (1969).

HSIAO, T.H.: Chemical basis of host selection and plant resistance in oligophagous insects. Entomologia exp. appl. 12, 777-788 (1969).

HSIAO, T.H.: Chemical feeding requirements of oligophagous insects. *In* "Insect and Mite Nutrition" (Ed., J.G. Rodriguez), pp. 225-240. Amsterdam: North-Holland Publ. Co. (1972).

HSIAO, T.H., FRAENKEL, G.: The influence of nutrient chemicals on the feeding behavior of the Colorado potato beetle, *Leptinotarsa decemlineata* (Coleoptera: Chrysomelidae). Ann. ent. Soc. Am. 61, 44-54 (1968a).

HSIAO, T.H., FRAENKEL, G.: Isolation of phagostimulative substances from the host plant of the Colorado potato beetle, *Leptinotarsa decemlineata* (Say). Ann. ent. Soc. Am. 61, 476-484 (1968b).

HSIAO, T.H., FRAENKEL, G.: The role of secondary plant substances in the food specificity of the Colorado potato beetle, *Leptinotarsa decemlineata* (Say). Ann. ent. Soc. Am. 61, 485-493 (1968c).

HSIAO, T.H., FRAENKEL, G.: Selection and specificity of the Colorado potato beetle for solanaceous and nonsolanaceous plants. Ann. ent. Soc. Am. 61, 493-503 (1968d).

ISHIKAWA, S., HIRAO, T., ARAI, N.: Chemosensory basis of host plant selection in the silkworm. Entomologia exp. appl. 12, 544-554 (1969).

JERMY, T.: On the nature of the oligophagy in *Leptinotarsa decemlineata* Say (Coleoptera: Chrysomelidae). Acta zool. hung. 7, 119-132 (1961).

JERMY, T.: Feeding inhibitors and food preference in chewing phytophagous insects. Entomologia exp. appl. 9, 1-12 (1966).

MA, W.C.: Dynamics of feeding responses in *Pieris brassicae* Linn. as a function of chemosensory input: a behavioural, ultrastructural and electrophysiological study. Meded. LandbHoogesch. Wageningen 72-11, 162 pp. (1972).

RODRIGUEZ, J.G. (Ed.): "Insect and Mite Nutrition". Amsterdam: North-Holland Publ. Co. (1972).

SCHOONHOVEN, L.M.: Chemosensory bases of host plant selection. A. Rev. Ent. 13, 115-136 (1968).

SCHOONHOVEN, L.M.: Gustation and foodplant selection in some lepidopterous larvae. Entomologia exp. appl. 12, 555-564 (1969).

SCHREIBER, K.: Über einige Inhaltsstoffe der Solanaceen und ihre Bedeutung für die Kartoffelkäferresistenz. Entomologia exp. appl. 1, 28-37 (1958).

SCHREIBER, K.: Steroid alkaloids: the Solanum group. *In* "The Alkaloids, Chemistry and Physiology" (Ed., R.H.F. Manski), Vol. 10, pp. 1-192. New York: Academic Press. (1968).

VAN EMDEN, H.G.: Insect/plant relationships. Symp. R. ent. Soc. Lond. No. 6. Oxford: Blackwell Scientific Publications. (1972).

WARDOJO, S.: Some factors relating to the larval growth of the Colorado potato beetle, *Leptinotarsa decemlineata* Say (Coleoptera: Chrysomelidae), on artificial diets. Meded. LandbHoogesch. Wageningen 69-16, 75 pp. (1969a).

WARDOJO, S.: Artificial diet without crude plant material for two oligophagous leaf feeders. Entomologia exp. appl. 12, 698-702 (1969b).

de WILDE, J., SCHOONHOVEN, L.M. (Ed.): Insect and Host Plant. Proc. 2nd Int. Symp. Wageningen, The Netherlands. Amsterdam-London: North-Holland Publ. Co. Entomologia exp. appl. 12, 471-810 (1969)

Journal paper No. 1458, Utah Agricultural Experiment Station, Logan, Utah.

THE ROLE OF THE HYDROCARBON α-FARNESENE IN THE BEHAVIOUR OF CODLING MOTH LARVAE AND ADULTS

O.R.W. Sutherland, R.F.N. Hutchins and C.H. Wearing

Entomology Division, Owairaka Research Centre,
Department of Scientific and Industrial Research,
Auckland, New Zealand

Over the past two decades, studies of the behaviour of phytophagous insects have increasingly emphasized the role of chemical cues in mediating the relationship between the insect and the food- or host-plant. Nevertheless, the isolation and identification of the plant-derived chemicals involved in various behavioural responses, particularly those perceived as odours, has proceeded slowly. This may be accounted for by two major factors. Firstly, in recent years much effort has been directed towards the elucidation of the chemical nature and behavioural activity of insect sex pheromones and sex attractants. Secondly, the task of designing critical bioassays for specific behavioural responses has become increasingly difficult as the variety of information the insect receives from the plant and the multiplicity of its responses to these have become more evident. Moreover, a single chemical can evoke several responses from a single insect depending very largely upon the physical state of the former and the physiological state of the latter.

A number of authors have classified chemical stimuli in terms of the responses they elicit from insects (e.g. Dethier, Barton Browne and Smith 1960; Beck 1965). However, the accuracy of any such classification of a particular chemical depends upon the success with which the various elements of insect behaviour are differentiated and evaluated. From the chemical viewpoint, the demands of chemoreception studies are no less exacting and have recently been explored by Bedoukian (1970) who emphasized the importance of purity, identity, and measurement of materials in such studies.

Thus, although knowledge of the nature of volatile plant constituents which affect pest behaviour is important in the understanding of that behaviour and of its possible manipulation, few plant attractants and volatile arrestants, oviposition stimulants and feeding stimulants or deterrents have been isolated and identified. Some notable exceptions are the cases of *Bombyx mori* (see review of Hamamura 1970), *Listroderes obliquus* (Sugiyama and Matsumoto 1957, 1959; Matsumoto and Sugiyama 1960), *Plutella maculipennis* (Gupta and Thorsteinson 1960a,b) and *Hylemya antiqua* (Matsumoto and Thorsteinson 1968a,b).

The codling moth *Laspeyresia pomonella* has been widely studied but remarkably few investigations of the behaviour of the moths and larvae have been undertaken. Female codling moths do not lay the major portion of their egg complement directly on fruit: most eggs are laid singly within a few centimetres of apples on leaves, twigs or stems (Wildbolz 1958; Putman 1962; Geier 1963). This pattern of egg distribution is particularly noticeable in the first, early summer, brood but it has also been found that most of the second generation eggs, although laid when apples are large and often ripe, are similarly placed (Putman 1962). Thus, most newly hatched larvae must locate the nearby fruit themselves. The successful penetration of fruit depends upon the behaviour of the gravid female and that of the newly hatched larva: in both cases a response to the presence of fruit is crucial. Gravid females and neonate larvae might locate fruit by undirected movement as suggested by Hall

(1934), but there is evidence to the contrary. Certainly, the limited reproductive capacity of each female casts doubt on the proposal that the movement of moths and larvae is entirely random.

The possibility that the movement of neonate larvae towards apple is directed was first raised by McIndoo (1928, 1929) and later by Garlick (quoted in Putman 1962) and Geier (1963), each of whom suggested that a positive response to odour was involved. The latter two authors showed that, in a semi-natural situation, newly hatched larvae moved more consistently toward fruit than would be expected if their locomotion was undirected, whereas McIndoo provided some laboratory evidence for olfactory attraction to apple but only over a few millimetres.

The oviposition behaviour of gravid codling moths is influenced by naturally occurring odorants. In reviewing earlier reports on moth "attractants and baits", Dethier (1947) concluded that essential oils such as citronellol, oil of cloves and pine tar oil attracted codling moths by acting as oviposition stimuli. In the field the oviposition of gravid moths is stimulated by the odour of apple (Wildbolz 1958) and in controlled laboratory tests apple volatiles stimulate not only oviposition but also flight activity of gravid female moths (Wearing, Connor and Ambler 1973).

The present paper concerns studies in which we aimed to confirm the influence of apple odour on some aspects of larval and adult behaviour of *L. pomonella* in the laboratory, and to isolate and identify the compounds responsible for this (Sutherland 1972; Sutherland and Hutchins 1972, 1973; Wearing and Hutchins 1973).

ATTRACTION OF NEWLY HATCHED LARVAE

Bioassay

The insects were reared on an artificial diet according to the method of Brinton, Proverbs and Carty (1969) at 25°C, RH 70-90%. Test larvae were always less than 12 hr old and had not had access to diet. Ripe "Sturmer Pippin" apples which had been in a cool store were used throughout.

Initially, a 2 x 1 cm piece of freshly cut apple skin was placed vertically on one side of a 5.2 cm diameter glass petri dish and a similarly sized piece of damp filter paper on the other side. In later experiments, a 2 x 1 cm piece of filter paper treated with approximately 100 ng of the test material in chloroform was placed on one side of the dish after evaporation of the solvent. A similar paper treated with pure chloroform was placed opposite. Ten vigorous larvae were released in the centre of the arena and the lid was replaced. The behaviour of the insects was then closely observed for 5 min and the number of larvae on or behind each filter paper or the apple skin recorded at 60-sec intervals. The data gave a somewhat conservative estimate of the response, for active larvae quite often left the treated filter paper temporarily after crawling over it for a minute or two. Each test was repeated five times with new larvae.

Olfactory response to apple odour

Neonate codling moth larvae were strongly attracted by the odour of apple skin. Within two minutes from the time of release more than half the larvae had reached the piece of apple skin and, at the conclusion of the 5-min test period, 42 of 50 larvae were on or behind it (Table 1).

Table 1. The response of newly hatched codling moth larvae to fresh apple skin and to a crude chloroform extract of whole apples. The number of larvae present on the skin or extract-treated paper and on appropriate blank papers were recorded at 60-sec intervals after release. Fifty larvae were used in each test.

Test material	Time elapsed (min)	Skin/ Test paper	Blank paper
Apple skin v blank	1	14	4
	2	28	0
	3	39	0
	4	41	0
	5	42	0
Extract v blank	1	23	1
	2	34	0
	3	34	0
	4	38	0
	5	41	0
Blank v blank	1	4	3
	2	5	2
	3	2	2
	4	2	4
	5	3	3

A series of photographs was taken of one such test which was conducted in darkness in order to eliminate a possible larval response to the edges of the apple skin or to its shape or colour. Photographs were taken with an electronic flash at the time of release of the larvae and at 60-sec intervals thereafter (Fig. 1). These photographs, together with numerous observations, indicated that orientation toward the skin was rapid and some larvae commenced moving in a direct path to the odour

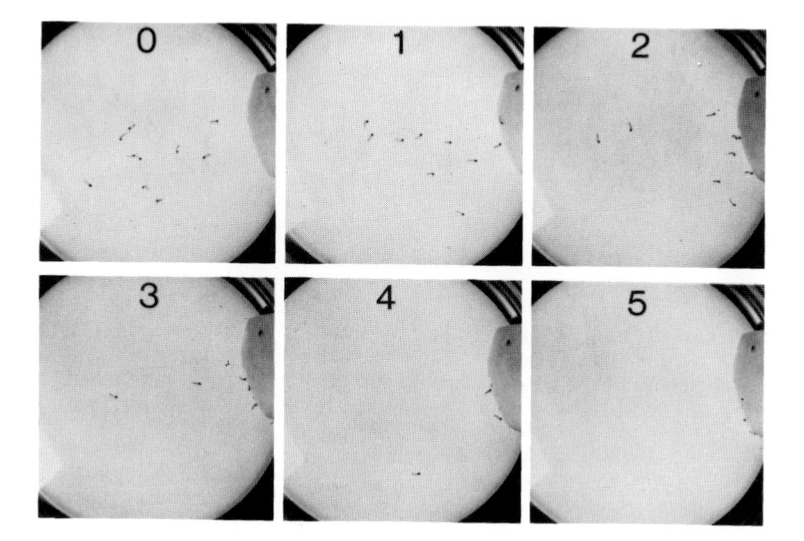

Fig. 1. The attraction of newly hatched codling moth larvae to apple skin in a closed 5.2 cm diameter Petri dish. (From Sutherland 1972)

source within a few seconds of release (e.g. photograph 1). The locomotion of the insects was characterised by brief pauses when the anterior half of the body was waved both vertically and horizontally before being lowered, usually in the direction of the odour source. Between such pauses for alteration in orientation the larvae typically moved forward along a straight path. This behaviour contained all the elements of klinotaxis and led most larvae rapidly to locate the odour source. In the test shown in Fig. 1 all larvae reached the apple skin within 5 min. The existence of an olfactory attractant (*sensu* Dethier, Barton Browne and Smith 1960) is clear but the bioassay gives no information on the ensuing stages of larval behaviour which culminate in the penetration of the fruit.

Isolation and identification of attractant

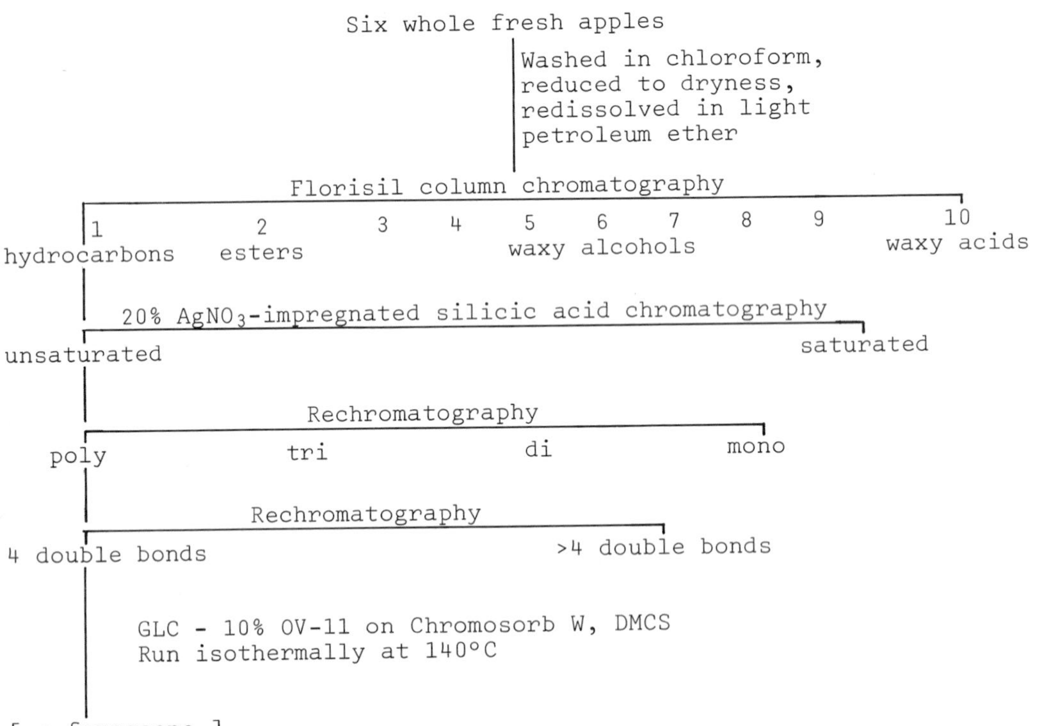

The isolation and identification of an attractant in the natural coating of "Sturmer Pippin" apples.

A crude chloroform extract of the outer coating of ripe fruit was prepared by washing six small "Sturmer Pippin" apples consecutively in 3 x 300 ml chloroform. The combined extracts were evaporated at reduced pressure at less than 40°C, and were made up to a final concentration of 6 apple equivalents/50 ml solvent. From this solution, 0.025 ml samples were drawn and placed on test filter papers. Larvae were as strongly attracted to this extract as they were to the fresh skin (Table 1). Close observation of the behaviour of the larvae in the closed dishes revealed no discernable differences in their response to skin or to the chloroform extract.

The crude extract of the outer coating of the fruit was chromatographed on Florisil (see scheme). Only the fraction containing the combined hydrocarbons was active. These were then separated by chromatography on silicic acid impregnated with 20% silver nitrate. Larvae responded strongly to the unsaturated hydrocarbon fraction but not at all to the saturated hydrocarbons. Further chromatography on silver nitrate-impregnated silicic acid eluted with increasing concentrations of diethyl ether in benzene, revealed that the active factor was a compound with more than three double bonds. This sample was then examined by gas-liquid chromatography. A Varian aerograph 1800 gas chromatograph was used and the samples were run isothermally at 140°C on 10% OV-11 on Chromosorb W, DMCS. One predominant peak appeared and one minor peak. These components were then separated by column chromatography as before, and the material comprising the predominant peak was found to be highly active in the bioassay. This was apparently a single compound with four double bonds. The minor fraction of hydrocarbons with more than four double bonds was without effect on larval behaviour.

At this stage, we learned that the acyclic sesquiterpene α-farnesene had been isolated from the natural coating of "Granny Smith" apples where it was the only sesquiterpene present (Murray, Huelin and Davenport 1964; Huelin and Murray 1966). We obtained a pure sample of α-farnesene and GLC of this, together with our active fraction, indicated that the two samples had identical retention times and that their peaks coincided when they were run together (Fig. 2A). Synthetic α-farnesene was then prepared by the dehydration of (*E*)-nerolidol with phosphoryl chloride (Anet 1970) and was found to have a retention time identical with that of the previous two samples (Fig. 2A).

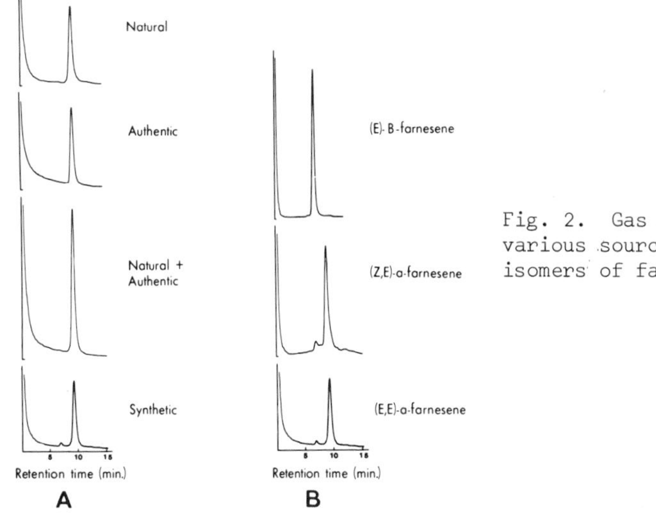

Fig. 2. Gas chromatograms of α-farnesene from various sources (A), and of three synthetic isomers of farnesene (B)

The three samples all proved to be equally active in the bioassay (Table 2). We therefore concluded that α-farnesene is a naturally occurring attractant for neonate codling moth larvae.

Activity of synthetic α-farnesene

A particular advantage of the standard bioassay procedure used in this study was its simplicity of design. An odour gradient was achieved by diffusion with a minimum of air movement. Under these conditions the concentration of odorant molecules encountered by the insect at any point in the arena is determined primarily by the physical properties of the

Table 2. The response of newly hatched codling moth larvae to natural, authentic and synthetic α-farnesene. The number of larvae present on each paper after 5 min was recorded. Fifty larvae were used in each test.

Hydrocarbon	Test paper	Blank paper
α-farnesene - natural*	31	1
α-farnesene - authentic	34	1
α-farnesene - synthetic	34	1

* Isolated from "Sturmer Pippin" apples by preparative gas chromatography.

compound. But also important are two further factors: the quantity of odorant at the source and the distance of larvae from the source.

In the previous tests a quantity of approximately 100 ng of test material was used. It was, however, unnecessary to expose larvae to this quantity of synthetic α-farnesene. A significant response could be obtained with as little as 15 ng of the attractant (Table 3). The lower limit is probably set by the high volatility of the compound and its rapid evaporation from a dry filter paper surface.

Table 3. The response of newly hatched codling moth larvae to various quantities of synthetic α-farnesene. The number of larvae present on each paper after 5 min was recorded. Fifty larvae were used in each test.

Quantity (nanogram)	Test paper	Blank paper
100	34	2
50	28	5
15	22	2
5	18	7

Table 4. The response of newly hatched codling moth larvae to 100 ng of synthetic attractant in closed arenas of various dimensions. The number of larvae present in each half of the arenas after 5 min was recorded. Fifty larvae were used in each test.

Distance of release from odour source (cm)	Test half	Blank half
2.50	49	1
5.00	47	3
5.75	38	12
6.75	31	19

Further evidence of the sensitivity of larvae to the attractant was provided by a series of tests in which larvae were released in closed petri dish arenas of various dimensions. In these tests the number of larvae present in each half of the dish at the end of a 5-min test period was recorded. We found that the insects were responsive to 100 ng of synthetic attractant located more than 6 cm away (Table 4).

Activity of stereo-isomers of farnesene

The sesquiterpene farnesene has three major geometric forms β-farnesene, α-farnesene, and allo-farnesene (I, II, III) (Naves 1966).

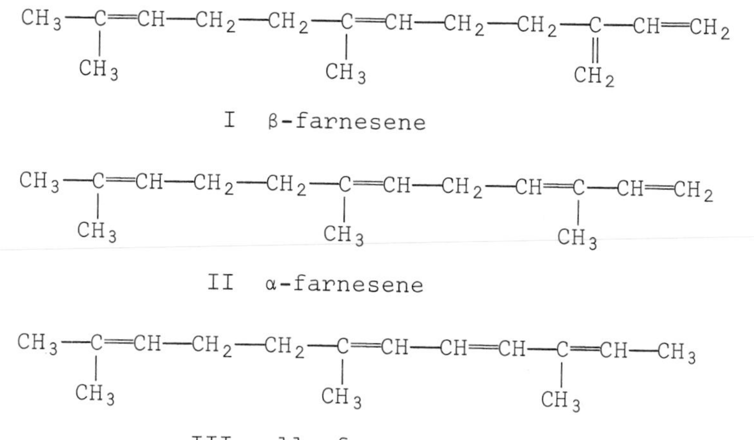

III allo-farnesene

But due to the unsaturation of the compound there are two possible β-isomers, four α-isomers and eight allo-isomers. Of these the two β-farnesenes and the four α-farnesenes were synthesized. These include three naturally occurring forms of farnesene: (*E*)-β-farnesene has been isolated from the oil of chamomilla, and (*E*,*E*)- and (*Z*,*E*)-α-farnesene constitute the farnesene found in the natural coating of apples (Anet 1970).

The compounds were obtained from a racemic mixture of (*E*)- and (*Z*)-nerolidol by the method of Anet (1970) and were purified by gas-liquid chromatography. Only two of the six synthetic isomers attracted codling moth larvae (Table 5). These were (*E*,*E*)-α-farnesene and (*Z*,*E*)-α-farnesene, the two isomers which occur naturally in apple.

Table 5. The response of newly hatched codling moth larvae to six synthetic stereo-isomers of farnesene. The number of larvae present on each paper after 5 min was recorded. Fifty larvae were used in each test.

Isomer	Test paper	Blank paper
(*E*)-β-farnesene	7	6
(*E*,*E*)-α-farnesene	34	2
(*Z*,*E*)-α-farnesene	34	1
(*Z*)-β-farnesene	5	10
(*E*,*Z*)-α-farnesene	8	10
(*Z*,*Z*)-α-farnesene	10	6

The larval response is unequivocal, but the reason for the differing activity of the various isomers is very much less clear. An examination of molecular models of the isomers, indicates that while there is little difference in molecular configuration between (*E*,*E*)- and (*Z*,*E*)-α-farnesene, between (*E*,*Z*)- and (*Z*,*Z*)-α-farnesene, or between the two β-farnesenes, the three pairs of molecules differ considerably from each other in shape (Fig. 3). Whereas the two active α-farnesenes have long relatively straight molecules, the profile of the inactive α-isomers is strongly bent. The inactivity of the β-farnesenes could probably be accounted for by the presence of two terminal double bonds, but their shorter and more compact molecular structure also contrasts with that of the active molecules.

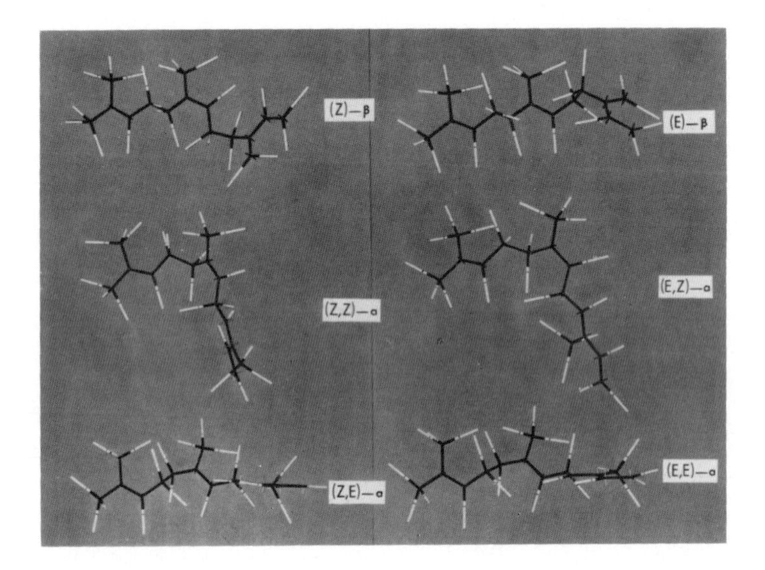

Fig. 3. Scale framework models of six synthetic stereo-isomers of farnesene constructed in the most probable configuration, i.e., that corresponding to the lowest energy conformation

The shape of an odorant molecule has long been held to influence the olfactory receptor response to that compound (Amoore 1964; Amoore *et al.* 1969). Although we have established that the response of codling moth larvae to the six farnesene isomers can be correlated with their molecular shape, the present results do not prove that this property alone is responsible for their activity in the bioassay. The receptor mechanism may be triggered by the appropriate juxta-position of active sites on the molecule and corresponding sites on the receptor surface (Roelofs and Comeau 1971), and, if so, the relative positions of the double bonds in the hydrocarbon molecule, rather than its overall shape, will be of prime importance

OVIPOSITION OF ADULT MOTHS

Bioassay

The moths were obtained from a laboratory population reared continuously on artificial diet (Brinton, Proverbs and Carty 1969) in a photoperiodic regime of 18 hr light : 6 hr dark at 24 ± 1°C. Gravid female moths were obtained by caging 20 newly emerged male and female moths in the dark for two days (see Wearing, Connor and Ambler 1973 for details).

The effect of the odour of whole apples on the flight activity and oviposition of the moths has been investigated by Wearing, Connor and Ambler (1973). Gravid moths were confined for 12 hr in chambers fitted with photoelectric activity detectors and lined on two sides with grooved waxed-paper upon which the moths could deposit their eggs. A constant air stream entered the control and test chambers on one side of each chamber through small holes in the waxed paper (the front paper) and exited similarly on the opposite side. Two whole mature "Sturmer Pippin" apples placed immediately behind the front paper provided the source of odour for the test chamber.

When several attempts to introduce α-farnesene at the appropriate concentration into the airstream failed to elicit a response, two different procedures were adopted in order to assess the response of female moths to the compound. A 3 cm diameter ball of cotton wool weighing about 0.1 g was treated with approximately 4 μg of synthetic (*E*,*E*)-α-farnesene in 0.1 ml chloroform. A blank cotton wool ball treated with an identical quantity of pure solvent was also prepared. After evaporation of the solvent the balls of cotton wool were transferred to glass petri dishes. In the first bioassay, the test and blank cotton balls were placed against the inner wall of separate 9.00 cm diameter petri dishes which already contained four female moths. The eggs laid in each dish after 30 min were counted and their position noted. In the second bioassay, 14.00 cm diameter petri dishes were used and the test and blank cotton wool balls were placed on opposite sides of the same dish. Eight moths were confined in each dish. Again the eggs were counted after 30 min and their distribution on the inner surface of each dish was plotted.

Preliminary work had revealed a periodicity in moth activity and oviposition, and since all tests were carried out in total darkness, the experiments were started shortly before onset of the usual dark period.

Response to apple odour

Those moths exposed to the odour of the whole fruit were more active and laid more eggs during the 12 hr test period than did the moths in the control chamber (Table 6). The possibility exists that, because of the duration of the experiment, the greater egg numbers laid in response to the odour could have resulted from more rapid maturation of eggs as well as from direct stimulation of oviposition itself. However, the extent of oviposition in trials of a shorter duration suggested that most eggs were laid during the early hours of the experiments and this was

Table 6. The oviposition and flight activity in an olfactometer of female codling moth in response to apple odour during a 12 hr period. Sum of 5 replicates each with 7 gravid moths per chamber.

Chamber	Total eggs laid	Mean % eggs on front paper	Detector activity count
Apple odour	502*	92.4	13310*
Blank	186	71.1	9949

* $P<0.001$ Wilcoxon signed rank test.

supported by later work.

Apple odour not only stimulated oviposition but also affected the distribution of eggs laid in the chambers. Compared with the moths in the control chambers, the moths exposed to the odour consistently laid a greater proportion of their eggs on the front paper through which the airstream entered the chamber.

Response to natural and synthetic α-farnesene

Both natural α-farnesene and synthetic (*E*,*E*)-α-farnesene stimulated oviposition by moths confined with a single cotton wool ball containing the material (Table 7). In most replicates, more eggs were laid in such dishes than in the control. When the α-farnesene-treated and blank cotton wool balls were placed in the same dish the moths laid more eggs on the side of the dish containing the treated ball (Table 8) and the density of eggs was greatest adjacent to the odour source (Fig. 4).

Table 7. The number of eggs laid by gravid female codling moths in 30 min in response to natural and synthetic α-farnesene. Sum of ten replicates each with 4 gravid moths per dish.

Test material	Test dish	Blank dish
Natural α-farnesene	149*	30
Synthetic (*E*,*E*)-α-farnesene	284*	113

* P<0.01 Wilcoxon signed rank test.

Table 8. The number of eggs laid by gravid female codling moths in 30 min in response to natural and synthetic α-farnesene. Test and control cotton wool balls on opposite sides of the same dish. Each replicate with 8 gravid moths per dish.

Test material	Test side	Control side
Natural α-farnesene	143*	34 (10 reps)
Synthetic (*E*,*E*)-α-farnesene	285*	128 (20 reps)

* P<0.001 cumulative binomial test.

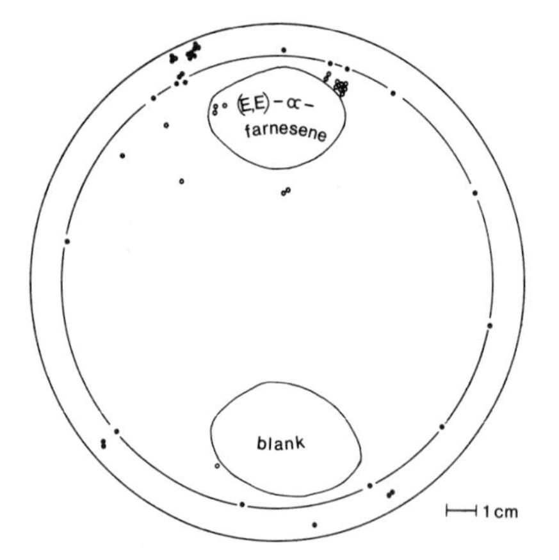

Fig. 4. The distribution of eggs laid in a 30-min period by 8 female codling moths confined in a closed petri dish with a ball of cotton wool treated with (*E*,*E*)-α-farnesene and a control ball. (From Wearing and Hutchins 1973).
0 - eggs on lid; ● - eggs on floor and sides

Wildbolz (1958) concluded from field experiments that apple odour induced oviposition by the codling moth. The results of the laboratory tests described here confirm the presence of one or more oviposition stimulant (*sensu* Dethier, Barton Browne and Smith 1960) in apple odour and establish that α-farnesene, a constituent of apple odour (Huelin and Coggiola 1968), is such a stimulant for this insect. All experiments in the present series have been characterised by a pattern of oviposition in which the density of eggs is greatest near the source of the odour. While a response to an oviposition stimulant could account for such a distribution, the location of the eggs suggests that α-farnesene, perhaps in combination with other apple volatiles, may be an attractant for the moths. However, it has not yet been possible to devise a bioassay which will demonstrate conclusively whether or not this is so.

Hillyer and Thorsteinson (1969) found that the ovarian development of *Plutella maculipennis* females was positively influenced by contact with the host-plant and suggested a possible role for olfactory stimuli. The female codling moth bears a full egg complement at emergence and during the oviposition period contains a constant number of ripe eggs (Geoffrion 1959, quoted in Deseö 1972). Therefore, it is conceivable that apple odour includes constituents which promote egg maturation and hence egg-laying. However, the results of the short 30 min bioassay (Table 7) leave no doubt that α-farnesene is an oviposition stimulant in the strict sense. Moreover, in another series of experiments in which this bioassay was continued for 12 hr, egg counts were made at half-hourly intervals for the first 4 hr. It was found that 70-80% of the eggs were deposited in the first 2 hr which suggests that the response of the moths to apple odour (Table 6) is at least largely behavioural.

A feature of all of the bioassays was the occurrence of oviposition in the absence of any known oviposition stimulant, a phenomenon also noted by Deseö (1970) and Wildbolz (1958). Some eggs were also laid at the maximum possible distance from the odour source (14 cm) in the closed petri dishes. This behaviour is reminiscent of oviposition in the field in which many eggs are not laid on the fruit, where the concentration of volatiles is greatest, but on leaves and twigs nearby.

FARNESENE AND THE BEHAVIOUR OF OTHER INSECTS

Naturally occurring terpenoids are widely implicated in insect behaviour as food- or host-plant attractants (Hamamura 1970; Becker, Petrowitz and Lenz 1971; Werner 1972; Fitzgerald and Nagel 1972; Kinzer, Ridgill and Reeves 1972; Camors and Payne 1972; Knopf and Pitman 1972), as feeding deterrents (Wada *et al.* 1970), as oviposition stimulants (Fitzgerald and Nagel 1972; Saxena and Sharma 1972), as sex and alarm pheromones (Blum 1969, 1970) and as repellents and defensive secretions (Weatherston and Percy 1970). Of these compounds some, for example citral, citronellol, limonene and α-pinene serve different functions in different species, and within a single species may induce more than one behaviour pattern when present at different concentrations.

There is increasing evidence that this may be true of farnesene. Besides functioning as an attractant for neonate *L. pomonella* larvae, synthetic (*E*,*E*)-α-farnesene attracts 5th instar larvae of this species in laboratory bioassays (Sutherland, unpublished data). Furthermore, the same isomer attracts newly hatched larvae of the apple leaf roller, *Epiphyas postvittana* (Sutherland, unpublished data).

The compound has also been implicated in ant behaviour. (*E*,*E*)-α-farnesene comprises the entire contents of the Dufours gland of *Aphaenogaster longiceps* (Cavill, Williams and Whitfield 1967; Anet 1970) and an unspecified isomer has been reported from the same gland of five other species of ant (Bergström and Löfqvist 1968, 1972). These findings led Blum (1969) and Gabba and Pavan (1970) to speculate that the compound may be an alarm pheromone for these species. An unspecified isomer of α-farnesene has since been identified from the contents of the Dufours gland of another ant, *Myrmica rubra* (Morgan and Wadhams 1972) while the acyclic sesquiterpenoid farnesal has been found in the mandibular glands of *Lasius fuliginosus* (Bernardi *et al.* 1967). However, an unequivocal demonstration of a role for farnesene in the behaviour of these insects is lacking.

On the other hand (*E*)-β-farnesene has been identified as the alarm pheromone of five species of aphid, and is effective against another five aphid species (Bowers *et al.* 1972; Edwards *et al.* 1973). Apparently no species which has been exposed to the compound has failed to respond, but it remains to be seen whether this is true of aphids other than the rather closely related species tested so far.

CONCLUSION

The bioassay procedures employed in this investigation were designed to illustrate and to characterise specific effects of some apple volatiles upon the behaviour of newly hatched larvae and gravid females of *L. pomonella*. One such compound, α-farnesene, has proven in our laboratory tests to be a potent attractant for newly hatched larvae and an oviposition stimulant for the moths. This is true both of naturally-derived and synthetic samples of the material. As α-farnesene is present in quite large quantities, not only in the natural coating of several varieties of apples (Meigh and Filmer 1967), but also in that of some other fruits attacked by *L. pomonella* (Murray 1969), we have no doubt that it plays a role in the events preceding the successful entry of fruit by young larvae. But whether this compound effects other responses, and whether other naturally occurring chemicals also have a role in the behaviour of this pest remain to be investigated.

There is already some indication that other chloroform-soluble components of the outer coating of apples may be involved in fruit location by larvae. Firstly, we have found that neonate larvae are consistently attracted more rapidly to a crude chloroform extract of the natural waxy apple coating than they are to pure α-farnesene isolated from such an extract. This suggests that, although we were unable to isolate any attractant other than α-farnesene from a crude extract, one or more other compounds may augment the larval response. Secondly, the insects tend to wander less frequently from a filter paper treated with the crude chloroform extract, which leads us to suspect the presence in the latter of an arrestant, a biting stimulant, or both.

Likewise, the oviposition response of gravid females to natural or synthetic α-farnesene is less intense than it is to the odour of whole apples and again the reason may be that normal oviposition is dependent upon the presence of more than one olfactory stimulant. In this respect the terpenoids geraniol and citral may be important. Already well-known as attractants for the Japanese beetle *Papilio japonica* (Fleming 1969) and *Bombyx mori* (Hamamura 1970) respectively, both compounds have a positive influence on the sexual behaviour of male codling moths (Butt *et al.* 1968; Skirkevicus and Tatjanskaite 1971). Since both geraniol and citral have been isolated and identified from apples (Power and Chesnut 1922; Langford, Muma and Cory 1943) the possibility that they may affect the behaviour of female moths or their larvae cannot be discounted.

REFERENCES

AMOORE, J.E.: Current status of the steric theory of odour. Ann. N.Y. Acad. Sci. 116, 457-476 (1964).

AMOORE, J.E., PALMIERI, G., WANKE, E., BLUM, M.S.: Ant alarm pheromone activity: correlation with molecular shape by scanning computer. Science, N.Y. 165, 1266-1269 (1969).

ANET, E.F.L.J.: Synthesis of (*E*,Z)-α-, (Z,Z)-α-, and (Z)-β-farnesene. Aust. J. Chem. 23, 2101-2108 (1970).

BECK, S.D.: Resistance of plants to insects. A. Rev. Ent. 10, 207-232 (1965).

BECKER, G., PETROWITZ, H.J., LENZ, M.: Über die Ursache der abschreckenden Wirkung von Kiefernholz auf Termiten. Z. angew. Ent. 68, 180-186 (1971).

BEDOUKIAN, P.Z.: Purity, identity and quantification of pheromones. Adv. Chemoreception 1, 19-34 (1970).

BERGSTRÖM, G., LÖFQVIST, J.: Odour similarities between the slave-keeping ants *Formica sanguinea* and *Polyergus rufescens* and their slaves *Formica fusca* and *Formica rufibarbis*. J. Insect Physiol. 14, 995-1011 (1968).

BERGSTRÖM, G., LÖFQVIST, J.: Similarities between the Dufour gland secretions of the ants *Camponotus ligniperda* (Latr.) and *Camponotus herculeanus* (L.) (Hym.). Ent. Scand. 3, 225-238 (1972).

BERNARDI, R., CARDANI, C., GHIRINGHELLI, D., SELVA, A., BAGGINI, A., PAVAN, M.: On the components of secretion of mandibular glands of the ant *Lasius (Dendrolasius) fuliginosus*. Tetrahedron Lett. 40, 3893-3896 (1967).

BLUM, M.S.: Alarm pheromones. A. Rev. Ent. 14, 57-80 (1969).

BLUM, M.S.: The chemical basis of insect sociality. *In* "Chemicals Controlling Insect Behaviour" (Ed., M. Beroza), 170 pp. London: Academic Press (1970).

BOWERS, W.S., NAULT, L.R., WEBB, R.E., DUTKY, S.R.: Aphid alarm pheromone: Isolation, identification and synthesis. Science, N.Y. 177, 1121-1122 (1972).

BRINTON, F.E., PROVERBS, M.D., CARTY, B.E.: Artificial diet for mass production of the codling moth *Carpocapsa pomonella* (Lepidoptera: Olethreutidae). Can. Ent. 101, 577-584 (1969).

BUTT, B.A., BEROZA, M., McGOVERN, T.P., FREEMAN, S.K.: Synthetic chemical sex stimulants for the codling moth. J. econ. Ent. 61, 570-572 (1968).

CAMORS, F.B., PAYNE, T.L.: Response of *Heydenia unica* (Hymenoptera: Pteromalidae) to *Dendroctonus frontalis* (Coleoptera: Scolytidae) pheromones and a host tree terpene. Ann. ent. Soc. Am. 65, 31-33 (1972).

CAVILL, G.W.K., WILLIAMS, P.J., WHITFIELD, F.B.: α-farnesene, Dufour's gland section in the ant *Aphaenogaster longiceps* (F.Sm.). Tetrahedron Lett. 23, 2201-2205 (1967).

DESEÖ, K.V.: The effect of olfactory stimuli on the oviposition behaviour and egg production of some micro-lepidopterous species. Colloques int. Cent. natn. Rech. scient. 189, 163-174 (1970).

DESEÖ, K.V.: The role of farnesylmethyl-ether applied on the male influencing the oviposition of the female codling moth (*Laspeyresia pomonella* L., Lepidopt.; Tortricidae). Acta phytopath. Hung. 7, 257-266 (1972).

DETHIER, V.G.: "Chemical Insect Attractants and Repellents", 289 pp. Philadelphia: The Blakiston Co. (1947).

DETHIER, V.G., BARTON BROWNE, L., SMITH, C.N.: The designation of chemicals in terms of responses they elicit from insects. J. econ. Ent. 53, 134-136 (1960).

EDWARDS, L.J., SIDDALL, J.B., DUNHAM, L.L., UDEN, P., KISLOW, C.J.: Trans-β-farnesene, alarm pheromone of the green peach aphid, *Myzus persicae*. Nature, Lond. 241, 126-127 (1973).

FITZGERALD, T.D., NAGEL, W.P.: Oviposition and larval bark-surface orientation of *Medetera aldrichii* (Diptera: Dolichopodidae). Response to a prey-liberated plant terpene. Ann. ent. Soc. Am. 65, 328-330 (1972).

FLEMING, W.E.: Attractants for the Japanese Beetle. Tech. Bull. U.S. Dep. Agric. 1399, 87 pp. (1969).

GABBA, A., PAVAN, M.: Researches on trail and alarm substances in ants. Adv. Chemoreception 1, 161-203 (1970).

GEIER, P.: The life history of codling moth *Cydia pomonella* (L.) (Lepidoptera: Tortricidae) in the Australian Capital Territory. Aust. J. Zool. 11, 323-367 (1963).

GUPTA, P.D., THORSTEINSON, A.J.: Food plant relationships of the diamond-back moth (*Plutella maculipennis* (Curt.)). I. Gustation and olfaction in relation to botanical specificity of the larva. Entomologia exp. appl. 3, 241-250 (1960a).

GUPTA, P.D., THORSTEINSON, A.J.: Food plant relationships of the diamond-back moth (*Plutella maculipennis* (Curt.)). II. Sensory regulation of oviposition of the adult female. Entomologia exp. appl. 3, 305-314 (1960b).

HALL, J.A.: Observations on the behaviour of newly hatched codling moth larvae. Can. Ent. 66, 100-102 (1934).

HAMAMURA, W.: The substances that control the feeding behaviour and the growth of the silkworm *Bombyx mori* L. *In* "Control of Insect Behaviour by Natural Products" (Eds., D.L. Wood, R.M. Silverstein, M. Nakajima), 345 pp. New York: Academic Press (1970).

HILLYER, R.J., THORSTEINSON, A.J.: The influence of the host plant or males on ovarian development or oviposition in the diamond back moth *Plutella maculipennis* (Curt.). Can. J. Zool. 47, 805-816 (1969).

HUELIN, F.E., COGGIOLA, I.M.: Superficial scald, a functional disorder of stored apples. IV. Effect of variety, maturity, oiled wraps and diphenylamine on the concentration of α-farnesene in the fruit. J. Sci. Fd Agric. 19, 297-301 (1968).

HUELIN, F.E., MURRAY, K.E.: α-farnesene in the natural coating of apples. Nature, Lond. 210, 1260-1261 (1966).

KINZER, H.S., RIDGILL, B.J., REEVES, J.M.: Response of walking *Conophthorus ponderosae* to volatile attractants. J. econ. Ent. 65, 726-729 (1972).

KNOPF, J.A.E., PITMAN, G.B.: Aggregation pheromone for manipulation of the Douglas-Fir beetle. J. econ. Ent. 65, 723-726 (1972).

LANGFORD, G.S., MUMA, M.H., CORY, E.N.: Attractiveness of certain plant constituents to the Japanese beetle. J. econ. Ent. 36, 248-252 (1943).

MATSUMOTO, Y., SUGIYAMA, S.: Attraction of leaf alcohol and some aliphatic alcohols to the adult and larvae of the vegetable weevil. Ber. Ōhara Inst. landw. Biol. 11, 359-361 (1960).

MATSUMOTO, Y., THORSTEINSON, A.J.: Effect of organic sulphur compounds on oviposition in onion maggot, *Hylemya antiqua* Meigen (Diptera: Anthomyiidae). Appl. Ent. Zool. 3, 5-12 (1968a).

MATSUMOTO, Y., THORSTEINSON, A.J.: Olfactory response of larvae of the onion maggot, *Hylemya antiqua* Meigen (Diptera: Anthomyiidae) to organic sulphur compounds. Appl. Ent. Zool. 3, 107-111 (1968b).

McINDOO, N.E.: Tropic responses of codling moth larvae. J. econ. Ent. 21, 631 (1928).

McINDOO, N.E.: Tropisms and sense organs of Lepidoptera. Smithson. misc. Collns 81, 1-59 (1929).

MEIGH, D.F., FILMER, A.A.E.: Chemical investigation of superficial scald in stored apples. Rep. agric. Res. Coun. (Ditton Laboratory) (1966-1967), 36-37 (1967).

MORGAN, E.D., WADHAMS, L.J.: Chemical constituents of Dufour's gland in the ant, *Myrmica rubra*. J. Insect Physiol. 18, 1125-1135 (1972).

MURRAY, K.E.: α-farnesene: isolation from the natural coating of apples. Aust. J. Chem. 22, 197-204 (1969).

MURRAY, K.E., HUELIN, F.E., DAVENPORT, J.B.: Occurrence of farnesene in the natural coating of apples. Nature, Lond. 204, 80 (1964).
NAVES, Y.-R.: Etudes sur les matières végétales volatiles CXCVIII (1) Contribution à la connaissance des farnésènes. Helv. chim. Acta 49, 1029-1041 (1966).
POWER, F.B., CHESNUT, V.K.: The odorous constituents of apples. II. Evidence of the presence of geraniol. J. Am. chem. Soc. 44, 2938-2942 (1922).
PUTMAN, W.L.: The codling moth *Carpocapsa pomonella* (L.) (Lepidoptera: Tortricidae): A review with special reference to Ontario. Proc. ent. Soc. Ont. 93, 22-60 (1962).
ROELOFS, W.L., COMEAU, A.: Sex pheromone perception: Synergists and inhibitors for the red banded leaf roller attractant. J. Insect Physiol. 17, 435-448 (1971).
SAXENA, K.N., SHARMA, R.N.: Embryonic inhibition and oviposition induction in *Aedes aegypti* by certain terpenoids. J. econ. Ent. 65, 1588-1591 (1972).
SKIRKEVICUS, A., TATJANSKAITE, L.: The sensitivity of the moth *Carpocapsa pomonella* L. to geraniol in connection with the different period of day. Proc. 1st All-Union Symp. on Insect Chemoreception, Vilnius, U.S.S.R. (1971): 133-138 (1971).
SUGIYAMA, S., MATSUMOTO, Y.: Olfactory responses of vegetable weevil larvae to allyl-, phenyl- mustard oils. Nôgaku kenkyû 45, 5-13 (1957).
SUGIYAMA, S., MATSUMOTO, Y.: Olfactory responses of the vegetable weevil larvae to various mustard oils. Nôgaku kenkyû 46, 150-157 (1959).
SUTHERLAND, O.R.W.: The attraction of newly hatched codling moth (*Laspeyresia pomonella*) larvae to apple. Entomologia exp. appl. 15, 481-487 (1972).
SUTHERLAND, O.R.W., HUTCHINS, R.F.N.: α-farnesene, a natural attractant for codling moth larvae. Nature, Lond. 239, 170 (1972).
SUTHERLAND, O.R.W., HUTCHINS, R.F.N.: Attraction of newly hatched codling moth larvae (*Laspeyresia pomonella*) to synthetic stereo-isomers of farnesene. J. Insect Physiol. 19, 723-727 (1973).
WADA, K., MATSUI, K., ENOMOTO, Y., OGISO, O., MUNAKATA, K.: Insect feeding inhibitors in plants. Part I. Isolation of three new sesquiterpenoids in *Parabenzoin trilobum* Nakai. Agric. biol. Chem. 34, 941-945 (1970).
WEARING, C.H., CONNOR, P.J., AMBLER, K.D.: The olfactory stimulation of oviposition and flight activity of the codling moth, *Laspeyresia pomonella* L., using apples in an automated olfactometer. N.Z. Jl Sci. 16, (1973).
WEARING, C.H., HUTCHINS, R.F.N.: α-farnesene, a naturally occurring oviposition stimulant for the codling moth, *Laspeyresia pomonella*. J. Insect Physiol. 19, 1251-1256 (1973).
WEATHERSTON, J., PERCY, J.E.: Arthropod defensive secretions. *In* "Chemicals Controlling Insect Behaviour" (Ed., M. Beroza), 170 pp. London: Academic Press (1970).
WERNER, R.A.: Response of the beetle, *Ips grandicollis*, to combinations of host and insect produced attractants. J. Insect Physiol. 18, 1403-1412 (1972).
WILDBOLZ, R.: Über die Orientierung des Apfelwicklers bei der Eiablage. Mitt. schweiz. ent. Ges. 31, 25-34 (1958).

NEUROSECRETORY AND CORPUS ALLATUM CONTROLLED EFFECTS ON MALE SEXUAL BEHAVIOUR IN ACRIDIDS

M.P. Pener

Department of Entomology, The Hebrew University, Jerusalem, Israel

This article sums up some advances made in the last few years in studies on the hormonal control of the mating behaviour of male locusts and grasshoppers. This subject was already treated to some extent in a recent paper (Pener 1972) related to the functions of the corpora allata (CA) in adult acridids. Further experiments have, however, revealed new facts, but increased the difficulties for making a synthesis.

THE CASE OF THE DESERT LOCUST AND THE RED LOCUST

In crowded desert locusts (*Schistocerca gregaria*), allatectomy of young adults results in the complete absence of male sexual behaviour (Loher 1961; Pener 1965, 1967a,b; Pener and Wajc 1971; Odhiambo 1966a; Cantacuzène 1967, 1968). Sectioning of the nervi corporum allatorum, or full denervation of the CA leads to delayed and subnormal mating behaviour (Pener 1965, 1967a). When allatectomy is performed on males which are already sexually mature, the intensity of mating behaviour decreases, but disappears completely only 4-6 weeks after the operation (Pener 1965, 1967a). The effect of these glands is not sex-specific; implantation of CA from females into previously allatectomized males induces mating behaviour about ten days later (Pener 1967b). The delayed reactions to these operations indicate that the influence of the CA on male sexual behaviour is probably indirect and that activation and/or stimulation of another organ may be involved (Pener 1965).

The maturation of young adult desert locusts is speeded up by the presence of already mature males (Norris 1954) which secrete a maturation accelerating pheromone (Loher 1961). Allatectomized males do not produce this pheromone (Loher 1961), either because they never become sexually mature, or because the CA hormone is directly responsible for pheromone production. On the other hand, the presence of very young males retards the maturation of young males (Norris 1954) and allatectomized adults exert a similar, but much more prolonged effect (Norris and Pener 1965). To account for this phenomenon Norris (1964, 1968) postulated that a maturation retarding pheromone is secreted by immature (and by allatectomized) locusts. If this is true, two pheromones, one retarding and one accelerating, seem to be responsible for the synchronous maturation of crowded desert locusts (Norris 1964), the switch over from inhibition to acceleration coinciding with the resumption of CA activity in the adult.

Production of these pheromones by solitary locusts has not yet been proved, although Norris (1968), citing her unpublished results and Thomas's personal communication, indicated that isolated males may produce at least the maturation accelerating pheromone. In any case, obviously, pheromones of fellow locusts cannot affect those which are kept under complete isolation. Norris (1952, 1954, 1964) demonstrated that isolated desert locusts mature more slowly than crowded ones (the situation is reversed in the migratory locust), possibly due to the lack of the effect of the pheromones and/or to a different endogenous "time table" of the maturation process. No studies have been performed on the endocrine factors controlling maturation and mating behaviour in isolated desert locusts. In the light of the findings concerning isolated migratory locusts such investigations seem to be desirable.

The CA exert a complete control over male sexual behaviour in crowded red locusts, *Nomadacris septemfasciata*, too. Allatectomized males do not exhibit sexual behaviour and reimplantation of active CA leads to the resumption of their mating activity (Pener 1968). Norris (1959) observed that maturation is somewhat more synchronous in crowded than in isolated *Nomadacris*. Her results suggest that if maturation regulating pheromones do exist in this species, their effect is considerably more limited than in *Schistocerca*.

Adults of both the desert and the red locust are able to exhibit a period of reproductive diapause (for references see Pener 1967a, 1968) during which the males do not display mating behaviour.

THE CASE OF THE GRASSHOPPER, *OEDIPODA MINIATA*

During the hot and dry summer in Israel, which lasts from May till October, *Oedipoda miniata* exhibits a reproductive diapause (Levy 1965; Broza and Pener 1969; Pener and Broza 1971). A study (Broza and Pener 1972) in which an investigation was made of the effect of the CA on the diapause in this grasshopper, as reflected by the effect on male mating behaviour, revealed the following results.

Implantation of active CA into diapausing males induces vigorous mating activity. The effect is temporary, the intensity of the sexual behaviour reaching a maximum during the second and third weeks following implantation, but later declining until by the fifth week the sexual behaviour scarcely exceeds that of the sham-operated or unoperated males which exhibit only extremely slight mating behaviour (see Broza and Pener 1972 for scoring system) during diapause. In fact, due to this "residual" sexual activity, the reproductive diapause may be regarded as being not absolutely complete in *O. miniata* males. It is complete, however, in the females and consequently reproduction does not take place during diapause.

On the other hand, allatectomized males also exhibit extremely slight mating behaviour even if they are kept under conditions of photoperiod and temperature which effectively terminate the diapause and induce intense sexual behaviour in non-allatectomized controls. The level of the mating behaviour shown by these allatectomized males is similar to that exhibited by normal diapausing males. The "residual" sexual activity of the diapausing males seems, therefore, to be independent of the CA.

In summary, these studies demonstrated that the CA exert almost, but not entirely, complete control over male sexual behaviour in the grasshopper, *Oedipoda miniata*, which is considered as a non-gregarious acridid.

THE CASE OF THE MIGRATORY LOCUST, *LOCUSTA MIGRATORIA*

Adults of the African migratory locust (*Locusta migratoria migratorioides*) do not exhibit reproductive diapause and photoperiod does not affect the timing or rate of their reproduction (Verdier 1969). However, other subspecies or strains of *Locusta migratoria* show different abilities for delaying sexual maturation under particular photoperiods (Verdier 1969; Perez, Verdier and Pener 1971); the delay being sometimes so marked that it resembles reproductive diapause.

Conflicting results regarding the effect of the CA on male mating behaviour have been reported for crowded migratory locusts. Girardie and Vogel (1966) and Girardie (1966) observed no sexual behaviour following allatectomy in *L.m. cinerascens*. Quo Fu (1965) recorded a lower copulation rate in allatectomized *L.m. manilensis*. Cantacuzène (1967) concluded that following allatectomy, males of *L.m. migratorioides* and of the "Kazalinsk" strain of *L. migratoria* are able to exhibit mating activity. She did not investigate, however, whether the operation affects the intensity of the sexual behaviour. Strong (1968) stated that male mating behaviour is not affected by the CA in *L.m. migratorioides*.

Using a comparative and quantitative method for recording the time spent on sexual behaviour, Wajc and Pener (1969) demonstrated that although allatectomized males of crowded *L.m. migratorioides* exhibited distinct sexual behaviour, the intensity of this behaviour was nevertheless considerably lower than that of the non-allatectomized controls. These findings fitted well with those of Quo Fu (1965) and did not contradict Cantacuzène's (1967) results.

Wajc and Pener's (1969) results disclosed that in the presence of normal (non-allatectomized) females, the percentage of time spent on sexual behaviour by allatectomized males was considerably lower than that of the controls. Furthermore, when allatectomized and control males were placed together in absence of females, the controls showed marked sexual behaviour toward the allatectomized males, but no mating behaviour of allatectomized males toward controls was recorded (males of some acridids, including *L. migratoria*, exhibit quite consistent homosexual behaviour, especially in absence of females). Finally, when observations were made on the sexual behaviour of males toward each other within a group of allatectomized males (that is, in absence of females and non-allatectomized males), the percentage of time spent on sexual behaviour was only slightly lower than that obtained for control males tested among fellow control males. The results obtained may partially be explained by the fact that both the urge to copulate and the effort made to repel copulation attempts of other males are reduced in allatectomized males, which seem to struggle even less against copulation attempts by other males than do normal females.

Wajc and Pener's (1969) results demonstrated that studies concerning hormonal effects on male sexual behaviour in insects must be based on comparative and quantitative methods. Furthermore, these findings explained the reason for Strong's (1968) inability to reveal the effect of allatectomy on male mating behaviour in *L.m. migratorioides*. Strong observed his males only in presence of similarly treated fellow males and did not test them toward females of differently treated (controls toward allatectomized, or *vice versa*) males.

In a joint paper, Pener, Girardie and Joly (1972) recently tried to reinvestigate the contradiction concerning Girardie and Vogel's (1966) and Girardie's (1966) findings as opposed to those of Wajc and Pener (1969). Since Girardie's original strain of *L.m. cinerascens* had unfortunately been lost, the study was performed on *L.m. migratorioides* using Girardie's (1966) surgical technique and Wajc and Pener's (1969) method of recording the percentage of time spent on sexual behaviour. Previous investigations had, however, disclosed that the median neurosecretory cells (MNSC) of the pars intercerebralis (PI) also affect male sexual behaviour in *L.m. cinerascens* (Girardie 1964; Girardie and Vogel 1966), at least through an activatory effect of the C cells of the MNSC on the CA (Girardie 1966, 1971). Pener, Girardie and Joly's study (1972) was extended, therefore, to cover the effect of both the MNSC and the CA.

In these experiments the sexual behaviour of allatectomized, PI electrocoagulated, sham-operated and unoperated controls, each treatment being performed a few days after the last moult, was observed during seven weeks. At the beginning of the fifth week, active CA were implanted into a proportion of the allatectomized and of the PI coagulated males. Average weekly percentages of sexual behaviour in presence of females (for details of the method and way of calculation see Pener 1967a; Wajc and Pener 1969) are summed up in Fig. 1. Consistently with Pener and Wajc's (1971) previous results, allatectomized males exhibited distinct sexual behaviour, but at a considerably lower intensity than did the controls. Reimplantation of CA increased their mating activity, which eventually reached the level shown by the controls. Electrocoagulation of the PI resulted in a practically complete absence of the sexual behaviour; during the whole experiment just one male exhibited a single attempt to mount a female. Implantation of CA into PI coagulated males did not induce mating behaviour (Fig. 1).

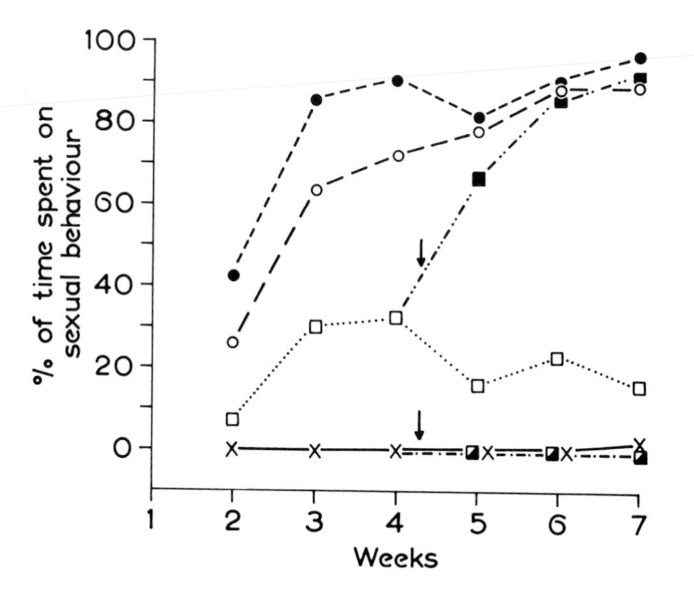

Fig. 1. Weekly average percentage of time spent on sexual behaviour by differently treated gregarious *Locusta migratoria migratorioides* males (after Pener, Girardie and Joly 1972).

●--------● - sham-operated males
O— — — — O - unoperated males
□·············□ - allatectomized males
■··-··-··-··■ - allatectomized males into which 3 pairs of CA were implanted
X————————X - PI coagulated males
◪·-·-·-·-◪ - PI coagulated males into which 3 pairs of CA were implanted

Vertical arrows show time of implantation of CA

The CA are responsible for the colour change toward yellow which coincides with the sexual maturation in some adult acridids, including crowded *L. migratoria* males (for references, see Pener 1972). Consistently with this concept, in Pener, Girardie and Joly's (1972) experiments, the allatectomized males did not show yellowing, whereas reimplantation of CA induced yellow colour. Electrocoagulation of the PI prevented the advance of yellow colouration almost to the same extent as did allatectomy. However, implantation of active CA into previously PI coagulated males, though it did not induce mating behaviour (see above), nevertheless promoted yellowing, a result which confirms that indeed the CA control the appearance of the yellow colouration. These findings showed that the absence of the yellow colour in the PI coagulated males was obtained because of the inactivity of the males' own CA and allow the conclusion that the PI activates the CA, thus exerting an allatotropic effect. In subsequent experiments, therefore, the advance of the yellow colouration was used as an indicator of CA activity and/or of the allatotropic effect of the PI.

Further experimental results from the same work demonstrated that implantation of PI into previously PI coagulated males induced distinct mating behaviour and promoted yellow colouration. Thus, both the effect of the PI on male sexual behaviour and its allatotropic effect was found to be humoral (neurosecretory).

Finally, using Girardie's (1966) method for selective electrocoagulation of the MNSC, Pener, Girardie and Joly (1972) destroyed either the C, or the A + B cells of the MNSC and for five weeks observed the sexual behaviour of the males so treated. The results indicated that the C cells are responsible for the control of the mating behaviour as well as for the allatotropic effect of the PI and the A + B cells have no effect in these respects.

Examinations of serial sections made from the PI at the end of the experiments provided additional substantial proof for these conclusions. Since in all experiments the males were individually marked, these sections made it possible to relate, *on the basis of the individual*

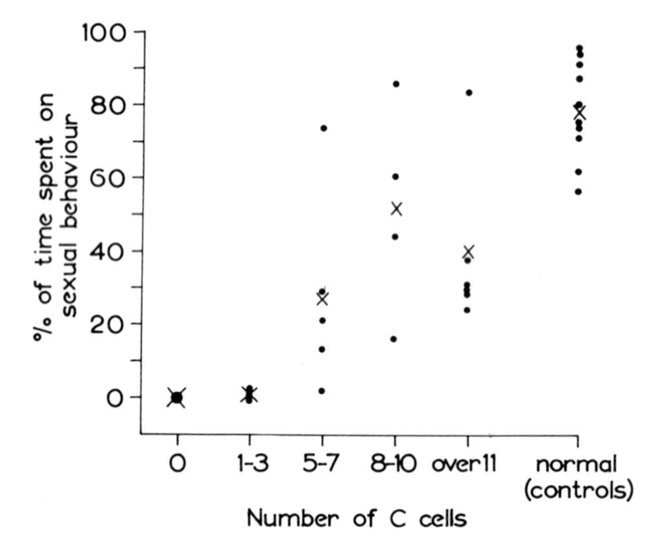

Fig. 2. Average percentage of time spent on sexual behaviour by individual gregarious *Locusta migratoria migratorioides* males during an entire period of experiments (5 or 7 weeks) as related to the number of C cells which remained undestroyed following selective (or complete) electrocoagulation of the pars intercerebralis. Unoperated control males (number of C cells is normal) are also included for comparison (after Pener, Girardie and Joly 1972).

● - result obtained from a single male

⬮ - results obtained from 12 males (11 males spent 0% of time on sexual behaviour and one male spent 0.46%; no graphical separation of these results is possible)

X - average value for all the males having similar number of C cells *males*, the number of C, or A + B cells which remained undestroyed to the intensity of the sexual activity. Fig. 2 clearly demonstrates that the intensity of the mating behaviour (expressed in average percentage of time spent on sexual behaviour during the whole period of observations) depends on the number of C cells. No such relationship was obtained for the A + B cells. Using similar procedures it was also found that the C, but not the A + B cells, are responsible for the allatotropic effect of the PI; a conclusion which was consistent with Girardie's (1966, 1967) findings.

In summary, the results of Pener, Girardie and Joly (1972) show the following effects in crowded adult *L.m. migratorioides* males (for schematic representation see Fig. 3): (1) The C cells of the PI completely control sexual behaviour inasmuch as no sexual behaviour is displayed after their destruction. Their effect is primary in the sense that it is not mediated through the CA. (2) The C cells activate the CA and in this way affect yellow colour and exert a secondary effect on mating behaviour. (3) The CA influence the intensity of sexual behaviour, but do not exercise a complete control since their removal does not completely inhibit mating behaviour. (4) The CA completely control yellow colouration. (5) The A + B cells do not affect mating behaviour or yellow colour; nor do they activate the CA.

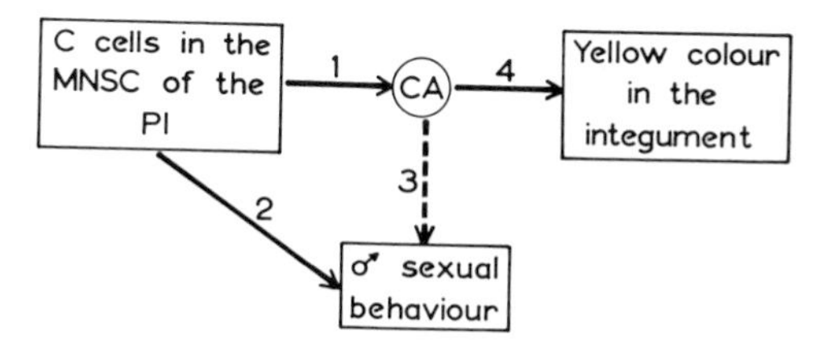

Fig. 3. Scheme of endocrine control of sexual behaviour and yellow colouration in gregarious adult *Locusta migratoria migratorioides* males. All arrows show positive (activatory) effects.
MNSC - median neurosecretory cells
PI - pars intercerebralis
CA - corpora allata
1 - allatotropic effect of the C cells
2 - complete control of male sexual behaviour by the C cells (not mediated through the CA)
3 - positive influence (but not complete control) of male sexual behaviour by the CA
4 - complete control of yellow colouration in the integument by the CA

All the studies hitherto cited relating to hormonal factors affecting male sexual behaviour in *Locusta migratoria*, were performed on crowded locusts only. However, while investigating the differential endocrine relations in different locust phases in the Centre for Overseas Pest Research, Pener (in prep.) recently compared the effect of the CA on mating behaviour in gregarious and solitary *L.m. migratorioides*.

In these investigations comparative experiments were set up using hatchlings from the same egg pods originating from crowded parents. Within 24 hr of hatching the nymphs were either placed in cages under conditions of heavy crowding, or separated one per glass jar. The colouration and certain morphological features of the isolated locusts showed considerable changes during nymphal development and following the last moult their colour and curvature of the pronotum were characteristic of the *solitaria* phase. In contrast, crowded locusts exhibited the typical colour and morphology of the *gregaria* phase (for comprehensive data on locust phases see Uvarov 1966; Albrecht 1967).

A comparative experiment was designed employing allatectomized, sham-operated, unoperated and extra CA implanted (3 pairs per recipient) males in each phase. Thus, in total, 8 experimental groups (4 treatments x 2 phases), each comprising at least 12 males, were set up. Surgical and control treatments were performed 2±1 days after the last moult and observations on mating behaviour started a few days later. Each *individual* male was placed with one solitary female 2-3 times per week and on each occasion observations were made for 2 hr. Thus, sexual behaviour of both gregarious and solitary males was tested toward solitary females, in order to standardize any effect which might have been exerted by the latter. Except when the actual observations (2x3 = 6 hr or 2x2 = 4 hr per week) were being made, each experimental group of gregarious males was kept in a separate small (12 litre) cage, and solitary males were maintained one per jar.

Fig. 4 shows the weekly average percentage of time spent on sexual behaviour. Consistently with the findings of Pener and Wajc (1971), Wajc and Pener (1969) and Pener, Girardie and Joly (1972), allatectomized gregarious males exhibited subnormal, though distinct mating behaviour. Implantation of extra CA accelerated maturation in gregarious males, inducing early and intense sexual activity which to some extent declined later; from the 5th week onwards the intensity of the mating behaviour shown by these males appearing to be slightly lower than that of the respective controls.

Regardless of which of the different operations and control treatments they received, the subsequent intensity of the mating behaviour was found to be roughly similar for every experimental group of solitary males. Judging from Fig. 4, the sole behavioural effect of the allatectomy appeared to be a slight delay in the onset of the mating activity. Implantation of extra CA affected neither the intensity nor the time of onset of the sexual behaviour.

A part of these experiments was repeated later in Jerusalem using a different strain of *L.m. migratorioides* and employing only unoperated and allatectomized solitary males. The intensity of the sexual behaviour was found to be well comparable with that obtained for solitary males in the previous experiments. The delayed onset of the mating activity in allatectomized males was reconfirmed. This time, however, the sexual behaviour of the allatectomized males appeared to be slightly more intense than that of the unoperated controls, indicating that the opposite situation in the previous experiment (Fig. 4) was probably obtained due to individual variation.

The surprising results of these experiments demonstrated that, except for having a limited influence on timing, the CA does not affect male mating behaviour in solitary *L.m. migratorioides*. With regard to the sexual characteristics of the two phases, the results revealed that sexual behaviour is more intense in normal (control) gregarious males than in solitary ones.

The most probable interpretation of these results would seem to be that the effect of the CA on mating behaviour is indeed different in gregarious and solitary locusts. According to this hypothesis, the organ or organs constituting the eventual target of the hormone (probably the nervous system), react differently in the different phases to the same endocrine factor. However, the following alternative explanation, based on different pheromone relations, may also be considered.

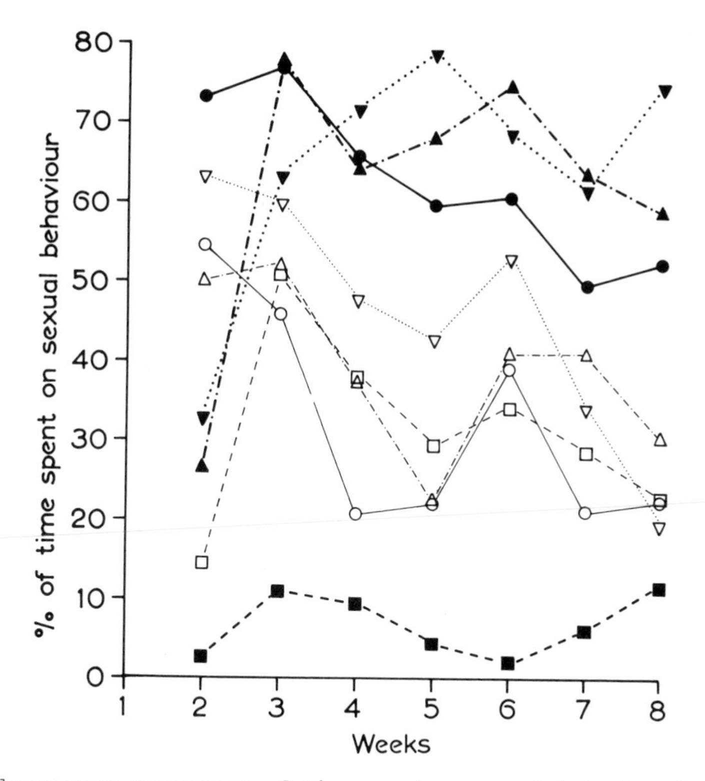

Fig. 4. Weekly average percentage of time spent on sexual behaviour by differently treated solitary *Locusta migratoria migratorioides* males as compared with that of their parallelly treated gregarious brothers in the same experiment. The averages for the 5th, 6th and 8th weeks are based on two observations per week; all the other averages are based on three observations per week.

□--------□ - allatectomized solitary males
△-·-·-·-·△ - sham-operated solitary males
▽·············▽ - unoperated solitary males
○————————○ - solitary males into which 3 pairs of extra CA (originating from sexually mature gregarious locusts) were implanted
■--------■ - allatectomised gregarious males
▲-·-·-·-·▲ - sham-operated gregarious males
▼·············▼ - unoperated gregarious males
●————————● - gregarious males into which 3 pairs of extra CA (originating from sexually mature gregarious locusts) were implanted.

Isolated adults of *L.m. migratorioides* mature more rapidly than crowded ones (Norris 1950). This situation is the reverse of that found in the desert locust. However, Norris (1964) demonstrated that in *both* species the presence of young adults inhibits maturation, while the presence of mature ones has an accelerating effect. She suggested that in *Locusta*, as in *Schistocerca*, the balance of two pheromones, one inhibiting and a second later accelerating development, regulates the synchronous maturation of crowded adults. Accordingly, the difference between the two species is merely in the timing of the switch over from one pheromonal effect to the other (Norris 1964, 1968).

Norris's suggestions can be related to the differential effect of the CA on mating behaviour in different phases by assuming that in *Locusta* (as in *Schistocerca*) the CA affects pheromone production and by further assuming that the pheromones do not merely regulate maturation, but also influence the intensity of male mating behaviour. According to this interpretation, crowded allatectomized males would secrete the maturation inhibiting pheromone for long periods (as happens in *S. gregaria*; cf. Norris and Pener 1965) and would inhibit each other's sexual behaviour. In contrast, crowded, non-allatectomized males would intensify each other's mating activity by persistently secreting the maturation accelerating pheromone. Finally, solitary males, which under complete isolation cannot be affected by either pheromone, would exhibit sexual behaviour of intermediate intensity which is not related to the activity of the CA. This theory, however, does not explain the delayed onset of the mating behaviour in allatectomized solitary males, thus, it is less feasible than the alternative one outlined above. This hypothesis could easily be proved or disproved by keeping crowded allatectomized and control males continuously together and testing their sexual behaviour toward females. Unfortunately, no such investigation has been reported; Wajc and Pener (1969), when observing mating behaviour of crowded allatectomized and control males each toward the other, placed such males together for only 4 hr per week.

Obviously, the two theories are not mutually exclusive. The CA may differentially affect the sexual behaviour in different phases of *L.m. migratorioides* and pheromonal effects may further amplify the differences. The "pheromonal theory", however, cannot by itself explain the effect of the CA on male mating behaviour in *Schistocerca gregaria*. In this species, Loher (1961) kept allatectomized males and controls continuously together with mature males, thus providing plenty of maturation accelerating pheromone. Despite the presence of this pheromone, the allatectomized males did not show mating behaviour.

SOME OTHER ACRIDIDS

Loher (1962) reported that the CA do not control male sexual behaviour in *Gomphocerus* (now *Gomphocerippus*) *rufus*. In this species, however, female sexual behaviour is affected by these glands. Allatectomized females of this species do not show "copulatory readiness", but persistently exhibit "primary defence" characteristic of immature females (Loher and Huber 1966). Hartmann (1971), studying the same species, has stated in the summary of a paper that: "Allatectomy, cauterization of the pars intercerebralis, castration, and the double operation allatectomy plus cauterization of the pars intercerebralis, do not influence the sexual behaviour of the male". However, in the text of the same paper Hartmann discloses that "Männchen von *Gomphocerus rufus*, die 2 bzw. 14 Tage nach der Imaginalhäutung allatektomiert wurden, oder deren Pars intercerebralis durch Kauterisierung zerstört worden war, zeigen in ihrem Sexualverhalten gegenüber aktiv kopulationsbereiten Weibchen keine oder nur graduelle Unterschiede im Vergleich mit Kontrollen. So behandelte Männchen sitzen in der Regel apathisch in der Beobachtungsarena und reagieren nur selten auf kopulationsbereite Weibchen." (Hartmann 1971, p. 213). It seems, therefore, that these operations may influence to some extent the *intensity* of the male sexual behaviour, even if they do not result in an all-encompassing effect.

Investigating the role of the CA in control of reproduction in female *Euthystira brachyptera*, Müller (1965) briefly mentioned that these glands do not affect male sexual behaviour in this species, since 3 allatectomized males sang and copulated normally. Obviously, no conclusion can be drawn from this data regarding the possible effect of the CA on the *intensity* of male mating behaviour.

SYNTHESIS

This article can be summed up in a single statement; for the time being no satisfying synthesis can be made. However, some explanations as to why generalization is so difficult may be considered.

Endocrine effects on male sexual behaviour in acridids have probably been studied most thoroughly in *Locusta migratoria migratorioides*. The main conclusions drawn for gregarious males of this subspecies are presented schematically in Fig. 3. This scheme fits the current factual knowledge, nevertheless it is probably imperfect and almost certainly incomplete. Every single arrow in the scheme may represent a full chain of causative events involving further organs and effects. Moreover, the possible existence of additional factors, which would have to be represented by new arrows in Fig. 3, cannot be excluded.

Furthermore, the scheme presented in Fig. 3 cannot be accepted as being fully applicable to any other acridid studied to date. An effect of the MNSC on male mating behaviour has not yet been investigated in *S. gregaria* and *N. septemfasciata*, thus, the existence of arrow No. 2 (in Fig. 3) and the exact nature of arrow No. 1 is doubtful in these species. Moreover, the situation in *Locusta* represented by arrow No. 3 differs from that existing in *Schistocerca* and *Nomadacris*: in both latter species removal of the CA completely inhibits male sexual behaviour. According to Hartmann's (1971) results the effect represented by arrow No. 2 is at least incomplete in *Gomphocerippus rufus*; both Loher (1962) and Hartmann (1971) suggest that in this species the influence shown by arrow No. 3 is absent (or slight ?). In diapausing *Oedipoda miniata* males implantation of active CA induces intense mating behaviour and temporarily terminates diapause. If we assume that the C cells are active during diapause, they cannot have an allatotropic effect (arrow No. 1), otherwise these cells would activate the male's own CA and the insect would not be in diapause. If, however, we assume that the C cells are inactive during diapause, they cannot have complete control over male mating behaviour (arrow No. 2); for if they exerted this kind of control, implantation of CA alone would not result in mating behaviour, as the implantation of active C cells would obviously be needed as well. *Ad absurdum*, the scheme as limited in Fig. 3 cannot be accepted for *O. miniata*.

The effect of the CA on male mating behaviour may be mediated solely through pheromonal effects in crowded *Locusta*, but this could not possibly be so in *Schistocerca*. It is probable that no pheromone relations similar to those existing in locusts affect maturation and/or regulate male mating behaviour in the non-gregarious grasshopper, *Oedipoda miniata*. Therefore, even if the "pheromone theory" should turn out to be true for *Locusta*, it cannot be extended for other species. Again, no general conclusion can be drawn.

Pener (1968, 1972) and Pener and Wajc (1971) related the differential effect of the CA on male mating behaviour to the ability of the species to exhibit reproductive diapause. These glands completely control male sexual behaviour in *Schistocerca* and *Nomadacris* which are able to exhibit complete reproductive diapause. In *Oedipoda miniata* males, the diapause is almost complete, and so is the effect of the CA. In *L.m. migratorioides* there seems to be no reproductive diapause and the CA affect male mating behaviour only partially (*gregaria*), or almost not at all (*solitaria*). Even different subspecies of (crowded) *L. migratoria* having different tendencies for exhibiting reproductive diapause, may show some *quantitative* differences in this effect of the CA (see Pener, Girardie and Joly 1972, for further details). It is difficult, however, to reconcile this theory with the complete control exerted by the C cells on male sexual behaviour in *L.m. migratorioides*. Why should this non-diapausing subspecies have an alternative all-encompassing mechanism? Or, if the C cells play a similar role in *Schistocerca* and *Nomadacris*, why do these species need the superimposed complete effect of the CA?

The differential effect of the CA on male mating behaviour may be correlated not only with presence or absence of reproductive diapause, but also with the taxonomic position of the species. According to Uvarov (1966), *Schistocerca* and *Nomadacris* belong to the subfamily Cyrthacanthacridinae, *Oedipoda* and *Locusta* to the Oedipodinae, while *Gomphocerippus* (and *Euthystira*) are classified as Gomphocerinae.

The mode of action of the CA in controlling male sexual behaviour is unknown, but the effect is probably indirect (Pener 1965). These glands affect locomotor activity in locusts (Odhiambo 1966b; Wajc and Pener 1971), perhaps by influencing the excitability and/or responsiveness of the nervous system and this may account for absence or reduction of the mating behaviour. This suggestion has already been discussed in detail (Pener 1972; Wajc and Pener 1971). However, even if we assume that the effect of the CA on sexual behaviour is a reflection of a more general effect on the nervous system, we still do not know whether this latter effect is direct or indirect. Even the possibility that there is a positive feed-back mechanism from the CA to the C cells and a subsequent effect of the C cells on the nervous system cannot be excluded.

Regardless of whether the effect of the CA on mating behaviour is direct or indirect, is specific or general, or is mediated or not through another organ or organs, as for example the neurosecretory cells of the last abdominal ganglion (cf. Delphin 1963), the fact remains that the effect is different in different species. The assumption that there is a general effect of these glands on the nervous system does not alter this conclusion. Either this general effect could differ between species, or a similar general effect could act differentially on male sexual behaviour, suppressing it completely in some species, but only reducing it in others.

Zdárek's (1968, 1971) studies on the bug, *Pyrrhocoris apterus*, and Foster's (1967) investigations on the dung fly, *Scatophaga stercoraria*, revealed that endocrine factors can affect male sexual behaviour in non-acridid insects, too. Engelmann (1970, p. 87-93), reviewing hormonal effects on female mating behaviour in several insect orders, has concluded that the CA exert complete or partial control in some species, but have no effect in some others. Thus, endocrine mechanisms involved in regulation of mating behaviour of insects seem to be complex and extremely flexible.

Perhaps this extraordinary flexibility constitutes a partial answer as to why and how insects are able to adapt themselves through relatively rapid evolutional processes to different or changing environments. This interpretation supposes that during the course of evolution, already existing hormones are easily and rapidly "employed" or "unemployed" for controlling and/or timing various behavioural (or even physiological) activities connected with the main events of the life cycle.

ACKNOWLEDGEMENTS

I am grateful to the Centre for Overseas Pest Research (formerly Anti-Locust Research Centre) London and its Director, Dr. P.T. Haskell for providing facilities and grant for the study on solitary *Locusta*. The experiments on the effect of the neurosecretory cells on male mating behaviour of crowded *Locusta* were performed in Prof. Joly's Laboratory in Strasbourg and were partially supported by a grant from the "Centre International des Stages du Ministère des Affaires Étrangers" of the French Government. Two former grants, one from the Centre for Overseas Pest Research, London, and another from the Authority of Research and Development of the Hebrew University, Jerusalem, supported some of the earlier studies. All of them are gratefully acknowledged.

REFERENCES

ALBRECHT, F.O.: "Polymorphisme Phasaire et Biologie des Acridiens Migrateurs". Paris: Masson and Cie. (1967).

BROZA, M., PENER, M.P.: Hormonal control of the reproductive diapause in the grasshopper, *Oedipoda miniata*. Experientia 25, 414-415 (1969).

BROZA, M., PENER, M.P.: The effect of the corpora allata on mating behaviour and reproductive diapause in adult males of the grasshopper, *Oedipoda miniata*. Acrida 1, 79-96 (1972).

CANTACUZÈNE, A.-M.: Effets comparés de l'allatectomie sur l'activité des glandes annexes mâles et le comportement sexuel de deux Acridiens: *Schistocerca gregaria* et *Locusta migratoria* (souches *migratorioides* et "Kazalinsk"). C. r. hebd. Séanc. Acad. Sci., Paris (Ser. D) 265, 224-227 (1967).

CANTACUZÈNE, A.-M.: Recherches morphologiques et physiologiques sur les glandes annexes males des Orthoptères. II. Les glandes annexes males des Acridiens aprés l'ablation des corps allates et au cours de la diapause imaginale. Bull. Soc. zool. Fr. 93, 545-557 (1968).

DELPHIN, F.: Histology and possible functions of neurosecretory cells in the ventral ganglia of *Schistocerca gregaria* Forsk. Nature, Lond. 200, 913-915 (1963).

ENGELMANN, F.: "The Physiology of Insect Reproduction". Oxford: Pergamon Press (1970).

FOSTER, W.: Hormone-mediated nutritional control of sexual behaviour in male dung flies. Science, N.Y. 158, 1596-1597 (1967).

GIRARDIE, A.: Fonction gonadotrope de la pars intercerebralis chez le mâle de *Locusta migratoria* L. (Orthoptère). C. r. hebd. Séanc. Acad. Sci., Paris (Groupe 12) 258, 2910-2911 (1964).

GIRARDIE, A.: Controle de l'activité génitale chez *Locusta migratoria*. Mise en évidence d'un facteur gonadotrope et d'un facteur allatotrope dans la pars intercerebralis. Bull. Soc. zool. Fr. 91, 423-439 (1966).

GIRARDIE, A.: Controle neuro-hormonal de la métamorphose et de la pigmentation chez *Locusta migratoria cinerascens* (Orthoptère). Bull. biol. Fr. Belg. 101, 79-114 (1967).

GIRARDIE, A.: Hormone et mécanismes endocrines contrôlant l'activité génitale de *Locusta migratoria* (Orthoptère). Archs Zool. exp. gen. 112, 635-648 (1971).

GIRARDIE, A., VOGEL, A.: Étude du contrôle neuro-humoral de l'activité sexuelle male de *Locusta migratoria* (L.). C. r. hebd. Séanc. Acad. Sci., Paris (D) 263, 543-546 (1966).

HARTMANN, R.: Der Einflus endokriner Faktoren auf die männlichen Akzessorischen Drüsen und die Ovarien bei der Keulenheuschrecke *Gomphocerus rufus* L. (Orthoptera, Acrididae). Z. vergl. Physiol. 74, 190-216 (1971).

LEVY, Y.: The life and biology of grasshoppers of the genus *Oedipoda* (Oedipodinae) in Israel with special reference to the development of the eggs. (In Hebrew). M.Sc. Thesis, The Hebrew University, Jerusalem. (1965).

LOHER, W.: The chemical acceleration of the maturation process and its hormonal control in the male of the desert locust. Proc. R. Soc. (B) 153, 380-397 (1961).

LOHER, W.: Die Kontrolle des Weibchengesanges von *Gomphorus rufus* L.

(Acridinae) durch die Corpora allata. Naturwissenschaften 49, 406 (1962).

LOHER, W., HUBER, F.: Nervous and endocrine control of sexual behaviour in a grasshopper (*Gomphocerus rufus* L., Acridinae). Symp. Soc. exp. Biol. 20, 381-400 (1966).

MÜLLER, H.P.: Zur Frage der Steuerung des Paarungsverhaltens und der Eireifung der Feldheuschrecke *Euthystira brachyptera* Ocsk. unter besonderer Berücksichtigung der Rolle der Corpora allata. Z. vergl. Physiol. 50, 447-497 (1965).

NORRIS, M.J.: Reproduction in the African migratory locust (*Locusta migratoria migratorioides* R. & F.) in relation to density and phase. Anti-Locust Bull. No. 6, 1-48 (1950).

NORRIS, M.J.: Reproduction in the desert locust (*Schistocerca gregaria* Forsk.) in relation to density and phase. Anti-Locust Bull. No.13, 1-49 (1952).

NORRIS, M.J.: Sexual maturation in the desert locust (*Schistocerca gregaria* Forskål) with special reference to the effects of grouping. Anti-Locust Bull. No.18, 1-44 (1954).

NORRIS, M.J.: Reproduction in the red locust (*Nomadacris septemfasciata* Serville) in the laboratory. Anti-Locust Bull. No.36, 1-46 (1959).

NORRIS, M.J.: Accelerating and inhibiting effects of crowding on sexual maturation in two species of locusts. Nature, Lond. 203, 784-785 (1964).

NORRIS, M.J.: Some group effects on reproduction in locusts. Colloques int. Cent. natn. Rech. scient. 173, 147-161 (1968).

NORRIS, M.J., PENER, M.P.: An inhibiting effect of allatectomized males and females on the sexual maturation of young male adults of *Schistocerca gregaria* (Forsk.) (Orthoptera: Acrididae). Nature, Lond. 208, 1122 (1965).

ODHIAMBO, T.R.: Growth and the hormonal control of sexual maturation in the male desert locust, *Schistocerca gregaria* (Forskål). Trans. R. ent. Soc. Lond. 118, 393-412 (1966a).

ODHIAMBO, T.R.: The metabolic effects of the corpus allatum hormone in the male desert locust. II. Spontaneous locomotor acitivity. J. exp. Biol. 45, 51-63 (1966b).

PENER, M.P.: On the influence of corpora allata on maturation and sexual behaviour of *Schistocerca gregaria*. J. Zool. 147, 119-136 (1965).

PENER, M.P.: Effects of allatectomy and sectioning of the nerves of the corpora allata on oöcyte growth, male sexual behaviour, and colour change in adults of *Schistocerca gregaria*. J. Insect Physiol. 13, 665-684 (1967a).

PENER, M.P.: Comparative studies on reciprocal interchange of the corpora allata between males and females of adult *Schistocerca gregaria* (Forskål) (Orthoptera: Acrididae). Proc. R. ent. Soc. (A) 42, 139-148 (1967b).

PENER, M.P.: The effect of corpora allata on sexual behaviour and "adult diapause" in males of the red locust. Entomologia exp. appl. 11, 94-100 (1968).

PENER, M.P.: The corpus allatum in adult acridids: the inter-relation of its functions and possible correlations with the life cycle. *In* "Proc. Int. Study Conf. Current and Future Problems of Acridology, London, 1970", pp. 135-147. Centre for Overseas Pest Research, London (1972).

PENER, M.P., BROZA, M.: The effect of implanted, active corpora allata on reproductive diapause in adult females of the grasshopper *Oedipoda miniata*. Entomologia exp. appl. 14, 190-202 (1971).

PENER, M.P., GIRARDIE, A., JOLY, P.: Neurosecretory and corpus allatum controlled effects on mating behavior and color change in adult *Locusta migratoria migratorioides* males. Gen. comp. Endocrinol. 19, 494-508 (1972).

PENER, M.P., WAJC, E.: The effect of allatectomy on male sexual behaviour in adult *Schistocerca gregaria* and *Locusta migratoria migratorioides*. *In* "Insect Endocrines" (Ed., V.J.A. Novák and K. Sláma) pp. 37-43. Academia, Praha. (Proc. int. Symp. Insect Endocrines, Brno, 1966) (1971).

PEREZ, Y., VERDIER, M., PENER, M.P.: The effect of photoperiod on male sexual behaviour in a North Adriatic strain of the migratory locust. Entomologia exp. appl. 14, 245-250 (1971).

QUO FU : Studies on the reproduction of the oriental migratory locust: the role of the corpora allata. (In Chinese with English Summary.) Acta ent. Sinica 14, 211-224 (1965).

STRONG, L.: Locomotor activity, sexual behaviour, and the corpus allatum hormone in males of *Locusta*. J. Insect Physiol. 14, 1685-1692 (1968).

UVAROV, B.P.: "Grasshoppers and Locusts. A Handbook of General Acridology." Vol. 1. London: Cambridge University Press. (1966).

VERDIER, M.: Diapauses d'une souche de *Locusta migratoria migratoria* L., en phase grégaire, et conditionnement photopériodique; comparaison avec d'autres Acridiens. I.- Réponse imaginale. Bull. Soc. zool. Fr. 94, 55-70 (1969).

WAJC, E., PENER, M.P.: The effect of the corpora allata on the mating behavior of the male migratory locust, *Locusta migratoria migratorioides* (R. & F.). Israel J. Zool. 18, 179-192 (1969).

WAJC, E., PENER, M.P.: The effect of the corpora allata on flight activity of the male African migratory locust, *Locusta migratoria migratorioides* (R. & F.). Gen. comp. Endocrinol. 17, 327-333 (1971).

ŽĎÁREK, J.: Le comportement d'accouplement à la fin de la diapause imaginale et son contrôle hormonal dans le cas de la punaise *Pyrrhocoris apterus* L. (Pyrrhocoridae, Heteroptera). Annls Endocr. 29, 703-707 (1968).

ŽĎÁREK, J.: Hormonal control of mating behaviour in *Pyrrocoris apterus* L. *In* "Insect Endocrines" (Eds., V.J.A. Novák and K. Sláma) pp. 51-61. Academia, Praha. (Proc. int. Symp. Insect Endocrines, Brno, 1966) (1971).

THE ROLE OF HORMONES IN THE REPRODUCTIVE BEHAVIOR OF FEMALE WILD SILKMOTHS

L.M. Riddiford

The Biological Laboratories, Harvard University, Cambridge, Mass., U.S.A.
Present address: *Department of Zoology, University of Washington, Seattle, Washington, U.S.A.*

Insects which do not feed in the imaginal stage must give a high priority to early reproduction. Therefore, mating and oviposition must occur shortly after adult emergence. The wild silkmoths show pronounced adaptations towards this end. The female emerges with nearly the full complement of mature eggs (Telfer and Rutberg 1960). Also, at the time of emergence she has a reservoir of sex pheromone (Preisner 1968) with which she can attract the male from distances as far as 2 miles (Rau and Rau 1929).

In order to maximize the chances of mating, the reproductive behaviors of these saturniid moths are coordinated by relatively well-defined environmental cues. To release pheromone, the female moth assumes the "calling" posture. The female hangs from a branch or twig and protrudes the terminal segments of the abdomen, thereby exposing the pheromone glands. The pheromone volatilizes from the exposed surfaces of the gland and is carried downwind. Fig. 1 shows that the calling behavior of the female moth occurs at a particular time in the photoperiodic cycle and that this time is characteristic of the species (Truman, Lounibos, and Riddiford, in preparation).

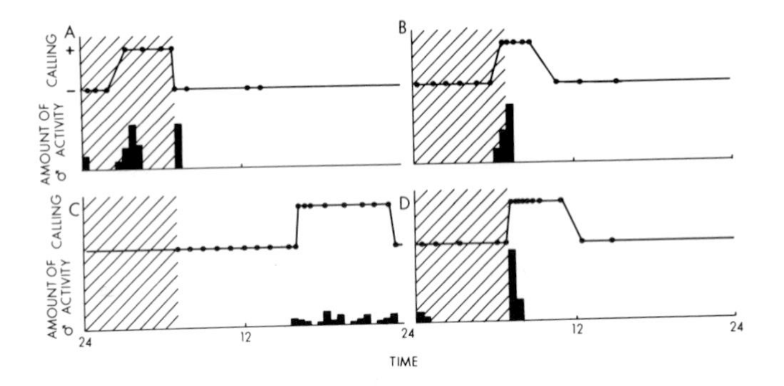

Fig. 1. The times of pheromone release ("calling") by the virgin female and of flight activity of the unmated male isolated from females in a 17L:7D photoperiod at 25°C. The • indicate time of observation of the females. The male activity was monitored continuously. (From Truman, Lounibos and Riddiford, in preparation).
A. *Samia cynthia*. B. *Antheraea pernyi*. C. *Callosamia promethea*. D. *Hyalophora cecropia*

The time of the flight activity of the male (Fig. 1) corresponds precisely with the timing of pheromone release in the female. The close temporal coordination between these complementary behaviors in the male and in the female maximizes the chances that a male will come in contact with the odor plume emitted by the virgin female (Wilson and Bossert 1963). Just as Payne, Shorey and Gaston (1970) showed for the cabbage looper moth, *Trichoplusia ni*, most of the saturniid males will respond behaviorally to the sex pheromone only at the time of day indicated by their flight activity. But the physiological response of the antennae as measured by the electroantennogram technique (Schneider 1957) remains constant throughout the photoperiod. Thus, the daily changes in responsiveness of the male to the female is not due to changes in antennal sensitivity but rather to alterations of responsiveness in the central nervous system.

Antheraea pernyi males are exceptions to this rule in that they will respond behaviorally to the female at any time of day (Riddiford 1970, and further unpublished data). In this species the female "calls" only during the late night (Fig. 1), but pheromone apparently is also emitted when she is not in the calling posture. Under dim illumination the male is excited and attracted by the pheromone at any time of day. The semi-domestication of this species by the Japanese for silk manufacture presumably contributes to this deviation from the typical saturniid behavior.

After mating, the behavior of the female moth changes. Calling behavior is no longer observed in the mated female (Falls 1933; Truman and Riddiford 1971). Moreover, the mated female shows greatly enhanced oviposition in comparison with that seen in the virgin (Truman and Riddiford 1971). The present paper is concerned with the controls over the reproductive behavior of the female moth.

BEHAVIOR OF THE VIRGIN FEMALE

As indicated in Fig. 1, most saturniid females "call" at a certain time of day in the absence of any other environmental cues. An exception to this rule is *Antheraea polyphemus*. In order to release sex pheromone, virgin females of the Polyphemus moth also require the presence of oak leaves (Riddiford and Williams 1967), the food plant of their larvae, or the volatile chemical from these leaves, *trans*-2-hexenal (Riddiford 1967). A similar behavior related to the larval food plant has also been reported for *Acrolepia assectella* (Rahn 1968).

In the case of Polyphemus, the induction of calling behavior is due to an interaction between chemical and temporal signals. Although surrounded with oak leaves, the female Polyphemus moth does not "call" during the photophase. Under a 17L:7D photoperiodic regime in an open cage, in the presence of oak or of *trans*-2-hexenal, Polyphemus females tend to begin "calling" somewhere between 3 to 6 hr after the lights-off signal (see Fig. 2). Since our stock has come from various regions of the country, this variation may be due to strain differences. In nature, Polyphemus have been reported as mating between 11.00 p.m. and midnight and at dawn (Rau and Rau 1929).

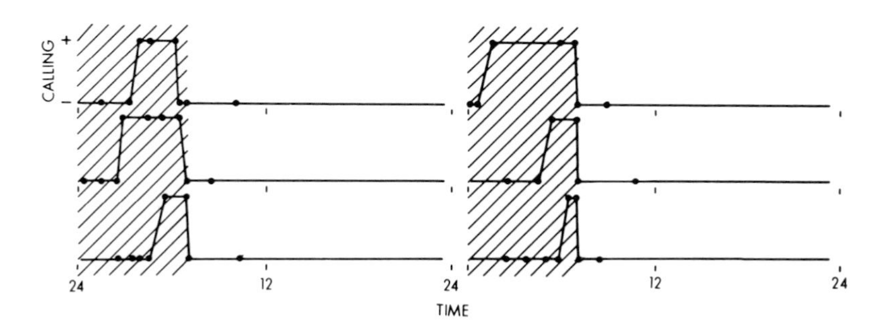

Fig. 2. The times of pheromone release ("calling") by virgin *Antheraea polyphemus* females in the presence of red oak (*Quercus rubrum*) leaves (left) and of vapors of 0.05% *trans*-2-hexenal (right) in an open cage. The • indicate time of observation of the females

In a closed system where air is passed over a 0.05% aqueous *trans*-2-hexenal solution and then through a jar containing females, the females usually begin calling 1½ to 3 hr after being placed in the dark (Riddiford and Williams 1971). This early onset of calling probably arises from a more chronic exposure to the *trans*-2-hexenal. Also, other environmental odors which may block olfactory reception or mask its effects (Riddiford 1967, 1970) are eliminated.

In spite of the more complex environmental control over the behavior of the virgin Polyphemus moth, the hormonal control over this behavior is the same as in *Hyalophora cecropia*, a species in which time within the photoperiodic cycle is the only influencing factor. In both Polyphemus and Cecropia females, the corpora cardiaca (CC) are necessary for calling behavior to occur (Riddiford and Williams 1971). The CC of these moths serve a dual function. It is a neurohaemal organ for the lateral and medial neurosecretory cells of the brain, and it contains its own intrinsic neurosecretory cells (Stumm-Zollinger 1957). The strictly endocrine corpora allata (CA) are attached to and lie posterior to the CC.

When the CA were removed from Cecropia or Polyphemus pupae, the resultant adult females displayed normal calling behavior. But removal of the CC-CA complex nearly completely abolished calling behavior in both species. Reimplantation of 3 pairs of CC-CA complexes into these females did not restore this behavior. Similarly, calling could be abolished by simply cutting the nerves from the brain to the CC. Since Stumm-Zollinger (1957) had shown that the neurohaemal portion of the CC in saturniid moths regenerated soon after removal of the CC, it was concluded that, in response to a nervous signal from the brain, the intrinsic neurosecretory cells of the CC secrete a hormone that elicits the calling behavior.

If the CC trigger "calling" by releasing a neurosecretory hormone, then this hormone should be found in the blood of a "calling" moth. To test this hypothesis, virgin Polyphemus females were exposed to *trans*-2-hexenal vapors in the dark. When they began "calling", their blood was collected (as described by Riddiford and Ashenhurst 1973), mixed with a few crystals of phenylthiourea and streptomycin, and injected into virgin Polyphemus females which had not been exposed to oak leaves or *trans*-2-hexenal. The recipient females were placed in the dark and checked at half-hour intervals for "calling". As seen in Table 1, 80% of the females injected with "calling" blood assumed the calling posture within 1.5 to 2.5 hr after injection, whereas none of the 10 females injected with virgin "noncalling" blood had released pheromone by 4 hr. Thus, the "calling" hormone can be recovered from the blood and presumably acts on the female nervous system to trigger the typical calling posture.

Table 1. Induction of pheromone release by *Antheraea polyphemus* virgin females by injection of blood from "calling" Polyphemus females.

Experimental situation	Number Tested	Number "Calling"*
Oak leaves or 0.05% *trans*-2-hexenal	40	31
No oak leaves or 0.05% *trans*-2-hexenal	14	0
Injected with blood from noncalling females	10	0
Injected with blood from calling females	10	8

* During scotophase or within 4 hr after injection.

BEHAVIOR OF THE MATED FEMALE

Once the female wild silkmoth mates, her behavior changes drastically. She no longer "calls" (Falls 1933; Riddiford and Williams 1971; Truman and Riddiford 1971), but she becomes more active (Truman, Lounibos and Riddiford, in preparation; Sweadner 1937), and lays eggs at a greatly accelerated rate (Truman and Riddiford 1971; N.M. Nijhout, unpublished studies in this laboratory). Of these behaviors, oviposition is most easily quantified and has served as an indicator of the switchover from virgin to mated behavior. As indicated in Fig. 3A, after mating a Cecropia female lays about 35% of her remaining eggs each night as compared with a virgin female of the same age which lays about 7% per night (Truman and Riddiford 1971).

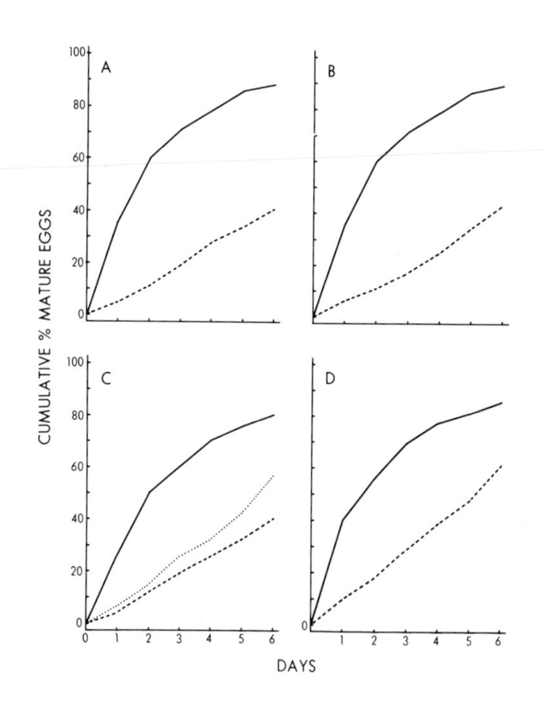

Fig. 3. The oviposition patterns of Cecropia females. The lines represent the average cumulative percent of eggs laid as a function of time after various treatments. Eggs were counted daily.

A. Mated (——) and virgin (---) females.

B. Females mated to normal males (——) and females mated to castrate males (---).

C. Virgin females each of which has received an implant of a bursa copulatrix from a mated female (——) or from a virgin female (---) or of a spermatheca from a mated female (...).

D. Mated allatectomized females (——) and mated cardiacectomized-allatectomized females (---).

(The data for A, B and D are from Truman and Riddiford 1971, and for C from Riddiford and Ashenhurst 1973)

In some insects, such as the grasshopper *Gomphocerus rufus* (Loher 1966) and the cockroach *Nauphoeta cinerea* (Roth 1962, 1964), the mechanical presence of the spermatophore in the bursa copulatrix provides the signal for the switchover from virgin to mated behavior. In other insects such as the mosquito *Aedes aegypti* (Gwadz, Craig and Hickey 1971; Gwadz 1972) and the housefly *Musca domestica* (Leopold, Terranova and Swilley 1971), the male accessory gland secretions enter the blood of the female from the bursa and act on the nervous system to effect the change in behavior. In the blood-sucking bug *Rhodnius prolixus* (Davey 1965) and in many lepidopterans (Norris 1933; Benz 1969, 1970) sperm seem to be necessary for the establishment of mated behavior. Likewise, as seen in Fig. 3B, the switch to mated behavior in the Cecropia moth is triggered by the reception of sperm; females mated to castrated males fail to show any change in behavior (Truman and Riddiford 1971).

In *Rhodnius*, a secretion from the sperm-filled spermatheca promotes egg-laying by activating the median neurosecretory cells of the brain to release a substance which stimulates the contraction of the oviducts (Davey 1965, 1967). In Cecropia, the story is slightly different. In this moth we have shown that the sperm apparently causes the bursa copulatrix to release a humoral factor which then elicits oviposition (Riddiford and Ashenhurst 1973). When implanted into a virgin female, the bursa from a mated female caused a switchover to the mated oviposition rate. A bursa from a virgin female did not have this effect (see Fig. 3C). In addition to increased oviposition, the mated bursa implant also caused the cessation of calling behavior. As indicated in Fig. 3C, this switchover to mated behavior was not caused by implantation of a mated spermatheca, irrespective of the presence or absence of sperm in the implant. Thus, prior to their migration to the spermatheca, the sperm apparently trigger secretory activity by the bursa. The bursa then continues to release its factor long after the sperm have left and can do so even when experimentally emptied of the spermatophore.

The bursa factor effects the change from virgin to mated behavior, but this change does not occur when the mated female has had her CC-CA complex removed (Fig. 3D; Truman and Riddiford 1971). Just as with calling behavior, the intrinsic neurosecretory cells of the CC are apparently involved. Also, connections between the brain and the CC are necessary for mated behavior to arise. The absence of mated oviposition behavior in a mated female whose CC-CA complex is detached from the brain suggests that the bursa factor does not act directly on the CC but rather through the central nervous system (CNS). This factor could modify the brain in such a way that when the incoming environmental signals are integrated, the nervous output to the CC stimulates those cells which release the oviposition-stimulating hormone rather than those cells which release the calling hormone. Alternatively, the bursa hormone could modify the CNS (presumably the abdominal nerve cord) in such a way that the same neurosecretory hormone from the CC elicits a different motor response. The second possibility seems more attractive. It keeps to a minimum the number of hormones made by the CC and thus on the number of different types of cells.

The blood from mated females stimulated increased oviposition of virgin Cecropia females for at least two days (Riddiford and Ashenhurst, 1973). Although blood immediately after mating did not show this activity, blood removed at any time after the first oviposition period was active. This blood presumably contains both the bursa factor and the CC hormone although the former is likely to be the major component. The bursa factor is probably released continuously into the blood as suggested by the continued enhanced oviposition elicited by bursa implants compared to the transient increase seen after blood injections.

By contrast, the oviposition-stimulating hormone from the CC is probably released just before oviposition in response to photoperiod signals and may also account for the enhanced activity of the mated female (Sweadner 1937; Truman, Lounibos and Riddiford, in preparation). This CC hormone may act directly on the ovarioles to cause contraction as does a neurosecretory substance in *Rhodnius* (Davey 1967).

Although the virgin Polyphemus moth requires oak leaves to elicit pheromone release, the mated Polyphemus female lays her eggs in the characteristic mated pattern even in the absence of this stimulus (M.M. Nijhout, unpublished data in this laboratory). The presence of leaves or of *trans*-2-hexenal does not significantly alter this mated rate of oviposition (J. Ashenhurst, unpublished data in this laboratory). Yet in the field female Polyphemus, just as female Cecropia, (Waldbauer and Sternburg 1967; Waldbauer, personal communication; Marsh 1941) lay eggs on the larval food plant. Thus, the hormonal milieu induced by mating prolongs the flight activity and stimulates oviposition. The odor of the larval food plant then serves to orient this oviposition activity. But the stimulus from the food plant is not essential for oviposition to occur. This is in contrast to the dependence on the food plant for oviposition by many feeding moths (Benz 1969, 1970; Yamamoto and Fraenkel 1960; Yamamoto, Jenkins and McClusky 1969).

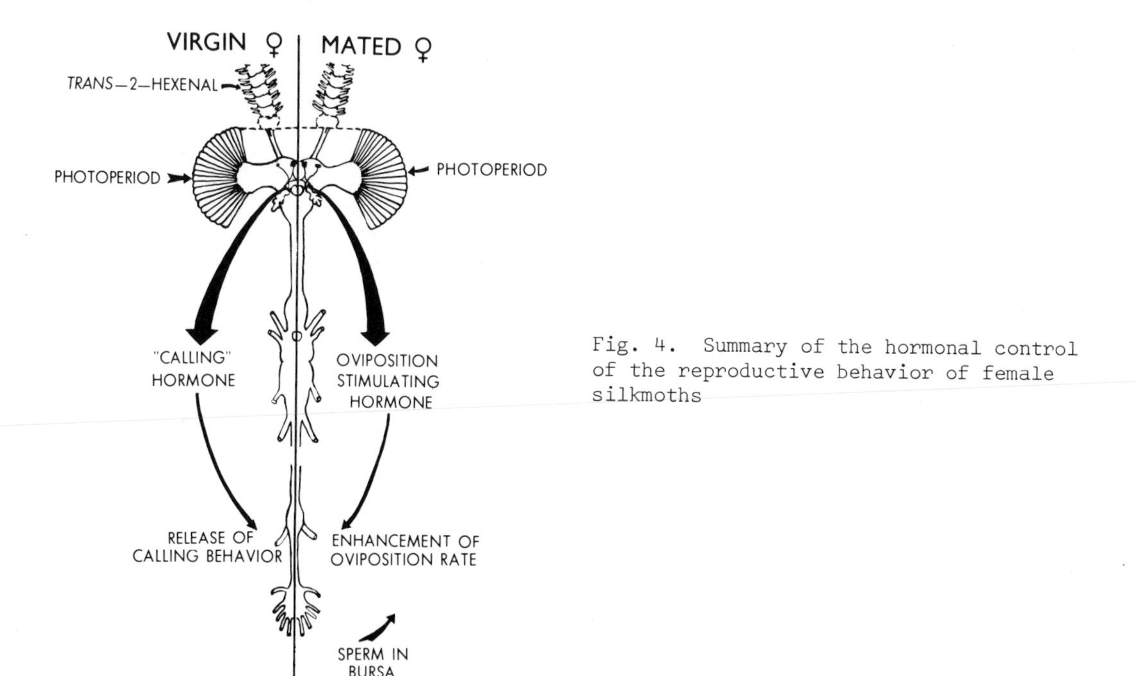

Fig. 4. Summary of the hormonal control of the reproductive behavior of female silkmoths

SUMMARY

As summarized in Fig. 4, a virgin female wild silkmoth responds to certain environmental cues by "calling" at a certain time of day and/or in a certain location that is propitious to mating. The environmental control is exercised through hormones from the intrinsic cells of the CC which regulate the type of behavior displayed. In a virgin female, the calling hormone is released and activates the calling posture, presumably via an action on the CNS. Once the female is mated, the transient presence of sperm in the bursa copulatrix apparently triggers secretion of a humoral factor from the bursa which causes the switch to "mated" behavior. This bursa factor probably acts via the CNS. Then, at a particular time in the photoperiodic regime, the intrinsic cells of the CC of the mated female release a neurosecretory hormone which both greatly increases the rate of oviposition and enhances flight activity.

ACKNOWLEDGEMENTS

I thank Dr. James Truman for a critical reading of the manuscript. The unpublished studies cited were supported by NSF grants GB-7966 and GB-24963 and a grant from the Rockefeller Foundation.

REFERENCES

BENZ, G.: Influence of mating, insemination, and other factors on oogenesis and oviposition in the moth *Zeiraphera diniana*. J. Insect Physiol. 15, 55-71 (1969).

BENZ, G.: The influence of the presence of individuals of the opposite sex, and some other stimuli on sexual activity, oogenesis, and oviposition in five lepidopterous species. *In* "L'influence des Stimuli Externes sur la Gamétogenèse des Insectes". Colloques int. Cent. natn. Rech. scient. 189, 174-206 (1970).
DAVEY, K.G.: Copulation and egg production in *Rhodnius prolixus*: the role of the spermathecae. J. exp. Biol. 42, 373-378 (1965).
DAVEY, K.G.: Some consequences of copulation in *Rhodnius prolixus*. J. Insect Physiol. 13, 1629-1636 (1967).
FALLS, O.: Sex attraction in *Samia cecropia*. Trans. Kans Acad. Sci. 36, 215-217 (1933).
GWADZ, R.W.: Neuro-hormonal regulation of sexual receptivity in female *Aedes aegypti*. J. Insect Physiol. 18, 259-266 (1972).
GWADZ, R.W., CRAIG, G.B.,Jr., HICKEY, W.A.: Female sexual behavior as the mechanism rendering *Aedes aegypti* refractory to insemination. Biol. Bull. mar. biol. Lab., Woods Hole 140, 201-214 (1971).
LEOPOLD, R.A., TERRANOVA, A.C., SWILLEY, E.M.: Mating refusal in *Musca domestica*: effects of repeated mating and decerebration upon frequency and duration of copulation. J. exp. Zool. 176, 353-360 (1971).
LOHER, W.: Die Steurung sexueller Verhaltensweisen und der Oocytenentwicklung bei *Gomphocerus rufus* L. Z. vergl. Physiol. 53, 277-316 (1966).
MARSH, F.L.: A few life history details of *Samia cecropia* within the southwestern limits of Chicago. Ecology 22, 331-337 (1941).
NORRIS, M.J.: Contributions towards the study of insect fertility. II. Experiments on the factors influencing fertility in *Ephestia kühniella* Z. (Lepidoptera, Phycitidae). Proc. zool. Soc. Lond. 4, 903-934 (1933).
PAYNE, T.L., SHOREY, H.H., GASTON, L.K.: Sex pheromones of noctuid moths: factors influencing antennal responsiveness in males of *Trichoplusia ni*. J. Insect Physiol. 16, 1043-1055 (1970).
PRIESNER, E.: Die interspezifischen Wirkungen der Sexuallockstoffe der Saturniidae (Lepidoptera). Z. vergl. Physiol. 61, 263-297 (1968).
RAHN, R.: Rôle de la plante-hôte sur l'attractivité sexuelle chez *Acrolepia assectella* Zeller (Lep. Plutellidae). C. r. hebd. Séanc. Acad. Sci., Paris 266, 2004-2006 (1968).
RAU, P., RAU, N.: The sex attraction and rhythmic periodicity in giant saturniid moths. Trans. Acad. Sci. St Louis 26, 83-221 (1929).
RIDDIFORD, L.M.: *Trans*-2-hexenal: mating stimulant for Polyphemus moths. Science, N.Y. 158, 139-141 (1967).
RIDDIFORD, L.M.: Antennal proteins of saturniid moths - their possible role in olfaction. J. Insect Physiol. 16, 653-660 (1970).
RIDDIFORD, L.M., ASHENHURST, J.B.: The switchover from virgin to mated behavior in female Cecropia moths: the role of the bursa copulatrix. Biol. Bull. mar. biol. Lab., Woods Hole 144, 162-171 (1973).
RIDDIFORD, L.M., WILLIAMS, C.M.: Volatile principle from oak leaves: role in sex life of the Polyphemus moth. Science, N.Y. 155, 589-590 (1967).
RIDDIFORD, L.M., WILLIAMS, C.M.: Role of the corpora cardiaca in the behavior of saturniid moths. I. Release of sex pheromone. Biol. Bull. mar. biol. Lab., Woods Hole 140, 1-7 (1971).
ROTH, L.M.: Hypersexual activity induced in females of the cockroach *Nauphoeta cinerea*. Science, N.Y. 138, 1267-1269 (1962).
ROTH, L.M.: Control of reproduction in female cockroaches with special reference to *Nauphoeta cinerea*. I. First preoviposition period. J. Insect Physiol. 10, 915-945 (1964).
SCHNEIDER, D.: Elektrophysiologische Untersuchungen von Chem- und Mechanorezeptoren der Antennae des Seidenspinners *Bombyx mori* L. Z. vergl. Physiol. 40, 8-41 (1957).
STUMM-ZOLLINGER, E.: Histological study of regenerative processes after transection of the nervi corporis cardiaci in transplanted brains

of the Cecropia silkworm (*Platysamia cecropia* L.). J. exp. Zool. 134, 315-326 (1957).

SWEADNER, W.R.: Hybridization and the phylogeny of the genus *Platysamia*. Ann. Carneg. Mus. 25, 163-242 (1937).

TELFER, W.H., RUTBERG, L.D.: The effects of blood protein depletion on the growth of the oocytes in the Cecropia moth. Biol. Bull. mar. biol. Lab., Woods Hole 119, 352-366 (1960).

TRUMAN, J.W., LOUNIBOS, L.P., RIDDIFORD, L.M.: Patterns of flight activity and pheromone release in saturniid moths. (In prep.)

TRUMAN, J.W., RIDDIFORD, L.M.: Role of the corpora cardiaca in the behavior of saturniid moths. II. Oviposition. Biol. Bull. mar. biol. Lab., Woods Hole 140, 8-14 (1971).

WALDBAUER, G.P., STERNBURG, J.G.: Host plants and the locations of the baggy and compact cocoons of *Hyalophora cecropia* (Lepidoptera: Saturniidae). Ann. ent. Soc. Am. 60, 97-101 (1967).

WILSON, E.O., BOSSERT, W.H.: Chemical communication among animals. Recent Prog. Horm. Res. 19, 673-716 (1963).

YAMAMOTO, R.T., FRAENKEL, G.: The physiological basis for the selection of plants for egg-laying in the tobacco hornworm, *Protoparce sexta* (Johan.). XIth Int. Congr. Ent. (Vienna, 1960) 3, 127-133 (1960).

YAMAMOTO, R.T., JENKINS, R.Y., McCLUSKY, R.K.: Factors determining the selection of plants for oviposition by the tobacco hornworm *Manduca sexta*. *In* "Insect and Host Plant" (Eds., J. de Wilde and L.M. Schoonhoven), pp. 504-508. Amsterdam: North-Holland Publishing Co. (1969).

HORMONES AND INSECT BEHAVIOR

L.M. Riddiford and J.W. Truman

The Biological Laboratories, Harvard University, Cambridge, Mass., U.S.A.
Present address: *Department of Zoology, University of Washington, Seattle, Washington, U.S.A.*

In insects, hormones are involved in the regulation of behavior to a greater extent than in other invertebrate or vertebrate groups. This extensive adoption of hormonal involvement in the function of the nervous system may be partially due to size restrictions and to the rich behavioral repertoires which insects characteristically show. Hormones, with their ability to reach every cell in the nervous system, are capable of causing widespread changes in responsiveness and thus of radically changing behavior. The use of this kind of hormonal control instead of more complex neural mechanisms may be a major factor in packing a broad array of behaviors into a small, simplified nervous system.

This paper is not meant as a comprehensive review since that is appearing elsewhere (Truman and Riddiford 1973), but rather it focuses on a few aspects of the action of the endocrine system on the insect nervous system. Here we have concentrated primarily on hormonal influences on the behavior of the giant silkmoths.

ROLE OF HORMONES IN BEHAVIOR

Wilson and Bossert (1963) classified pheromones as either releasers or primers. A releaser pheromone, such as the alarm pheromone of ants, causes an immediate behavioral response in the receiver. Primer pheromones are more subtle in their action and typically show a delayed effect on behavior. Primers work through the endocrine system of the receiver to alter its physiological state. A classic example of the latter is the suppression of ovarian development of worker bees by the queen substance. The effects of hormones on behavior are analogous to those outlined for pheromones (Truman and Riddiford 1973). Thus, we have separated these effects into two broad categories - releasers and modifiers (Fig. 1).

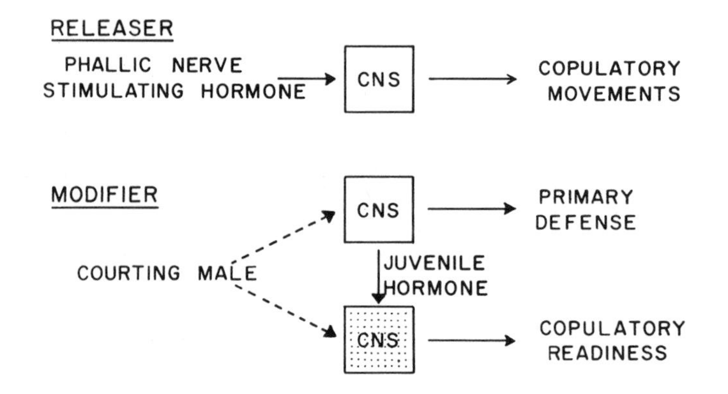

Fig. 1. Schematic representation of the releaser and the modifier effects of hormones on insect behavior. The examples illustrate the effect of the phallic nerve-stimulating hormone on the behavior of male *Periplaneta* and the action of juvenile hormone (JH) on female *Gomphocerus*

A hormone may directly trigger a specific piece of behavior and thus act as a releaser. This effect on behavior is relatively rapid - usually arising within a matter of minutes - and does not require additional environmental stimuli. Thus far, hormones of the releaser type are neurosecretory in nature.

A modifier hormone is analogous to a primer pheromone in that it does not elicit an immediate behavioral reaction. Rather, it alters the behavioral responsiveness of the animal such that a given environmental signal evokes a new behavior. Modifiers thus regulate behavior so that the animal's responses are appropriate for its internal and external environment.

The releaser effects of hormones

The first demonstration that complex pieces of behavior could be directly released by hormones is seen in the study of the phallic nerve stimulating hormone of cockroaches (Milburn, Weiant and Roeder 1960; Milburn and Roeder 1962). When injected into male cockroaches, extracts of the corpora cardiaca (CC) caused rhythmic twisting movements of the abdomen which resembled movements observed during copulation. Similarly, when applied to the nerve cord of a semi-dissected individual, these extracts promoted rhythmic bursting in the phallic nerve (Fig. 2).

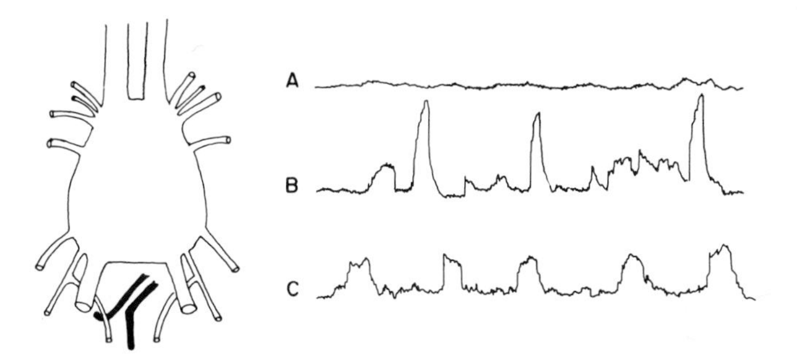

Fig. 2. Effect of the phallic nerve-stimulating hormone on the motor activity in the phallic nerve. A - before treatment; B - 10 min after adding CC extract; C - 35 min after. Records are representations of integrated neural activity given in Milburn and Roeder (1962). Diagram on the left shows the placement of the recording and the indifferent electrodes

The motor responses to the phallic nerve stimulating hormone are identical to those induced in male cockroaches by decapitation (Roeder, Tozian and Weiant 1960). The latter maneuver removes the subesophageal centers which normally inhibit the motor program for copulatory movements; therefore, this behavior becomes expressed. Similarly, the phallic nerve stimulating hormone apparently acts on the subesophageal ganglion to remove this inhibition of the abdominal motor centers.

The reason for the existence of releaser hormones is a bit puzzling. As is shown below, these hormones are released in response to well-defined environmental stimuli and act on discrete portions of the nervous system to trigger a specific motor output. Thus, they perform the same function as would neural pathways from the integrative centers of the brain to the motor centers elsewhere in the nervous system. Moreover, this hormonal transmission is not as efficient as neural transmission since minutes are required for hormonal messengers, as compared to the milliseconds needed for classical neural mechanisms. One benefit of hormonal links may lie in the maintenance of sustained pieces of behavior. Thus, the persistence of the "calling hormone" in the blood of female moths (see below) may be well suited for the tonic maintenance of the "calling posture" which is often held for a number of hours. Also, a hormone bathes all parts of the nervous system and thus may be more efficient in eliciting a coordinated behavioral pattern than is a complex neural network. But, at this time, theories on the relative benefit of releaser hormones can be only speculative.

The modifier effects of hormones

The more commonly reported action of hormones on behavior is that of changing behavioral states. In insects, the most familiar example is the action of juvenile hormone (JH) in causing the maturation of sexual behavior in the adult. Engelmann (1960) first demonstrated the necessity of the corpora allata (CA) and thus of JH for the onset of sexual receptivity in females of the cockroach *Leucophaea madrae*. When females were allatectomized shortly after ecdysis, 70% never displayed sexual receptivity. Normal behavior was restored by implantation of loose CA.

Similarly, in the grasshopper *Gomphocerus rufus*, the CA are necessary not only for the development of female receptivity, but also for its maintenance (Loher 1962; Loher and Huber 1964, 1966). The young immature female responds to the courting behavior of the male by kicking him away and by escape reactions - "primary defense" behavior. Under the influence of the CA, the maturation of sexual behavior then occurs so that she displays "copulatory readiness" six days after emergence. In response to the same stimulus of a courting male, the female now stridulates and moves towards him. Mounting and copulation then ensue.

When female *Gomphocerus* were allatectomized shortly after ecdysis, there was no switchover from primary defense to copulatory readiness. Similarly, after excision of these glands, sexually mature females reverted to defensive behavior. Thus, the continued presence of JH is required for the maintenance of the receptive condition. Indeed, the relationship between JH and receptive behavior in female grasshoppers is analogous to that between the sex steroids and estrous behavior in female rats and mice.

Modifier hormones are used in various contexts. Primer pheromones may change behavior by causing the secretion of a modifier hormone. The action of the maturation pheromone which speeds sexual maturation in males of *Schistocerca* is mediated through the release of JH (Loher 1961). Long-term effects of changing environmental stimuli are also often accomplished through modifier hormones. For example, in *Oncopeltus*, the onset of migratory behavior which occurs in response to decreasing day lengths is caused by increased titers of JH (Rankin, this volume). Finally, modifier hormones also serve to coordinate behavior with developmental or physiological changes within the insect. The dual action of JH in promoting egg maturation and the onset of receptive behavior in some female insects is a prime example. Also, in larval locusts ecdysone acts on the nervous system to suppress locomotor activity during the vulnerable period of the ecdysone-induced molt (Haskell and Moorhouse 1963). Indeed, in these cases the nervous system has evidently captured the use of pre-existing hormones so as to establish a close coupling between physiology and behavior.

THE INFLUENCE OF HORMONES IN THE BEHAVIOR OF WILD SILKMOTHS

The metamorphosis of the caterpillar into the moth is accompanied by an extensive reorganization of the nervous system. This newly formed nervous system of the adult contains a number of latent behavioral programs which at first cannot be utilized. Then, in response to the changing physiological states of the moth, these programs become successively revealed. Concurrently, others become shut-off. The switching on and off of these behaviors is accomplished through the action of a number of insect hormones.

The activation of adult behavior

The time of adult emergence signals the most dramatic change in behavior which has been recorded in insects. The developmental processes of metamorphosis fashion the adult nervous system out of that of the pupa in approximately three weeks. But, at the end of this period, the pharate moth, which is encased in the pupal skin, still shows only pupal movements. This lack of adult behavior is not due to confinement by the pupal cuticle. When this cuticle is removed before the normal time of eclosion, the behavior of the moth remains like that of a pupa (Blest 1960; Truman 1971).

Complex adult motor patterns, such as flight or locomotion, cannot be elicited in a peeled, pharate *Antheraea pernyi* moth. In the normal adult male, exposure to the female sex pheromone under dim illumination causes excitation and intense flight activity. But the peeled pharate male, exposed to the same stimulus, shows no response. This deficiency is not at the sensory level since a normal electroantennogram response (Schneider 1957) is elicited from the antennae of the pharate male by the sex pheromone (Riddiford, unpublished). Thus, the inhibition of adult behavior is most probably a central phenomenon.

The lack of adult behavior extends beyond complex motor acts. The posture of the peeled moth is abnormal and shows a distinct lack of tonus. Even simple reflexes such as the righting reflex, seen when the animal is placed on its back, are absent. The behavioral repertoire of the peeled moth includes occasional spasmodic movements of the legs, various abdominal twitches, and the pupal-like rotary movements of the abdomen.

At the time of day when eclosion normally occurs, the peeled moth begins to show changes in its behavior. The first sign of this transition is the appearance of the pre-eclosion behavior - a 1.25 hr long program of abdominal movements which normally precedes eclosion (Truman 1971). This is followed rapidly by eclosion and by the spreading of the wings. Within a span of a few hours the full adult behavioral repertoire has been "turned on".

This dramatic assumption of adult behavior is hormonally controlled (Fig. 3). Evidence of this control was first obtained from experiments designed to locate the biological clock which was responsible for the daily "gating" of eclosion (Truman and Riddiford 1970). Removal of the brains from the pupae resulted in moths which emerged randomly throughout the day and night and which showed aberrant pre-eclosion behaviors. When a brain was implanted into the abdomens of similarly debrained pupae, the resulting moths emerged at their characteristic time of day and followed the proper sequence of behaviors.

An "eclosion hormone" was found in the brain and CC of pharate moths. When homogenates of these organs were injected into pharate animals,

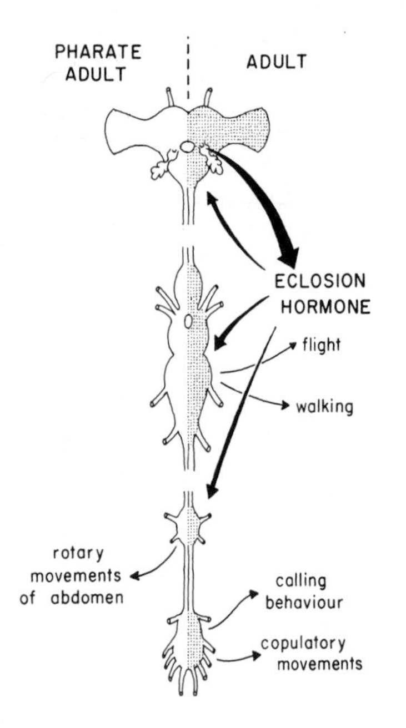

Fig. 3. The modifier effect of the eclosion hormone. Diagram shows some of the behavior which can be elicited before and after exposure to the eclosion hormone

normal eclosion and the precocious assumption of adult behavior followed (Truman 1971). Eclosion hormone activity is essentially absent from the brain and CC of larvae or pupae, but it appears in these structures during the latter two-thirds of adult development (Truman 1973). At the time of eclosion it appears in the blood. Thus, this hormone is made only during adult development for the purpose of "turning on" the adult nervous system.

Besides "turning on" some parts of the nervous system, the eclosion hormone also "turns off" others. The abdomen of the newly emerged moth is supplied with dense bands of muscle which run longitudinally along the dorsal and lateral sides of the abdomen. These intersegmental muscles are used during eclosion and then rapidly degenerate (Finlayson 1956; Lockshin and Williams 1965a). The signal for these muscles to degenerate is apparently given by the motor neurons which supply them because shortly after eclosion these neurons become silent. Moreover, if one forces these nerves to remain active by either electrical or pharmacological means, the corresponding muscles are preserved (Lockshin and Williams 1965b,c). Consequently, the absence or presence of these muscles in the moth 3 to 4 days after eclosion can be used as an indicator of whether or not the motor neurons had "shut off".

Lockshin (1969) later reported that when the abdomen of a pharate *Antheraea polyphemus* moth was isolated prior to eclosion, the muscles were routinely retained. Apparently, in the absence of an additional influence from the anterior end, the motor neurons remain active. This signal is supplied by the eclosion hormone (Truman 1970). When injected into isolated abdomens from pharate Polyphemus moths, the hormone triggers the muscle degeneration on schedule. Thus, one specific aspect of eclosion hormone action appears to be the "turning off" of the motor neurons which supply the intersegmental muscles.

The main function of the eclosion hormone is to ensure a smooth coordinated activation of adult behavior. In a debrained moth, which is exposed to the eclosion hormone, some of the neural centers nevertheless eventually turn on. But other behaviors are seldom or never seen, and there is no coordination between the activation of the various parts.

The eclosion hormone also has a releaser action in that it directly triggers the pre-eclosion behavior and eclosion. In *Hyalophora cecropia*, the pre-eclosion behavior has three distinct parts which differ as to the frequency and type of abdominal movements. The first half hour consists of fairly rapid rotatory movements of the abdomen followed by a half hour of relative quiescence with an occasional rotatory movement. The final 15 min consists of rapid, anteriorly-directed peristaltic movements of the abdomen which assist the moth in emerging. This distinctive motor pattern is the first overt sign of eclosion hormone action and begins 10 to 15 min after injection. The neural circuitry necessary for this response to the eclosion hormone is restricted to the chain of 4 abdominal ganglia (Truman 1971). Manifestly, injection of the hormone into isolated pharate abdomens typically results in the full and complete pre-eclosion behavior followed by eclosion.

The relative importance of sensory input and of prepatterned information for the performance of the pre-eclosion behavior was examined by electrophysiological recordings from semidissected isolated abdomens (Truman and Sokolove 1972). Cecropia abdomens were opened ventrally to expose the chain of abdominal ganglia and all nerves to the periphery were severed. Efferent activity was then recorded from the dorsal nerves of the second and third abdominal ganglia. These nerves carry the axons which supply the intersegmental muscles - the main muscle groups which are responsible for the movements seen during the pre-eclosion behavior.

In response to the addition of the eclosion hormone, these semi-isolated nerve cords generated the complete pre-eclosion behavior. As seen in Fig. 4, the temporal arrangement of efferent bursts recorded from the dorsal nerve mimicked the temporal sequence of movements seen in the intact animal. An initial period of frequent bursting was followed by a quiescent period which ended abruptly with a second bout of intense bursting. The patterning of the motor bursts during these three phases was that which could generate the movements typically seen in the intact insect. During the first hyperactive period and the quiet phase, the bursts typically showed a rotational patterning. Each burst was composed of several volleys which occurred synchronously in the ipsilateral roots of the second and third ganglia. These efferent volleys, however, showed a clear right-left alternation (Fig. 4). This pattern thus generates the rotational movements of the abdomen which are observed during the first two phases of the pre-eclosion behavior. With the abrupt onset of the third phase, a new burst pattern was seen. It consisted of one major volley which occurred synchronously in both roots of a given ganglion. These bilateral bursts began in the more posterior ganglion and then proceeded anteriorly (Fig. 4). This pattern of motor activity generates a wave of contraction moving in an anterior direction along the abdomen. Thus, the pre-eclosion behavior is due to a prepatterned motor program which is built into the abdominal ganglia. This program consists of the motor information for the rotational and peristaltic movements and also of information as to the relative frequency of bursting and the timing of the switchover from one burst type to the next. The entire program is then activated and read off in response to the hormonal command from the brain.

In causing the pre-eclosion behavior, the hormone probably serves a triggering function and need not be present during the complete display of the behavior. This conclusion arises from the fact that a small number of debrained moths (16%) nevertheless performed the pre-eclosion behavior (Truman 1971). Thus, in the absence of the hormone, the circuitry which generates the pre-eclosion behavior can occasionally become "activated" and the proper behavior then ensues.

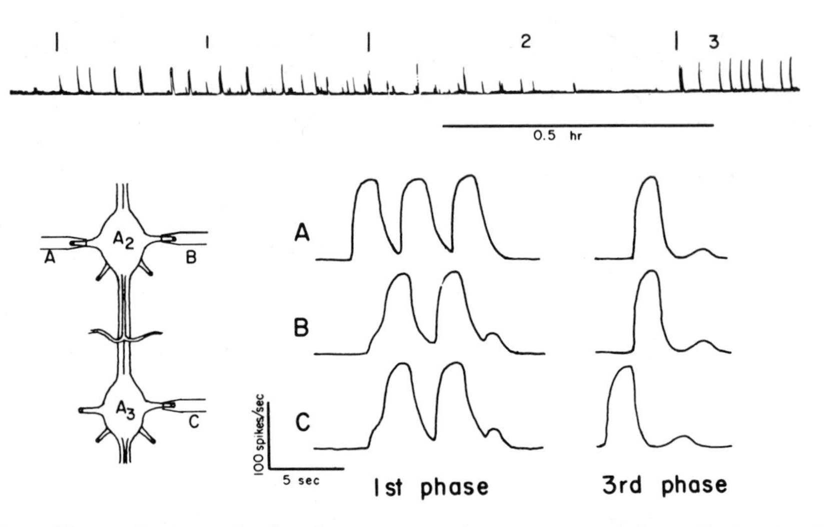

Fig. 4. The effect of the eclosion hormone on the motor activity of the deafferented abdominal nerve cord of pharate Cecropia moths.
(Top) Integrated efferent activity from the right dorsal nerve of ganglion A_2. Hormone was added to the preparation approximately 40 min before the onset of the first burst. (1) The first hyperactive period, (2) the quiet period, and (3) the second hyperactive period.
(Bottom) The "fine-structure" of the bursts which were typically recorded during the first and the last phases of the response to the eclosion hormone. Records are representations of the integrated neural activity which was recorded from the deafferented abdominal nerve cord. The letters refer to the electrode placements shown on the diagram to the left (Truman and Riddiford 1973)

The hormonal influences in the behavior of adult silkmoths

In the post-eclosion behavior of the adult moth, hormones continue to play a prominent role, but only in the case of the female. Some male insects, such as *Schistocerca gregaria*, require JH for the proper onset of male behavior (Odhiambo 1966; Pener 1967), but the CA have no role in the saturniids (Riddiford, unpublished). Moreover, the CC of the female moth are required for proper regulation of reproductive behavior (Riddiford and Williams 1971; Truman and Riddiford 1971), but these glands are not involved in the mating response of the male, at least in the case of Pernyi moths (Riddiford, unpublished).

The lack of hormonal control in male behavior is not entirely surprising. The male silkmoth displays full sexual behavior immediately after emergence, and this behavior remains unaltered throughout the remainder of his life. Thus, there is no opportunity for a hormonally induced switch from one behavioral state to another.

The reproductive behavior of the female moth is more complex; but unlike many female insects (Engelmann 1970; Truman and Riddiford 1973), she does not rely on the CA for the control of her behavior. Instead, her CC play a prominent role.

At certain times of day, or in response to other environmental cues, virgin female silkmoths assume a characteristic "calling" posture. The moth suspends herself from a support and protrudes the terminal segments of the abdomen, thus exposing the pheromone glands. The assumption of this posture is due to the daily release of a "calling hormone" from the intrinsic neurosecretory cells of the CC (Riddiford and Williams 1971; Riddiford, this volume). The importance of the CC to this behavior was demonstrated by removing the glands from pupae and then observing the behavior of the resulting moths. This operation severely impaired the ability of the virgin females to call. Evidence for a hormonal control was further strengthened by showing that blood from calling females could induce this behavior when it was injected into non-calling recipients. Blood from non-calling females had no effect.

The calling hormone appears to have only a releaser action, but its effects differ somewhat from the releaser effects of the eclosion hormone. In response to the latter hormone, the insect behaves as an automaton showing little or no response to changes in environmental conditions until after eclosion is completed. By contrast, while under the influence of the calling hormone, the female remains responsive to other stimuli. Thus, disturbances or changes in the level of illumination can cause an abrupt termination of calling even though the hormone is in the blood.

Mating causes a change in behavior of the female moth. The virgin female "calls", displays a low level of locomotor activity, and, as she ages, lays eggs at a low rate. After mating, calling ceases and locomotion and oviposition are enhanced. These changes can be attributed to the action of two hormones - one from the CC and one from the copulatory bursa.

The stimulus for the switch from virgin to mated behavior is supplied by the sperm which are deposited by the male (Truman and Riddiford 1971). Matings which involve the transfer of a spermatophore and accessory gland secretions but not sperm are completely ineffective in causing the change in the female's behavior. The sperm apparently induce the bursa copulatrix to secrete a "bursa factor" into the blood (Riddiford and Ashenhurst 1973). Once the secretion is induced, it continues in the absence of sperm, as was demonstrated by implanting an emptied bursa from a mated female into a virgin female. The recipient laid eggs in a mated fashion. Neither the bursa from a virgin female nor the spermatheca from a mated female could cause this behavior when implanted into a virgin host.

But there is a second endocrine step which is necessary for the assumption of mated behavior. Female Cecropia, which lack their CC, show normal virgin behavior except that they do not call. When these females are subsequently mated, their behavior remains typically virgin - the female continues to display a low rate of oviposition. Thus the CC appear to be involved in the enhancement of oviposition rate which is a result of insemination (Truman and Riddiford 1971). As with calling, this involvement of the CC is most certainly hormonal.

The mode of action of the two hormones and the relationship between the bursa hormone and the oviposition-stimulating hormone from the CC are unclear at this time. The bursa factor may act exclusively on the brain centers which regulate the function of the CC, thereby turning off calling hormone secretion and turning on the release of the oviposition hormone. The other extreme interpretation is that the bursa factor is acting on the rest of the nervous system to alter its response to a single hormone from the CC. Then, depending on the state of the nervous system, the same hormone would promote either calling or oviposition. The relative role of the hormones probably would be somewhere between these two extremes.

Therefore, as summarized in Fig. 5, in the silkmoth the adult behavioral patterns are first turned on through the action of the eclosion hormone. The female then has an additional switch within her adult behavioral repertoire - that of going from the virgin to the mated state. This latter switch is apparently accomplished through the action of the bursa hormone. Incorporated in these various behavioral states, one also finds hormones which release specific pieces of behavior: the eclosion

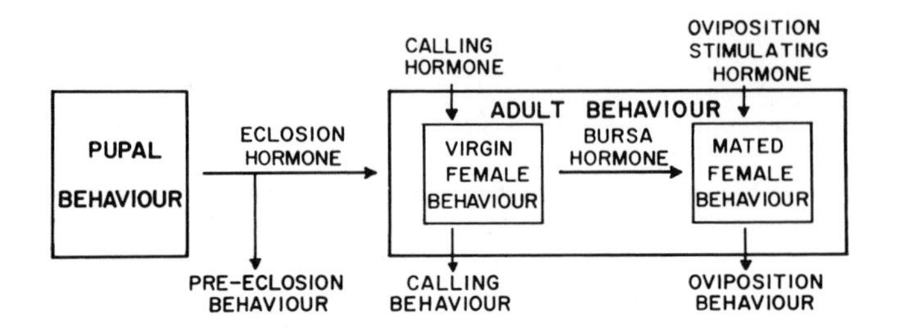

Fig. 5. Schematic representation of the hormonal influences which are involved in the behavior of an adult female silkmoth. Modifier effects occur along the horizontal axis; releaser effects along the vertical axis

hormone triggers the pre-eclosion behavior, the calling hormone releases calling, and the oviposition-stimulating hormone causes egg laying. The neuroendocrine and endocrine systems therefore have a central role in the regulation of the behavioral responses of these moths.

EPILOGUE

In this paper we have attempted to show that hormones play an important role in the control of insect behavior and, therefore, that their action is intimately involved in the function of the insect nervous system. The behavioral results of hormone action seem to be of two broad types - either a specific piece of behavior is triggered, as in the case of the releaser effects, or the responsiveness of the animal is modified such that new patterns of behavior can be elicited. The mechanistic significance of this division of effects has yet to be explored. Do modifier and releaser effects become manifest through basically different neural mechanisms? Do releasers act as long-distance transmitters by causing the depolarization of specific groups of interneurons which then generate the appropriate behaviors? How do modifiers change the responsiveness of neural circuits? Do they induce the formation of new synapses or change levels of transmitter synthesis or alter resting potential levels of postsynaptic neurons? Many questions as to hormone-neuron interactions are posed. Their answers should lead us to a better understanding of how behavior is controlled and of how the nervous system functions.

REFERENCES

BLEST, A.D.: The evolution, ontogeny, and quantitative control of settling movements of some New World saturniid moths, with some comments on distance communication by honey bees. Behaviour 16, 188-253 (1960).

ENGELMANN, F.: Hormonal control of mating behaviour in an insect. Experientia 16, 69-70 (1960).

ENGELMANN, F.: "The Physiology of Insect Reproduction", 307 pp. New York: Pergamon Press (1970).

FINLAYSON, L.H.: Normal and induced degeneration of abdominal muscles during metamorphosis in the Lepidoptera. Q. Jl microsc. Sci. 97, 215-234 (1956).

HASKELL, P.T., MOORHOUSE, J.E.: A blood-borne factor influencing the activity of the central nervous system of the desert locust. Nature, Lond. 197, 56-58 (1963).

LOCKSHIN, R.A.: Programmed cell death. Activation of lysis by a mechanism involving the synthesis of protein. J. Insect Physiol. 15,

1505-1516 (1969).
LOCKSHIN, R.A., WILLIAMS, C.M.: Programmed cell death. I. Cytology of degeneration in the intersegmental muscles of the Pernyi silkmoth. J. Insect Physiol. 11, 123-133 (1965a).
LOCKSHIN, R.A., WILLIAMS, C.M.: Programmed cell death. III. Neural control of the breakdown of the intersegmental muscles of the silkmoths. J. Insect Physiol. 11, 601-610 (1965b).
LOCKSHIN, R.A., WILLIAMS, C.M.: Programmed cell death. IV. The influence of drugs on the breakdown of the intersegmental muscles of silkmoths. J. Insect Physiol. 11, 803-809 (1965c).
LOHER, W.: The chemical acceleration of the maturation process and its hormonal control in the male of the desert locust. Proc. R. Soc. (B) 153, 380-397 (1961).
LOHER, W.: Die Kontrolle des Weibchengesanges von *Gomphocerus rufus* L. (Acridinae) durch die Corpora allata. Naturwissenschaften 49, 406 (1962).
LOHER, W., HUBER, F.: Experimentelle Untersuchungen am Sexualverhalten des Weibchens der Heuschrecke *Gomphocerus rufus* L. (Acridinae). J. Insect Physiol. 10, 13-36 (1964).
LOHER, W., HUBER, F.: Nervous and endocrine control of sexual behavior in a grasshopper (*Gomphocerus rufus* L., Acridinae). Symp. Soc. exp. Biol. 20, 381-400 (1966).
MILBURN, N.S., ROEDER, K.D.: Control of efferent activity in the cockroach terminal abdominal ganglion by extracts of the corpora cardiaca. Gen. comp. Endocr. 2, 70-76 (1962).
MILBURN, N., WEIANT, E.A., ROEDER, K.D.: The release of efferent nerve activity in the roach, *Periplaneta americana*, by extracts of the corpus cardiacum. Biol. Bull. mar. biol. Lab., Woods Hole 118, 111-119 (1960).
ODHIAMBO, T.R.: Growth and the hormonal control of sexual maturation in the male desert locust, *Schistocerca gregaria* (Forskål). Trans. R. ent. Soc. Lond. 118, 393-412 (1966).
PENER, M.P.: Effects of allatectomy and sectioning of the nerves of the corpora allata on oocyte growth, male sexual behaviour, and colour change in adults of *Schistocerca gregaria*. J. Insect Physiol. 13, 665-684 (1967).
RIDDIFORD, L.M., ASHENHURST, J.B.: The switchover from virgin to mated behavior in female Cecropia moths: the role of the bursa copulatrix. Biol. Bull. mar. biol. Lab., Woods Hole 144, 162-171 (1973).
RIDDIFORD, L.M., WILLIAMS, C.M.: Role of the corpora cardiaca in the behavior of saturniid moths. I. Release of sex pheromone. Biol. Bull. mar. biol. Lab., Woods Hole 140, 1-7 (1971)
ROEDER, K.D., TOZIAN, L., WEIANT, E.A.: Endogenous nerve activity and behaviour in the mantis and cockroach. J. Insect Physiol. 4, 45-62 (1960).
SCHNEIDER, D.: Electrophysiologische Untersuchungen von Chemo- und Mechanoreceptoren der Antenne der Seidenspinners, *Bombyx mori* L. Z. vergl. Physiol. 40, 8-41 (1957).
TRUMAN, J.W.: The eclosion hormone: its release by the brain and its action on the central nervous system of silkmoths. Am. Zoologist 10, 511-512 (1970).
TRUMAN, J.W.: Physiology of insect ecdysis. I. The eclosion behaviour of saturniid moths and its hormonal release. J. exp. Biol. 54, 805-814 (1971).
TRUMAN, J.W.: Physiology of insect ecdysis. II. The assay and occurrence of the eclosion hormone in the Chinese oak silkmoth, *Antheraea pernyi*. Biol. Bull. mar. biol. Lab., Woods Hole 144, 200-211 (1973).
TRUMAN, J.W., RIDDIFORD, L.M.: Neuroendocrine control of ecdysis in silkmoths. Science, N.Y. 167, 1624-1626 (1970).
TRUMAN, J.W., RIDDIFORD, L.M.: Role of the corpora cardiaca in the behavior of saturniid moths. II. Oviposition. Biol. Bull. mar.

biol. Lab., Woods Hole 140, 8-14 (1971).
TRUMAN, J.W., RIDDIFORD, L.M.: Hormonal mechanisms underlying insect behavior. Adv. Insect Physiol. 10, (1973).
TRUMAN, J.W., SOKOLOVE, P.G.: Silkmoth eclosion: hormonal triggering of a centrally programmed pattern of behavior. Science, N.Y. 175, 1491-1493 (1972).
WILSON, E.O., BOSSERT, W.H.: Chemical communication among animals. Recent Prog. Horm. Res. 19, 673-716 (1963).

MIGRATORY BEHAVIOUR OF THE FEMALE OF THE COMMON COCKCHAFER *MELOLONTHA MELOLONTHA* L. AND ITS NEUROENDOCRINE REGULATION

M.M.C. Stengel

Station de Zoologie, I.N.R.A., Centre de Recherches de Colmar, 68021 Colmar, France

During her life above ground, the female adult cockchafer passes through two or three ovarian cycles, each of which is characterized by orientated migrations which lead her towards feeding areas and thence back to egg laying sites.

Upon leaving the soil, the female cockchafer makes an orientated flight towards its place of feeding (the edge of a forest, a thicket or an isolated tree). This is the "pre-feeding flight". At this time the ovaries are undeveloped. Approximately thirteen days later, after mating and feeding, and when the eggs of the first ovarian cycle are fully developed, the female flies in the reverse direction and returns to lay eggs in the fields whence she came. This flight is called the "oviposition flight". Then, after having layed her eggs, she returns to the forest in the "post-oviposition flight".

Some females pass through as many as three ovarian cycles, that is to say, they make three complete migrations to the feeding site and back again to the oviposition site (Couturier and Robert 1958, 1962). The periodicity of these migratory cycles has not been precisely determined but it is known to be linked to the ovarian cycle, which is, itself, dependent not only on the age of the females, but also on the climatic and feeding conditions to which they have been subjected. The mean duration of the first ovarian cycle is 13 days, that of the second 11 days, and of the third 10 days.

The migratory activity of the female cockchafer is therefore characterized by reversal of the sense of the direction of flight, which is linked to the state of maturation of the oocytes. I have set out to study the physiological bases of this important phase of the female's activity.

The connection between oogenesis and the reversal of flight led me to study the neuroendocrine regulation of the cyclical migratory behaviour. In the first section, I will discuss the migratory behaviour, and examine evidence concerning the senses involved, in order to lead up to recent work on its neuroendocrine regulation.

MIGRATORY BEHAVIOUR

Behaviour of females

I will limit myself to discussing the main flights and will neglect the secondary flights which take place within the feeding area.

The main characteristic of these flights is their orientation. The experiments performed in unfamiliar country, and which are detailed below, have shown that we are dealing with truly oriented flight and that the mechanism may be described as 'photohorotaxis' (Fig. 1). The forest edge or thicket towards which the cockchafers fly constitutes the attractive site, the direction of flight being determined and main-

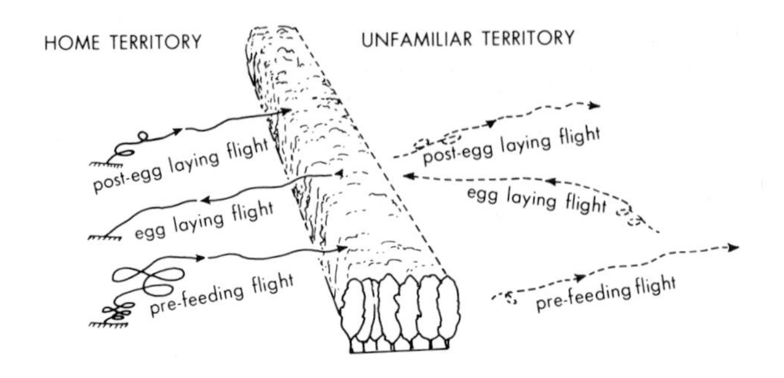

Fig. 1. Schematic diagram showing (solid arrow) the paths taken by a female in the course of different flights during her life above ground in her home territory and (broken arrows) the paths followed when the flights take place in unfamiliar territory (from Couturier and Robert 1958)

tained by the sharp contrast provided by the dark object. More precisely, the insects are attracted by the silhouette which, of those in their environment, subtends the greatest angle. Orientated flight may take place from as far away as 3 km. Cockchafers are capable of orientating themselves towards a silhouette only when in flight. Wingless cockchafers do not do so by walking. It seems that the possibility of orientation depends on the existence of the increased nervous activity which accompanies flight.

The oviposition flight is characterized by a reversal of the sense of the direction of flight; that is to say, females return to the fields on a flight path parallel to the pre-feeding flight, but in the reverse direction.

The post-oviposition flight is parallel to the two others and has the same direction as that of the pre-feeding flight which determines the directions of all subsequent flights. In its first pre-feeding flight the cockchafer uses guide marks and steers towards a silhouette, that is to say, it takes into consideration the countryside which surrounds it. This, however, is not true of the other principal flights, for, if a cockchafer is captured in the course of one of these flights, and is freed in another place at another time of day, it takes up the geographical orientation of the flight it was on at the time of its capture. Furthermore, a female captured in the course of its pre-feeding flight, and which is fed in the laboratory until her eggs are ready for laying, takes a direction of flight which is the reverse of that of the pre-feeding flight, regardless of where she is released. Similarly, females which have laid in the laboratory, and are then released in the field, make their post-oviposition flight in the same direction as that of their congeners which have remained at the original site.

After having established the direction of its initial flight by the perception of a silhouette, the female cockchafer re-orientates herself by using as yet undetermined astronomical guide marks independently of features of the countryside.

Couturier and Robert (1958, 1962) have studied the external factors likely to play a role in the re-establishment and maintenance of orientation. Neither changes in the photoperiodic regime, the interposition of a smoke screen in the axis of flight, chilling, nor narcosis, seem to have any influence on the migratory behaviour. The sun has been shown to play a part in the re-establishment of orientation of daytime flight in unfamiliar territory. On days on which the sun is obscured by cloud, cockchafers do not take flight. Couturier and Robert have shown that the cockchafers use the sun to re-orient themselves even though the main flights occur in the evening, after sunset.

Schneider (1961, 1963) thought that cockchafers could perceive magnetic fields but this idea has had to be abandoned. The same author believes he has evidence of a gravitational sense in the cockchafer. If this hypothesis is verified, a great step forward in the explanation of the maintenance of orientation will, without doubt, have been made.

The passage of time has an influence on the orientation sense of the cockchafer. Females take 28 days to lose all sense of orientation but males take only 12 days.

The above ground life of the female cockchafer is therefore marked by a migratory and orientated behaviour which manifests itself as cyclical displacements between the feeding site and the oviposition site. This behaviour is, above all, characterized by reversal in the flight direction, which is linked to cycles of egg maturation.

Behaviour of males

Upon leaving the soil, the male, like the female, is attracted by and makes an orientated flight towards a silhouette and is capable of again taking up his original orientation if released in strange territory. Unlike the female, however, the male never shows reversal in the sense of direction of flight, but moves only within the adult feeding area where it feeds and mates.

NEUROENDOCRINE REGULATION

The link between egg development and the reversal of flight direction has led to the study of the neuroendocrine system and of its relationships to oogenesis and behaviour.

Neurosecretory system

Current work on the neurosecretory system of the female of *Melolontha melolontha* shows that it is very little different from the classical model met with in other insects. One thing which should be noted, however, is the existence of a connection between the neurosecretory cells of the pars intercerebralis (PI) and the frontal ganglion. This current study has revealed that changes which parallel oogenesis are just as clearly apparent in the neurosecretory cells of the PI as in the corpora cardiaca (CC) and the corpora allata (CA).

In the course of oogenesis, the neurosecretory cells increase in size and the density of the nuclei of the cells likewise increases. The CC and the allato-cardiaca nerves become charged with neurosecretory material.

The role of the corpora allata in migratory behaviour

Migratory behaviour of the female

The first study of the relationship between the neuroendocrine system and migratory behaviour consisted of investigating the role of the CA.

The action of the CA on the migratory behaviour, especially on the reversal of flight of females which are ready to lay eggs (pre-oviposition females), was studied by making transplants of CA. The CA were removed with fine forceps from a sacrificed female and transplanted into the haemocoel of another female through a window cut in the third last visible abdominal sternite. A crystal of mixed antibiotic (penicillin and streptomycin) and a crystal of phenylthiourea were introduced together with the CA. The transplants were made from and to females of known physiological age, captured when in the course of a pre-feeding flight or in the course of an oviposition flight. The captured females were kept at a temperature of 0°C until the day of the operation.

The CA of pre-feeding females were implanted into pre-feeding and into pre-oviposition females, and the CA of pre-oviposition females into pre-oviposition females and into pre-feeding females. For each group of 50 transplants, 3 control groups each consisting of 50 individuals were used; 1 group of pre-oviposition females, 1 group of pre-feeding females, and 1 group of pre-feeding females which were sham-operated and received antibiotic and phenylthiourea but no CA. After the operation, the females were kept without males and without food at a temperature of 12°C for 48 hr to permit the healing of the wound. The control groups were kept under the same conditions. After this interval the females were released individually in the field. For each, the orientation of flight was noted. Releases can only be made in the morning, in the following meteorological conditions: temperature above 10°C, wind speed of less than 90 m/min, mainly clear sky, and absence of precipitation.

Pre-feeding females which received CA from other pre-feeding females did not change their migratory behaviour at all, i.e. they retained the orientation of pre-feeding flight. Pre-oviposition females which received CA from similar females or from pre-feeding females also kept their initial flight direction. Seventy per cent of 200 pre-feeding females which received CA from pre-oviposition females showed reversal of the direction of flight, such that their flight direction was identical to that of pre-oviposition females (Fig. 2) (Stengel and Schubert 1970). Thus, the implantation of CA, taken from pre-oviposition females, into pre-feeding females causes the latter to take the reversed flight

RESULTS OF RELEASES ON 25.5.1969

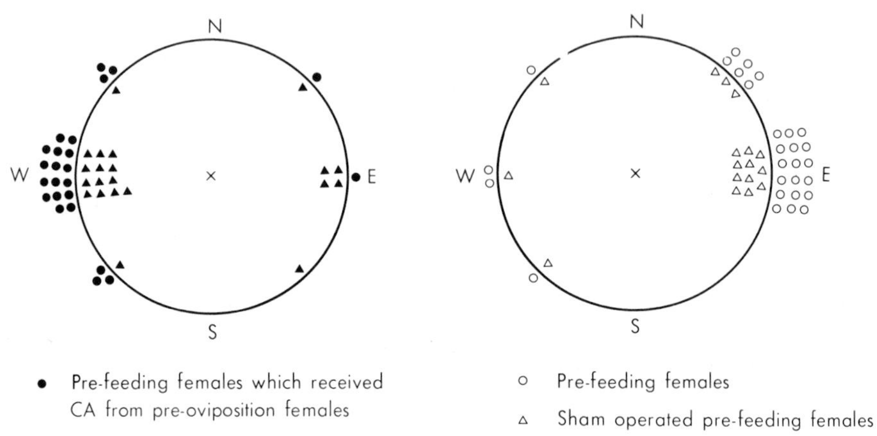

Fig. 2. Orientations taken by different categories of females. The pre-feeding females having received CA from pre-oviposition females reversed their direction of flight. Each point represents the flight orientation of one individual

direction which is characteristic of the migratory behaviour of the pre-oviposition female cockchafer. The CA therefore play a role in the mechanism which regulates the migratory behaviour of the female of *Melolontha melolontha*.

Migratory behaviour of the male

In 1971, we studied the influence of the CA from pre-laying females on the migratory behaviour of the male. The techniques used were the same as those used in studying the role of the CA in the migratory behaviour of the female.

Males which underwent a sham operation did not change their flight direction. Seventy-three per cent of 216 males which received CA from pre-oviposition females showed reversal of their flight direction, this new direction being identical with that of the pre-oviposition females (Fig. 3) (Stengel and Schubert 1972a).

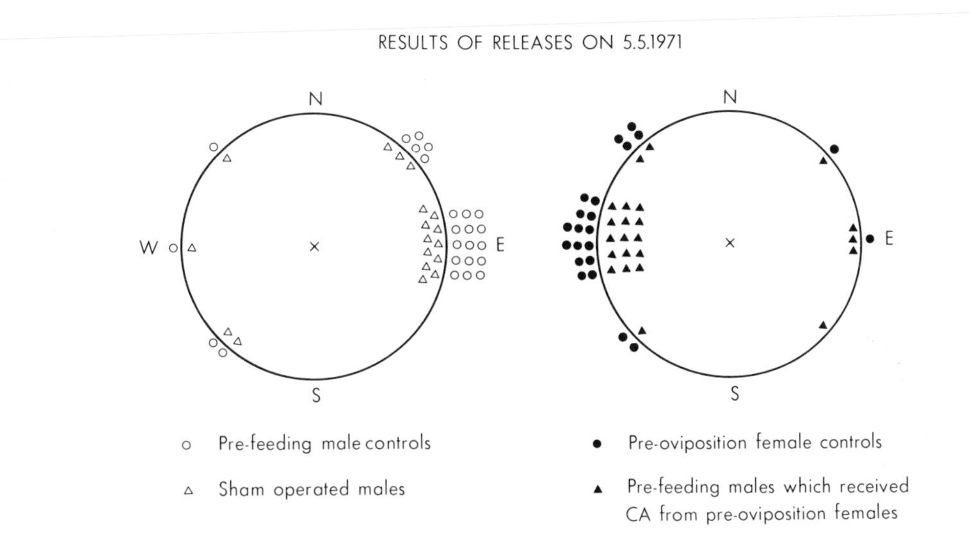

Fig. 3. The influence of CA from pre-oviposition females on the migratory behaviour of males. The pre-feeding males which received CA from pre-oviposition females reversed their direction of flight. Each point represents the flight orientation of one individual

The implantation of the CA from pre-oviposition females to pre-feeding males caused the latter to reverse their flight direction to that of the donor females, even though the male never shows this reversal in nature. The CA of pre-oviposition females are therefore equally capable of eliciting a reversal of flight direction in the two sexes.

Influence of the corpora allata of the pre-oviposition female on oogenesis in the pre-feeding female

In order to study the influence of the CA on oogenesis, the operative techniques described above were again employed. The effects of implantation were determined by comparing the size and the number of the oocytes in the operated females sacrificed after 5 or 10 days of feeding with those in control females (Stengel and Schubert 1972b). The results showed that the CA of pre-oviposition females, which had fully developed eggs in their ovaries, blocked oogenesis (vitellogenesis and the emission of oocytes at the level of the germarium) in the pre-feeding recipient females, even though they still possessed their own functional CA which normally regulate and accelerate oogenesis (Table 1).

Table 1. Length of the oocytes of experimental groups of females which had fed for 5 or 10 days after receiving different treatments. The lengths are given in hundredths of millimetres.

		1st oocyte	2nd oocyte	3rd oocyte	4th oocyte	5th oocyte
0 days feeding	Untreated	157	80	50		
5 days feeding	Untreated	218	132	85	64	32
	Control operated	219	130	82	62	31
	CA from pre-oviposition females implanted	157	84	52	30% 32	
10 days feeding	Untreated	362	50% 362 rest 261-325	231	162	93
	Control operated	362	50% 362 rest 261-325	240	156	92
	CA from pre-oviposition females implanted	224	124	81	71	38

CONCLUSION

The adult life of the female cockchafer is characterized by a cyclical migratory behaviour which leads the female from the field where she emerged and where she will go to lay her eggs, to the forest where she feeds, and vice versa. This activity is linked to the maturation of eggs and parallels changes in the neuroendocrine system. The main characteristic of this migratory behaviour is the reversal of the sense of direction of flight in females which are ready to lay. The CA play an active role in bringing about this reversal. At this time the CA release one or more factors which release the reversal mechanism for the sense of flight direction and at the same time block oogenesis.

The CA of the pre-oviposition female are capable of releasing the reversal mechanism even in males which do not normally reverse their flight direction. The CA of the pre-oviposition females, when implanted into pre-feeding females, the CA of which are actively fulfilling their gonadotrophic function, block oogenesis and thus are capable of countering the effect of the recipient's own CA. We thus have demonstrated two types of hormonal activity of the CA, occurring at different times: the gonadotrophic activity in the pre-feeding female, and the oostatic activity in the pre-oviposition female, in which it also releases the mechanism which brings about the reversal of flight direction.

It does not seem that we have here a case of the simple diminution of the rate of release of juvenile hormone, as this could not alone explain the fact that the CA of a pre-oviposition female can mask, or even counteract, the activity of the CA of a pre-feeding female. We have phenomena which can reconcile the descriptions by Adams, Hintz and Pomonis (1968) concerning the house fly, and by Dixon and Moser (1972) concerning the larvae of bees.

Recent studies by Stengel and Schubert (1972c,d) show that the neurosecretory cells of the PI have actions identical with those of the CA. It seems, therefore, that the neurosecretory cells of the PI secrete the hormones which will be released by the CA. But what is the origin of this double hormonal activity? One can envisage a rhythm of activity which is specific to the neurosecretory cells of the PI which is more or less linked to oogenesis. Also one can conceive that the oostasis and flight reversal are the result of an activity specific to the CA. This activity, which is governed by oogenesis, will be influenced directly by the hormonal activity of the PI.

The migratory behaviour of the cockchafer provides us with a very good example of neurohormonal regulation of behaviour which depends on the existence of two types of neurosecretory activity which are separated in time.

REFERENCES

ADAMS, T.S., HINTZ, A.M., POMONIS, J.G.: Oöstatic hormone production in houseflies, *Musca domestica*, with developing ovaries. J. Insect Physiol. 14, 983-993 (1968).

COUTURIER, A., ROBERT, P.: Recherches sur les migrations du Hanneton commun (*Melolontha melolontha* L.). Annls Épiphyt. 9, 257-329 (1958).

COUTURIER, A., ROBERT, P.: Observations sur le comportement du Hanneton commun (*Melolontha melolontha* L.) (Coléoptère Scarabidae). Revue Zool. agric. 3(7-9), 99-108 (1962).

DIXON, S.E., MOSER, E.: Duality in function in *Corpora allata* of honeybee larvae. Can. J. Zool. 50, 593-595 (1972).

SCHNEIDER, F.: Beeinflussung der Aktivität des Maikäfers durch Veränderung der gegenseitige Lage magnetischer und elektrischer Felder. Mitt. schweiz. ent. Ges. 33, 225-237 (1961).

SCHNEIDER, F.: Orientierung und Aktivität des Maikäfers unter dem Einfluss richtungsvariabler künstlicher elektrischer Felder und weiterer ultraoptischer Bezugssysteme. Mitt. schweiz. ent. Ges. 36, 1-26 (1963).

STENGEL, M., SCHUBERT, G.: Rôle des *Corpora allata* dans le comportement migrateur de la femelle de *Melolontha melolontha* L. (Coléoptère Scarabidae). C. r. hebd. Séanc. Acad. Sci., Paris (D) 270, 181-184 (1970).

STENGEL, M., SCHUBERT, G.: Influence des *Corpora allata* de la femelle pondeuse de *Melolontha melolontha* L. (Coléoptère Scarabidae) sur l'ovogénèse de la femelle préalimentaire. C. r. hebd. Séanc. Acad. Sci., Paris (D) 274, 426-428 (1972a).

STENGEL, M., SCHUBERT, G.: Influence des *Corpora allata* de la femelle pondeuse de *Melolontha melolontha* L. (Coléoptère Scarabidae) sur le comportement migrateur du mâle. C. r. hebd. Séanc. Acad. Sci., Paris (D) 274, 568-570 (1972b).

STENGEL, M., SCHUBERT, G.: Influence de la *Pars intercerebralis* et des *Corpora cardiaca* de la femelle pondeuse sur l'ovogénèse de la femelle préalimentaire de *Melolontha melolontha* L. (Coléoptère Scarabidae). C. r. hebd. Séanc. Acad. Sci., Paris (D) 275, 1653-1654 (1972c).

STENGEL, M., SCHUBERT, G.: Rôle de la *Pars intercerebralis* et des *Corpora cardiaca* de la femelle pondeuse de *Melolontha melolontha* L. (Coléoptère Scarabidae) dans le comportement migratoire de la femelle préalimentaire. C. r. hebd. Séanc. Acad, Sci., Paris (D) 275, 2161-2162 (1972d).

A COMPARISON OF THE MIGRATORY STRATEGIES OF TWO MILKWEED BUGS, *ONCOPELTUS FASCIATUS* AND *LYGAEUS KALMII*

R.L. Caldwell

Department of Zoology, University of California, Berkeley, California, U.S.A.

Views on the nature of migration and dispersal in insects have changed markedly in the past 25 years. Williams (1957) in the 1940s and 1950s was largely responsible for calling attention to the large number of insect species which migrate. However, he stressed that insect migrations were mass occurrences, often triggered by unfavorable conditions, in which flight was unidirectional and under the control of the individual. Dispersal was considered to be mainly a passive and accidental process in which weak-flying insects were scattered by the wind. In the late 1950s, Rainey (1963) demonstrated that movements of African migratory locusts were correlated with prevailing wind patterns and suggested that the direction of migration in these strong fliers was independent of their orientation and thus also passively determined. This led to the questioning of whether or not there was a valid distinction which could be drawn, on the basis of directionality, between the so-called active migrants and passive dispersants. The problem was further complicated when considering the vertical and horizontal components of orientation and the fact that there was then, as there still is, so little data on the orientation of flying insects (Kennedy 1961). Waloff (1972) has recently presented evidence that locusts do in fact orient their direction of flight down wind, thus pointing to the difficulties in attempting to distinguish among dispersal strategies using orientation as a criterion.

Alternative approaches to the characterization of migration and dispersal were suggested in the early 1960s by Johnson (1963, 1965), Kennedy (1961), and Southwood (1960, 1962); for recent reviews see Johnson (1969) and Dingle (1972 and this volume). While each viewed migration with a slightly different bias, all seemed to agree that such flights may be characterized as a distinct behavioral, physiological, and ecological syndrome. Long range directional flight was no longer considered essential to a definition of migration, but rather a specialized adaptation enhancing the probability of locating appropriate habitats, as is known to occur in some Lepidoptera. Also, the distinction between passive and active flight was no longer made, since both types of flight have evolved, leading to the most efficient possible displacement of populations given the physiological and morphological constraints of the species involved, regardless of whether vectors such as wind or self-determined directional flight are used. It was the initiation and timing of flight, not its strength or direction, that was considered important.

Behaviorally, Kennedy (1961) stressed that migration involved the occurrence of persistent, straightened-out flight. He did not distinguish between low altitude directional flight where the individual determined its orientation and high altitude flight where directionality is determined by the prevailing winds. Indeed, many migrants demonstrate both at different times during migratory flight. What is important is that mechanisms have evolved to ensure that flight is unidirectional, leading to efficient displacement. Equally important to Kennedy's definition of migration was the accentuation of locomotion with a depression of 'vegetative' activities such as mating, feeding, and oviposition. Thus, during migration, the individual's responses to stimuli normally eliciting vegetative activities will be inhibited.

Ecologically, migration involves leaving a habitat and has the effect of increasing the mean distances between individuals of a population. Pre- and post-diapause flights are sometimes exceptions, since they tend to concentrate individuals or merely displace entire populations. It was often held that the major importance of migration was to provide a safety valve to alleviate population pressure under unfavourable conditions; large numbers of insects leaving the habitat would reduce population pressures and increase the probability of survival of remaining individuals (Elton 1927). However, due to the risk during exodus and to the lack of evidence, in many cases, for a return flight, such a process is of questionable value since emigrants would be selected against (Williams 1958). Southwood (1960, 1962) proposed that the major evolutionary significance of migration lies in that it allows insects to keep pace with temporary habitats. Thus, the less permanent a habitat, the greater the probability that species utilizing it would engage in migratory flights.

Since migration and dispersal both lead to a scattering of individuals and both tend to remove individuals from where they developed, Johnson, Kennedy, and Southwood do not distinguish between these two processes, except to consider migration as a set of adaptations to ensure that dispersal occurs under specified conditions.

The milkweed bugs of the family Lygaeidae appear to be ideally suited for a study of the behavioral and physiological mechanisms effecting migration. Milkweed bugs breed almost exclusively on milkweeds (usually of the genus *Asclepias*) which, in most cases, are members of early successional communities and are thus temporary in nature. Therefore, milkweed bugs might be expected to make migratory flights at some time in their life cycle in order to keep pace with changes in habitat location.

Sufficient data are currently available for only two species of milkweed bugs, *Oncopeltus fasciatus* and *Lygaeus kalmii*, to allow for a detailed analysis of migratory strategies. Both species have almost identical ranges over the North American continent, extending from southern Canada to Central America and from the Atlantic to the Pacific (Slater 1964). Within this range, both lygaeids are often found on the same stands of milkweed and display few apparent differences in resource utilization. However, there is one major difference between the two. *O. fasciatus* is a known long distance migrant (Dingle 1967) and is not capable of overwintering in north temperate regions (Essig 1929). From its winter range of sub-tropical and tropical North America, migrants spread northward each spring and summer, re-establishing breeding populations in the northern United States and southern Canada, producing 2-4 generations. In the fall, the northern populations, which have greatly proliferated, again migrate and presumably some individuals, at least, make their way south.

L. kalmii, on the other hand, is capable of overwintering over all of its range by becoming dormant (Blatchley 1926). In Iowa, this species emerges from winter hibernacula near old milkweed patches in early spring and is observed flying on the first warm days (above 19°C). However, there is no evidence that it travels over long distances and all flight ceases within a few weeks. During the late spring, summer, and early fall, two to three generations are produced and in the late fall, newly emerged adults enter winter hibernacula (Caldwell 1969).

When *O. fasciatus* populations re-invade northern regions each year, they effectively scan the available habitats of *Asclepias*. However, *L. kalmii* populations, by remaining within their habitat year round also must have evolved means of tracking changes in habitat location, finding new stands of milkweed as they form, and leaving old ones as they become senescent. The spring flights observed in *L. kalmii* serve this purpose. There are, then, two different patterns of movement represented by these two species; one producing seasonal extensions of the population range and the other causing a shuffling of populations within the range. Both effectively utilize new habitats as they form. These two patterns of dispersal require different types of flight. Colonizing *O. fasciatus* must cover annually distances of up to several hundred miles if the population is to exploit the food resources to the north, whereas *L. kalmii* needs only to search the immediate area to locate new stands of milkweed.

ONCOPELTUS FASCIATUS

As mentioned above, *O. fasciatus* populations annually invade northern latitudes in the spring and summer and return to the tropics in the fall (Dingle 1967). The direction of population displacement is apparently determined by the prevailing wind patterns at the time of migration. In the spring, on warm sunny days when flight is most likely to occur, the prevailing winds, at least for the central United States, are southerly moving up the Mississippi Valley. In the fall, sunny days are more likely to occur during periods of high pressure with winds out of the north. Similar north-south movements, up and down the Mississippi River valley, correlated with wind direction have been documented for several other Hemiptera (see Dingle 1972). Glick (1939) reports taking *O. fasciatus* in aerial plankton nets at several hundred feet and my own field observations indicate that this species often takes off flying strongly upwards until out of sight (80 to 100 m), a pattern of take-off similar to that found in many wind-assisted migrants. Thus *O. fasciatus* has evolved a pattern of flight where the primary directional component is provided by the prevailing winds. The question then may be asked, what mechanisms has this species evolved to ensure that it flies at the appropriate times to effect a north-south migration?

In the laboratory, when flight-tested using a tethered, still air technique (Dingle 1965), females make long flights (arbitrarily defined as flights over 30 min) at a specific stage in the life-cycle; post-tenerally and pre-reproductively. For example, females reared at 23°C on a 16L-8D photoperiod make most long flights 8 to 12 days after the adult molt. In *O. fasciatus* the end of the teneral period may be objectively defined by the cessation of the deposition of daily cuticular growth rings (Dingle 1966; Dingle, Caldwell and Haskell 1969). The length of the teneral period is temperature- but not photoperiod-sensitive, varying from approximately 5 days at 31°C to 10 days at 19°C. Bugs tested in my laboratory at 23°, 25°, and 27°C indicated that the age at which they made their first long flight closely corresponds to the cessation of cuticle deposition. The termination of the migratory period in the female is determined by the onset of oviposition (Caldwell and Rankin 1972; Rankin, this volume). For example, on a 16L-8D, 23°C photoperiod-temperature regimen, the mean age of first oviposition is 14 days (Dingle 1965), which is just after peak flight activity has subsided. Males also do not begin to undertake long flights until after the teneral period, but may fly at any time thereafter. However, there does seem to be some correlation between age of first copulation and peak flight activity (Caldwell, unpublished).

Field observations on the age at which flights are initiated correlate quite well with the laboratory studies. In the field, young adult *O. fasciatus* remain in aggregations on milkweed pods until approximately 10 days after the adult molt, at which time the aggregations break up and individuals disperse. During the fall at northern latitudes, individuals have been observed initiating migratory flights shortly after the aggregations dissolve (Caldwell and Rankin, in press).

The proportion of a population of *O. fasciatus* which migrates is determined by two factors: (1) by the number of potential migrants which are genetically programmed to fly at a particular age, and (2) in females, by the length of time available for migration between the end of the teneral period and the onset of oviposition. In the latter case, the time of the onset of oviposition is both environmentally and genetically determined (Dingle, this volume). When Dingle (1968a) flight-tested *O. fasciatus* in the laboratory, only 30% of the females and 20% of the males of a "wild type" population flew at least once for over 30 min. These bugs were reared at 16L-8D, 23°C. However, 60% of both male and female offspring of long flying parents flew for longer than 30 min. When *O. fasciatus* are maintained under conditions which delay the onset of reproduction and thus extend the time available for migration (discussed below), 61% of the males (N = 43) and 87% of the females (N = 30) tested every 5 days for 40 days made at least one flight of over 30 min. Some individuals flew early in the life cycle, others later, but none made long flights throughout the entire period (Caldwell and Rankin 1972; Caldwell, unpublished). Experiments under way in this laboratory indicate that the age of first long flight is also under genetic control and is amenable to selection. For example, we have been able to produce a strain of bugs which make their first long flights four days after the adult molt rather than the usual eight. Thus, the proportion of potential migrants in a population and the age at which they fly appears to be genetically determined.

Whether or not potential migrants ever migrate is determined by several environmental factors, most of which seem to operate, at least in females, by determining the age at which first oviposition occurs, although some may also have a more direct effect on flight mechanisms. Temperature determines the length of the teneral period, as mentioned above. At 23°C, first long flights occur at 8-10 days (Dingle 1965), while at 25°C first long flights are recorded at 6 days (Caldwell, unpublished). At 27°C and above, flight should begin at 5-6 days, but Dingle (1968a) found that very few individuals at these high temperatures ever flew. At least in females, this can be attributed to the fact that at these temperatures the onset of oviposition occurs soon after the onset of the post-teneral period and inhibits flight (see below). At temperatures below 23°C, there is a considerable period of time between the end of the teneral period and the onset of reproduction, but Dingle (1968a) observed little flight activity. Low temperatures probably inhibit the expression of the migratory tendency in some as yet undetermined manner.

O. fasciatus delays the onset of reproduction under the influence of short photoperiods (Dingle 1968b). Photoperiod has no effect on the length of the teneral period, but a 12L-12D, 23°C regimen, will delay the onset of oviposition to a mean of 63.5 days after the adult molt for bugs maintained as single pairs (Dingle, this volume), whereas on a 16L-8D, 23°C regimen under the same rearing conditions mean first oviposition occurs at 14 days (Caldwell, unpublished). At higher temperatures, the effects of short photoperiods are attenuated so that above 27°C, very little delay is evident. Below 23°C, the effects of short days have yet to be determined, although an extrapolation of the temperature-response curve indicates that oviposition should be delayed for at least several months. While Dingle (1968a) was unable to demonstrate any effect of photoperiod on the proportion of bugs making long flights shortly after the end of the teneral period, he found that significantly more females reared at 12L-12D, 23°C and tested on day 25 flew than did those reared at 16L-8D, 23°C and tested at the same age. He did not test 25 day old males. Note that short day females have not yet begun to oviposit by day 25 while the long day females have. Delays in the onset of oviposition have two major effects. Firstly, individuals have a greater number of days available for flight before being inhibited by oviposition, and secondly, those individuals which were programmed to fly at an older age will have a chance to migrate before oviposition is initiated. The situation is less clear for males, although a similar relationship between photoperiod and the onset of reproduction seems to exist (Caldwell and Rankin 1972).

The quality and availability of food also affects flight in *O. fasciatus*. When females reared at 16L-8D, 23°C were starved from day 20 to 28 following the adult molt and then flight-tested, significantly more made long flights (>1 hr) than did fed controls of the same age (Dingle 1968a). Starved females began to curtail egg production and hence lifted the inhibition of flight by oviposition. Furthermore, about the same proportion of females (30%) made long flights after starvation as flew pre-ovipositionally. It would be interesting to know if the individuals that flew earlier were the same as those which flew after starvation, or if different individuals were programmed to fly at that specific age and, with the cessation of oviposition, began to fly. Unfortunately, these data are not available. Rankin (this volume) reports that starvation lasting for several weeks eventually reduces the number of individuals making long flights and suggests that this is due to the inhibition of juvenile hormone secretion which she has shown to be of primary importance in promoting flight in *O. fasciatus*. She hypothesizes that starvation initially reduces the juvenile hormone (JH) titers necessary to maintain egg production but that titers are still sufficient to promote flight, thus producing the initial burst of flight activity with starvation. Eventually, with the continually falling JH titers, there is a cessation of flight as well. Caldwell and Rankin (1972) have suggested that the continued northward migration of *O. fasciatus* in the late spring and early summer, when long photoperiods and high temperatures would be expected to inhibit flight, is maintained because of the lack of milkweed seed necessary for oviposition. This lack of seed delays the onset of oviposition and thereby extends the migratory period, although poor quality food (plant juices) are sufficient to sustain flight.

Two other factors delay the onset of reproduction in *O. fasciatus*. First, Dingle (1968b) demonstrated that crowding delays the onset of oviposition from 9 to 34 days, depending on the temperature and photoperiod used. Although he did not flight-test bugs reared under different crowding conditions, it is reasonable to assume that such delays in oviposition would lengthen the period during which flights could occur just as in the other cases cited above. No significant difference was found between the proportion of bugs, reared in this laboratory as single pairs, which made long flights and the proportion of long flying individuals obtained by Dingle (1965) who kept 20 pairs of bugs per container (both groups were tested under the same conditions; flight-tested on days 8-10, and reared at 16L-8D, 23°C). In both cases approximately 30% of the females and 20% of the males made long flights. This indicates that crowding does not directly affect the proportion of bugs flying at a given age as long as the tests are carried out pre-ovipositionally.

Second, the lack of mating partners may affect migration. Mated females oviposit at an earlier age and produce large clutches more frequently than do virgins (Abbott 1967; Gordon and Bandal 1967; Gordon and Loher 1968; Loher and Gordon 1968; Caldwell, unpublished). Dingle (1965, 1966) reports that both virgin males and females made long flights at an older age than did mated bugs, although he did not determine if the virgin females were ovipositing. Results from my laboratory indicate that virgin females do fly at an older age because of the delay in the onset of oviposition and also fly occasionally after oviposition has begun; in the latter case, such flights are usually on days when oviposition does not occur. The manner in which mating affects males remains unclear.

Most daily activities of this species, including feeding, drinking, mating, oviposition, flight, and general locomotor activity, are governed by circadian rhythms which are set by the daily light cycle (Caldwell and Dingle 1967; Caldwell and Rankin, in prep.). The integration of these rhythmic activities produces a pattern of temporal segregation that determines which behaviors are more likely to occur at various times of the day. Before oviposition, peak flight activity occurs 8 hr after light onset, whereas mating and feeding reach a maximum in the evening, 12-16 hr after light onset. The segregation of mating and feeding into a period of the day when flight does not occur is similar in effect to the inhibition of vegetative activities in relation to migratory flight discussed by Kennedy (1961). In *O. fasciatus*, which may migrate over a period of several days, such temporal segregation allows bugs to mate and take in energy and water during the long migration but still fly during the middle of the day when temperatures are favorable for flight.

Recent observations in this laboratory indicate that during long flights, *O. fasciatus* passes through various states of responsiveness to landing stimuli. Upon take-off, there is a period marked by a low threshold of the landing reflex (extension of the legs) to contrasting visual stimuli. Within a few minutes, individuals which are going to make a long flight become unresponsive to landing stimuli. This period may last from a few minutes to several hours. Prior to the cessation of flight, the landing reflex returns. Again, this is reminiscent of Kennedy's proposed inhibition of responsiveness to vegetative stimuli.

Thus, *O. fasciatus* is a migrant species with specific genetic, physiological and environmental control mechanisms to ensure that, at an appropriate time in the life cycle, flight will be initiated and sustained in a certain proportion of the population. Factors signalling either current or approaching conditions unfavorable for population growth, such as short photoperiods, low temperatures, poor quality food, crowding, and the lack of sex partners, all act to delay the onset of reproduction and thus extend the time available for individuals to express their genetically set migratory tendencies. Conditions favoring population growth such as abundant food, high temperatures, and long photoperiods hasten reproductive development and prevent most individuals from migrating. The temporal segregation of vegetative activities to a time during the day when flight does not occur and the inhibition of landing responses during the flight maximize the opportunity for migratory flights to be initiated and sustained.

LYGAEUS KALMII

The biology of *Lygaeus kalmii* is not well known. The only mention of flight in this species other than in my own work (Caldwell 1969; Caldwell and Hegmann 1969) is by Townsend (1887) in which he reported capturing one flying female in Michigan on March 3. The work reported below was conducted at the University of Iowa and therefore is only applicable to Iowa populations of *L. kalmii* unless otherwise stated

(Caldwell 1969).

As mentioned above, *L. kalmii* was observed to fly in the field only on the first warm (>19°C) days of March and early April, shortly after emerging from winter hibernacula. In both 1967 and 1968, the migration period lasted for approximately 12 days. Bugs did not fly on days during this period when temperatures were below 19°C or when wind velocities were greater than 7-10 m/sec. No flights were observed before 1000 hr or after 1700 hr; temperatures below 19°C in the morning and evening precluded flight. *L. kalmii* were not observed flying in the field at other times of the year except for occasional short flits from one plant to another.

Observations on several hundred flying *L. kalmii* in the field indicated that flights fell into two categories, trivial and straight-line flights. Trivial flights were short, usually 1-10 m, from one plant to another. Bugs seldom flew higher than 2 m and the flight direction was erratic with many mid-course deviations. Such flights were more common at temperatures above 23°C and were more frequently undertaken by males. There seemed to be no particular orientation to sun or wind direction.

Straight-line flights were the most common during the spring migration and were undertaken by more females than males. Flight was in a straight line at a height of 2-3 m following the contours of the earth. Most flights were down wind, but there were exceptions and no consistent orientation to the sun could be determined. Bugs were often observed to deviate from their flight paths to land on the white floss of old milkweed pods. They also were attracted to white cotton balls placed in the field. The presence of white floss in milkweeds is usually associated with broken pods containing seed and is probably the cue used by *L. kalmii* to locate milkweed patches. Since most milkweeds in Iowa are perennials, the presence of milkweed floss is a good indicator of both a current food source and of developing plants during the coming year. The maximum distance travelled in this mode of flight was not determined, but several bugs were followed for up to 400 m. (The inability of the observer to run faster or farther precluded further pursuit.) Laboratory evidence that bugs will fly non-stop for up to 30 hr suggests that if appropriate stimuli are not encountered to arrest flight, they could continue on for many kilometers.

The major mode of dispersal in *L. kalmii* seems to be the spring straight-line flights. These flights are uni-directional, a prerequisite for most definitions of migration (Kennedy 1961), but they do not appear to include inhibition of vegetative reflexes which Kennedy predicts should accompany migration. However, since these flights are unidirectional and seem to be an adaptation to ensure that dispersal occurs under certain conditions (discussed below), I shall refer to them as migratory.

Field caught *L. kalmii* were flight-tested in the laboratory using the same tethered technique as described for *O. fasciatus* the day after capture. It was found that 11.1% (N = 144) of the bugs captured in March flew for over one hour whereas only 3.4% (N = 116) of those taken in June through August made similar long flights. Of bugs captured in November and December, 22.8% (N = 136) made long flights. The large proportion of fall bugs flying in the laboratory does not agree with the field observations. However, since field temperatures at this time of the year in Iowa rarely reach the 19°C required for flight, these results are not surprising. The lack of flight in the summer can be attributed to reproductive activity which will be discussed below.

Since it is apparent that not all *L. kalmii* in a population migrate, and that the proportion of bugs flying under various conditions differs, laboratory experiments were conducted to determine what factors affect flight. Laboratory stocks derived from Iowa field populations were used. *L. kalmii* will not mature at constant temperatures below 25°C, so stocks were maintained at 35°C, which seems to be optimum for growth. Most experiments were conducted with bugs reared at this temperature, although similar results were usually obtained from 25°C stocks. Differences will be pointed out. All flight-tests were conducted at 23°C.

In *O. fasciatus*, flight is post-teneral and pre-reproductive. *L. kalmii* were repeatedly flight-tested every two days following adult eclosion to determine if this was also true for this species. At 35°C, peak flight activity occurred 4-10 days after the adult molt; at 25°C peak activity came at 12-15 days, although some individuals continued to make long flights after several weeks. The deposition of cuticular growth rings is not temperature or photoperiod sensitive in *L. kalmii* as in *O. fasciatus*, and the teneral period as measured by this index lasted 7-9 days at both 25° and 35°C. At 35°C flight can occur as early as one day after adult ecdysis; peak activity occurs before the end of the teneral period, which indicates that the correlation between the end of the teneral period and the beginning of migration is not as strong in *L. kalmii* as in *O. fasciatus*.

The question of whether or not flight is pre-reproductive is even more complicated in *L. kalmii*. Females first oviposit at a mean age of 4 days at 35°C and 12 days at 25°C. Whether or not they have mated has no effect on the age of first oviposition or on the number of eggs produced. Therefore, flight, unlike that in *O. fasciatus*, is not pre-ovipositional. However, mating has a considerable effect on flight performance. When bugs were maintained from the adult molt either as virgins or as potential mated pairs (6 males or females or 3 pairs per container), 34.8% of the virgin females and 4.8% of the mated females flew for over 30 min whereas 61.9% of the virgin males and 6.0% of the mated males made long flights ($N \simeq 20$ for all groups). Similar results were obtained for different densities and for field captured immature adults. At 25°C, the same trend was demonstrated, but the differences were not as marked, which suggests that the inhibition of flight by mating is not as great at low temperatures. When bugs are maintained as pairs they mate several times a day. However, even a single mating reduced flight activity. A group of virgins was flight-tested on day 6, allowed one mating on day 7, and retested on day 8. The proportion of bugs flying over 30 minutes was reduced 20% for females ($N = 26$) and 25% for males ($N = 29$). First mating occurs 3 days after the adult molt at 35°C and at 8 days at 25°C, which is well before peak flight activity in both cases.

The quality and abundance of food also affects flight in *L. kalmii*. When newly emerged adults were given only flowers of *Asclepias verticillata* in contrast to controls fed seed, and both groups were maintained as virgins and flight-tested on day 4, 62% of the females and 51% of the males fed flowers flew for longer than 30 min, whereas 25% of the female and 24% of the male controls made similar long flights ($N \simeq 20$ in all groups). When these two groups were tested at later ages, no difference was found in the proportions flying. Thus poor quality food seems to initiate flight at an earlier age. Complete starvation had a similar effect. Poor quality food also delayed the onset of mating by several days so that the bugs flew before they mated.

Other factors which might affect flight in *L. kalmii* were also examined. Crowding had no effect, except that increased densities of cultures containing both sexes produced more frequent mating and therefore a greater inhibition of flight. *L. kalmii* did not respond to photoperiod, either with respect to reproduction or flight. Also, flight and other daily activities such as mating were not governed by circadian rhythms and individuals remained responsive to landing stimuli. The genetics of migration will be discussed below.

These various laboratory results may be used to interpret the pattern of migration seen in the field. Bugs developing in the late spring, summer, and early fall are exposed to high temperatures and are usually able to locate adequate food. (Unlike *O. fasciatus*, *L. kalmii* normally feeds on the ground where seed is usually available.) Under these conditions, mating occurs prior to the onset of flight and thus migration is inhibited. If for some reason food or mates are not available, flight occurs and more desirable habitats are located. In the late fall, temperatures drop below the levels required for flight. By sunning themselves bugs can elevate their body temperatures sufficiently to molt to adults, but do not fly or reproduce. When brought into the laboratory and warmed above 19°C, they immediately fly, which indicates that the flight response has matured. Below approximately 22°C, mating behavior does not develop so that, in the field, bugs are still sexually immature. As temperatures drop still lower, adults move into winter hibernacula. Nymphs and sexually mature adults do not survive the winter, but in the early spring when temperatures reach 12-15°C, the young adults emerge. They begin feeding and also spend a considerable amount of time sunning themselves. Measurements indicate that a bug in still air and bright sun can elevate its internal temperature as much as 10°C. Under these conditions they soon begin to copulate, since body temperatures are brought above the 22°C required for mating to develop. However, air temperatures are still below the 19°C required for flight. As soon as air temperatures reach the required levels, those individuals which have not yet mated migrate. If on emergence from winter hibernacula, scarcity of food and/or non-availability of males indicates that the milkweed patch presently occupied is probably undergoing senescence, the probability of flight is increased and thus the location of new habitats. However, when conditions are favorable for further population growth, as indicated by abundant food and a successful production of individuals the preceding year, few individuals will migrate due to mating activity in the population. By using temperature as a cue, *L. kalmii* populations are able to grow rapidly as soon as favorable conditions occur. The initiation of flight due to a lack of food and the suppression of flight by mating allows the population to assess the probability of reproductive success over the coming year and respond appropriately.

The proportion of a population which is committed, even under favorable conditions, to migrate will be a balance between the losses of reproductive potential in the population and the gains achieved through the possibilities of colonization. This will depend upon many factors: the risk of death during migration due to predation and accident as opposed to the risk of staying in the present habitat, the probability of locating suitable habitats due both to the biological properties of the migrant and the relative permanence and distribution of the habitat, and the productivity of the species in a particular habitat. In *L. kalmii*, differences occur in the proportion of a population which is genetically programmed to migrate under appropriate conditions. In Iowa, milkweed is ubiquitous and in most areas one patch is never more than several hundred meters from another. Patches usually last from 5 to 10 years, so that while a population in a particular patch must move every 15 to 30 generations, the probability of finding a new habitat is very high. Although the new patch may have been colonized already, the individuals locating it will not die of starvation or be exposed to serious risk while covering long distances. Thus the cost of migration in this area is probably not great. In the laboratory under optimum conditions for flight, 12.8% of the offspring (males and females combined) of field captured adults made flights of over 1 hr (N = 646). The distribution of flight times in this population is highly skewed with most bugs flying only a few minutes, but a few flew for over 12 hr. In contrast to the Iowa population, offspring from parents captured in western Colorado were tested also under conditions determined for that population to be optimum for flight and only 3.4% (N = 146) made flights of over one hour. On the western slopes of the Rocky Mountains milkweed is relatively rare and is found primarily in mountain valleys. One patch may be several kilometers from another, making the location of other milkweed by a migrant unlikely. Apparently, this factor has been weighed evolutionarily and the commitment to migration reduced in these mountain populations. One further example may serve to emphasize this point. In the hills around San Francisco Bay, milkweed is rare. However, a small patch has been planted in the Botanical Garden of the University of California, Berkeley. It measures approximately 3 by 4 m and has been there for several years. At some time in the past, *L. kalmii* colonized it and has maintained a stable population for at least the past five years. I have sampled this population repeatedly and flight-tested offspring reared in the laboratory under conditions optimum for flight in other Northern California populations. I have yet to find a single bug that flew for more than a few seconds. Apparently, a non-migrant "island" population has been selected. Selection studies on Iowa and Colorado populations and an estimate of the heritability of flight (Caldwell and Hegmann 1969) indicate that there is considerable genetic variation in *L. kalmii* populations for migratory potential and that it is amenable to selection. This further suggests that the differences in the proportions of migrants found in various populations discussed above are due to genetic causes.

COMPARISON OF MIGRATORY STRATEGIES OF *O. FASCIATUS* AND *L. KALMII*

The two migratory strategies described for *O. fasciatus* and *L. kalmii* are markedly different. If *O. fasciatus* is to colonize northern habitats, it must have a mechanism which ensures that individuals return south in the fall or flight will be selected against. However, all bugs cannot be programmed to fly every generation, since during the summer it is more opportune to reproduce, and apparently the energetics of reproduction and migration are incompatible in this species. The solution has been to use photoperiod and food availability to predict the times when flight should occur to take the species either north or south on prevailing winds. A system has evolved where individuals are genetically programmed to fly at specific ages and where oviposition (in females) inhibits flight. Factors such as short photoperiod and poor quality food delay oviposition and thereby allow individuals to express their migratory tendency. I have attempted to summarize this in Fig. 1a. The lines for short photoperiod could just as easily be labelled poor quality food, since both factors delay the onset of reproduction. Note that it is oviposition which sets the time available for migration. Mating generally occurs prior to or during flight, which is a good strategy for a colonizer since it ensures that females arrive in new habitats carrying sperm and therefore need not delay reproduction until the arrival of a male.

On the other hand, *L. kalmii* is more opportunistic. It migrates only when its habitat deteriorates or when it exceeds the carrying capacity of the environment (various factors, particularly parasitism, usually preclude the latter). This species tracks the environment by relying on cues such as food and the presence of mates to assess the status of the habitat. Fig. 1b partially depicts this strategy but is slightly misleading since it represents data collected in the laboratory from

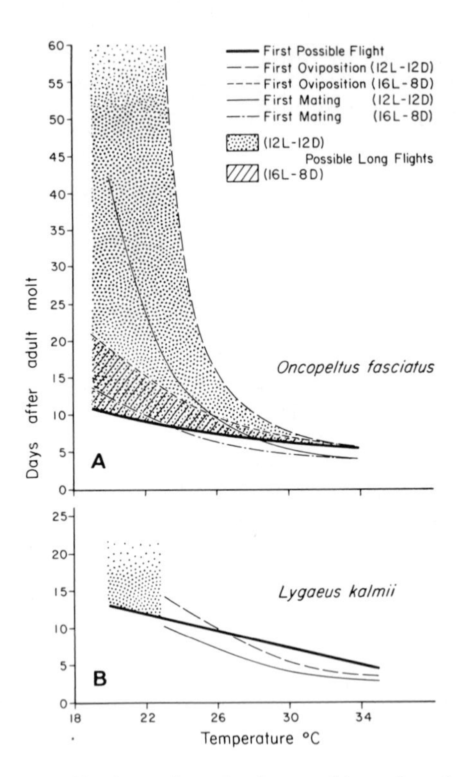

Fig. 1. A. Potential age of migration in *Oncopeltus fasciatus* females under different temperatures and photoperiods. Oviposition, but not mating, inhibits flight. Values were obtained from Iowa populations reared in the laboratory, from the egg, at the specified condition and maintained as adults as single mated pairs. 16L-8D and 12L-12D photoperiods and 19, 23, 25, 27 and 31°C temperatures were used. The age of first possible flight (over 30 min) was obtained from flight-testing individuals at 23, 25, and 27°C and was estimated from data on the age of cessation of cuticle deposition plus one day at 19 and 31°C. Photoperiod does not affect age of first possible flight. (Data from Dingle 1968a, 1972, this volume; Dingle, Caldwell and Haskell 1969; and Caldwell, unpublished.)

B. Potential age of migration in *Lygaeus kalmii* males and females under different temperatures. Photoperiod has no effect in this species. Mating, but not oviposition, inhibits flight. Values for first mating and first oviposition were obtained from Iowa populations reared from the egg at 25, 27, 31 and 35°C and maintained as adults as single mated pairs except for bugs brought from the field as newly emergent adults and maintained at 23°C. Age of first possible flight (over 10 min) was determined by flight-testing virgin bugs reared from the egg at 25, 27, 31 and 35°C and from field captured, newly emergent virgin adults maintained at 19 and 23°C. (Data from Caldwell 1969)

bugs reared at constant temperatures. Still, it indicates that as long as mating occurs prior to flight, no migration will occur. If mating does not take place (lack of mates or food), the inhibition of flight is not realized and a genetically determined proportion of the population migrates. Oviposition has no effect on flight; females reproduce as soon as males are located.

The modes of flight used by these two species are also quite different. *O. fasciatus* must traverse long distances by riding the winds at high altitudes. The associated inhibition of various vegetative reflexes ensures that flight is initiated and sustained. *L. kalmii* need not fly so far. It uses a low altitude, unidirectional flight but remains responsive to stimuli which signal an appropriate habitat. Both strategies effectively screen available habitats, but the former is obligatory while the latter is facultative.

In both species, the proportion of potential migrants is genetically determined. In *O. fasciatus*, at least for populations that colonize northern latitudes, all individuals must migrate with a potential for returning south. Genetic variation seems to be primarily concerned with the age at which flight may occur, a mechanism which, when coupled to the inhibition of flight by oviposition, is sensitive to environmental factors predicting the future and allows the population to migrate at the appropriate time. In *L. kalmii*, individual populations vary with respect to what proportion is committed to colonization when it is probable that the current habitat is degenerating. In this case a balance is struck between the risk of not locating a suitable habitat and the risk of staying. Thus we have two species utilizing the same resources, but displaying two different migratory strategies, each with its own set of underlying control mechanisms which ensure that the population will disperse when necessary to carry out the strategy it has adopted.

REFERENCES

ABBOTT, C.E.: Male influence on ovarian development in large milkweed bug, *Oncopeltus fasciatus* (Hemiptera: Lygaeidae). Ann. ent. Soc. Am. 60, 344-347 (1967).

BLATCHLEY, W.S.: "Heteroptera or true bugs of Eastern North America". Indianapolis: Nature Publishing Co. (1926).

CALDWELL, R.L.: A comparison of the dispersal strategies of two milkweed bugs, *Oncopeltus fasciatus* and *Lygaeus kalmii*. Ph.D. Thesis, University of Iowa. (1969).

CALDWELL, R.L., DINGLE, H.: The regulation of cyclic reproduction and feeding activity in the milkweed bug *Oncopeltus* by temperature and photoperiod. Biol. Bull. mar. biol. Lab., Woods Hole 133, 510-525 (1967).

CALDWELL, R.L., HEGMANN, J.P.: Heritability of flight duration in the milkweed bug, *Lygaeus kalmii*. Nature, Lond. 223, 91-92 (1969).

CALDWELL, R.L., RANKIN, M.A.: The effect of a juvenile hormone mimic on flight in the milkweed bug, *Oncopeltus fasciatus*. Gen. Comp. Endocr. 19, 601-605 (1972).

CALDWELL, R.L., RANKIN, M.A.: The separation of migratory from vegetative behavior in *Oncopeltus fasciatus*. J. comp. Physiol. (In press).

DINGLE, H.: The relation between age and flight activity in the milkweed bug, *Oncopeltus*. J. exp. Biol. 42, 269-283 (1965).

DINGLE, H.: Some factors affecting flight activity in individual milkweed bugs (*Oncopeltus*). J. exp. Biol. 44, 335-343 (1966).

DINGLE, H.: A probable interaction between behavioral and meteorological events in the migration and distribution of the milkweed bug (*Oncopeltus*). Proc. 4th Int. Congr. Biometerol., New Brunswick, N.J., 1966 (Abstr.) (1967).

DINGLE, H.: The influence of environment and heredity on flight activity in the milkweed bug, *Oncopeltus*. J. exp. Biol. 48, 175-184 (1968a).

DINGLE, H.: Life history and population consequences of density, photoperiod, and temperature in a migrant insect, the milkweed bug, *Oncopeltus*. Am. Nat. 102, 149-163 (1968b).

DINGLE, H.: Migration strategies of insects. Science, N.Y. 175, 1327-1335 (1972).

DINGLE, H., CALDWELL, R.L., HASKELL, J.B.: Temperature and circadian control of cuticle growth in the bug, *Oncopeltus fasciatus*.

J. Insect Physiol. 15, 373-378 (1969).
ELTON, C.S.: "Animal Ecology". London: Sidgwick and Jackson. (1927).
ESSIG, E.O.: "Insects of Western North America". New York: Macmillan (1926).
GLICK, P.A.: The distribution of insects, spiders and mites in the air. Tech. Bull. U.S. Dep. Agric. 673, 1-150 (1939).
GORDON, H.T., BANDAL, S.K.: Effect of mating on egg production by the large milkweed bug, *Oncopeltus fasciatus* (Hemiptera: Lygaeidae). Ann. ent. Soc. Am. 60, 1099-1102 (1967).
GORDON, H.T., LOHER, W.: Egg production and male activation in new laboratory strains of the large milkweed bug, *Oncopeltus fasciatus*. Ann. ent. Soc. Am. 61, 1573-1578 (1968).
JOHNSON, C.G.: Physiological factors in insect migration by flight. Nature, Lond. 198, 423-427 (1963).
JOHNSON, C.G.: Migration. *In* "The Physiology of Insecta" (Ed. M. Rockstein), pp. 187-226. New York: Academic Press (1965).
JOHNSON, C.G.: "Migration and Dispersal of Insects by Flight". London: Methuen (1969).
KENNEDY, J.S.: A turning point in the study of insect migration. Nature, Lond. 189, 785-791 (1961).
LOHER, W., GORDON, H.T.: The maturation of sexual behavior in a new strain of the large milkweed bug, *Oncopeltus fasciatus*. Ann. ent. Soc. Am. 61, 1566-1572 (1968).
RAINEY, R.C.: Meteorology and the migration of desert locusts. Tech. Notes Wld met. Org. No. 54,(Anti-Locust Mem. No. 7) 115 pp.(1963).
SLATER, J.A.: "A catalogue of the Lygaeidae of the world", Vol. 1. Connecticut: University of Connecticut, Storrs. (1964).
SOUTHWOOD, T.R.E.: The flight activity of Heteroptera. Trans. R. ent. Soc. Lond. 112, 173-220 (1960).
SOUTHWOOD, T.R.E.: Migration of terrestrial arthropods in relation to habitat. Biol. Rev. 37, 171-214 (1962).
TOWNSEND, C.H.T.: On the life history of *Lygaeus turcicus* Fab. Entomologica am. 3, 53-55 (1887).
WALOFF, Z.: Orientation of flying locusts, *Schistocerca gregaria* (Forsk.), in migrating swarms. Bull. ent. Res. 62, 1-72 (1972).
WILLIAMS, C.B.: Insect migrations. A. Rev. Ent. 2, 163-180 (1957).

THE HORMONAL CONTROL OF FLIGHT IN THE MILKWEED BUG, *ONCOPELTUS FASCIATUS*

M.A. Rankin

Harvard University Biological Laboratories, Cambridge, Massachusetts, U.S.A.

Insect migration has been described as a distinct physiological, behavioral and ecological syndrome (Kennedy 1961; Johnson 1963; Southwood 1962; Dingle 1972). It is usually post-teneral and pre-reproductive, most migrants being characterized by immature ovaries, hypertrophied fat bodies and often also by a positive phototaxis. Kennedy (1961) in his analysis of migratory behavior, indicated that response to migratory stimuli is generally accompanied by and dependent upon an internal inhibition of the 'vegetative' reflexes (such as feeding, mating, or oviposition) that will eventually arrest the movement. A migratory flight then, may be distinguished from a purely appetitive flight by a concomitant suppression of these vegetative activities.

Since juvenile hormone (JH) and the corpus allatum (CA) have been implicated in the control of reproductive development and behavior in many insects, Kennedy (1956) and later Johnson (1966, 1969) proposed that a deficiency of this hormone might characterize the migratory period. However, evidence from several sources suggests that in fact just the reverse may be true. Cassier (1963, 1964) found that implantation of CA from mature male *Locusta migratoria* into a male of the same age led to an increase in the speed of walking associated with a stronger phototactic response.

According to Odhiambo (1966), removal of the CA in male *Schistocerca gregaria* resulted in a decrease in locomotor activity which could be reversed by implanting active CA into allatectomized males. Odhiambo went so far as to propose a direct effect of the CA hormone on the level of excitability of the central nervous system. Also, Lebrun (1969) reported that the readiness of *Calotermes flavicolis* to embark on "nuptial swarming flights" depended upon high titers of JH, and Stengel and Schubert (1970), and Stengel (this volume) found that the state of activity of the CA affected orientation of flight in female *Melolontha melolontha*. Most recently Wajc and Pener (1971) have demonstrated that allatectomized adult *Locusta migratoria migratorioides* males flew significantly less than did sham-operated controls. It follows then, from the above results, that an increase in JH, if it produced increased locomotor activity and phototactic responses, could result in stimulation of migratory behavior.

Dingle (1965, 1966, 1968) has shown that the milkweed bug, *Oncopeltus fasciatus*, fits the Johnson - Kennedy characterization of a migrant insect very well. Flight is post-teneral and pre-reproductive. In response to short photoperiods, this species undergoes an adult reproductive diapause and a much greater time is available for long flights. Long photoperiods indicate favorable breeding conditions and bring on reproduction very quickly whereas the combination of long photoperiod and high temperature inhibits flight almost entirely. Therefore, to explore the role of the CA in controlling the onset and timing of insect migratory behavior, an attempt was made to alter the amount of JH activity in a population of *O. fasciatus* and to measure the subsequent effects on flight.

JUVENILE HORMONE MIMIC APPLICATIONS

Thirty-two pairs of males and females received 10 μg of a juvenile hormone mimic (Calbiochem, JH synthetic B grade) 5 days after the adult molt. Animals were flight tested on day 8 (using a tethered flight technique (Dingle 1965)) and every three days thereafter until day 20 and then on days 25 and 30. All were maintained under a 12L-12D 24°C (diapause inducing) regimen and were checked daily for reproductive activity. Forty-three control pairs were similarly treated but no hormone was applied. In a second series of experiments juvenile hormone mimic (JHM) (20 μg) was applied as above to 30 pairs of animals reared in a 16L-8D photoperiod two days after the adult molt while a group of 26 pairs of control animals was again treated similarly without receiving the hormone mimic. Subjects were flight tested only on day 5 (Caldwell and Rankin 1972).

JHM application significantly increased the proportion of males making long flights. When JHM was applied on day 5, between 35 and 60% of the treated males flew for over one hour on any given test day while only 5 to 21% of the control males made such flights. Eighty-seven percent of the treated males, in contrast to only 42% of the control animals, made at least one long flight ($P < 0.001$) (Fig. 1). Similarly,

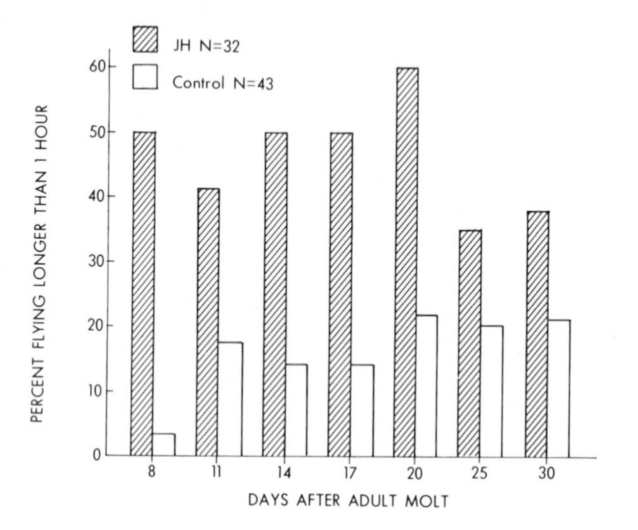

Fig. 1. Effect of JHM on flight in male *O. fasciatus*. Treated males were flight-tested on the days indicated. Controls received no hormone. (From Caldwell and Rankin 1972.)

when applied to long-day males on day 2, in the flight test on day 5, a significantly greater proportion of the treated animals made flights of over one hour than did controls (36.7% as compared to 7.7% of the controls) (Fig. 2) (Caldwell and Rankin 1972).

Unlike treated males, females receiving JHM on day 5 showed no significant increase in the proportion of individuals making flights of over one hour. However, when the hormone mimic was applied very early (on day 2) and flight tests administered on day 5, a significant increase in flight activity was observed among treated individuals (70% as compared to 35% in the control group, $P < 0.01$) (Fig. 2).

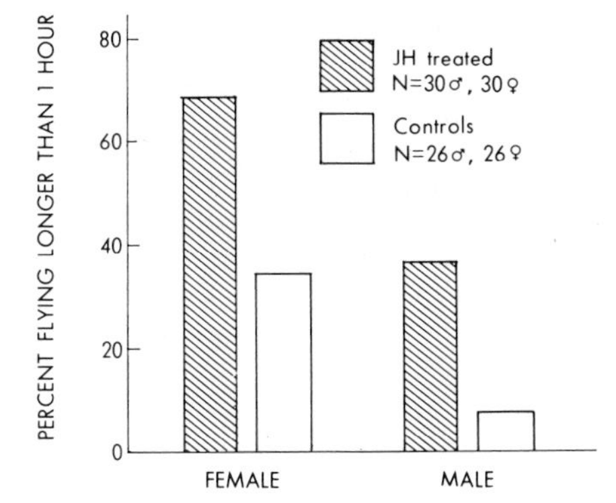

Fig. 2. Effect of early application of JHM on male and female *O. fasciatus*. Controls were untreated. All bugs were flight-tested on day 5. (From Caldwell and Rankin 1972.)

Hormonal applications also induced reproductive activity (Table 1); the mean age of first mating in the control group was day 23, mean age of first oviposition was day 38 while in the hormone treated group these ages were day 10 and 14 respectively. Thus the conflicting results among the group of JHM-treated older females in which flight activity decreased rather than increased in response to hormone treatment can be resolved. Hormone application had broken the reproductive diapause produced by the exposure to short photoperiod and since oviposition inhibits long flight behavior in *O. fasciatus* females, the effect of the applied hormone on egg development obscured any possible effect on flight.

Table 1. Mean age to first mating of JHM-treated and untreated single pairs of *O. fasciatus* (12L-12D 24°C).

Treatment	n	Mean age to first mating	Mean age to first egg
JHM 10 μg (day 5)	17	10	14
Control (untreated)	27	23	38

To remove this complicating effect of the hormone on egg development, groups of females were ovariectomized and given hormone treatment (10 μg, as above) on day 5 or 10 after the adult molt. In flight tests on day 10, 20, and 30, hormone applications produced a significant increase in flight activity over untreated ovariectomized controls, JHM treated intact females, or untreated intact controls. Thus juvenile hormone mimic was shown to affect flight of older females when its effect on ovarian development was eliminated (Fig. 3). It is interesting to note that Caldwell (1971) has stated that topically applied JHM affects pre-flight behavior in *O. fasciatus*, resulting in a significant decrease in length of time before free take-off.

EFFECTS OF STARVATION AND ALLATAL IMPLANTATION ON FLIGHT

To duplicate the effects of the synthetic hormone mimic using a natural hormone source, a group of 32 individuals maintained under short-day conditions (12L-12D, 23°C) received a single CA implant on day 2 and were flight tested on days 8 and 15. These implants had no significant effect on flight nor on reproductive development and were judged to be ineffective as a hormone source. A second series of implants was there-

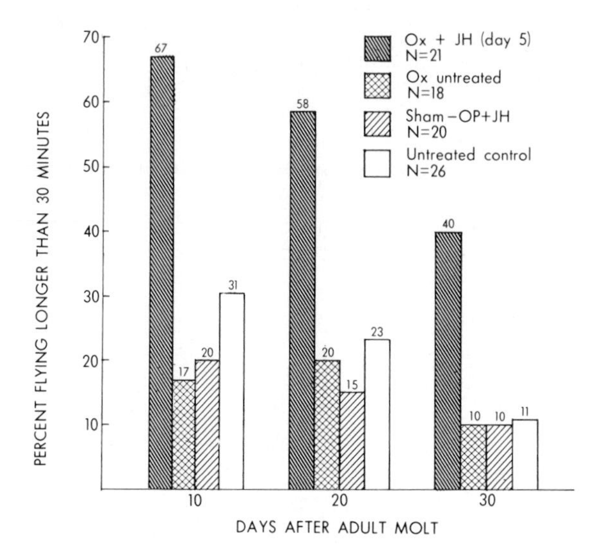

Fig. 3. Effect of JHM treatment on ovariectomized and intact female *O. fasciatus*. Controls were untreated ovariectomized and intact females. Hormone was administered on day 5

fore done using three glands per host. Both intact and ovariectomized females as well as intact males received implants. Flight tests were administered on days 10, 15, 20 and 30. On day 10 all implanted groups showed a significant increase in flight activity over control groups (Figs. 4 and 5). The initial high level of flight behavior was maintained in subsequent flight tests in males and ovariectomized females implanted with CA, but not in intact allatal implanted females where flight levels had declined by day 15 and were not significantly greater

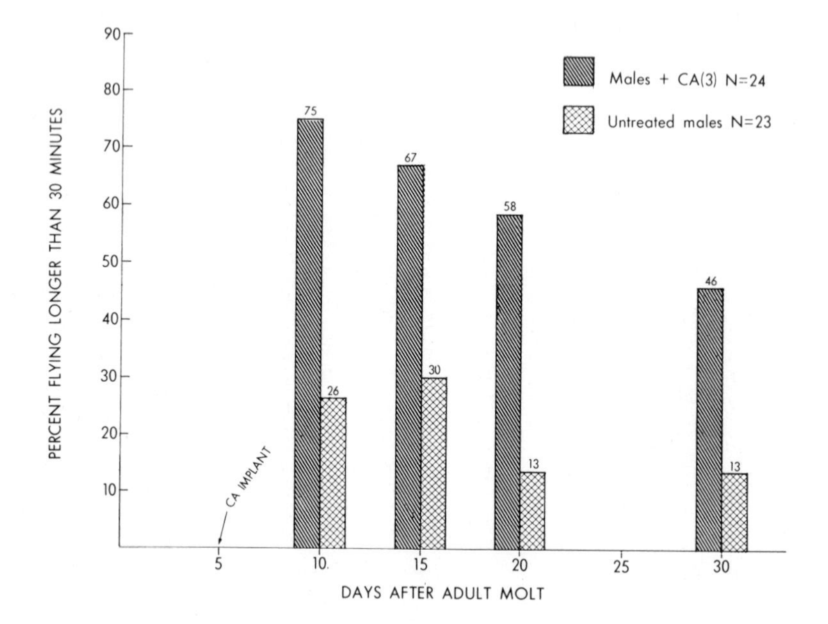

Fig. 4. Effect of implantation of three CA on flight in male *O. fasciatus*

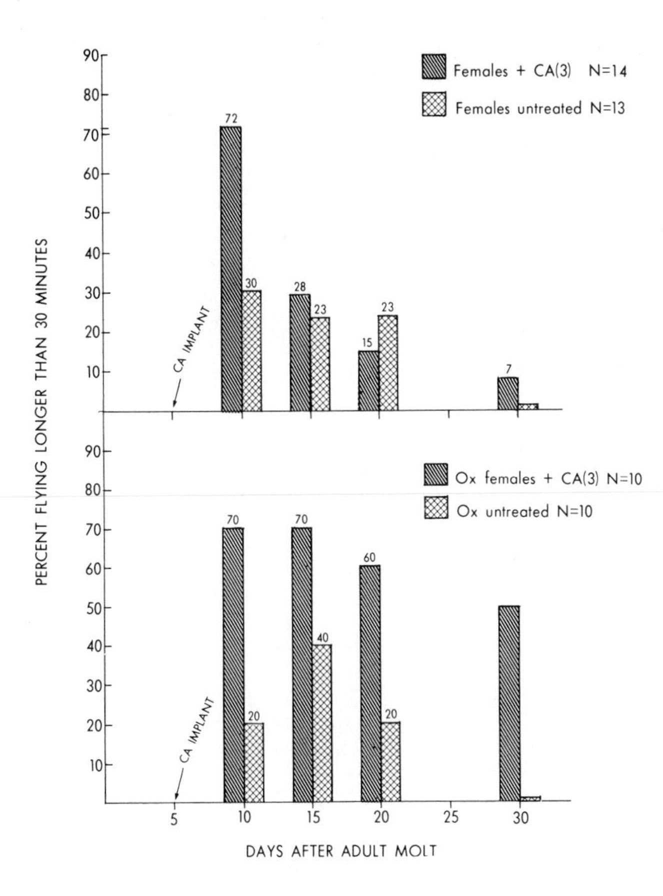

Fig. 5. Effect of implantation of three CA on flight of intact (A) and ovariectomized (B) female *O. fasciatus*. Controls were non-implanted, intact (A) or ovariectomized (B) females

than controls. This was likely due once again to the onset of oviposition in these females by this time and the resultant inhibition of further flight. These differences in response to allatal implantation or JHM application between ovariectomized and intact females clearly demonstrate the existence of an oogenesis-flight syndrome in this species.

To do the reciprocal experiment and investigate the effects of a decreased level of allatal hormone in the blood, it was necessary to attempt to remove the gland and flight test allatectomized insects. If this operation resulted in a decrease in flight activity, reimplantation of a CA or application of JHM as replacement therapy would be expected to restore the flight response. Due to extremely high post-operative mortality, it did not prove feasible to flight test allatectomized individuals. An alternative to removal of the CA as a method of reducing JH levels was therefore investigated.

The possibility that starvation might be analogous to surgical allatectomy was suggested by Johansson's (1958) investigation* of the relationship of nutrients to reproductive activity in *O. fasciatus*. He concluded that the CA is inhibited via nervous pathways from the brain

* Johansson's (1958) relevant findings included: (1) The CA is necessary for egg development in *O. fasciatus* but the brain neurosecretory cells are not, though fecundity is reduced in their absence. (2) Starvation from the time of the adult molt results in a complete inhibition of egg production. (3) Implanting one CA or cutting the nervous connectives from the brain to the gland but leaving the gland *in situ* resulted in egg production in starved females.

during starvation and termed this condition "pseudoallatectomy". Starved (pseudoallatectomized) bugs were therefore flight tested, were then given allatal implants or JHM application as replacement therapy and were flight tested again.

When a group of 46 *O. fasciatus* from a stock that had been artificially selected for high flight for five generations ("superfliers") were starved from day 5 and flight tested on days 10, 15, and 20, flight activity was observed to increase rather than diminish immediately after food deprivation. By day 20, however, the flight response had declined in the starved group to a level significantly below that of the fed controls. Allatal implants and JHM applications were administered to starved individuals on day 20 after flight levels had diminished. JHM application to one portion of the starved population significantly increased the number of individuals making long flights ($P < 0.03$). Fifty-six percent of the individuals in this treatment group made long flights on day 25 as compared with 14% of the untreated starved control group (Fig. 6). The effect persisted in the 30 day flight test but was not as great. Evidence that JHM was acting as the natural hormone could be seen in its effect on egg production. Four of the 8 females receiving hormone laid some eggs while none of the starved controls did so. Allatal implantation was less effective as replacement therapy. Long flights were not significantly affected and only one female laid eggs.

A second starvation-implant experiment was done which was generally similar in design and regimen to the previous one. In this, starved insects and insects which had access to food were flight tested on day 10. Fifteen pairs of starved animals received implants of three CA per individual on day 11 and were flight tested on days 15, 20, and 30. Their flight performances were compared with a control group consisting of 12 pairs of untreated starved individuals. Both groups were observed at 3 hr intervals during the light-on period of each day and checked for mating and oviposition behavior.

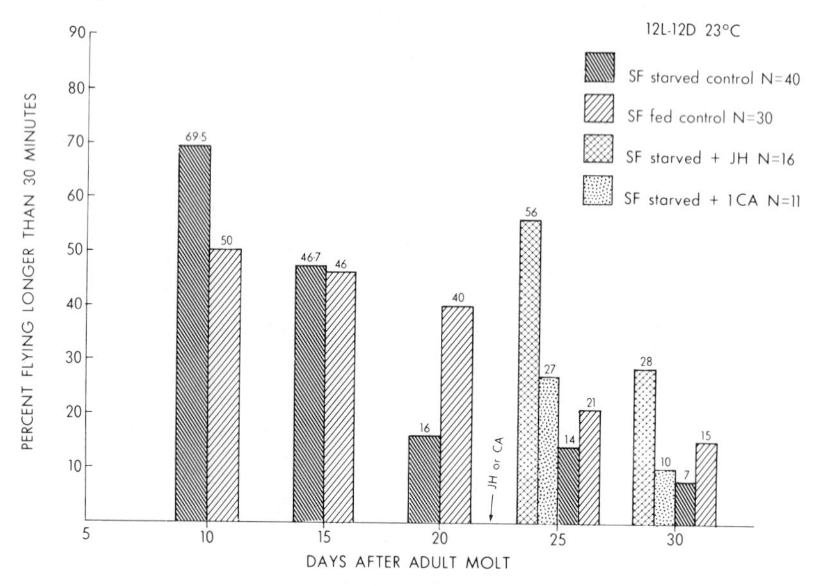

Fig. 6. Effect of application of JHM or implantation of one CA on flight in starved *O. fasciatus*. Controls were superflier (SF) starved and fed individuals. All groups were of superflier stock. After flight tests on day 20 the starved group was divided into three parts, one receiving a single implanted CA, a second receiving 10 μg JHM and a third which remained untreated and starved

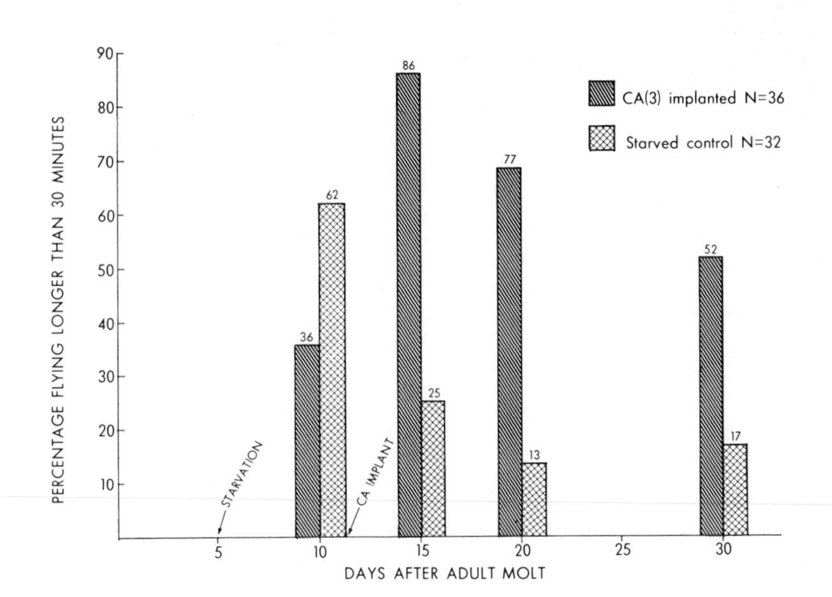

Fig. 7. Effect of implantation of three CA on flight of a starved superflier population of *O. fasciatus*. Controls were non-implanted, starved superfliers. (The entire starved population was flight tested together on day 10 and, without reference to flight performance, animals were chosen at random to receive allatal implants. The initial flight performances of the two groups thus formed were recorded separately to show initial levels of flight before treatment because by chance the flight performance of the implanted group was significantly lower than that of the control group in this first test.)

The percentage of individuals in the allatal triple-implanted group making long flights at each flight test was significantly greater than in the control group. On day 15, 86% of the starved-implanted group as compared to 25% of the starved untreated controls made a long flight ($P < 0.001$) (Fig. 7).

Reproductive activity was also stimulated by the implanted glands; seven of 18 starved-implanted females laid eggs at least once and of these, three laid several large clutches while none of the control females had oviposited. Flight tests on days 20 and 30 indicated that females, having laid eggs, were much less willing to make long flights; in fact, none of these females made a flight greater than 10 min, though all had done so on day 15. Eighty-nine percent ($^8/_9$) of the females tested on day 30 that had not oviposited made flights greater than 30 min. Most (70%) of the triple-implanted males tested on day 30 made long flights while only 17% of the unimplanted controls did so (both males and females).

These results indicate that the CA itself can stimulate flight behavior in this species and the effect can be duplicated by application of synthetic JHM. Flight and ovarian development can be restored among starved individuals by allatal implants, indicating that the eventual decline in flight as well as lack of ovarian development after starvation is indeed the result of allatal inhibition.

The inter-relationship between egg development and flight was again illustrated by the results of later flight tests. In every instance starved implanted females that had oviposited made no further long flights while most starved-implanted females that did not oviposit continued to make long flights for the duration of the experiment.

The increase in the flight response of superflier strains after starvation suggests a possible explanation for a paradox in the theory of *Oncopeltus* migration strategy. The basic hypothesis (mainly from Dingle (1968)) is that, in response to short photoperiods, reproduction is delayed and a portion of the resident southern population migrates northward carried by prevailing winds. Long photoperiods and high temperatures indicate favorable conditions, hasten the onset of reproduction, and result in a reduction in flight activity in terms of both the proportion of individuals flying and the duration of flights. The difficulty in this hypothesis is that migrant *O. fasciatus* usually appear in Iowa and similar latitudes in late June or early July (Dingle 1965), the longest days of the year. The increased flight response of the superfliers, after starvation in the allatal implantation experiments, suggests a possible explanation for this contradiction. Food availability may influence flight activity in the field as it appears to do in laboratory flight tests. Caldwell (1969) in field studies of *O. fasciatus* has pointed out that milkweed in northern areas of the U.S. (above the 40th parallel) is just beginning to flower in late June and early July. Therefore, food available at this time would be of poor quality, primarily flowers or green pods. *O. fasciatus* fed on green pods or flowers have been shown (Caldwell 1969) to delay reproduction as though they were malnourished even though high rearing temperatures and long photoperiods should have brought on oviposition very rapidly.

Although the population of superfliers used in the allatal implantation experiments had been selected for long flight for five generations, its full flight potential had apparently not been realized under normal laboratory culture conditions. Starvation resulted in a phenotypic modification of the flight response to a higher level. In nature then, the effect of poor quality food on a migrant would presumably be to delay reproduction and stimulate flight behavior regardless of photoperiod. Such a response would be highly adaptive since it would effectively maximize both the utilization of available northern habitats and the probability of escaping unfavorable conditions which would result in a lack of food.

DISCUSSION

If during the adult diapause the flight period is to be pre-reproductive and is of some duration, it is difficult to explain how the allatal hormone could induce migratory behavior without stimulating concomitant ovarian development, which would then inhibit flight in the female. Similarly, the involvement of the CA in the flight response after starvation is somewhat ambiguous. Though starvation does apparently inhibit the CA, it does not immediately reduce flight and, in fact, appears to stimulate that behavior for at least a short time after food is removed. The question once again arises as to how flight can be enhanced while ovarian development is inhibited, if both are stimulated by the same hormone.

One explanation might be that the flight system responds to a lower titer of hormone than does the reproductive system. It may be postulated that when the CA becomes reactivated after the imaginal molt, its degree of activity is determined by temperature and photoperiod, and the flight system is sensitive to very low titers of hormone while the ovaries require higher levels to begin development. According to this hypothesis, under diapause conditions, levels of hormone high enough for ovarian development would not be reached until late in the diapause period, whereas under non-diapause conditions, long photoperiods and high temperatures would result in immediate high levels of allatal activity, egg development and little flight in the female. If this were the case, one would expect to be able to produce a correlated response in age at first flight by applying artificial selection for either very early or very late reproduction to a population (under diapause-inducing conditions) for several generations. Similarly, directional selection for either early or late age of first migratory flight should produce a correlated shift in the same direction of the mean age of first reproduction in the population. According to this model, one would not expect to be able to select for both early flight and late reproduction (or *vice versa*) at the same time in a single genetic strain.

Starvation at day 5 apparently does inhibit the CA and results in a suppression of egg production. There is, however, an enhancement of the initial flight response which is possibly due to lower levels of hormone than will stimulate the ovaries. Flight levels then decline after about day 20 to very low levels and, if oviposition has not begun, starvation prevents it from occurring. Starvation then may stop egg production almost immediately after the allatum is suppressed whereas the flight system, responsive to low blood titers, would be stimulated by the hormone still present in the system until it dropped below the low threshold necessary for flight. The fact that applications of JH and allatal implants restore first flight and then ovarian development supports the above suggestion and indicates that it is actually the CA itself which is responsible for the post-starvation effects.

Another observation relevant to this question was made in the course of flight testing untreated diapausing control animals. Dingle (1968) has reported that photoperiod does not affect the overall proportion of a population making long flights, but rather that conditions which delay reproduction merely extend the time available for flights. Having tested a short-day population 10 days after the adult molt and again at 25 days, he assumed that the 30% making long flights were the same individuals at each test. Our data obtained by repeatedly testing individual control animals does not support this interpretation since 60.5% of such males made long flights at least once in the diapause period. Although the proportion of bugs flying at each test after day 8 was always 15% to 20%, it was not the same individuals flying each time. Apparently some animals characteristically fly soon after the imaginal molt and others, if a short photoperiod delays reproduction, fly later. In this experiment, none continued to make long flights throughout the entire diapause period. It is obvious therefore that the delay in reproduction in response to appropriate photoperiod and temperature provides not only more time for migratory flight, as Dingle (1968) indicated, but also results in a higher total percentage of the population making such flights. Possibly the time of flight during the diapause period differs for each individual and corresponds to a genetically determined time at which the CA becomes reactivated and begins secreting hormone at the low levels necessary for stimulation of flight.

One might propose, as an alternative to the above model, that when the CA is reactivated after the imaginal molt, it produces a high titer of hormone which stimulates the flight system but not the reproductive system and then diminishes to a low level, gradually rising again to stimulate egg production. It is possible that the reproductive system could not respond rapidly enough to utilize a brief increase in hormone titer while the flight system could. Such a fluctuation in hemolymph JH levels has been demonstrated (de Wilde *et al.* 1968) in *Leptinotarsa*. This is the only species in which JH titers have been examined over the life cycle of an insect. Using the *Galleria* bioassay for JH these workers showed a brief period of high JH activity associated with prediapause behavior after the adult molt. This is followed by a period of low activity and an adult reproductive diapause. The initial high titers are associated with pre-diapause behavior which may actually constitute a type of migration.

Whitmore, Whitmore and Gilbert (1972) have recently demonstrated that, in *Hyalophora glovi*, administration of JH resulted in the appearance of 6 carboxylesterases in the hemolymph which presumably are produced by the target tissues as a result of their response to the hormone. These enzymes break down JH in the hemolymph and act as a feedback mechanism controlling hemolymph titers of JH. If this kind of mechanism were operative in *O. fasciatus*, it would explain how an early high level of JH secretion could stimulate flight without also inducing ovarian development. If the flight system responded more quickly to the hormone than the ovaries, the esterase production would remove high levels of JH from the blood and ovarian activation would not occur. A great deal more secretion by the CA would then be required to bring the level of JH in the hemolymph up to a constant high level presumably necessary to stimulate oogenesis.

The demonstration of an effect of juvenile hormone itself on flight in the female migrant *O. fasciatus* is particularly significant since, if mating can occur before migration, as it does in this species, the female can found a colony without the presence of a male. Further, it is significant that JH is known to control egg development in this species. Pre-reproductive migration in the female results in eggs being deposited in a newly invaded habitat where there is likely to be abundant food and little competition. Thus, control of both flight and ovarian development by the CA increases the probability of the successful colonization of the new habitat by assuring a close coordination of these two activities in response to photoperiod, temperature changes, food shortages and any other environmental cue, be it proximal or ultimate, which affects the activity of the gland.

Caldwell (personal communication) has shown that application of JHM to *Lygaeus* results in no significant increase in flight activity. This species is a year-round resident at temperate latitudes; some members of the population are capable of long flights but apparently this species uses a different strategy to escape unfavorable environmental conditions. *Lygaeus* has probably not evolved an allatal control mechanism to coordinate flight with ovarian development, since in this species mating is the stimulus which inhibits further flight while egg development and oviposition in a virgin female have no effect on this behavior. The CA is probably primarily concerned with the control of egg production as is true in most species. The influence of JH on flight behavior is perhaps an added function which has evolved only in migratory species in which close coordination of flight and reproduction is highly advantageous.

In *Dysdercus*, where starvation stimulates flight, whereas feeding results in ovarian development and wing muscle hystolysis (Dingle 1972, and this volume), Edwards (1970) has concluded that flight muscle degeneration and vitellogenesis are induced by the CA which is apparently activated by feeding. It should be noted however that his conclusions are based on implantations of retrocerebral complexes rather than isolated CA. Here is a situation, just opposite to that in *Oncopeltus*, which seems to fit the Johnson-Kennedy JH deficiency syndrome theory of migration. The allatum apparently controls the initiation of "vegetative" functions and when activated, terminates the flight period irrevocably by initiating wing muscle hystolysis. Borden and Slater (1968) have shown, as well, that topically applied synthetic JH induced wing muscle degeneration in *Ips confusus* within two days. Yet Stegwee *et al.* (1963) found that in *Leptinotarsa decemlineata*, the CA are necessary to maintain the flight muscles, degeneration occurring after allatectomy. The reimplanting of CA or the application of JH results in regeneration of these muscles. It seems likely that species such as *Dysdercus* and *Ips* are essentially opportunists, in that the life span is relatively short and reproduction ensues quickly after finding an adequate food source and mate. No energy is diverted from reproduction for maintaining other options in terms of further flight. Species employing such a strategy are likely to evolve a mechanism to separate migratory from vegetative phenomena, in order to maximize the latter. *Oncopeltus, Leptinotarsa*, and similar species are somewhat longer lived, not so opportunistic in terms of immediate reproduction, and tend to maximize their options in fairly complex responses to different environmental contingencies. Here the CA is perhaps used to coordinate an animal's total response to changing environmental conditions and maximize the probability of utilizing a temporary habitat and effecting successful colonization.

ACKNOWLEDGEMENTS

Appreciation is expressed to Drs. Hugh Dingle, Roy L. Caldwell and Joseph Frankel for helpful discussions of this work and to Dr. John Rankin for technical assistance. This research was supported by an N.S.F. predoctoral fellowship.

REFERENCES

BORDEN, J.H., SLATER, C.E.: Induction of flight muscle degeneration by synthetic juvenile hormone in *Ips confusus* (Coleoptera: Scolytidae). Z. vergl. Physiol. 61, 366-368 (1968).

CALDWELL, R.: A comparison of dispersal strategies in two species of milkweed bugs, *Oncopeltus fasciatus* and *Lygaeus kalmii*. Ph.D. Thesis, University of Iowa. (1969).

CALDWELL, R.: Juvenile hormone mimic effects on pre-migratory behavior in the milkweed bug, *Oncopeltus fasciatus*. (Abstract). Am. Zool. 11, 643 (1971).

CALDWELL, R.L., RANKIN, M.A.: Effects of a juvenile hormone mimic on flight in the milkweed bug, *Oncopeltus fasciatus*. Gen. comp. Endocrinol. 19, 601-605 (1972).

CASSIER, P.: Action des implantations de corps allates sur la réactivité phototropique de *Locusta migratoria migratorioides* (R.et F.), phase gregaria. C. r. hebd. Séanc. Acad. Sci., Paris 257, 4048-4049 (1963).

CASSIER, P.: Etude et interpretation des effets a long terme, des implantations abdominales de corps allates, sur la réactivité phototropique de *Locusta migratoria migratorioides* (R.et F.), phase gregaire. C. r. hebd. Séanc. Acad. Sci., Paris 258, 723-725 (1964).

DINGLE, H.: The relation between age and flight activity in the milkweed bug, *Oncopeltus*. J. exp. Biol. 42, 269-283 (1965).

DINGLE, H.: Some factors affecting flight activity in individual milkweed bugs (*Oncopeltus*). J. exp. Biol. 44, 335-343 (1966).

DINGLE, H.: The influence of environment and heredity on flight activity in the milkweed bug, *Oncopeltus*. J. exp. Biol. 48, 175-184 (1968).

DINGLE, H.: Migration strategies of insects. Science, N.Y. 175, 1327-1335 (1972).

EDWARDS, F.J.: Endocrine control of flight muscle histolysis in *Dysdercus intermedius*. J. Insect Physiol. 16, 2027-2031 (1970).

JOHANSSON, A.S.: Relation of nutrition to endocrine-reproductive functions in the milkweed bug, *Oncopeltus fasciatus* (Dallas). Nytt Mag. Zool. 7, 1-132 (1958).

JOHNSON, C.G.: Physiological factors in insect migration by flight. Nature, Lond. 198, 423-427 (1963).

JOHNSON, C.G.: A functional system of adaptive dispersal by flight. A. Rev. Ent. 11, 233-260 (1966).

JOHNSON, C.G.: "Migration and Dispersal of Insects by Flight". London: Methuen. (1969).

KENNEDY, J.S.: Phase transformation in locust biology. Biol. Rev. 31, 349-370 (1956).

KENNEDY, J.S.: A turning point in the study of insect migration. Nature, Lond. 189, 785-791 (1961).

LEBRUN, D.: Corps allates et instinct genesique de *Calotermes flavicollis* Fabr. Le déclenchement de l'activité sexuelle des jeunes imagos ailées de *Calotermes flavicollis* Fabr. nécessite la présence dans l'organisme d'un taux élevé d'hormone juvénile. C. r. hebd. Seanc. Acad. Sci., Paris 269, 632-634 (1969).

ODHIAMBO, T.R.: The metabolic effects of the corpus allatum hormone in the male desert locust. J. exp. Biol. 45, 51-63 (1966).

SOUTHWOOD, T.R.E.: Migration of terrestrial arthropods in relation to habitat. Biol. Rev. 37, 171-214 (1962).

STEGWEE, D., KIMMEL, E.C., deBOER, J.A., HENSTRA, S.: Hormonal control of reversible degeneration of flight muscle in the Colorado potato beetle, *Leptinotarsa decemlineata* Say (Coleoptera). J. Cell Biol. 19, 519-527 (1963).

STENGEL, M., SCHUBERT, G.: Rôle des corpora allata dans le comportement migrateur de la femalle de *Melolontha melolontha* L. (Coléop. Scarabaeidae). C. r. hebd. Séanc. Acad. Sci., Paris 270, 181-184 (1970).

WAJC, E., PENER, M.P.: The effect of the corpora allata on the flight activity of the male African migratory locust, *Locusta migratoria migratorioides* (R. and F.). Gen. comp. Endocrinol. 17, 327-333 (1971).

WHITMORE, D., WHITMORE, E., GILBERT, L.: Juvenile hormone induction of esterases; a mechanism for the regulation of juvenile hormone titer. Proc. natn. Acad. Sci., U.S.A. 69, 1592-1595 (1972).

de WILDE, J., STAAL, G., de KORT, C., DeLOOF, A., BAARD, G.: Juvenile hormone titer in the hemolymph as a function of photoperiodic treatment in the adult Colorado beetle (*Leptinotarsa decemlineata* Say). Proc. K. ned. Akad. Wet. (C) 71, 321-326 (1968).

THE EXPERIMENTAL ANALYSIS OF MIGRATION AND LIFE-HISTORY STRATEGIES IN INSECTS

H. Dingle

Department of Zoology, University of Iowa, Iowa City, Iowa, U.S.A.

The introduction of experimental methods into the study of insect migration is a relatively recent phenomenon. Earlier studies based on field observations, light trapping, and the compilation of arrival and breeding dates had clearly shown that migration in insects was widespread (Williams 1958), but it was not until migrant insects were studied in both laboratory and field that the nature of migration as a physiological syndrome was understood. Not that field observations were rendered less important, far from it (e.g. Baker 1969), but the data from experimental approaches could now be combined with the field data to provide an integrated view of the whole migration process (Johnson 1969; Dingle 1972). Pioneering the experimental approach was J.S. Kennedy whose now classic papers on the flight and settling of *Aphis fabae* (Kennedy 1958, 1965, 1966, this volume; Kennedy and Booth 1963a,b, 1964) showed that migration is characterized by the enhancement of flight behaviour, the inhibition of settling and reproductive behaviour, and complex interactions among the three.

Studies of the relation between migration and reproduction in a variety of insects indicate that for the most part reproduction begins following the migratory flight, although in some species such flight may occur between bouts of reproductive activity. In females of many species, migration occurs prior to egg development, a fact which lead Johnson to refer to the relation between migration and reproduction as the "oogenesis-flight syndrome" (Johnson 1969) during which wing muscle development and activity is maximized and reproductive development is minimized. This concept of migration has been criticized because the role of the males was not clear, and indeed males often retain flight capacity after it has been lost in females (e.g. Dingle and Arora 1973) and may therefore have different strategies. However, during copulation males of some species are tied to females for long periods and are effectively prevented from migrating (Dingle and Arora 1973). It is perhaps preferable to refer to a "reproduction-flight syndrome" for both physiological and ecological reasons to be developed below (see also Caldwell and Rankin, this volume), but the chief tenets of the Johnson model, at least for females, remain basically sound (Dingle 1972).

The timing of reproduction is of course important to the role of migrant insects as colonizers of new habitats (Dingle 1972), and it is this aspect of the study of migration that has profited from recent theoretical advances in colonization and demographic theory (e.g. MacArthur and Wilson 1967). If a migrant is to be a successful colonizer, it must be able to reproduce and leave descendants. Its ability to do so is indicated by its "reproductive value" which is the expected contribution of an individual to future population growth (Fisher 1958; Dingle 1972). Reproductive value in most species reaches a maximum just as reproduction begins; the significance of pre-reproductive migration is thus obvious. Colonization potential is also enhanced by rapid population growth which occurs when r, the so-called intrinsic rate of increase, is high. Migrants thus tend to be "r strategists" emphasizing high reproductive rates adapted to unstable habitats rather than "K strategists" emphasizing lower rates, but maximum survival of offspring in environments near the carrying capacity, K (MacArthur and

Wilson 1967). Indeed a summary of the available evidence shows that migrant insects are most likely to be denizens of temporary or unstable habitats (Southwood 1962).

What is becoming increasingly obvious from both the physiological and ecological data is that migration is an integral part of the life history of any given species and cannot be meaningfully separated from it. As such, migration is intimately related to other aspects of the life history such as growth, maturation, timing of reproduction, and total reproductive effort. These relationships are outlined in Fig. 1 which gives a generalized model of the relation between migration and life histories.

As Fig. 1 illustrates, following the adult moult there is a period of maturation prior to migration. This period can vary from a few hours (e.g. aphids, some butterflies) to a few (bugs) or several (locusts) days; during this maturation a variety of internal and external stimuli interact to influence both rate and end product in terms of the individual's tendency to migrate. At the end of the maturation period there is a "choice point" where individuals may settle down and begin reproducing or migrate before doing so. The bell shaped curve across the choice point indicates that depending on species, habitat, and previous individual history a variable proportion of any given population will be migrants ranging from zero (apterous or brachypterous forms of species with alary polymorphism) to virtually 100% (gregarious phase of migratory locusts). In the case of populations proceeding directly to reproduction, later unfavourable environmental circumstances may result in an inter-reproductive migration. In the case of migration, there may be a temporary halt for a stay at a feeding or diapause site. But whichever track is followed, reproduction and population growth eventually ensue. The above points are taken up in more detail below.

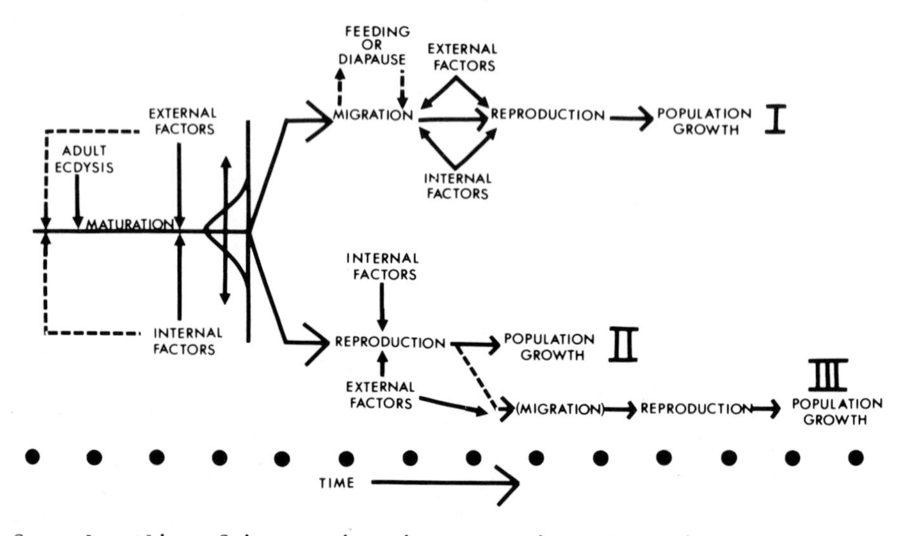

Fig. 1. General outline of insect migration strategies. Following adult ecdysis is a period of maturation influenced by various internal (hormones, nervous programming) and external (photoperiod, temperature, food supply) factors. Following this period the population contains a varying proportion of migrants indicated by the arrow on the distribution curve. Track I (migration) may include a delay for diapause or feeding. Track II (no migration) leads directly to population growth but external factors may result in inter-reproductive migration (Track III). Track II may also involve pre-reproductive diapause, omitted for simplicity

MATURATION

There is now a great deal of evidence from experimental studies of migrant insects demonstrating that flight is a function of age. In typical examples using tethered insects maximum flight duration occurs early in adult life following a maturation period (Williams, Barness and Sawyer 1943; Atkins 1960; Dingle 1965; Rygg 1966; Johnson 1969; Caldwell, this volume). During this period of maturation the cuticle hardens and may thicken via the deposition of growth rings (Neville 1967; Dingle, Caldwell and Haskell 1969). Various internal changes, such as growth of fat body and reproductive organs, also occur and in the milkweed bug, *Oncopeltus fasciatus*, are indicated by more feeding in early post-eclosion individuals than in older bugs (Caldwell and Rankin, in press; Caldwell, this volume).

In *O. fasciatus* feeding in the teneral adult occurs throughout the day with approximately the same frequency. During the migratory period, however, mid-day feeding levels decline. Feeding is restricted primarily to evening hours, whereas flight is maximum during the middle of the day. Copulation also appears following the teneral period, but it too reaches its highest frequency in the evening when it may proceed simultaneously with feeding. Oviposition, like flight, occurs maximally at midday but is postmigratory. Changes in responsiveness to "vegetative" stimuli therefore segregate such activities from flight either by age or by time of day (Caldwell and Rankin, in press).

Neurophysiological investigations of the flight system in locusts also indicate a period of maturation in the nervous programme controlling flight (Kutsch 1971, this volume). On the first day following imaginal ecdysis a locust will fly with loss of tarsal contact, but with a wing-beat frequency of only 10 Hz, roughly half that of a mature adult in flight. Interval histograms of the spikes of single neuromuscular units indicate prolonged interspike intervals relative to those in the fully mature adult. The wingbeat frequency increases in exponential fashion as the animal matures reaching asymptote of *c.* 20 Hz at 3 weeks which coincides with the termination of cuticle deposition (Neville 1967). Thus maturation is, at least in this respect, an integrated process. The role of hormones in this process is considered by Rankin elsewhere in this volume.

MIGRANTS AND NON-MIGRANTS

The most conspicuous alternation of migrant and non-migrant pathways (Fig. 1) occurs in those insects with alary polymorphism such as aphids and gerrids. In general, those environmental inputs which signal that conditions are favourable to reproduction result in apterous or brachypterous morphs. In aphids the relevant stimuli appear to be tactile. With low population densities and hence little contact between individuals, apterous individuals are produced whereas crowded conditions with increased contact result in the production of alatae (Lees 1966, 1967) and increased migratory flight (Dixon 1969). In *Megoura viciae* high temperature suppresses alate production and in other species photoperiod may also exert an influence, but in either case are subordinate to tactile inputs. They do, however, have an important role in determining whether morphs will be sexual or parthenogenetic. In some species of Gerroidea wing polymorphism is apparently a function of temperature with low temperatures during vitellogenesis resulting in an increase in the number of brachypterous or micropterous forms observed in summer. However, genetic factors may also be involved, and the situation is far from clear (Brinkhurst 1959, 1963; Guthrie 1959). An experimental analysis of wing polymorphism in *Gerris odontogaster* revealed no genetic differences between morphs; rather, decreasing daylengths during or prior to the fourth instar induce a long-winged diapause adult, a finding which indicates that photoperiod operates a switch mechanism (Vepsäläinen 1971). Carefully controlled experimental studies of other species are clearly needed.

Polymorphism may also occur in the flight musculature with no externally visible result in the wings themselves. In various British corixids low temperatures will result in flightless forms and high in normal; younger larvae of early summer are most sensitive to high temperature, but the effect can still be seen in older, late summer larvae or in newly moulted adults (Young 1965). Changes in food during the summer are perhaps also a contributing influence.

Food is certainly the critical factor in the wing muscle polymorphism of various African cotton stainer bugs of the genus *Dysdercus* (Edwards 1969; Arora 1971; Dingle and Arora 1973). In females of *D. cardinalis*, *D. fasciatus*, *D. intermedius*, *D. nigrofasciatus* and *D. superstitiosus*, wing muscle histolysis is induced by feeding. In the absence of food, the wing muscles remain intact and migration occurs. In fed *D. cardinalis*, *D. nigrofasciatus* and *D. superstitiosus* females, a brief period of flight may occur prior to histolysis, but fed *D. fasciatus* females never fly. Results of flight tests on *D. fasciatus* and *D. nigrofasciatus* are shown for comparison in Fig. 2. *D. fasciatus* feeds on the large fallen fruits of the baobab tree (*Adansonia*) which are abundant seasonally. When fruits are present, females feed, histolyse wing muscles, and reproduce; in the absence of food, they migrate. For the remaining species, food plants are usually scattered, but generally available, annuals and perennials. Here a short flight to a new host plant is advantageous; under conditions of starvation, such as might occur during severe drought, long distance migration is still possible. An additional factor in *D. intermedius* is copulation which leads to

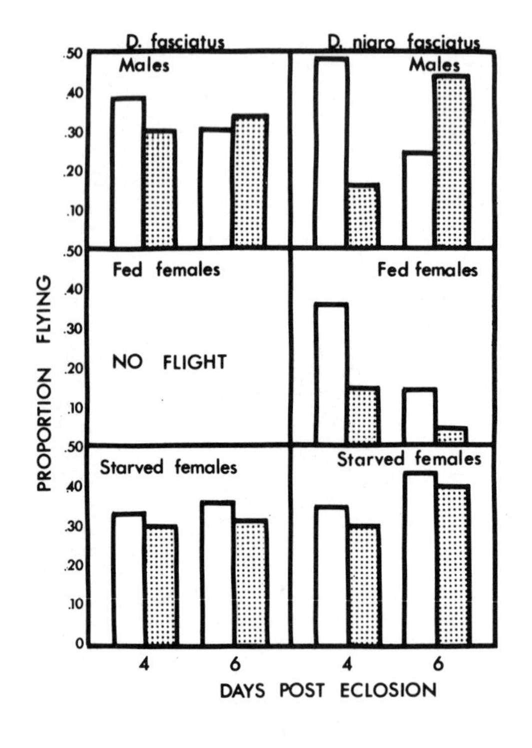

Fig. 2. Comparison of proportions of bugs flying for less than 30 min (open bars) and more than 30 min (stippled bars) in *Dysdercus fasciatus* and *D. nigrofasciatus* following adult eclosion. *D. superstitiosus* and *D. cardinalis* are similar to *D. nigrofasciatus*

muscle histolysis even in starved females; this is not the case for the other species.

Finally, the proportion of migrants in a population may vary independently of such gross morphological differences as alary or flight muscle polymorphism. Here it is useful to distinguish between long duration undistracted migratory flight and shorter appetitive "trivial" flights in search of food or mate (Southwood 1962; Johnson 1969). In a population of the milkweed bug *Oncopeltus fasciatus* maintained under an LD16:8 photoperiod of 23°C, for example, only 30% of females and 18% of males flew in tethered flight for 30 min or longer which is a reasonable operational criterion for migration (Dingle 1966) whereas many more flew for briefer periods. Raising the temperature to 27°C reduced the amount of migratory flight whereas decreasing the photoperiod to LD12:12 increased it (Dingle 1968a; Caldwell and Rankin 1972). Selection for long flyers increased the proportion flying for over 30 min to about 65% in both sexes in one generation. The proportion of long flyers in another milkweed bug, *Lygaeus kalmii*, varied with season and in different parts of the range (Caldwell 1969). Heritability (see below) of flight duration in *L. kalmii* varied from 0.2 to 0.4 clearly indicating the presence of genetic variance (Caldwell and Hegmann 1969). Genetic and environmental influences are thus involved in both species.

MIGRATION AND DIAPAUSE

The physiological and ecological similarities between migration and diapause are considerable (Kennedy 1961); both, for example, involve the temporary suppression of reproductive development. Indeed the two phenomena may occur together in the adult insect and migratory flights to and from diapause sites are common (Johnson 1969). Well known examples include the bogong moth, *Agrotis infusa*, in Australia (Common 1954), the heteropteran *Eurygaster integriceps* in the Middle East (Brown 1965), and numerous species of coccinellids (Hodek 1967; Johnson 1969). The entire life cycle of the heteropteran *Lygaeus equestris* in Sweden was studied by Solbreck (1972) who found an increase in fat body size prior to the autumn pre-diapause flight and a decrease over the course of the winter, little or no ovarian development during autumn, winter, and early spring, and rapid ovarian development during May coinciding with spring migration to breeding sites and increased mating. The species thus fits Johnson's (1969) model of a reproduction-flight syndrome. Similarly *L. kalmii* also undergoes reproductive development in the spring after overwintering, and adults which begin reproducing in the autumn fail to survive the winter (Caldwell 1969, and this volume).

Migration and diapause are also intimately related in the milkweed bug *Oncopeltus fasciatus*. In North America this species migrates north in the spring and colonizes stands of developing milkweed. Migration proceeds northward in a series of steps with each succeeding generation advancing a few hundred miles farther (Dingle 1972). When the photoperiod decreases in the autumn to LD12:12, the bugs enter ovarian diapause and field populations are observed to decrease as reproduction ceases and adults leave as they mature (Caldwell 1969; Dingle 1972). Presumably the emigrating bugs move south on the generally southwards directed winds of autumn, and the diapause allows several weeks for the bugs to make the journey before reproducing. Reproduction does take place at various locations in the southern U.S., the West Indies, and Mexico throughout the winter. In the laboratory we have shown that the diapause results both in more bugs flying for long periods (Caldwell and Rankin 1972; Rankin, this volume) and in a longer period of adult life during which such flights can take place (Dingle 1968a; Caldwell and Rankin 1972; Rankin 1972).

Table 1. Selection for age at first oviposition in females of *Oncopeltus fasciatus*. Change in mean age at first reproduction ($\bar{\alpha}$) with selection and calculation of realized heritability from $R = h^2S$. Generation 1 transferred from LD16:8, 23°C to LD12:12, 23°C as eggs. Remaining generations reared at LD12:12.

Generation	$\bar{\alpha}$	Selection Differential (S)	Response to Selection (R)	Heritability (h^2)
1	63.5			
2	61.3	6.6	2.2	0.33
3	50.9	9.6	10.3	1.08
4	37.8	7.3	13.1	1.79
5	35.3	5.3	2.5	0.47
6	21.7	10.3	13.6	1.30
7	16.7	4.4	5.0	1.14
8	14.9	2.7	1.8	0.67

This ovarian diapause in *O. fasciatus* has proven to be remarkably flexible. Selection experiments in the laboratory have shifted the age at first oviposition from 63.5 days, i.e. following ovarian diapause, to 16.7 days, a time typical of non-diapausing insects, after 7 generations (Table 1). In these experiments bugs were initially transferred as eggs from LD16:8 at 23°C to LD12:12 at 23°C. These bugs are indicated as the first generation in Table 1. The 5 earliest pairs to reproduce were then selected as the parents of the next generation and so on through the remaining generations. Several offspring from generation 7 (i.e. generation 8) were transferred as eggs to LD11:13, and 30 pairs from these eggs again displayed reproductive diapause, not reproducing until past 60 days old. Selection for earlier reproduction, therefore, has not eliminated diapause, but rather has shifted the critical photoperiod for its induction.

Selection experiments with other species of insect have also indicated the flexibility of diapause (Slifer and King 1961; Barry and Adkisson 1966; Pickford and Randell 1969; Honěk 1972). There is also geographic variation in critical photoperiod within species (Danilevskii 1965; Tauber and Tauber 1972) and crosses between different geographical races produce offspring with intermediate characteristics (Danilevskii 1965). On the basis of these and other data, it is tacitly assumed that diapause depends on the operation of several different genes (Lees 1955; Barry and Adkisson 1966; Morris and Fulton 1970; Honěk 1972).

The appropriate methods of analyzing such polygenic characters are available from quantitative genetics (Falconer 1960). The basic models of quantitative genetics assume that individual differences in continuously varying traits are the result of both genetic and environmental differences. The total (phenotypic) variance for any trait is expressed by:

$$V_P = V_A + V_N + V_E$$

where V_A is the additive genetic variance, V_N is the non-additive genetic variance (due to dominance and epistasis), and V_E is the environmental variance. The additive genetic variance arises from the summation of effects across all genes contributing to a trait and is the variance contributing specifically to parent-offspring resemblance (as in human height, for example) as opposed to the resemblance between all members of a species (as in the number of fingers, for example). The relationship between additive genetic variance and phenotypic variance is expressed by the heritability (h^2) of a trait where $h^2 = V_A/V_P$. Heritability is important because it predicts sensitivity to selection and hence the rate at which evolution can occur.

The heritability of a trait can be estimated in several ways (Falconer 1960). The one used here for diapause is to determine the "realized heritability" from the selection experiments described above; it is illustrated in Fig. 3. Basically the method involves selecting

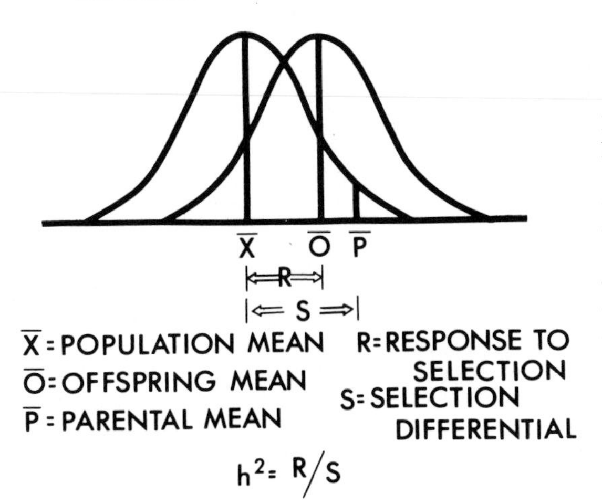

Fig. 3. Determination of realized heritability. The selection differential is the difference between the mean score of the selected parents ($\bar{P}$) and the mean of the population from which they were taken ($\bar{X}$). Response to selection is the difference between the mean score of the offspring of the selected parents ($\bar{O}$) and the population mean

parents whose mean score, in this case days to first oviposition, differs in the desired direction from the mean of the population; the difference between the parental mean and the population mean is the selection differential, S. The mean score of the offspring of these parents is then calculated, and the difference between offspring mean and population mean is the response to selection, R. On the assumption that the response to selection is the result of additive genetic variance, $h^2 = R/S$. Obviously any environmental effects unaccounted for will bias the estimate. In this case the environment is controlled in so far as possible, but unknown transient environmental factors could still introduce bias. One final point should also be noted, namely that the above experiments have been done without an *unselected* control line, i.e. a line free of both artificial and natural selection. Such a control line could also control for environmental transients. The reasons for lack of a control involve the nature of the character, age at first reproduction, which is extremely sensitive to selection as the heritability estimates indicate, making the establishment of effective controls difficult. Attempts are in progress to establish an unselected line although two trials so far have failed to demonstrate complete freedom from the effects of natural selection.

Response to selection, selection differential, and the resultant realized heritabilities derived from selection for age at first reproduction in *O. fasciatus* at LD12:12 are shown in Table 1. The mean heritability from all generations is 1.18 indicating, since additive genetic variance cannot be greater than total variance, that the heritability for this trait must be close to 1.0. In other words, the response to selection is essentially equal to the selection differential; the trait thus exhibits maximum genetic flexibility.

Such flexibility is unexpected in a trait as important to survival as the critical photoperiod determining diapause. This is because this trait should be closely related to Darwinian fitness and such traits usually undergo strong directional selection and display little additive genetic variance and much non-additive variance. Heritability is thus apt to be low while conversely the coefficient of genetic determination is higher (Fisher 1958; Falconer 1960; Dobzhansky 1970). Although non-additive genetic effects have not been eliminated in the selection experiments with *O. fasciatus*, the high heritability estimates suggest that variance due to such non-additive effects must be low. Therefore, these high heritability estimates remain to be explained.

In this connection, it is important to note that natural selection for critical photoperiod in *O. fasciatus* is not unidirectional. In the spring, as the bug moves north, daylength is increasing, and it is an advantage to reproduce early. This is because early reproduction leads to rapid population growth (see below) utilizing the seasonally abundant milkweed crop. The bug is unable, however, to survive the winter so continued reproduction is highly disadvantageous. Rather the bug enters diapause, thus permitting escape through migration. This is, of course, triggered by the shorter photoperiods of late summer and early autumn. The direction of selection thus reverses itself seasonally and could be the factor maintaining high additive genetic variance in the population. For a trait closely associated with diapause, namely the amount of heat necessary to break diapause in the fall webworm *Hyphantria cunea*, Morris and Fulton (1970) also obtained a high heritability estimate. Further, the fact that critical photoperiods for hybrid offspring of different geographical races of several Russian Lepidoptera tend to be intermediate between those of the parents suggests high heritabilities here as well. In this case selection presumably differs with latitude. Finally, the selection experiments of Honěk (1972) on diapause in the pentatomid bug *Aelia* also indicate high heritabilities. The available data thus support the notion that additive genetic variance for diapause or diapause related traits is indeed high, indicating considerable genetic as well as environmental influence.

TERMINATION OF MIGRATION

The stimuli which cause arrestment of migratory flight are frequently associated with the food plant either directly through food intake or indirectly through some substance which the plant produces. Feeding may induce wing muscle degeneration thus effectively preventing further migration or may induce reproduction which inhibits all but short distance (trivial) flights to new host plants or to seek mates. Examples of forms with wing muscle degeneration following feeding are aphids (Kennedy 1958 et seq.), scolytid beetles (Bhakthan, Borden and Nair 1970), and females of several species of *Dysdercus* (Edwards 1969; Duviard 1972; Dingle and Arora 1973). The scolytids may regenerate the muscles for a later flight (Bhakthan, Nair and Borden 1971), but in aphids and *Dysdercus* the process of degeneration is apparently irreversible. In *Aphis fabae* and the scolytids *Dendroctonus pseudotsugae* and *Trypodendron lineatum*, the initial response to the host plant and subsequent settling behaviour are primed by the previous migratory flight; with no flight, the responses either have a high threshold or fail to appear (Kennedy 1958 et seq.; Bennett and Borden 1971). In *A. fabae* there may also be a searching flight prior to settling.

Food induced reproduction which suppresses further migration is seen in locusts and in *Oncopeltus fasciatus*. Locusts (*Schistocerca*) delay reproduction when fed senescent leaves, but mature reproductively when fed fresh. The active substance is apparently the plant growth hormone gibberellin A_3. Terpenoids from young leaves of various aromatic shrubs may have the same effect (Ellis, Carlisle and Osborne 1965). The relevant factors in *O. fasciatus* are not clear, but abdominal stretch from ripening eggs may be involved (Rankin 1972). Reproduction acts directly in the bugs *Lygaeus kalmii* (Caldwell 1969 and this volume) and *Dysdercus intermedius* (Edwards 1969). In the former no migration occurs after mating, and in the latter mating induces wing muscle histolysis. No matter what the precise stimulus, however, it is clear that inputs signalling a favourable environment for reproduction and population growth induce flight arrestment.

REPRODUCTION AND POPULATION GROWTH

Population growth rate is measured by r, the intrinsic rate of increase, such that in a population of a given size, N, growing in an unlimited universe and approaching a stable age distribution, growth will occur according to the relation $dN/dt = rN$. This value is approximated by trial and error substitution in the equation

$$\sum_{\alpha}^{\omega} l_x m_x e^{-rx} = 1$$

where α and ω are, respectively, the ages at first and last reproduction, l_x is survival to age x, m_x is the number of (female) offspring produced by a female of age x, and e is the base of natural logarithms. It is characteristic of r that it is increased especially by earlier reproduction (α) and also by higher fecundity (m_x); prolonging reproduction has relatively little effect on r (Cole 1954; Lewontin 1965). In general it is advantageous for a species which like a migrant insect is colonizing new habitats, to maximize fecundity and reproduce early thus increasing r. An exception of course is found when it becomes disadvantageous to reproduce early, as when offspring will be killed by an oncoming winter, and reproduction may be delayed for diapause or migration thus reducing r. An example is *O. fasciatus* which on a long day (LD16:8) at warm temperature (27°C) has an r of .0810 per individual per day, while under diapause inducing conditions of LD12:12 and 23°C, it has an r of .0369 (Dingle 1968b, 1972). High temperature also reduces the amount of migratory flight (Dingle 1968a). There is, therefore, considerable environmentally induced within species variation in r.

The value of r also varies between species as a function of habitat and the resultant strategies. An example occurs in 3 species of *Dysdercus* from East Africa. *Dysdercus fasciatus* feeds on a seasonally rich food source, the fallen fruits of the baobab tree (*Adansonia*). This situation is exploited by maximizing early reproduction and fecundity producing an r of .0939 per individual per day, the highest of the 3 species. *Dysdercus nigrofasciatus* is intermediate with an r of .0878. It feeds on a variety of annual and perennial Malvales which are generally available, but never as abundant as baobab fruits in season. *Dysdercus superstitiosus* also feeds on annuals and perennials and can utilize a variety of species of plant other than Malvales, the only *Dysdercus* to do so. It thus faces the least environmental fluctuation, and has the lowest r, .0616. The reproductive strategies of these bugs thus reflect the type of environmental resources utilized and are consistent with the migration strategies discussed above.

The relation between demographic factors and colonizing ability is contained in the concept of reproductive value (Fisher 1958; Dingle 1972). This is analogous to compound interest on a bank account and indicates the expected contribution of an individual of specified age to future population growth (i.e. in terms of the "compound interest" generated both by its own future reproduction and that of its descendants). The higher the value the more likely the founding of a successful colony. Mathematically, the reproductive value, V_x, of an individual of age x in a growing population is defined relative to its value at birth, V_o. Thus where t is time and the other terms are as previously defined

$$\frac{V_x}{V_o} = \frac{e^{rx}}{l_x} \int_x^{\infty} l_t m_t e^{-rt} dt$$

Usually the integral can be replaced by a summation sign and the time by age for greater simplicity of calculation. Integration or summation is from x to infinity (or ω) so that only future births are included, and since V_x is relative to V_o, we can set $V_o = 1$. Note that rate of increase, survivorship, and fecundity all enter into the calculation of V_x.

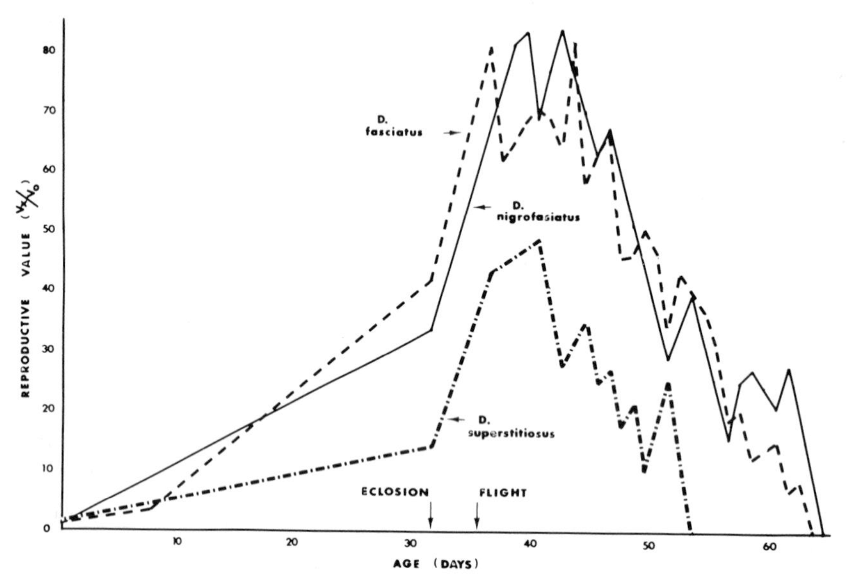

Fig. 4. Reproductive values for 3 species of *Dysdercus* plotted as a function of age from birth. Note that maximum values are reached immediately following flight. Delays in reproduction allowing migratory flight would displace curves to the right

Since reproductive value is a measure of colonizing ability, it is an important statistic for migrant insects. In Fig. 4 it is plotted as a function of age for 3 species of *Dysdercus*; on the same graph is indicated the age at which flight occurs. What is clear from the figure is that flight appears in the life history just prior to the time of maximum reproductive value, indicating that upon arrival in a new habitat colonization potential is high. Since maximum reproductive value usually coincides with the onset of reproduction, starvation, which increases the likelihood of long distance migration, displaces the curves to the right, and V_x is still maximal at the termination of flight. The advantages of delaying reproduction until the end of migration and the invasion of a new habitat are reflected in the reproductive-flight syndrome of Johnson (1969) again indicating the integration of ecological and physiological strategies. Similar relationships between reproductive value and migratory flight have been obtained for *Oncopeltus fasciatus* and *Aphis fabae*

(Dingle 1965, 1968a). Maximum reproductive values for diapausing species such as *Lygaeus kalmii* and *L. equestris* would occur at the time of post-diapause migration and reproduction in the spring.

PROSPECT

Data from numerous migrant insects clearly indicate the co-evolution of migration and life-history strategies. Returning to the scheme outlined in Fig. 1, there is indeed evidence for a maturation phase prior to flight, for a distribution of migrants and non-migrants in a population as a consequence of both genetic and environmental factors, for a close linkage in many species between diapause and migration, and for the integration of migratory and colonizing abilities. Migration and life-history strategies are therefore best considered together.

Clearly experimental methods have increased our understanding in an area which until not long ago was considered largely a problem for field observations. These experimental methods have been drawn not only from physiology, but also from quantitative genetics and population biology, with the result that we now know more about both mechanisms and strategies in insect migration and the integration of the physiological and ecological characteristics of migrants. In my laboratory, we are currently pursuing our studies of the genetics of diapause as well as of various life history parameters, such as fecundity, longevity, and growth rate, and of flight itself. At the same time we are continuing ecological studies of field populations and physiological and demographic studies of migrants in the laboratory. Such an experimental and interdisciplinary approach, we feel, is at the heart of an analysis of a behaviour as complex and with as many ramifications as migration.

Classically, the "experimental analysis of behaviour" has meant physiological analysis. Perhaps, however, we are now ready for a broadening of the traditional conception of "experimental". I am not advocating that approaches should be abandoned but, rather, I am asking whether they could be profitably combined with the large body of experimental and analytical methodology available from, for example, population biology and quantitative genetics. My own belief is that they can, as exemplified by the studies of migration. Life history strategies influence behaviour and may determine constraints imposed upon the nervous system. Further, most behaviours are complex and graded, and their variances usually involve allele differences at many loci. The techniques appropriate to the analysis of these differences therefore derive from quantitative genetics rather than the analysis of single gene differences traditional in biology, although the latter can also illuminate behaviour and are encompassed by quantitative genetic analysis (DeFries and Hegmann 1970). It should be possible with quantitative techniques to construct organisms (e.g. by selection) whose behavioural differences are amenable to the determination of underlying nervous mechanisms and to determine the extent to which gene differences underlying behaviour also impose modifications on the structure and function of the nervous system (Hegmann 1972; Kater and Hegmann 1973). The interrelationships between behaviour, life histories, and natural selection should also be analyzable. It is thus by the incorporation of powerful methodological and analytical tools available from other disciplines, that the experimental analysis of behaviour seems at the threshold of significant advances.

ACKNOWLEDGEMENTS

Drs. Roy L. Caldwell, Mary Ann Rankin, and Christer Solbreck have given freely of their time to discuss many of the problems of insect migration. All have read the manuscript. Dr. Joseph P. Hegmann patiently introduced me to quantitative genetics and discussed the selection experiments at length; he should not be held responsible for any errors of interpretation on my part. Mrs. Toni Yeager has devoted far more of her time than could reasonably be expected to giving excellent care to the bugs and maintaining the selected lines. Finally, my work on migration has been supported by NSF Grants GB-2949, 6444, and 8705 and a Special Fellowship from the USPHS.

REFERENCES

ARORA, G.: A study of oocyte development in the adult pyrrhocorid bugs of the genus *Dysdercus*. M.Sc. Thesis, University of Nairobi. (1971)

ATKINS, M.D.: A study of the flight of the Douglas-fir beetle *Dendroctonus pseudotsugae* Hopk. (Coleoptera: Scolytidae). II. Flight movements. Can. Ent. 42, 941-954 (1960).

BAKER, R.R.: The evolution of the migratory habit in butterflies. J. anim. Ecol. 38, 703-746 (1969).

BARRY, B.D., ADKISSON, P.L.: Certain aspects of the genetic factors involved in the control of the larval diapause of the pink bollworm. Ann. ent. Soc. Am. 59, 122-125 (1966).

BENNETT, R.B., BORDEN, J.H.: Flight arrestment of tethered *Dendroctonus pseudotsugae* and *Trypodendron lineatum* (Coleoptera: Scolytidae) in response to olfactory stimuli. Ann. ent. Soc. Am. 64, 1273-1286 (1971).

BHAKTHAN, N.M.G., BORDEN, J.H., NAIR, K.K.: Fine structure of degenerating and regenerating flight muscles in a bark beetle, *Ips confusus*. I. Degeneration. J. Cell Sci. 6, 807-820 (1970).

BHAKTHAN, N.M.G., NAIR, K.K., BORDEN, J.H.: Fine structure of degenerating and regenerating flight muscles in a bark beetle, *Ips confusus*. II. Regeneration. Can. J. Zool. 49, 85-89 (1971).

BRINKHURST, R.O.: Alary polymorphism in the Gerroidea (Hemiptera-Heteroptera). J. anim. Ecol. 28, 211-230 (1959).

BRINKHURST, R.O.: Observations on wing polymorphism in the Heteroptera. Proc. R. ent. Soc. Lond (A) 38, 15-22 (1963).

BROWN, E.S.: Notes on the migration and direction of flight of *Eurygaster* and *Aelia* species (Hemiptera, Pentatomoidea) and their possible bearing on invasions of cereal crops. J. anim. Ecol. 34, 93-107 (1965).

CALDWELL, R.L.: A comparison of the dispersal strategies of two milkweed bugs, *Oncopeltus fasciatus* and *Lygaeus kalmii*. Ph.D. Thesis, University of Iowa. (1969)

CALDWELL, R.L., HEGMANN, J.P.: Heritability of flight duration in the milkweed bug, *Lygaeus kalmii*. Nature, Lond. 223, 91-92 (1969).

CALDWELL, R.L., RANKIN, M.A.: Effects of a juvenile hormone mimic on flight in the milkweed bug, *Oncopeltus fasciatus*. Gen. comp. Endocrin. 19, 601-605 (1972).

CALDWELL, R.L., RANKIN, M.A.: The separation of migratory from vegetative behavior in *Oncopeltus fasciatus*. J. comp. Physiol. (In press.).

COLE, L.C.: The population consequences of life history phenomena. Q. Rev. Biol. 29, 103-137 (1954).

COMMON, I.F.B.: A study of the ecology of the adult bogong moth, *Agrotis infusa* (Boisd.) (Lepidoptera: Noctuidae) with special reference to its behaviour during migration and aestivation. Aust. J. Zool. 2, 223-263 (1954).

DANILEVSKII, A.S.: "Photoperiodism and Seasonal Development of Insects". Edinburgh: Oliver and Boyd. (1965)

DeFRIES, J.C., HEGMANN, J.P.: Genetic analysis of open-field behavior. *In* "Contributions to Behavior-Genetic Analysis - The Mouse as a Prototype" (Eds., G. Lindzey and D.D. Thiessen), pp. 23-56. New

York: Appleton-Century-Crofts. (1970)
DINGLE, H.: The relation between age and flight activity in the milkweed bug, *Oncopeltus*. J. exp. Biol. 42, 269-283 (1965).
DINGLE, H.: Some factors affecting flight activity in individual milkweed bugs (*Oncopeltus*). J. exp. Biol. 44, 335-343 (1966).
DINGLE, H.: The influence of environment and heredity on flight activity in the milkweed bug *Oncopeltus*. J. exp. Biol. 48, 175-184 (1968a).
DINGLE, H.: Life history and population consequences of density, photoperiod, and temperature in a migrant insect, the milkweed bug *Oncopeltus*. Am. Nat. 102, 149-163 (1968b).
DINGLE, H.: Migration strategies of insects. Science, N.Y. 175, 1327-1335 (1972).
DINGLE, H., ARORA, G.: Experimental studies of migration in bugs of the genus *Dysdercus*. Oecologia 12, 119-140 (1973).
DINGLE, H., CALDWELL, R.L., HASKELL, J.B.: Temperature and circadian control of cuticle growth in the bug *Oncopeltus fasciatus*. J. Insect Physiol. 15, 373-378 (1969).
DIXON, A.F.G.: Population dynamics of the sycamore aphid *Drepanosiphum platanoides* (Schr.) (Hemiptera: Aphididae): Migratory and trivial flight activity. J. anim. Ecol. 38, 585-606 (1969).
DOBZHANSKY, T.: "Genetics of the Evolutionary Process". New York: Columbia University Press. (1970)
DUVIARD, D.: Les vols migratories de *Dysdercus voelkeri* Schmidt (Hemiptera: Pyrrhocoridae) en Cote d'Ivoire. I. Le syndrome "oogenese vol migratoire". O.R.S.T.O.M., Centre d'Adiopodoume, Cote d'Ivoire, 13 pp. (1972).
EDWARDS, F.J.: Environmental control of flight muscle histolysis in the bug *Dysdercus intermedius*. J. Insect Physiol. 15, 2013-2020 (1969).
ELLIS, P.E., CARLISLE, D.B., OSBORNE, D.J.: Desert locusts: sexual maturation delayed by feeding on senescent vegetation. Science, N.Y. 149, 546-547 (1965).
FALCONER, D.S.: "Introduction to Quantitative Genetics". Edinburgh: Oliver and Boyd. (1960)
FISHER, R.A.: "The Genetical Theory of Natural Selection". New York: Dover. (1958).
GUTHRIE, D.M.: Polymorphism in the surface water-bugs (Hemipt.-Heteropt.: Gerroidea). J. anim. Ecol. 28, 141-152 (1959).
HEGMANN, J.P.: Physiological function and behavioral genetics. I. Genetic variance for peripheral conduction velocity in mice. Behav. Gen. 2, 55-67 (1972).
HODEK, I.: Bionomics and ecology of predaceous Coccinellidae. A. Rev. Ent. 12, 79-104 (1967).
HONĚK, A.: Selection for non-diapause in *Aelia acuminata* and *A. rostrata* (Heteroptera, Pentatomidae) under various selective pressures. Acta ent. bohemoslov. 69, 73-77 (1972).
JOHNSON, C.G.: "The Migration and Dispersal of Insects by Flight". London: Methuen. (1969)
KATER, S.B., HEGMANN, J.P.: The snail and the mouse: Neurophysiological and genetic analysis of behavior. *In* "Colorado Symposium on Behavior Genetics: Simple Systems" (Ed., J.R. Wilson). Boulder: Colorado Associated University Press. (1973)
KENNEDY, J.S.: The experimental analysis of aphid behavior and its bearing on current theories of instinct. Proc. Xth Int. Congr. Ent. (Montreal, 1956) 2, 397-404 (1958).
KENNEDY, J.S.: A turning point in the study of insect migration. Nature, Lond. 189, 785-791 (1961).
KENNEDY, J.S.: Co-ordination of successive activities in an aphid. Reciprocal effects of settling on flight. J. exp. Biol. 43, 489-509 (1965).
KENNEDY, J.S.: The balance between antagonistic induction and depression of flight activity in *Aphis fabae* Scopoli. J. exp. Biol. 45, 215-

228 (1966).
KENNEDY, J.S., BOOTH, C.O.: Free flight of aphids in the laboratory. J. exp. Biol. 40, 67-85 (1963a).
KENNEDY, J.S., BOOTH, C.O.: Co-ordination of successive activities in an aphid. The effect of flight on the settling responses. J. exp. Biol. 40, 351-369 (1963b).
KENNEDY, J.S., BOOTH, C.O.: Co-ordination of successive activities in an aphid. Depression of settling after flight. J. exp. Biol. 41, 805-824 (1964).
KUTSCH, W.: The development of the flight pattern in the desert locust, *Schistocerca gregaria*. Z. vergl. Physiol. 74, 156-168 (1971).
LEES, A.D.: "The Physiology of Diapause in Arthropods". London: Cambridge University Press. (1955)
LEES, A.D.: The control of polymorphism in aphids. Adv. Insect Physiol. 3, 207-277 (1966).
LEES, A.D.: The production of the apterous and alate forms in the aphid *Megoura viciae* Buckton, with special reference to the role of crowding. J. Insect Physiol. 13, 289-318 (1967).
LEWONTIN, R.C.: Selection for colonizing ability. *In* "The Genetics of Colonizing Species" (Eds., H.G. Baker and G.L. Stebbins). New York: Academic Press, pp. 77-94 (1965).
MacARTHUR, R.H., WILSON, E.O.: "The Theory of Island Biogeography". Princeton: Princeton University Press. (1967).
MORRIS, R.F., FULTON, W.C.: Heritability of diapause intensity in *Hyphantria cunea* and correlated fitness responses. Can. Ent. 102, 927-938 (1970).
NEVILLE, A.C.: Chitin orientation in cuticle and its control. Adv. Insect Physiol. 4, 213-286 (1967).
PICKFORD, R., RANDELL, R.L.: A non-diapause strain of the migratory grasshopper, *Melanoplus sanguinipes* (Orthoptera: Acrididae). Can. Ent. 101, 894-896 (1969).
RANKIN, M.A.: The inter-relationship and physiological control of flight and reproduction in *Oncopeltus fasciatus* (Heteroptera: Lygaeidae). Ph.D. Thesis, University of Iowa. (1972)
RYGG, T.D.: Flight of *Oscinella frit* L. (Diptera, Chloropidae) females in relation to age and ovary development. Entomologia exp. appl. 9, 74-84 (1966).
SLIFER, E.H., KING, R.L.: The inheritance of diapause in grasshopper eggs. J. Hered. 52, 39-44 (1961).
SOLBRECK, C.: Sexual cycle, and changes in feeding activity and fat body size in relation to migration in *Lygaeus equestris* (L.) (Het., Lygaeidae). Ent. Scand. 3, 267-274 (1972).
SOUTHWOOD, T.R.E.: Migration of terrestrial arthropods in relation to habitat. Biol. Rev. 37, 171-214 (1962).
TAUBER, M.J., TAUBER, C.A.: Geographic variation in critical photoperiod and in diapause intensity of *Chrysopa carnea* (Neuroptera). J. Insect Physiol. 18, 25-29 (1972).
VEPSÄLÄINEN, K.: The role of gradually changing daylength in determination of wing length, alary dimorphism and diapause in a *Gerris odontogaster* (Zett.) population (Gerridae, Heteroptera) in South Finland. Ann. Acad. Sci. fenn. (A., IV Biologica) 183, 1-25 (1971).
WILLIAMS, C.B.: "Insect Migration". London: Collins. (1958)
WILLIAMS, C.M., BARNESS, L.A., SAWYER, W.H.: The utilization of glycogen by flies during flight and some aspects of the physiological ageing of *Drosophila*. Biol. Bull. mar. biol. Lab., Woods Hole 84, 263-272 (1943).
YOUNG, E.C.: Flight muscle polymorphism in British Corixidae: ecological observations. J. anim. Ecol. 34, 353-389 (1965).

INDEX TO AUTHORS

INDEX TO INSECTA AND OTHER ANIMALS

SUBJECT INDEX

ALLE ZEIT WACH
1842